Vegetable Grafting: Principles and Practices

Vegetable Grafting: Principles and Practices

Edited by Myles Spencer

STATES
ACADEMIC PRESS
www.statesacademicpress.com

States Academic Press,
109 South 5th Street,
Brooklyn, NY 11249, USA

Visit us on the World Wide Web at:
www.statesacademicpress.com

ISBN: 978-1-63989-751-3

Cataloging-in-Publication Data

Vegetable grafting : principles and practices / edited by Myles Spencer.
p. cm.
Includes bibliographical references and index.
ISBN 978-1-63989-751-3
1. Grafting. 2. Vegetables. 3. Vegetative propagation. 4. Rootstocks. I. Spencer, Myles.
SB123.65 .V44 2023
631.541--dc23

Table of Contents

Preface

Grafting is a vegetative propagation technique wherein the tissues of two plants are joined in order to continue their growth together. There are different techniques of grafting, namely, approach grafting, side grafting, splice grafting, saddle grafting, cleft grafting and flat grafting. The advantages of grafting include fast and large-scale production of desirable varieties of fruiting trees and flowering shrubs. Propagation methods such as layering and cutting are usually not successful for growing commercially significant plants but they can be grown easily using grafting. This horticulture technique is also used to overcome the issue of sexual incompatibility between two different species. It can be used for cloning a scion variety to maintain consistency of characters such as color, size and flavors. This book outlines the principles and practices of vegetable grafting. It will also provide interesting topics for research, which interested readers can take up. With the state-of-the-art inputs by acclaimed experts of this field, this book targets students and professionals.

This book is the end result of constructive efforts and intensive research done by experts in this field. The aim of this book is to enlighten the readers with recent information in this area of research. The information provided in this profound book would serve as a valuable reference to students and researchers in this field.

At the end, I would like to thank all the authors for devoting their precious time and providing their valuable contribution to this book. I would also like to express my gratitude to my fellow colleagues who encouraged me throughout the process.

Editor

1

The Role of Grafting in the Resistance of Tomato to Viruses

Roberta Spanò [1,*], Massimo Ferrara [2], Donato Gallitelli [1] and Tiziana Mascia [1]

1 Department of Soil, Plant and Food Sciences, University of Bari "Aldo Moro", 70126 Bari, Italy; donato.gallitelli@uniba.it (D.G.); tiziana.mascia@uniba.it (T.M.)
2 Institute of Sciences of Food Production (ISPA)—CNR, 70126 Bari, Italy; massimo.ferrara@ispa.cnr.it
* Correspondence: r.spanomail@gmail.com

Abstract: Grafting is routinely implemented in modern agriculture to manage soilborne pathogens such as fungi, oomycetes, bacteria, and viruses of solanaceous crops in a sustainable and environmentally friendly approach. Some rootstock/scion combinations use specific genetic resistance mechanisms to impact also some foliar and airborne pathogens, including arthropod or contact-transmitted viruses. These approaches resulted in poor efficiency in the management of plant viruses with superior virulence such as the strains of tomato spotted wilt virus breaking the Sw5 resistance, strains of cucumber mosaic virus carrying necrogenic satellite RNAs, and necrogenic strains of potato virus Y. Three different studies from our lab documented that suitable levels of resistance/tolerance can be obtained by grafting commercial tomato varieties onto the tomato ecotype Manduria (Ma) rescued in the framework of an Apulian (southern Italy) regional program on biodiversity. Here we review the main approaches, methods, and results of the three case studies and propose some mechanisms leading to the tolerance/resistance observed in susceptible tomato varieties grafted onto Ma as well as in self-grafted plants. The proposed mechanisms include virus movement in plants, RNA interference, genes involved in graft wound response, resilience, and tolerance to virus infection.

Keywords: vegetable grafting; RNAi; wound and pathogen response; plant viruses; disease tolerance/resistance; tomato ecotype

1. Introduction

The tomato virome consists of at least 136 characterized virus species threatening tomato production worldwide plus a number of new viruses that are being progressively identified by next-generation sequencing [1]. The role of these not yet characterized viruses in pathogenesis, disease development, and crop losses is unknown. Characterized and economically relevant viruses infecting tomato include the strains of tomato spotted wilt virus (TSWV) breaking the resistance conferred by the SW5 gene (Sw5-RB) (TSWV), strains of cucumber mosaic virus (CMV) harboring harmful variants of their satellite RNAs (satRNA), and isolates of potato virus Y (PVY) necrogenic to tomato, which appeared since the 1980s, and quickly became one of the most important tomato diseases in the Mediterranean countries. TSWV, CMV, and PVY are transmitted by arthropods and are included among the top ten economically important plant viruses [2] for which efficient and environmentally friendly methods of control are not yet available. Means of natural transmission, extended host range, and appearance of new or recombinant strains with increased virulence may account for control failures. CMV infects approximately 1300 monocots and dicots within 500 genera of more than 100 botanical families [3] including crop and wild species, which are important for virus persistence in open field [2,4]. TSWV host range is also large, consisting of at least 1090 species among crop and wild plants in 84 botanical families [2,5,6]. Finally, PVY has a relatively narrower host range compared to TSWV and CMV but, similar to them, new and emerging strains may cause severe epidemics in pepper, potato, tobacco, and tomato [2,7,8]. In spite of several attempts to obtain suitable levels of resistance by classical breeding or transgenesis [4,8–11], field control of these viruses is still based on routine pesticide sprays against vectors, which proved scantily efficient because of the stylet-borne nature of the non-persistent transmission of CMV and PVY mediated by aphids and the complex persistent propagative relationship between thrips and TSWV, which replicates in both plants and thrips, thus leading to the continuous emergence of resistance-breaking strains and new species [12]. On the other hand, introgression of potential resistance genes into commercial crops proved ineffective against virus strains with high genome plasticity and superior virulence. Several recessive (sw2, sw3, and sw4) and dominant (Sw1a and Sw1b) resistance genes against TSWV have been reported and incorporated into commercial tomato cultivars (see for review [9] and references quoted therein) but their resistance was isolate-specific and, therefore, it was quickly overcome by TSWV infection. Only the Sw-5 gene cluster originated from *Solanum peruvianum* provided durable and stable resistance against TSWV isolates from different geographic areas. However, large deployment of such Sw5-TSWV-resistant tomato varieties has resulted in the onset of RB isolates overcoming the resistance by single-amino acid substitutions in the TSWV NSM protein [13,14]. Resistance sources against CMV have been identified within the genus *Solanum*, but all were strain specific and/or quantitative [15]. Thus, CMV-resistant tomato varieties are not yet commercially available. Transgenic resistance of tomato varieties expressing the CMV coat protein or beneficial satRNAs has been shown to be partially [16,17] or highly efficient [18,19], respectively, in field trials. However, implementation of satRNA-mediated protection in agriculture is still a matter of debate for potential risks [20]. Necrogenic strains of PVY induce necrotic lesions on tomato leaves followed by a streak on stems with all the commercial hybrids and varieties commonly grown in greenhouses and open field. Therefore, unlike what is described in potato, pepper, and tobacco, PVY did not reveal any strain specialization to infect tomato [21]. This, in theory, should have facilitated the search for PVY resistance genes to be introgressed in tomato. According to Parrella et al. [22], resistance against PVY and another potyvirus tobacco etch virus has been identified in the wild tomato accession "PI 247087." The resistance gene corresponds to a eukaryotic translation initiation factor that interacts with viral VPg and prevents virus accumulation in inoculated tissues [23]. In spite of these results, PVY-resistant tomato varieties are not yet available on the market. Thus, introgression of potential resistance genes into commercial crops proved ineffective against virus strains with high genome plasticity and superior virulence.

Therefore, alternative integrated management strategies were exploited.

Suitable levels of resistance to abiotic and biotic stresses have been successfully obtained in vegetable annual crops by grafting two genotypes selected as scion and rootstock. Actually, grafting offers a rapid surgical alternative to the time-consuming classical breeding or transgenesis to combine desired characteristics of rootstock and scion in vegetable crops [24,25].

Ideally, scion must cope with crop performance and yield required by farmers and fruit quality and nutritional value required by consumers [26]. The rootstock should guarantee cultivation under adverse environments such as salinity, nutrient deficiency, drought, pollutants, and soil-borne pathogens such as fungi, oomycetes, bacteria, and viruses [27–33]. Among vegetable crops, tomato, eggplant, sweet pepper, watermelon, melon, and cucumber are commonly and economically grafted in Asia, Europe, and North America. To give some examples (reviwed by [34]), watermelon, melon, and cucumber are prevalently grafted onto *Cucurbita maxima* × *C. moschata* rootstocks such as "Shintoza," "Ps1313," "TZ 148" or onto *C. moschata* "Jinxinzhen," *C. pepo* "Brava," and *C. ficifolia* in the case of cucumber. Tomato has been grafted mainly onto tomato genotypes and interspecific hybrids such as the *Solanum lycopersicum* × *S. habrochaites* "Maxifort" and "Beaufort" rootstocks, which improve crop yield, tolerance to soilborne pathogens and fruit quality, favoring higher accumulation of phenolic compounds, vitamin C, lycopene, and flavonoids in fruits of grafted plants [35]. Eggplant has been grafted traditionally on *Solanum* spp. wild species such as *S. integrifolium*, *S. torvum*, and *S. sisymbriifolium*, which provide high level of resistance to bacterial wilt, Fusarium wilt, Verticillium wilt, and root-knot nematodes. Compared to tomato and eggplant, grafted sweet peppers are not typically commercialized. Fruit shape is important in pepper crops but certain rootstocks can modulate gene expression and fruit development of the scion. Tsaballa et al. [36] demonstrated that phenotypic change in sweet pepper can be inherited for up to two generations of seed produced by these progenies. Nonetheless, suitable levels of resistance against root-knot nematodes and *Phytophthora capsici* were attained in sweet pepper grafted on *Capsicum annuum* accessions "AR96023" and "AF2638," respectively. Finally, this list could not be exhaustive without including the control of Verticillium wilts obtained by grafting seed-propagated globe artichoke hybrids onto cardoon [37] as well as the list of potential rootstocks with special characteristics to manage biotic and abiotic stresses in tomato, eggplant, chili, potato, cucumber, muskmelon, pumpkin, and wax gourd reported by Kumar et al. [38].

Grafted tomato plants were implemented in the early 1960s but their use aroused little interest because farmers were particularly discouraged by the higher price of grafted plants compared to non-grafted seedlings. Recent surveys report that the price of grafted tomato seedlings for fresh market tomato production is between 0.4 to 1.2 USD per grafted seedling in the USA and some Asian countries, including Japan and Korea, and between 0.6 to 1.2 EUR per grafted seedling in Spain and some European countries [39]. Nonetheless, actual estimates suggest between 20% and 40% of tomatoes are grafted [39] and their implementation is rapidly increasing all over the world in organic and environmentally friendly crops for the control of soil-borne pathogens as an alternative to the banned methyl bromide and to mitigate effects of abiotic stresses.

The higher cost of the grafted plantlets is now compensated by a number of advantages, which include production increase and earliness of the harvest with a reduced number of plants, rational use of irrigation water, use of fewer fertilizers, and elongation of the crop cycle. A good rootstock/scion combination usually guarantees a robust root system and the maintenance of good vegetative vigor and resistance to deal with abiotic and biotic stresses until the end of the farming cycle [27]. All these aspects guarantee net returns large enough to make grafted tomatoes significantly more profitable than non-grafted seedlings [40]. Grafting is a viable approach also to limit damage by viruses transmitted through soil such as melon necrotic spot virus (MNSV) vectored by *Olpidium* sp. or the contact-transmissible tobacco mosaic virus (TMV), which reaches the soil in infected plant debris or as seed-coat contaminant. Rootstocks with specific characteristics have been developed and deployed for the control of MNSV in muskmelon and watermelon [41–43] and of TMV in tomato [38]. However, the majority of plant viruses is airborne by arthropods and pollen or transmitted by contact or aerosol [44,45] (Table 1).

Some promising attempts were recently made to reduce incidence of tomato yellow leaf curl virus, TSWV, and pepino mosaic virus in different scion-rootstocks combinations [46–49].

Table 1. Virus species recorded in Italian tomato crops.

Virus	Acronim	Transmission
Alfalfa mosaic virus	AMV	aphids
Cucumber mosaic virus	CMV	aphids
Parietaria mottle virus	PMoV	thrips
Pelargonium zonate spot virus	PZSV	pollen, thrips
Pepino mosaic virus	PepMoV	contact
Potato virus Y	ToMV	aphids
Tobacco mosaic virus	TMV	contact
Tomato mosaic virus	ToMV	contact/aerosol
Tomato brown fruit rugose virus	ToBFRV	contact
Tomato cholorosis virus	ToCV	aleyrodids
Tomato infectious chlorosis virus	TICV	aleyrodids
Tomato spotted wilt virus	TSWV	thrips
Tomato yellow leaf curl virus	TYLCV	aleyrodids
Tomato yellow leaf curl Sardinia virus	TYLCSV	aleyrodis

2. Why Grafting?

Samuel [50] described a general pattern of translocation of tobacco mosaic virus (TMV) into plant tissues suggesting that after entering tobacco leaf tissue, TMV particles primarily reach roots through phloem sieve tubes and between 2 to 3 days post-inoculation (dpi) they follow phloem sap in the reverse way up, bypass the old leaves, and invade systemically the young plant tissues [51]. Andrianifahanana et al. [52] reported a similar pattern of virus translocation in pepper of the potyvirus pepper mottle virus. The authors documented virus movement from the inoculated leaf down the stem toward the roots via the external phloem. Then the virus entered internal phloem at the cotyledon node and rapidly spread upward to the young tissues. A phloem-translocation pattern has been described also for cauliflower mosaic virus [53], cherry leafroll virus [54], and TMV [55] (reviewed in [56–58]). Internal and external phloem is present in the stem of *Solanaceae* and of some other plant families such as *Cucurbitaceae* and corresponds to the adaxial and abaxial phloem strands of leaf veins and petioles. In tomato leaflets the xylem is in the center of the vein with the phloem distributed on both the adaxial and abaxial sides of the bundle. The abaxial phloem exports sugars from the leaf in a descending direction to roots. Thus, a virus entering abaxial leaf phloem is expected to follow transport of photoassimilates down to roots, according to the source-sink relationship [59]. For potyviruses, virus loading in the external phloem depends on the virus-host combination [56] and for potato virus A it is controlled by the virus-coded Vpg [60]. On the contrary, a complex structure of CMV containing viral RNA, coat protein, and the 3a movement protein loads into sieve elements of minor veins of *Nicotiana clevelandii*. The complex successively assembles into complete virus particles within minor vein sieve elements [58,61]. Unfortunately no such data are available for TSWV.

Results from these studies were seminal to our idea to exploit grafting as a strategy to control airborne viral infections in tomato.

This review gives a brief account of the results and proposes mechanisms probably involved in the resistance/tolerance observed in grafted plants of the three cases-study carried out in our laboratory to manage one Sw5-RB strain of TSWV [62], two CMV strains supporting stunting and necrogenic satRNAs [63], and one recombinant PVY strain necrogenic to tomato [64].

3. Screening of *Solanum* spp. Germplasm

Local tomato ecotypes and commercial hybrids and *Solanum* spp. genotypes were screened for resistance/tolerance to challenge inoculation of specific strains of TSWV, CMV, and PVY characterized by superior virulence.

Screening for potential rootstocks privileged local tomato ecotypes found and characterized in the framework of regional actions aimed at preserving biodiversity of vegetable crops in Apulia (southern Italy). Selected ecotypes were Fiaschetto (Fi), Giallo invernale (Gi), Manduria (Ma), Morciano (Mor), Racalino (R), and Regina (Re). Screenings included also *Solanum integrifolium*, *S. nigrum*, *S. torvum*, eggplant cv. Molfettese (Mo), and commercial tomato hybrids, Faino (Fa), Messapico (Me), Taylor (Ta) or not Pullrex (Pu), carrying the Sw5-resistance gene to TSWV and the tomato variety UC82 (UC) as highly susceptible control. Tests were carried out with cleft-grafted and non-grafted plants rub-inoculated on the first leaf above the graft junction with sap extracted from infected plants and grown under glasshouse [62–64] or exposed to field inoculum (this review). According to Kumar et al. [38], complete repair of graft wound would be achieved between 7 and 10 days after grafting, thus we were confident that by the time of inoculation, continuous cambial connections between scion and rootstock had been restored. In glasshouse tests, plants were grown and maintained at 24 ± 2 °C with 16 h photoperiod and monitored daily for disease symptoms. Accumulation of viral RNA was estimated by dot blot or tissue print hybridization in samples collected from inoculated and mock-inoculated plants at 14, 21, and 28 days post-inoculation (dpi). Plant tissues were ground in the presence of an alkaline solution and spotted onto positively charged nylon membranes that were hybridized overnight with virus-specific DIG-labeled RNA or DNA probes. Chemiluminescent hybridization signals were used to detect and quantify the accumulation of viral RNA by using *Glyceraldehyde 3-phosphate dehydrogenase* (*GAPDH*) as housekeeping gene for normalization [65]. Viral genomic RNAs and virus-specific small interfering RNAs were separated on agarose or polyacrylamide gels, respectively, and detected by northern blot hybridization with specific radiolabeled or DIG-labeled probes [62,63] or by NGS sequencing [64].

For field tests, three biological replicates of each plant at first branching stage were transplanted according to a fully randomized block scheme. In 2016 and 2017, two experimental fields were set up between March and April using 270 and 180 plants per field, respectively. Data were collected before the middle of August, at the end of vegetative season.

3.1. Results with TSWV

S. nigrum and UC showed systemic symptoms ranging from severe mosaic, to leaf and stem necrosis and plant death in response to infection of the Sw5-RB strain TSWV-CiPz, whereas Ma, Mo, and *S. integrifolium* showed mild mosaic and, more interestingly, recovered from viral disease symptoms between 21 and 28 dpi (Table 2). The lowest levels of viral RNA were detected in Ma, Mo, and *S. integrifolium*, therefore these genotypes were selected as rootstocks to prepare graft combinations. Self-grafted UC/UC, Ma/Ma, Fa/Fa, Me/Me, and the Me and Fa scions grafted onto Ma showed mild mosaic. All these plants recovered from disease symptoms and displayed increased growth, abundant leaf canopy, and root development, compared to non-grafted infected plants. Oneweek later, also UC/Mo, and Pu/Pu recovered from disease symptoms. Recovery was not observed with non-grafted UC and Pu. Viral RNA load was markedly reduced in the Me/Ma graft combination as well as in all scions of the self-grafted combinations (Table 2).

In a field test set up in 2016, the only virus detected was a Sw5-RB TSWV strain identified by targeting NS_m sequences flanking the *Mae*I restriction site with RT-PCR RFLP [66] and rub-inoculation onto the commercial tomato hybrids Diaz and York carrying the Sw5 resistance gene [67]. Field infection damaged 7–8% of the plants with necrotic symptoms on leaves and fruits. Among the tomato ecotypes evaluated, Regina (Re) proved the most susceptible (Figure 1) but despite the overall low incidence of viral infection in the field plants self-grafted or grafted onto Ma showed a productive advantage compared to the non-grafted counterparts. In a second field test set up in 2017, we exposed to field inoculum the Sw5 commercial tomato hybrid Taylor self-grafted or grafted onto Ma. Non-grafted Taylor plants served as control. Again, the only virus detected was an Sw5-RB TSWV strain. Graft combinations Taylor/Taylor and Taylor/Ma produced between 6 and 9 kg/plant, whereas the non-grafted Taylor counterparts produced between 4 and 5 kg/plant, which is thought to be the standard production of this hybrid. In addition, the non-grafted Taylor plants, showed sporadic symptoms of Sw5-RB TSWV infection and about ten days delay in fruit ripening, compared to the grafted plants.

Table 2. Response of *Solanum* spp. genotypes, tomato varieties, and graft combinations to challenge inoculation of Sw5-RB TSWV isolate CiPz.

	14 *	**21 ***	**21 ***	**28 ***
Non-Grafted Plants	**Symptoms**	**Symptoms**	**Viral RNA Load ****	**Symptoms**
UC82 (UC)	VN, LE	LY	2.63 ± 0.08 [a]	PD
Manduria (Ma)	mLD	R	0.85 ± 0.02 [e]	R
Regina (Re)	VN, LE	LY	1.85 ± 0.01 [c]	PD
Sw5-Messapico (Me)	VN, LE	LY	1.91 ± 0.01 [c]	PD
Sw5-Faino (Fa)	VN, LE	LY	2.0 ± 0.03 [bc]	PD
Pullrex (Pu)	LY	D	2.1 ± 0.04 [b]	PD
Molfettese (Mo)	mMos	R	0.78 ± 0.02 [e]	R
Solanum integrifolium	mMos	R	0.87 ± 0.03 [e]	R
S. nigrum	sMos, sLD	N, sLD, LE	1.9 ± 0.09 [c]	N
S. torvum	mMos	mMos	1.25 ± 0.06 [d]	mMos

Table 2. *Cont.*

Grafted Plants ***	14 * Symptoms	21 * Symptoms	21 * Viral RNA Load **	28 * Symptoms
UC/Mo	Mos	R	0.19 ± 0.02 [c,d,e]	R
UC/*S. integrifolium*	sMos, LE	sLD, LY	0.35 ± 0.06 [a]	N
UC/Ma	Mos	R	0.10 ± 0.06 [e,f,g]	R
UC/UC	mMos	R	0.04 ± 0.008 [f,g]	R
Ma/Ma	mMos	R	0.01 ± 0.003 [g]	R
Me/Mo	sMos,	sMos, LY	0.30 ± 0.07 [a,b,c,d]	LY, sLD
Me/*S. integrifolium*	sMos,	sMos, LY	0.31 ± 0.07 [a,b,c]	LY, sLD
Me/Ma	mMos	R	0.01 ± 0.004 [g]	R
Me/Me	mMos	R	0.02 ± 0.006 [g]	R
Fa/Mo	Mos	Mos	0.22 ± 0.04 [b,c,d,e]	R
Fa/*S. integrifolium*	LE, VN	LY, VN	0.38 ± 0.04 [a]	N
Fa/Ma	mMos	R	0.12 ± 0.04 [e,f,g]	R
Fa/Fa	mMos	R	0.12 ± 0.06 [e,f,g]	R
Pu/Mo	sMos	sLD, VN	0.34 ± 0.02 [a,b]	N
Pu/*S. integrifolium*	LE, VN	LY, VN	0.36 ± 0.03 [a]	SLD, VN
Pu/Ma	Mos	LE, VN	0.18 ± 0.04 [d,e]	LY, VN
Pu/Pu	Mos	R	0.16 ± 0.03 [e,f]	R

* = days post-inoculation; ** = estimates of accumulation of viral RNA by quantitative dot blot hybridization at 21 dpi. Chemiluminescent hybridization signals were used to detect and quantify the accumulation of viral RNA by using *Glyceraldehyde 3-phosphate dehydrogenase* (*GAPDH*) as housekeeping gene for normalization. Each value represents average of three biological replicates ± standard error among replicates. Letters represent statistically significant differences values according to ANOVA analysis for non-grafted and grafted plants using Tukey's test ($p \leq 0.05$); *** = scion/rootstock; LE = leaf epinasty; LY = leaf yellowing; mLD = mild leaf distortion; mMos = mild mosaic; Mos = mosaic; N = whole leaf necrosis; PD = plant death; R = Recovery; sLD = severe leaf distortion; sMos = severe mosaic; VN = vein necrosis. Sw5 = plant expressing Sw5 resistance gene.

Figure 1. Severe symptoms of a Sw5-RB strain of TSWV on tomato cv Regina (**A**) compared to the same variety grafted onto Manduria (**B**) during 2016 field tests.

3.2. Results with CMV

CMV infections in tomato crops became suddenly relevant around the 1980s after the introduction of the so-called "Asian" strains characterized by superior aggressiveness compared to endemic CMV strains [68]. After initial severe outbreaks between 1988 and 1992, CMV has been detected sporadically in tomato fields although outbreaks may still occur after very mild winters, which may lead to huge increases in overwintering aphid populations that spread the virus from infected foci in wild plants communities into the crops [68,69]. Therefore we limited our study to glasshouse tests. Infection of CMV-TTS carrying a tomato top stunting satRNA induced leaf distortion and apical stunting in UC and Ma (Table 3), which persisted in UC until 30 dpi but not in Ma plants that recovered from disease symptoms. Symptoms elicited in non-grafted plants were very similar to those of grafted plants but, by 21 dpi, the latter fully recovered from disease symptoms (Table 3). Viral RNA loads in Ma were about 1.6-fold lower than in UC whereas no significative differences were estimated among grafted and self-grafted plants. Systemic infection of CMV-77 carrying a necrogenic satRNA induced stunting and severe stem and leaf necrosis in both Ma and UC, between 15 and 18 dpi (Table 3). Ma plants transiently recovered from necrosis between 18 and 21 dpi but by 30 dpi, these plants also died (Table 3). Non-grafted genotypes showed a more severe and rapid disease progression than grafted plants. Necrosis developed in all UC self-grafted plants and in one out of the four UC/Ma plants but none of the plants died. Interestingly, self-grafted Ma plants recovered from necrosis by 21 dpi (Table 3). Because of the occurrence of a necrotic phenotype, we monitored the accumulation of viral RNA at 15 dpi. At this time-point, viral RNA load in Ma was 8-fold lower than in UC and this ratio was substantially similar among the three graft combinations tested (Table 3).

Table 3. Response of tomato varieties UC82 (UC) and Manduria (Ma) and their graft combinations to infection of CMV-TTS and CMV-77.

CMV-TTS *	**15 *****	**21 *****	**21 *****	**30 *****
Plants	**Symptoms**	**Symptoms**	**Viral RNA Load †**	**Symptoms**
UC	TS, LD	sTS	0.16 ± 0.008 [b]	sTS
Ma	TS, LD	R	0.25 ± 0.001 [a,b]	R
UC/Ma §	TS	R	0.33 ± 0.021 [a]	R
UC/UC §	TS	R	0.31 ± 0.025 [a]	R
Ma/Ma §	TS	R	0.3 ± 0.007 [a]	R
CMV-77 **	**15 *****	**15 *****	**21 *****	**30 *****
Plants	**Symptoms**	**Viral RNA Load †**	**Symptoms**	**Symptoms**
UC	SN, LN	0.04 ± 0.001 [c]	PD	PD
Ma	mS	0.33 ± 0.025 [a,b]	tR	PD
UC/Ma §	SN, LN	0.24 ± 0.07 [b]	SN, LN ¥	SN, LN
UC/UC §	SN, LN	0.43 ± 0.06 [a]	SN, LN	SN, LN
Ma/Ma §	YBL	0.03 ± 0.001 [c]	R	R

* = supporting TTS satRNA; ** = supporting 77 satRNA; *** = days post-inoculation (dpi); † = ng of viral RNA estimated by quantitative dot blot hybridization at 15 and 21 dpi. Chemiluminescent hybridization signals were used to detect and quantify the accumulation of viral RNA by using *GAPDH* as housekeeping gene for normalization. Each value represents average of three biological replicates ± standard error among replicates. Letters represent statistically significant differences values according to ANOVA analysis for non-grafted and grafted plants using Tukey's test ($p \leq 0.05$); § = scion/rootstock; LD = leaf distortion; LN = leaf necrosis; mS = moderate stunting; PD = plant death; SN = stem necrosis; TS = top stunting; sTS = severe top stunting; R = recovery; tR = transitory recovery; YBL = yellowing of basal leaves; ¥ = necrosis developed in 1 out of 4 challenged plants.

3.3. Results with PVY

PVY isolates infecting tomato show high genetic diversity, including genetic recombination. One of such recombinants, denoted PVY^{C}-to, was characterized as an interlineage recombinant isolate of the PVY^{C} group and associated with necrotic spots and vein necrosis on the leaflets and pale yellow spots scattered on fruit skin of a table tomato variety [70].

In two independent glasshouse experiments, PVY^{C}-to infected systemically UC and Ma. UC infection was characterized by mosaic, leaf blade reduction, and twisting with some necrotic spots scattered on the leaf surface. Conversely, infection in Ma was substantially asymptomatic but viral RNA loads at 14 dpi did not differ significantly between UC and Ma, despite the differences in symptoms. Infection of PVY^{C}-to in the three graft combinations induced a mild distortion of leaf margin and a slightly reduced growth. Viral RNA loads at 14 dpi did not differ significantly among the three graft combinations but was approximately 2.5-fold higher than in non-grafted plants. Despite the difference in viral RNA loads, non-grafted Ma and all grafted plants recovered from disease symptoms by 21 dpi with a mean of 3-fold reduction in viral RNA loads compared with the estimates at 14 dpi sampling time. Non-grafted UC plants did not recover from disease symptoms; rather, they suffered increased disease severity and a further 2-fold increase in viral RNA loads by passing from 14 to 21 dpi (Table 4).

Table 4. Response of tomato varieties UC82 (UC) and Manduria (Ma) and their graft combinations to the infection of PVY^{C}-to.

	14 *	14 *	21 *	21 *
Plants	**Symptoms**	**Viral RNA Load *****	**Symptoms**	**Viral RNA Load *****
UC	Mos, LD	5.72 ± 1.34 ^{b}	sMos, sLD	16.07 ± 1.96 ^{a}
Ma	mLD	8.03 ± 0.93 ^{b}	R	2.22 ± 0.68 ^{c}
UC/Ma **	mS	17.92 ± 1.84 ^{a}	R	4.69 ± 1.16 ^{bc}
UC/UC **	A	13.68 ± 2.4 ^{a}	A	6.03 ± 0.44 ^{b}
Ma/Ma **	mLD	16.07 ± 1.96 ^{a}	R	2.39 ± 1.30 ^{c}

* = days post inoculation (dpi); ** = scion/rootstock; *** = pg of viral RNA estimated by quantitative dot blot hybridization at 14 and 21 dpi. Chemiluminescent hybridization signals were used to detect and quantify the accumulation of viral RNA by using *GAPDH* as housekeeping gene for normalization. Each value represents average of three biological replicates ± standard error among replicates. Letters represent statistically significant differences values according to ANOVA analysis for non-grafted and grafted plants using Tukey's test ($p \leq 0.05$); Mos = mosaic LD = leaf distortion; mLD = moderate leaf distortion; mS = moderate stunting; sMos = severe mosaic; sLD = severe leaf distortion; R recovery; A = asymptomatic.

4. Proposed Models

4.1. Virus Movement in Grafted Plants

Given the airborne nature of TSWV, we studied virus movement in grafted plants of *Nicotiana benthamiana* and tomato (Figure 2). At 5 dpi, TSWV-CiPz was detected in the inoculated and in the apical leaf of *N. benthamiana* whereas downward virus movement was evident only at 8 dpi by high virus accumulation above and below the graft union. Increased virus-specific signal at 13 dpi in the petiole of the apical leaf suggested virus movement toward the plant apex. By analogy with potyviruses [52] and TMV [57], the interval of 5 dpi between the inoculation and detection of TSWV-CiPz in the inoculated leaf might mirror initial viral replication and protein expression in rub-inoculated epidermal cells. Cell-to-cell movement then, may have led infection outwards to the immediately adjacent cells, with a relatively slow process estimated in one cell/2 h so that infection boundary advanced only a small distance per day [57]. Leaves on the same side of the stem having phloem traces adjacent to one another and carrying upward plant material and virus [59] may account for the simultaneous detection of TSWV-CiPz in the apex and in the inoculated leaf of *N. benthamiana*. Tissue-print hybridization of UC and Ma plants showed a virus movement pattern similar to that observed in *N. benthamiana*. The virus did not move out from the inoculated leaf until 4 dpi to descend quickly through the petiole down to the graft and roots. Distribution and accumulation of TSWV-CiPz RNA in non-grafted and grafted tomato was analyzed at 19 dpi (Figure 2). Viral RNA was detected in all sections but the hybridization signal was much weaker in the apex and inoculated leaf than in the roots, suggesting a reduced viral RNA load. The reduction was particularly marked in self-grafted Ma plants.

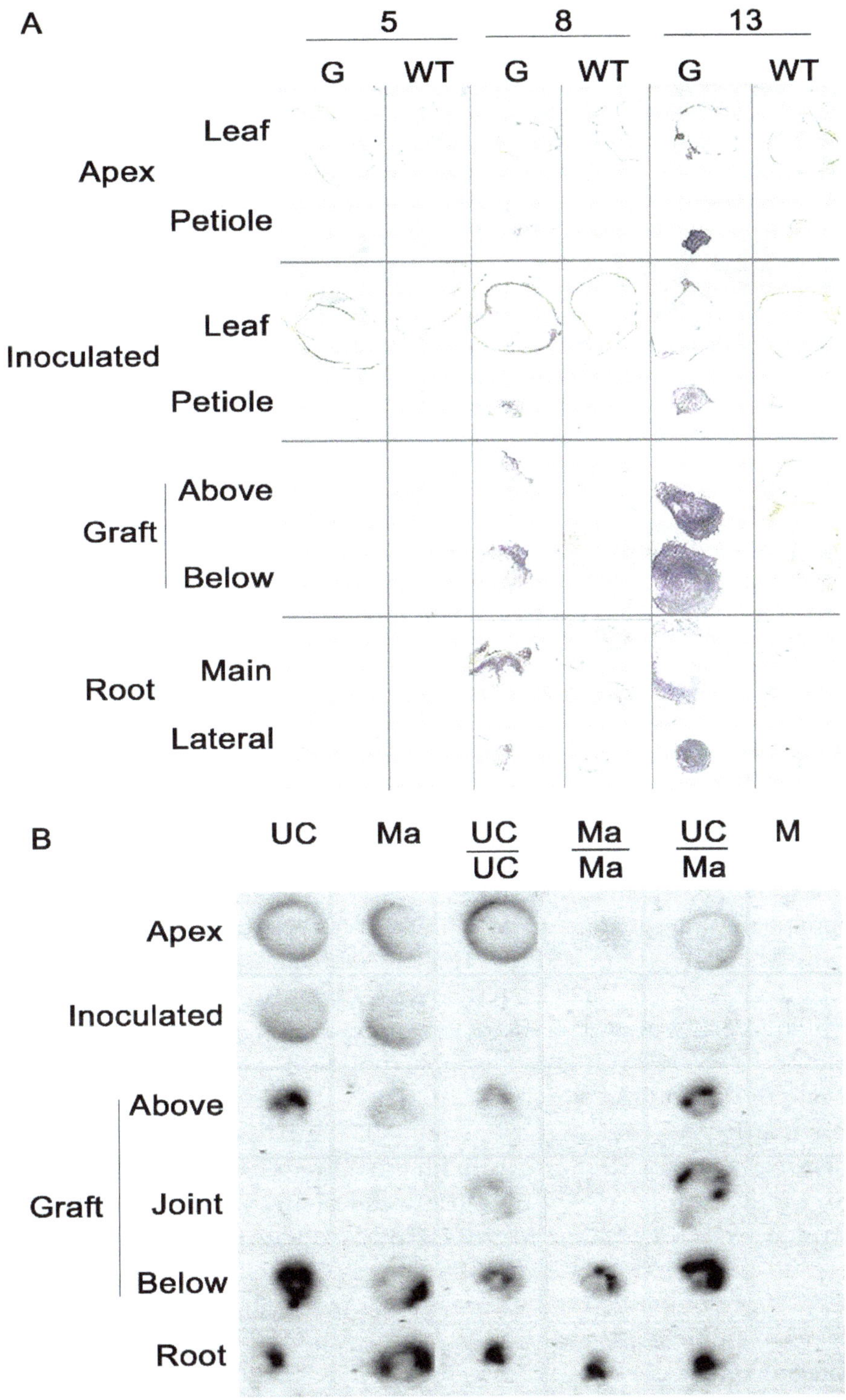

Figure 2. Detection of TSWV in sections of self-grafted (G) and non-grafted (WT) *N. benthamiana* plants at 5, 8, and 13 dpi with TSWV-CiPz. (**A**). Cross sections were prepared from apex, inoculated leaf, above and below graft junction, and from main and lateral roots. Virus was detected by using an antiserum against the NSs protein. (**B**). Detection of TSWV-CiPz RNA in UC, Ma, UC/UC, Ma/Ma, and UC/Ma at 19 dpi. Viral RNA was detected by tissue print hybridization with a DIG-labelled RNA probe for TSWV M RNA. Cross sections were prepared from apex, stem junction of the inoculated leaf, above, at and below the graft junction and from principal root. M = mock-inoculated UC plant.

4.2. RNA Interference

RNA interference (RNAi) is a defensive reaction against biotic stresses conserved in eukaryotic organisms. The reaction includes the synthesis of invasive double-stranded RNAs (dsRNA), such as the replicative intermediates or secondary structures of single stranded RNA (ssRNA) viruses, which are recognized and diced into 21–24 small RNA (sRNA) duplexes by conserved type III ribonucleases of the Dicer-like (DCL) protein family. The small dsRNA fragments denoted primary viral short interfering RNAs (primary vsiRNAs) are then loaded onto proteins of the Argonaute (AGO) family to build up RNA-induced silencing complexes (RISCs) that are the effectors of the process. RISCs removes one of the two strands from primary vsiRNAs and uses the other to detect and degrade viral sequences complementary to the RISC-retained vsiRNA strand [71,72]. After interference initiation in one or few cells, plant RNA-dependent RNA polymerases (RDR1 or RDR6) increase the amount of dsRNA as a DCL substrate for the production of secondary vsiRNAs and amplification of the RNAi signal. Thus, initial RNAi signal spreads progressively to adjacent cells through plasmodesmata and over long distance through plant phloem [73–75]. Interestingly, systemic movement of RNAi signal was demonstrated by grafting and in studies with viruses and viroids [76–79]. Another point in favor of a tolerance/resistance mechanism mediated by RNAi is the occurrence of a recovery phenotype prevalently in grafted plants and independently from the challenging virus. Recent evidences suggested that recovery from virus-induced disease symptoms might be the consequence of a robust delivery of secondary vsiRNAs produced by the DCL4/RDR6/SGS3 RNAi pathway to saturate the viral suppressor of RNA silencing (VSRs) activity involved in determining disease symptom with the formation of a virus tolerant condition in infected tissues [80]. Therefore, we evaluated whether a differential response based on RNAi could be involved in the different susceptibility to TSWV-CiPz, CMV-TTS and CMV-77 observed among the *Solanum* spp. genotypes tested. Increase in the population of small RNAs (sRNAs) purified at 21 dpi was clearly seen in non-grafted UC and Ma plants, upon TSWV-CiPz infection [62]. Results from hybridization of the three L, M, and S genomic RNAs of TSWV-CiPz with radiolabeled sRNA probes suggested that the majority of sRNAs in the total pool corresponded to vsiRNAs produced from genomic RNA S whereas those corresponding to genomic segments L and M were apparently less represented [62]. The TSWV S RNA encodes the non-structural protein NSs that has VSRs activity [81–83]. Quantitative results from hybridization suggested also a stronger accumulation of vsiRNAs in Ma than in UC plants in agreement with the 3-fold reduction of viral RNA and the recovery phenotype elicited by Ma plants compared to UC plants (Table 2). Relative abundance of transcripts of *AGO*1, *AGO*4, *DCL*1, *DCL*2, *DCL*4, *PAZ*, *RDR*1, and *RDR*6 (orthologs of *Arabidopsis thaliana*) in roots and leaves of Ma plants at 21 dpi with TSWV-CiPz suggested that Ma plants could elicit a stronger RNAi response compared to UC even when used as rootstock. For example, *AGO*4 and *RDR*6, were up-regulated in leaves of UC and Me when grafted onto Ma [62]. Thus, the symptom attenuation and reduced accumulation of TSWV-CiPz RNA estimated in UC and Me when grafted onto Ma strengthens the involvement of an RNAi-based response through the up-regulation of *AGO*4 and *RDR*6 in Ma roots.

Results with CMV-TTS and CMV-77 in non-grafted plants showed significant differences only in the relative abundance of *DCL*2 transcripts in UC and of *AGO* genes in Ma. On the contrary, viral infection stimulated up-regulation of *DCL*2, *AGO*2, and *RDR*1 in self-grafted UC and of *DCL*2, *AGO*2 and *RDR*6 in the UC/Ma graft combination. Hence, vsiRNAs specific for CMV-TTS and its satRNA were searched in UC leaves in a time-course from 14 to 35 dpi. VsiRNAs derived from the genomic RNA of CMV-TTS were detected only at 14 dpi, i.e., concomitantly with the recovery from diseases symptoms (Table 3). On the contrary, siRNAs against TTS-satRNA were detected at all time points suggesting that host RNAi-triggered degradation of genomic CMV-TTS RNAs was enhanced by its satRNA. This is not surprising, since CMV satRNAs behave as molecular parasites of their helper viruses regardless of whether they attenuate or exacerbate virus-induced symptoms [84]. This conclusion is also compatible with the recovery from disease symptoms shown by Ma and all the graft combination by 30 dpi (Table 3). Estimates of the relative abundance of *DCL*, *AGO*, and *RDR* transcripts were obtained from non-grafted and tomato plants challenged by CMV-77 whereas the early development of necrosis in UC prevented the analysis for this tomato genotype. To partially overcome this limitation, we used the transgenic tomato line UCTC5.9.2 (UCTC) expressing constitutively the Tfn benign variant of CMV satRNAs (Tfn-satRNA), which confers tolerance against CMV-77 [85]. Non-grafted UCTC plants responded to infection of CMV-77 by up-regulation of *DCL*2, *AGO*1, *AGO*2, and *RDR*1 transcripts whereas only *AGO*1 was up-regulated in Ma. On the contrary, a remarkable up-regulation was recorded in all UC/UCTC and Ma/UCTC plants upon infection with CMV-77. RQ values evidenced the up-regulation of *AGO*2 (between 12- and 20-fold), *RDR*1 (between 12- and 20-fold) and *RDR*6 (between 4.5- and 15-fold). *AGO*2 and *RDR*6 showed an up-regulation of 15- and 6.3-fold, respectively, in self-grafted UCTC infected by CMV-77, compared to mock-inoculated controls [63]. A general scheme showing the pattern of RNAi-based response in grafted and non-grafted plants is shown in Figure 3.

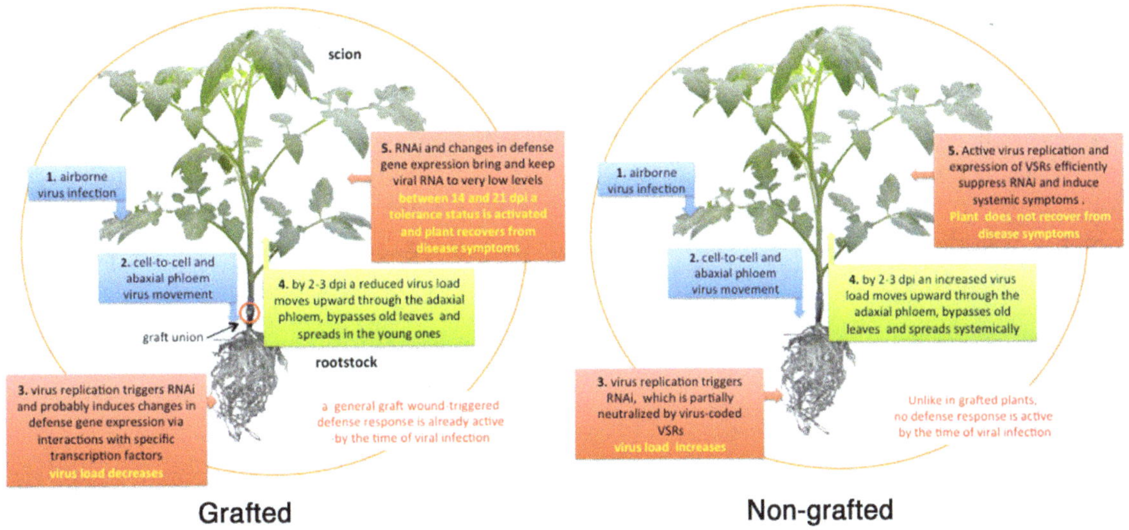

Figure 3. Proposed flow diagram of an airborne virus infection in grafted and non-grafted tomato plants. By the time of virus infection, a general graft wound-response is already active only in grafted plants (pink oval in the background). After cell-to-cell movement in epidermal cells, the virus enters the abaxial phloem, bypasses graft union (circled in red), and accumulates in roots following the source-to-sink phloem sap movement. In rootstock tissues, heavy loads of replicating viral RNA trigger RNAi, probably induces changes in defense gene expression, and signals molecules via interactions with specific transcription factors. Reduced virus loads move from roots into the scion bypassing the graft union and old leaves and spreads into young leaves where it may induce symptoms. Virus load is still sufficient to trigger RNAi and changes in defense gene expression. Between 14-21 days post infection (dpi), virus replication and load are very low, a tolerant state is activated and plant recovers from disease symptoms. In roots of non-grafted plants, the virus replicates actively and accumulates in root tissues by suppressing RNAi via the expression of its VSRs. Then, the virus spreads upward to infect systemically tomato leaves. Active virus replication and expression of its VSRs induce disease symptoms and hinders recovery.

4.3. *Response to Graft Wound and Viral Infection*

We used the Ion Torrent RNAseq analysis to identify variations in the transcriptome of grafted and non-grafted Ma and UC tomato plants, exposed or not to the infection of PVY^{C}-to. Results provided unique information on gene differentially expressed (DEGs) in the two tomato genotypes in response to graft wound and virus infection. Genes were considered significantly differentially expressed only when log2 of their fold change (FC) expression (log2FC) was $\geq$1 with a false discovery rate (FDR) $\leq$ 0.05 [64]. Beyond the use of resistant rootstocks to manage specific soil-borne pathogens of tomato such as fungi, oomycetes, bacteria, and viruses [27,86], it has been suggested that grafting *per se* activates systemic defense mechanisms in response to the graft wound [87,88]. In our study, the graft wound induced 1991 unique DEGs, which accounted for the 5.88% of the total genes

(33,810) annotated in tomato genome (Solyc). This percentage was close to the 8% DEGs of the total genes annotated in *A. thaliana* involved in plant wound healing process [89,90]. Most of the graft wound responsive DEGs encoded signal molecules that overlapped their role also in response to abiotic stress and pathogen attack [64,87,90,91]. Table 5 show a selection of such graft wound-responsive genes detected in mock-inoculated and infected tomato plants. Among them APETALA2/ethylene-responsive transcription factor (RAP2/ERF), xyloglucan hydrolase (XTH) and late embryogenesis abundant protein14/light stress-regulated protein3 (LEA14/LSR3) were specifically up-regulated only in Ma/Ma grafted plants. RAP2/ERF are key regulators in various abiotic stress responses, including wounding repair mechanisms and hormone responses such as ethylene and jasmonic acid (JA) [92,93]. Some defense-related genes that are induced by ethylene contain a cis-acting sequence denoted GCC box, which is the core of the ethylene-responsive element and is essential for the expression of several pathogensis-related genes [94]. LEA14 is expressed in response to wound whereas LSR3 is expressed against dehydration that may occur at graft interface before the callus formation at stage 3 of graft union formation [38]. Other genes coding for wound-responsive protein family or maternal effect embryo arrest (MEE) are up-regulated in all graft combinations of mock-inoculated and PVY^{C}-to infected plants. MEE14 and MEE59 are involved in blocking the endosperm development [95] and, interestingly, fruits of Ma/Ma and UC/Ma grafted tomato plants did develop very few or no seeds in tests carried out for two consecutive years. The causes leading to this aspermy need a more detailed investigation.

Table 5. Significative differential expression of graft-responsive genes in mock-inoculated and infected tomato plants.

	Mock			Infected		
Locus Name	Ma/Ma *	UC/UC *	UC/Ma *	Ma/Ma *	UC/UC *	UC/Ma *
Solyc11g072600.1 (RAP2/ERF4) ***	3.20 **			3.63		
Solyc04g080700.2 (wound-responsive family protein)	3.57	3.26	2.55	2.51	2.86	2.61
Solyc07g056000.2 (XTH)			1.74		2.55	
Solyc07g052980.2 (XTH)	1.27			2.24		
Solyc02g091920.2 (XTH)	−1.58	−1.94	−1.61	−1.44		
Solyc09g075210.2 (AtLEA5/SAGS21)	−3.29	−4.33	−3.41	−4.52	−1.93	−2.13
Solyc06g006000.2 (MEE59)	5.03	5.51	5.49	3.75	4.23	4.25
Solyc01g108910.2 (MEE14)	5.41	5.03	5.00	5.00	4.54	4.64
Solyc01g095140.2 (LEA14, LSR3)	2.30			3.42	2.43	2.21

* = Scion/rootstock graft combination; ** = significantly differentially expressed genes (DEGs) with *p* value adjusted for multiple testing with the Benjamini–Hochberg procedure which controls false discovery rate (FDR) ≤ 0.05; *** = Arabi name in parentheses; RAP5/ERF4 = APETALA5/ethylene-responsive transcription factor 4; XTH = xyloglucan endotransglucosilase/hydrolase; AtLEA5/SAGS21 = late embryogenesis abundant protein 5/senescence-associated gene 21; MEE14,MEE59 = maternal effect embryo arrest; LEA14/LSR3 = late embryogenesis abundant protein/light stress protein.

Upon infection of PVY^{C}-to, the number of unique DEGs in grafted plants was reduced from 1991 to 1075 with a shift from 5.88% to 3.17% of total Solyc annotated genes, suggesting a diverging contribution of graft wound and viral infection in the number of DEGs. For example, Ma/Ma grafted plants modulated 74.4% and 80% of the unique DEGs in response, respectively, to graft wound and PVY^{C}-to. Additionally, the Ma rootstock reduced by 1.5-fold the number of unique DEGs in UC scion compared to self-grafted plants, whereas it did not make any other significant modulation in the transcripts of infected UC/UC and UC/Ma grafted plants. Recent evidences on VSRs coded by PVY^{C}-to could account for the decrease in the number of unique DEGs induced by graft wound in grafted plants mock-inoculated. PVY Hc-Pro binds in vivo to 21 nt vsiRNAs during infection in *N. benthamiana* [96] whereas the transgenic expression of Hc-Pro in tobacco altered the expression of hormone-responsive and defense-related genes involved in modifications of cell wall, transcriptional regulation, protein processing, and photosynthesis [97]. In *A. thaliana*, turnip mosaic virus Hc-Pro down-regulated a salicylic acid-mediated response by binding to the salicylic acid (SA)-binding protein SABP3. SABP3 is a tobacco chloroplastic carbonic anhydrase (CA1, SABP3) that was down-regulated in infected UC, Ma, Ma/Ma, and UC/Ma [64].

PVY codes Vpg as an additional VSRs. This protein is involved in the suppression of RNA silencing via the degradation of suppressor of gene silencing 3 (SGS3), which is crucial for the synthesis of dsRNAs derived from viral replication and, in turn, necessary for vsiRNAs production. DCL4/RDR6/SGS3 are thought to be involved in the appearance of the recovery phenotype in virus-infected plants [80]. However, our analysis could not specifically associate a significative up-/down-regulation of *SGS3* with the presence/absence of the recovery.

Thus in this study, the number of graft wound-responsive DEGs were reduced by either the viral infection or the Ma rootstock through mechanisms that will be discussed in the sub-heading 4.4. Most of the DEGs induced by grafting and potyviral infection were involved in the functional categories of signaling/response to stimulus and metabolic processes. PVY^{C}-to infection did not exclusively modulated any gene in the photosynthesis functional category. Conversely, 5 genes modulated only in grafted plants (Solyc01g105050.2, Solyc07g054210.2, Solyc07g063600.2, Solyc09g059640.1, Solyc10g007690.2) and 5 genes modulated by graft wound and potyviral infection (Solyc01g105030.2, Solyc06g060310.2, Solyc08g067320.1, Solyc08g067330.1, Solyc09g011080.2) were involved in the photosynthesis functional category and down-regulated.

4.4. Oxidative Stress and Antioxidant Enzymes

The production of reactive oxygen species (ROS) is a general event following grafting although it has been proposed that excess of ROS production in grafted cucurbits could be an indication of some graft incompatibility [98,99]. Beyond the implications in graft wound response, rapid production and accumulation of ROS such as O^{2-} and hydrogen peroxide is an important component of the plant response to viral infections both in the infected and non-infected tissues [100]. In infected plants, ROS toxicity can serve as direct protective agent driving the cross-linking of the cell wall or triggering the collapse of challenged host cell at the activation of the hypersensitive response (HR) to generate apoptopic-like signals [101]. In the PVY infection in tomato, the R/*avr* HR may be initiated by the interaction of tomato *pot*-1 resistance gene with the potyviral Vpg [102]. Thus, an up-regulation of genes coding for ROS scavenger enzymes is expected to mitigate ROS toxicity. Indeed, both graft-wounded and infected plants produce antioxidant enzymes such as catalase (CAT), peroxidase (POD) monooxygenase (MO), and manganese or copper/zinc superoxide dismutase (SOD) as ROS scavengers to mitigate ROS toxicity. In compatible grafted plants, the level of such antioxidants is higher than in non-grafted plants, is governed by the rootstock and enables better and strong root development in contrast to low level of antioxidants detected in the incompatible graft interactions [103]. In our study, Solyc03g11780.2 (respiratory burst oxidase homolog D (RbohD)) involved in ROS production was up-regulated in all graft combinations both in mock-inoculated and infected plants, with a seemingly superior up-regulation in mock-inoculated Ma/Ma and in all infected grafted plants (Table 6). Similarly, Solyc11g018800.1 (POD), Solyc12g094620.1 (CAT2) and Solyc04g025650.2 (monooxigenase 1, MO1) were overexpressed in almost all graft combinations both in mock-inoculated and infected plants, whereas Solyc02g082760.2 (CAT2) and Solyc06g049080.2 (MSD1) were significantly down-regulated only in mock-inoculated grafted plants, with the exception of Ma/Ma infected plants (Table 6). These data are in agreement with the notion that in compatible grafted plants, genes coding for ROS scavenger enzymes are up-regulated to compensate for an high ROS production in response to both graft and virus infection.

Table 6. Significative differential expression of graft-responsive antioxidant genes in mock-inoculated and infected tomato plants.

Locus Name	GO *	Mock			Infected		
		Ma/Ma **	UC/UC	UC/Ma	Ma/Ma	UC/UC	UC/Ma
Genes Involved in ROS Production							
Solyc03g117980.2 (RbohD) ***	GO:0004601	2.79 †	1.51	1.56	2.97	2.17	2.21
Genes Involved in ROS Scavenging							
Solyc01g006300.2 (POD2)	GO:0004601 GO:0006979 GO:0042744	3.38	2.45	2.54			
Solyc11g018800.1 (POD)	GO:0004601 GO:0006979 GO:0042744	4.21	3.10	2.89	4.05	4.04	4.18
Solyc02g082760.2 (CAT2)	GO:0004096 GO:0006979 GO:0042744	−1.76	−2.05	−2.22	−1.69		
Solyc12g094620.1 (CAT2)	GO:0004096 GO:0042542 GO:0042744	3.88	2.59	2.05		1.95	2.12
Solyc04g025650.2 (MO1)	GO:0004497	1.69	2.13	2.07	3.32	3.22	3.12
Solyc06g049080.2 (MSD1)	GO:0004784	−1.51	−1.48	−1.49			

* = Gene ontology involved in ROS response; ** = Scion/rootstock graft combination compared to non-grafted plants; *** = Arabi name in parentheses; RbohD = Respiratory burst oxidase homologue D; POD = Peroxidase 2; CAT = Catalase; MO = Monooxygenase; MSD = Manganese superoxide dismutase; † = significantly differentially expressed genes (DEGs) with p value adjusted for multiple testing with the Benjamini–Hochberg procedure which controls false discovery rate (FDR) ≤ 0.05.

4.5. Resilience and Tolerance

Response to graft wound and viral infection was clearly different in Ma and UC. Our study recorded 91 DEGs in UC plants inoculated with PVY^{C}-to and a huge accumulation of viral RNA between 14 and 21 dpi, which probably interfered with recovery from disease symptoms. Conversely, we recorded only 2 DEGs (*DCL2* and *CSD2*) in plants of the Ma ecotype, a reduced accumulation of viral RNA and recovery from disease symptom upon infection of the three taxonomically distinct viruses used in our studies. Thus, Ma was refractory to changes in gene expression and resilient to disease symptoms. *DCL2* and *CSD2* were up-regulated in Ma, suggesting that it employed few resources in response to virus infection but specific enough to recover from disease symptoms. According to Kørner et al. [80], recovered tissues may be in a "tolerant state," characterized by reduced levels of viral RNA and accumulation of vsiRNAs.

Maintenance of a virus infection in recovered leaves has a beneficial impact also on the resilience of plants to abiotic stresses and drought [104]; a characteristic that we did not test. Conversely, the up-regulation of resistance genes requires energy resources, which are generally recruited from primary metabolism and plant growth and development [105]. For example, leaves are generally source tissues but upon infection of airborne pathogens, such as viruses, they strengthen cell wall and down-regulate genes involved in photosynthesis [105]. In UC, we detected 18 down-regulated genes involved in photosynthesis in response to both PVY^{C}-to infection and graft wound. Therefore, we propose that the Ma response to the infection of TSWV, CMV, and PVY could be a case of tolerance, which is now reconsidered as "a mitigation of the impact of virus infection irrespective of the pathogen load" [104–107].

5. Conclusions

Severe strains of TSWV, CMV, and PVY are arthropod-borne pathogens that induce devastating crop diseases worldwide. Current control methods are unsustainable whereas grafting offers a promising environmentally friendly alternative. The results of our studies suggest that grafting may represent a rapid and sustainable approach to deal quickly and mitigate the negative effects of airborne viral infections in tomatoes when useful resistance genes are not available or have not yet been identified. The Manduria tomato ecotype has proven to be an adaptive rootstock, considering its tolerance to high virus loads and resilience to the appearance of symptoms induced by three taxonomically distinct arthropod-borne viruses. However, the graft wound itself activates defense responses that impact infection as demonstrated by susceptible varieties which, when grafted on themselves, showed levels of tolerance to viral infection and resilience to the appearance of symptoms very similar to those found in the Manduria ecotype. Thus grafted plants probably operate with higher energy resources compared to non-grafted counterparts and necessary to respond to wounding. At the same time, the energy resources contribute to prevent over-accumulation and expression of viral RNAs. This equilibrium limits the cost for the host while saving resources by reducing the number of DEGs necessary to respond to viral infection.

In principle, the adoption of vegetable grafting in agriculture requires an accurate selection of the rootstock to cope with specific biotic and abiotic stresses. Collectively our studies offer an unprecedented view of grafting based on tolerance rather than the resistance that might be fruitfully applied to a wide range of pathogens such as fungi, oomycetes, and bacteria, as well as other virus species.

Author Contributions: Conceptualization, D.G.; Formal Analysis, R.S. and M.F.; Data Curation, R.S., M.F. and T.M.; Writing—Original Draft Preparation, D.G.; Writing—Review and Editing, R.S., M.F., D.G. and T.M. All authors have read and agreed to the published version of the manuscript.

References

1. Xu, C.; Sun, X.; Taylor, A.; Jiao, C.; Xu, Y.; Cai, X.; Wang, X.; Ge, C.; Pan, G.; Wang, Q.; et al. Diversity, distribution, and evolution of tomato viruses in China uncovered by small RNA sequencing. *J. Virol.* **2017**, *91*, e00173–17. [CrossRef] [PubMed]
2. Scholthof, K.B.G.; Adkins, S.; Czosnek, H.; Palukaitis, P.; Jacquot, E.; Hohn, T.; Hohn, B.; Saunders, K.; Candresse, T.; Ahlquist, P.; et al. Top 10 plant viruses in molecular plant pathology. *Mol. Plant Pathol.* **2011**, *12*, 938–954. [CrossRef] [PubMed]
3. García-Arenal, F.; Palukaitis, P. Cucumber mosaic virus. In *Desk Encyclopedia of Plant and Fungal Virology*; Mahy, B.W.J., van Regenmortel, M.H.V., Eds.; Elsevier: Amsterdam, The Netherlands, 2008; pp. 171–176.
4. Jacquemond, M. Cucumber mosaic virus. *Adv. Virus Res.* **2012**, *84*, 440–505.
5. Parrella, G.; Gognalons, P.; Gebre-Selassiè, K.; Vovlas, C. An update of the host range of Tomato spotted wilt virus. *J. Plant Pathol.* **2003**, *85*, 227–264.
6. Turina, M.; Tavella, L.; Ciuffo, M. Tospoviruses in the Medirranean area. *Adv. Virus Res.* **2012**, *84*, 404–439.
7. Moury, B.; Verdin, E. Viruses of pepper crop in the Mediterranean basin: A remarkable stasis. *Adv. Virus Res.* **2012**, *84*, 128–163.
8. Loebenstein, G.; Gaba, V. Viruses of Potato. *Adv. Virus Res.* **2012**, *84*, 210–248.
9. Turina, M.; Kormelink, R.; Resende, T. Resistance to tospoviruses in vegetable crops: Epidemiological and molecular aspects. *Annu. Rev. Phytopathol.* **2016**, *54*, 347–371. [CrossRef]
10. de Oliveira, A.S.; Boiteux, L.S.; Kormelink, R.; Resende, R.O. The Sw-5 gene cluster: Tomato breeding and research toward orthotospovirus disease control. *Front. Plant Sci.* **2018**, *9*, 1055. [CrossRef]
11. Oliveira, R.M.; Dianese, É.C.; Lima, M.F.; Resende, R.O.; Inoue-Nagata, A.I.; Boiteux, L.S. Sources of resistance to Potato virus Yand Pepper yellow mosaic virus in *Solanum* (section *Lycopersicon*) germplasm. *Eur. J. Plant Pathol.* **2017**, *150*, 691–699. [CrossRef]
12. Pappu, H.R.; Jones, R.A.C.; Jain, R.K. Global status of Tospovirus epidemics in diverse cropping systems: Successes achieved and challenges ahead. *Virus Res.* **2009**, *141*, 219–236. [CrossRef] [PubMed]

13. López, C.; Aramburu, J.; Galipienso, L.; Soler, S.; Nuez, F.; Rubio, L. Evolutionary analysis of tomato Sw-5 resistance-breaking isolates of Tomato spotted wilt virus. *J. Gen. Virol.* **2011**, *92*, 210–215. [CrossRef] [PubMed]
14. Peiró, A.; Cañizares, M.C.; Rubio, L.; López, C.; Moriones, E.; Aramburu, J.; Navarro, J.S. The movement protein (NSM) of Tomato spotted wilt virus is the avirulence determinant in the tomato Sw-5 gene-based resistance. *Mol. Plant Pathol.* **2014**, *15*, 802–813. [CrossRef]
15. Abad, J.; Anastasio, G.; Fraile, A.; García-Arenal, F. A search for resistance to Cucumber mosaic virus in the genus Lycopersicon. *J. Plant Pathol.* **2000**, *82*, 39–48.
16. Murphy, J.F.; Sikora, E.J.; Sammons, B.; Kaniewski, W.K. Performance of transgenic tomatoes expressing cucumber mosaic virus CP gene under epidemic conditions. *Hort. Sci.* **1998**, *33*, 1032–1035. [CrossRef]
17. Tomassoli, L.; Ilardi, V.; Barba, M.; Kaniewski, W. Resistance of transgenic tomato to Cucumber mosaic cucumovirus under field conditions. *Mol. Breed.* **1999**, *5*, 121–130. [CrossRef]
18. Gallitelli, D.; Vovlas, C.; Martelli, G.P.; Montasser, M.S.; Tousignant, M.E.; Kaper, J.M. Satellite-mediated protection of tomato against cucumber mosaic virus: II. Field test under natural epidemic conditions in southern Italy. *Plant Dis.* **1991**, *75*, 93–95. [CrossRef]
19. Sayama, H.; Sato, T.; Kominato, M.; Natsuaki, T.; Kaper, J.M. Field testing of a satellite-containing attenuated strain of cucumber mosaic virus for tomato protection in Japan. *Phytopathology* **1993**, *83*, 405–410. [CrossRef]
20. Mascia, T.; Gallitelli, D. Satellites as Viral Biocontrol Agents. In *Viroids and Satellites*; Hadidi, A., Flores., R., Randles, J.W., Palukaitis, P., Eds.; Academic Press: London, UK, 2017; pp. 681–688.
21. Aramburu, J.; Galipienso, L.; Matas, M. Characterization of potato virus Y isolates from tomato crops in northeast Spain. *Eur. J. Plant Pathol.* **2006**, *115*, 247–258. [CrossRef]
22. Parrella, G.; Ruffel, S.; Moretti, A.; Morel, C.; Palloix, A.; Caranta, C. Recessive resistance genes against potyviruses are localized in collinear genomic regions of the tomato (*Lycopersicon* spp.) and pepper (*Capsicum* spp.) genomes. *Theor. Appl. Genet.* **2002**, *105*, 855–861. [CrossRef]
23. Kang, B.C.; Yeam, I.; Frantz, J.D.; Murphy, J.F.; Jahn, M.M. The *pvr1* locus in pepper encodes a translation initiation factor eIF4E that interacts with Tobacco etch virus VPg. *Plant J.* **2005**, *41*, 392–405. [CrossRef] [PubMed]
24. Albacete, A.; Martinez-Andujar, C.; Martinez-Pérez, A.; Thompson, A.J.; Dodd, I.C.; Pérez-Alfocea, F. Unravelling rootstock x scion interactions to improve food security. *J. Exp. Bot.* **2015**, *66*, 2211–2226. [CrossRef] [PubMed]
25. Rouphael, Y.; Kyriacou, M.C.; Colla, G. Vegetable Grafting: A Toolbox for Securing Yield Stability under Multiple Stress Conditions. *Front. Plant Sci.* **2018**, *8*, 2255. [CrossRef] [PubMed]
26. Kyriacou, M.C.; Rouphael, Y. Towards a new definition of quality for fresh fruits and vegetables. *Sci. Hortic.* **2018**, *234*, 463–469. [CrossRef]
27. Louws, F.J.; Rivard, C.L.; Kubota, C. Grafting fruiting vegetables to manage soilborne pathogens, foliar pathogens, arthropods and weeds. *Sci. Hortic.* **2010**, *127*, 127–146. [CrossRef]
28. Colla, G.; Rouphael, Y.; Leonardi, C.; Bie, Z. Role of grafting in vegetable crops grown under saline conditions. *Sci. Hort.* **2010**, *127*, 147–155. [CrossRef]
29. Colla, G.; Rouphael, Y.; Mirabelli, C.; Cardarelli, M. Nitrogenus efficiency traits of mini-watermelon in response to grafting and nitrogen-fertilization doses. *J. Plant Nutr. Soil Sci.* **2011**, *174*, 933–941. [CrossRef]
30. Savvas, D.; Colla, G.; Rouphael, Y.; Schwarz, D. Amelioration of heavy metal and nutrient stress in fruit vegetables by grafting. *Sci. Hortic.* **2010**, *127*, 156–161. [CrossRef]
31. Rouphael, Y.; Cardarelli, M.; Colla, G. Yield, mineral composition, water relations, and water use efficiency of grafted mini-watermelon plants under deficit irrigation. *HortScience* **2008**, *43*, 730–736. [CrossRef]
32. Schwarz, D.; Rouphael, Y.; Colla, G.; Venema, J.H. Grafting as a tool to improve tolerance of vegetables to abiotic stresses: Thermal stress, water stress and organic pollutants. *Sci. Hortic.* **2010**, *127*, 162–171. [CrossRef]
33. Borgognone, D.; Colla, G.; Rouphael, Y.; Cardarelli, M.; Rea, E.; Schwarz, D. (Effect of nitrogen form and nutrient solution pH on growth and mineral composition of self-grafted and grafted tomatoes. *Sci. Hortic.* **2013**, *149*, 61–69. [CrossRef]
34. Gaion, L.A.; Braz, L.T.; Carvalho, R.F. Grafting in vegetable crops: A great technique for Agriculture. *Int. J. Veg. Sci.* **2018**, *24*, 85–102. [CrossRef]

35. Riga, P.; Benedicto, L.; García-Flores, L.; Villaño, D.; Medina, S.; Gil-Izquierdo, A. Rootstock effect on serotonin and nutritional quality of tomatoes produced under low temperature and light conditions. *J. Food Comp. Anal.* **2016**, *46*, 50–59. [CrossRef]
36. Tsaballa, A.; Athanasiadis, C.; Pasentsis, K.; Ganopoulos, I.; Nianiou-Obeidat, I.; Tsaftaris, A. Molecular studies of inheritable grafting induced changes in Pepper (*Capsicum annuum*) fruit shape. *Sci. Hort.* **2013**, *149*, 2–8. [CrossRef]
37. Temperini, O.; Calabrese, N.; Temperini, A.; Rouphael, Y.; Tesid, R.; Lenzid, A.; Carito, A.; Colla, G. Grafting artichoke onto cardoon rootstocks: Graft compatibility, yield and Verticillium wilt incidence. *Sci. Hort.* **2013**, *149*, 22–27. [CrossRef]
38. Kumar, S.; Bharti, N.; Saravaiya, S.N. Vegetable Grafting: A Surgical Approach to combat biotic and abiotic stresses: A review. *Agric. Rev.* **2018**, *39*, 1–11. [CrossRef]
39. Lee, J.M.; Kubota, C.; Tsao, S.J.; Bie, Z.; Echevarria, P.H.; Morra, L.; Oda, M. Current status of vegetable grafting: Diffusion, grafting techniques, automation. *Sci. Hortic.* **2010**, *127*, 93–105. [CrossRef]
40. Singh, H.; Kumar, P.; Chaudhari, S.; Edelstei, M. Tomato grafting: A global perspective. *Hortscience* **2017**, *52*, 1328–1336. [CrossRef]
41. Avgelis, A.D. Watermelon necrosis caused by a strain of melon necrotic spot virus. *Plant Pathol.* **1989**, *38*, 618–622. [CrossRef]
42. Gomez, J.; Cuadrado, I.; Velasco, V. Melon necrotic spot virus (MNSV) in Almeria. III. The effect of melon grafting on MNSV control. *Bol. San. Veg. Plagas* **1993**, *19*, 197–219.
43. Hirai, G.; Nakazumi, H.; Yagi, R.; Nakano, M.; Shiga, Y.; Dohi, H. A new MNSV resistant melon rootstock variety "Sorachi Dai Ko 3". *Bull. Hokkaido Prefect. Agricult. Exp. Stn.* **2003**, *84*, 47–54.
44. Banttari, E.E.; Vanette, J.R. Aerosol spread of plant viruses: Potential role in disease outbreaks. *Ann. N. Y. Acad. Sci.* **1980**, *353*, 167–173. [CrossRef]
45. Fillhart, R.C.; Bachand, G.D.; Castello, J.D. Airborne transmission of tomato mosaic Tobamovirus and its occurrence in red spruce in the northeastern United States. *Can J. Forest Res.* **1997**, *27*, 1176–1181. [CrossRef]
46. Rivero, R.M.; Ruiz, J.M.; Romero, L. Role of grafting in horticultural plants under stress conditions. *Food. Agric. Environ.* **2003**, *1*, 70–74.
47. Rivard, C.L.; Louws, F.J. Grafting to manage soilborne diseases in heirloom tomato production. *Hortscience* **2008**, *43*, 2104–2111. [CrossRef]
48. Miguel, A. Use of grafted plants and IPM methods for the production of tomatoes in the Mediterranean region. In *Proceedings of the International Conference Alternatives to Methyl Bromide, Lisbon, Portugal, 27–30 September 2004*; Batchelor, T., Flávia, A., Eds.; European Commission: Brussels, Belgium; pp. 141–145.
49. Schwarz, D.; Beuch, U.; Bandte, M.; Fakhro, A.; Buttner, C.; Obermeieret, C. Spread and interaction of Pepino mosaic virus (PepMV) and *Pythium aphanidermatum* in a closed nutrient solution recirculation system: Effects on tomato growth and yield. *Plant Pathol.* **2010**, *59*, 443–452. [CrossRef]
50. Samuel, G. The movement of tobacco mosaic virus within the plant. *Ann. Appl. Biol.* **1934**, *21*, 90–111. [CrossRef]
51. Hull, R. *Plant Virology*, 5th ed.; Academic Press: London, UK, 2014.
52. Andrianifahanana, M.; Lovins, K.; Dute, R.; Sikora, E.; Murphy, J.F. Pathway for phloem-dependent movement of pepper mottle potyvirus in the stem of *Capsicum annuum*. *Phytopathology* **1997**, *87*, 892–898. [CrossRef]
53. Leisner, S.M.; Turgeon, R.; Howell, S.H. Long distance movement of cauliflower mosaic virus in infected turnip plants. *Mol. Plant Microbe Interact* **1992**, *5*, 41–47. [CrossRef]
54. Mas, P.; Pallas, V. Long-distance movement of cherry leaf roll virus in infected tobacco plants. *J. Gen. Virol.* **1996**, *77*, 531–540. [CrossRef]
55. Cheng, N.H.; Lin, C.L.; Carter, S.A.; Nelson, R.S. Vascular invasion routes and systemic accumulation patterns of tobacco mosaic virus in Nicotiana benthamiana. *Plant J.* **2000**, *23*, 349–362. [CrossRef] [PubMed]
56. Ding, X.S.; Carter, S.A.; Deom, C.M.; Nelson, R.S. Tobamovirus and potyvirus accumulation in minor veins of inoculated leaves from representatives of the Solanaceae and Fabaceae. *Plant Physiol.* **1998**, *116*, 125–136. [CrossRef]
57. Vuorinen, A.L.; Kelloniemia, J.; Valkonen, J.P.T. Why do viruses need phloem for systemic invasion of plants? *Plant Sci.* **2011**, *181*, 355–363. [CrossRef]
58. Navarro, J.A.; Sanchez-Navarro, J.A.; Pallas, V. Key checkpoints in the movement of plant viruses through the host. *Adv. Virus Res.* **2019**, *104*, 1–64.

59. Turgeon, R. The sink–source transition in leaves. *Annu. Rev. Plant Physiol. Plant Mol. Biol.* **1989**, *40*, 119–138. [CrossRef]
60. Rajamäki, M.-L.; Valkonen, J.P.T. Viral genome-linked protein (VPg) controls accumulation and phloem-loading of a potyvirus in inoculated potato leaves. *Mol. Plant Microbe Interact.* **2002**, *15*, 138–149.
61. Blackman, L.M.; Boevink, P.; Cruz, S.S.; Palukaitis, P.; Oparka, K.J. The movement protein of cucumber mosaic virus traffics into sieve elements in minor veins of *Nicotiana clevelandii*. *Plant Cell* **1998**, *10*, 525–538. [CrossRef]
62. Spanò, R.; Mascia, T.; Kormelink, R.; Gallitelli, D. Grafting on a Non-Transgenic Tolerant Tomato Variety Confers Resistance to the Infection of a Sw5- Breaking Strain of Tomato spotted wilt virus via RNA Silencing. *PLoS ONE* **2015**, *10*, e0141319. [CrossRef]
63. Spanò, R.; Gallitelli, D.; Mascia, T. Grafting to manage infections of top stunting and necrogenic strains of cucucmber mosaic virus in tomato. *Ann. App. Biol.* **2017**, *171*, 393–404. [CrossRef]
64. Spanò, R.; Ferrara, M.; Montemurro, C.; Mulè, G.; Gallitelli, D.; Mascia, T. Grafting alters tomato transcriptome and enhances tolerance to an airborne virus infection. *Sci. Rep.* **2020**, *10*, 2538. [CrossRef]
65. Mascia, T.; Santovito, E.; Gallitelli, D.; Cillo, F. Evaluation of reference genes for quantitative reverse transcription polymerase chain reaction normalization in infected tomato plants. *Mol. Plant Pathol.* **2010**, *11*, 805–816. [CrossRef] [PubMed]
66. Finetti-Sialer, M.M.; Lanave, C.; Padula, M.; Vovlas, C.; Gallitelli, D. Occurrence of two distinct tomato spotted wilt virus subgroups in Southern Italy. *J. Plant Pathol.* **2002**, *84*, 145–152.
67. Ciuffo, M.; Finetti Sialer, M.M.; Gallitelli, D.; Turina, M. First report in Italy of a resistance breaking strain of infecting tomato cultivars carrying the resistance gene. *New Dis. Rep.* **2005**, *10*, 48.
68. Gallitelli, D. The ecology of cucumber mosaic virus and sustainable agriculture. *Virus Res.* **2000**, *71*, 9–21. [CrossRef]
69. Aramburu, J.; Galipienso, L.; Lopez, C. Reappearance of cucumber mosaic virus isolates belonging to subgroup IB in tomato plants in north-eastern Spain. *J. Phytopathol.* **2007**, *155*, 513–518. [CrossRef]
70. Mascia, T.; Finetti-Sialer, M.M.; Cillo, F.; Gallitelli, D. Biological and molecular characterization of a recombinant isolate of Potato virus Y associated with a tomato necrotic disease occurring in Italy. *J. Plant Pathol.* **2010**, *92*, 131–138.
71. Ruiz, M.T.; Voinnet, O.; Baulcombe, D.C. Initiation and maintenance of virus-induced gene silencing. *Plant Cell* **1998**, *10*, 937–946. [CrossRef]
72. Ding, S.W.; Voinnet, O. Antiviral immunity directed by small RNAs. *Cell* **2007**, *130*, 413–426. [CrossRef]
73. Kalantidis, K.; Schumacher, H.T.; Alexi-adis, T.; Helm, J.M. RNA silencing movement in plants. *Biol. Cell* **2008**, *100*, 13–26. [CrossRef]
74. Molnar, A.; Melnyk, C.; Baulcombe, D.C. Silencing signals in plants: A long journey for small RNAs. *Genome Biol.* **2011**, *12*, 215. [CrossRef]
75. Mermigka, G.; Verret, F.; Kalantidis, K. RNA silencing movement in plants. *J. Integr. Plant Biol.* **2016**, *58*, 328–342. [CrossRef] [PubMed]
76. Palauqui, J.C.; Elmayan, T.; Pollien, J.M.; Vaucheret, H. Systemic acquired silencing: Transgene-specific post-transcriptional silencing is transmitted by grafting from silenced stocks to non-silenced scions. *EMBO J.* **1997**, *16*, 4738–4745. [CrossRef] [PubMed]
77. Han, Y.; Grierson, D. Enhancement of post-transcriptional gene silencing by grafting. *Plant Signal Behav.* **2008**, *3*, 30–33. [CrossRef] [PubMed]
78. Ali, E.M.; Kobayashi, K.; Yamaoka, N.; Ishikawa, M.; Nishiguchi, M. Graft Transmission of RNA Silencing to Non-Transgenic Scions for Conferring Virus Resistance in Tobacco. *PLoS ONE* **2013**, *8*, e63257.
79. Kasai, A.; Sano, T.; Harada, T. Scion on a Stock Producing siRNAs of Potato Spindle Tuber Viroid (PSTVd) Attenuates Accumulation of the Viroid. *PLoS ONE* **2013** *8*, e57736. [CrossRef]
80. Kørner, C.J.; Pitzalis, N.; Peña, E.J.; Erhardt, M.; Vazquez, F.; Heinlein, M. Crosstalk between PTGS and TGS pathways in natural antiviral immunity and disease recovery. *Nat. Plants* **2018**, *4*, 157–164. [CrossRef]
81. Takeda, A.; Sugiyama, K.; Nagano, H.; Mori, M.; Kaido, M.; Mise, K.; Tsuda, S.; Okuno, T. Identification of a novel RNA silencing suppressor, NSs protein of Tomato spotted wilt virus. *FEBS Lett.* **2002**, *532*, 75–79. [CrossRef]
82. Bucher, E.; Sijen, T.; de Haan, P.; Goldbach, R.; Prins, M. Negative-strand tospoviruses and tenuiviruses carry a gene for a suppressor of gene silencing at analogous genomic positions. *J. Virol.* **2003**, *77*, 1329–1336. [CrossRef]

83. Hedil, M.; De Ronde, D.; Kormelink, R. Biochemical analysis of NSs from different tospoviruses. *Virus Res.* **2017**, *242*, 149–155. [CrossRef]
84. Rubino, L. Biology of satellites. In *Viroids and Satellites*; Hadidi, A., Flores, R., Randles, J.W., Palukaitis, P., Eds.; Academic Press: London, UK, 2017; pp. 567–575.
85. Cillo, F.; Finetti Sialer, M.M.; Papanice, M.A.; Gallitelli, D. Analysis of mechanisms involved in the cucumber mosaic virus satellite RNA-mediated transgenic resistance in tomato plants. *Mol. Plant Microbe Interact.* **2004**, *17*, 98–108. [CrossRef]
86. Gilardi, G.; Gullino, M.L.; Garibaldi, A. Critical aspects of grafting as a possible strategy to manage soil-borne pathogens. *Sci. Hortic.* **2013**, *149*, 19–21. [CrossRef]
87. Guan, W.J.; Zhao, X.; Hassell, R.; Thies, J. Defense mechanisms involved in disease resistance of grafted vegetables. *Hortscience* **2012**, *47*, 164–170. [CrossRef]
88. Cookson, S.J.; Clemente Moreno, M.J.; Hevin, C.; Nyamba Mendome, L.Z.; Delrot, S.; Magnin, N.; Trossat-Magnin, C.; Ollat, N. Heterografting with nonself rootstocks induces genes involved in stress responses at the graft interface when compared with autografted controls. *J. Exp. Bot.* **2014**, *65*, 2473–2481. [CrossRef]
89. Melnyk, C.W. Plant grafting: Insights into tissue regeneration. *Regeneration* **2016**, *4*, 3–14. [CrossRef] [PubMed]
90. Melnyk, C.W.; Gabel, A.; Hardcastle, T.J.; Robinson, S.; MIyashima, S.; Grosse, I.; Meyerowitz, E.M. Transcriptome dynamics at *Arabidopsis* graft junctions reveal an inter tissue recognition mechanism that activates vascular regeneration. *Proc. Natl. Acad. Sci. USA* **2019**, *115*, E2447–E2456. [CrossRef]
91. Cheong, Y.H.; Chang, H.-S.; Gupta, R.; Wang, X.; Zhu, T.; Luan, S. Transcriptional profiling reveals novel interactions between wounding, pathogen, abiotic stress, and hormonal responses in Arabidopsis. *Plant Physiol.* **2002**, *129*, 661–677. [CrossRef]
92. Heyman, J.; Canher, B.; Bisht, A.; Christiaens, F.; De Veylder, L. Emerging role of the plant ERF transcription factors in coordinating wound defense responses and repair. *J. Cell Sci.* **2018**, *131*, jcs208215. [CrossRef]
93. Xie, Z.; Nolan, T.M.; Jiang, H.; Yin, Y. AP2/ERF Transcription Factor Regulatory Networks in Hormone and Abiotic Stress Responses in Arabidopsis. *Front. Plant Sci.* **2019**, *10*, 228. [CrossRef]
94. Fujimoto, S.Y.; Ohta, M.; Usui, A.; Shinshi, H.; Ohme-Takagi, M. Arabidopsis ethylene-responsive element binding factors act as transcriptional activators or repressors of GCC box-mediated gene expression. *Plant Cell* **2000**, *12*, 393–404.
95. Pagnussat, G.C.; Yu, H.-J.; Quy, A.; Ngo, Q.A.; Rajani, S.; Mayalagu, S.; Johnson, C.S.; Capron, A.; Xie, L.-F.; Ye, D.; et al. Genetic and molecular identification of genes required for female gametophyte development and function. *Arab. Dev.* **2005**, *132*, 603–614. [CrossRef]
96. Del Toro, F.J.; Donaire, L.; Aguilar, E.; Chung, B.-N.; Tenllado, F.; Canto, T. Potato virus Y HCPro suppression of antiviral silencing in *Nicotiana benthamiana* plants correlates with its ability to bind in vivo to 21- and 22-nucleotide small RNAs of viral sequence. *J. Virol.* **2017**. [CrossRef] [PubMed]
97. Soitamo, A.J.; Jada, B.; Lehto, K. HC-Pro silencing suppressor significantly alters the gene expression profile in tobacco leaves and flowers. *BMC Plant Biol.* **2011**, *11*, 68. [CrossRef] [PubMed]
98. Aloni, B.; Karni, L.; Deventurero, G.; Levin, Z.; Cohen, R.; Katzir, N.; Lotan-Pompan, M.; Edelstein, M.; Aktas, H.; Turhan, E. Physiological and biochemical changes at the rootstock-scion interface in graft combinations between Cucurbita rootstocks and a melon scion. *J. Hortic. Sci. Biotech.* **2008**, *83*, 777–783. [CrossRef]
99. Xu, Q.; Guo, S.R.; Li, H.; Du, N.S.; Shu, S.; Sun, J. Physiological aspects of compatibility and incompatibility in grafted cucumber seedlings. *J. Am. Soc. Hort. Sci.* **2015**, *140*, 299–307. [CrossRef]
100. Mandadi, K.K.; Sholthof, K.-B.G. Plant Immune Responses Against Viruses: How Does a Virus Cause Disease? *Plant Cell* **2013**, *25*, 1489–1505. [CrossRef]
101. Sagi, M.; Davydov, O.; Orazova, S.; Yesbergenova, Z.; Ophir, R.; Stratmann, J.W.; Fluhr, R. Plant respiratory burst oxidase homologs impinge on wound responsiveness and development in *Lycopersicon esculentum*. *Plant Cell* **2004**, *16*, 616–628. [CrossRef]
102. Rouxel, T.; Balesdent, M.-H. Avirulence Genes. In *Encyclopedia of Life Sciences (ELS)*; John Wiley & Sons, Ltd.: Chichester, UK, 2010.
103. Aloni, B.; Cohen, R.; Karni, L.; Aktas, H.; Edelstein, M. Hormonal signaling in rootstock-scion interactions. *Sci. Hort.* **2010**, *127*, 119–126. [CrossRef]

104. Paudel, D.B.; Sanfaçon, H. Exploring the Diversity of Mechanisms Associated with Plant Tolerance to Virus Infection. *Front. Plant Sci.* **2018**, *9*, 1575. [CrossRef]
105. Berger, S.; Sinha, A.K.; Roitsch, T. Plant physiology meets phytopathology: Plant primary metabolism and plant–pathogen interactions. *J. Exp. Bot.* **2017**, *58*, 4019–4026. [CrossRef]
106. Cooper, J.I.; Jones, A.T. Responses of plants to viruses: Proposals for the use of terms. *Phytopathology* **1983**, *73*, 127–128. [CrossRef]
107. Pagan, I.; Garcia-Arenal, F. Tolerance to plant pathogens: Theory and experimental evidence. *Int. J. Mol. Sci.* **2018**, *19*, 810. [CrossRef] [PubMed]

2

Effects of Grafting on Morphophysiological and Yield Characteristic of Eggplant (*Solanum melongena* L.) Grafted onto Wild Relative Rootstocks

Ibrahim Musa [1,2], Mohd Y. Rafii [1,3,*], Khairulmazmi Ahmad [4], Shairul Izan Ramlee [3], Muhammad Asyraf Md Hatta [5], Yusuff Oladosu [1], Ismáila Muhammad [1], Samuel Chibuike Chukwu [1], Nur Nadzirah Mat Sulaiman [1], Arolu Fatai Ayanda [1] and Jamilu Halidu [3]

1 Institute of Tropical Agriculture and Food Security, Universiti Putra Malaysia, Serdang 43400, Selangor, Malaysia; ibrahimmusa@fukashere.edu.ng (I.M.); oladosuy@upm.edu.my (Y.O.); ismuha@gsu.edu.ng (I.M.); chukwu.samuel@ebsu.edu.ng (S.C.C.); gs50487@student.upm.edu.my (N.N.M.S.); gs53735@student.upm.edu.my (A.F.A.)

2 Department of Agronomy, Faculty of Agriculture, Federal University of Kashere, Gombe 0182, Gombe State, Nigeria

3 Department of Crop Science, Faculty of Agriculture, Universiti Putra Malaysia, Serdang 43400, Selangor, Malaysia; shairul@upm.edu.my (S.I.R.); gs51813@student.upm.edu.my (J.H.)

4 Department of Plant Protection, Faculty of Agriculture, Universiti Putra Malaysia, Serdang 43400, Selangor, Malaysia; khairulmazmi@upm.edu.my

5 Department of Agriculture Technology, Faculty of Agriculture, Universiti Putra Malaysia, Serdang 43400, Selangor, Malaysia; m.asyraf@upm.edu.my

* Correspondence: mrafii@upm.edu.my

Abstract: Grafting is regarded as an integral component of sustainable vegetable production. It is important in the management of soil-borne diseases, and reports suggest that grafting with viable rootstocks can enhance crop growth and yield. This research was conducted using splices and cleft grafting techniques to investigate graft compatibility among varieties of high yielding eggplant scion (MCV1, MCV2, CCV1, CCV2, CCV3, NCV, and TCV) grafted onto wild rootstocks (MWR, BWR, and TWR) to study their morphophysiological and yield characteristics. High yielding scions grafted onto wild relative rootstocks were compared with two controls including self-grafted and non-grafted. All the scion had a high rate of germination (≥95%) and remarkable graft success (100%) was recorded in MCV1, MCV2, and TCV using the cleft techniques. Generally, the use of rootstocks resulted in higher total and marketable fruit yield compared to the non-grafted and self-grafted scion plants, respectively. In particular, MWR and TWR rootstock conferred the highest vigour to the scion, resulting in the highest values recorded for total and marketable fruit yield, number of fruits per plant and average fruit weight. A similar result was obtained in fruit length and diameter, where long and wide fruits were observed in scions grafted onto MWR and TWR rootstocks, respectively. Grafting of high yielding eggplant scion onto resistant MWR, BWR and TWR eggplant rootstock was found to be beneficial for eggplant cultivation. The remarkable compatibility and vigour of the rootstock with scion led to the improvement in total and marketable yield of the fruits. As such, it can be concluded that the use of wild relative rootstocks of eggplant species can be a valuable method of improving eggplant production.

Keywords: eggplant; splice grafting; cleft grafting; scion-rootstock combinations; *Solanum torvum*

1. Introduction

Eggplant (*Solanum melongena* L.) is one of the most cultivated vegetables in temperate and tropical regions of the world. It differs in colour, shape and fruit size. The most widely cultivated type is the purple cultivars [1]. The colour depends on content of anthocyanins present in the fruit skin. Anthocyanins are powerful antioxidants group that are located in vacuoles of plant cell belonging to phenolic flavonoids. Eggplant skin extracts have shown high capacity for inhibiting inflammation that leads to atherosclerosis and suppressing metastasis and tumours growth in the blood vessels by developing scavenging free radicals that can damage proteins and lipids [2]. As reported by Yang [3], eggplant was ranked in the top 10 for SOS activity (superoxide scavenging) among 120 vegetable species evaluated based on their antioxidant activity. Additionally, flavonoids isolated from eggplant have shown a great antioxidant activity and their presence in diet is associated with lower risk of heart disease, lung cancer and stroke [2].

Despite the high economic importance of this crop, it is susceptible to many pest and diseases particularly insects, nematodes, bacterial wilt, Verticillium and Fusarium wilts [4]. As reported by Moncada et al. [2], more than 78% yield losses were recorded due to soil-borne pests and pathogens such as *Meloidogyne spp.*, Fusarium and Verticillium stroke [2]. For the past few decades, the primary objectives in crop production have been to increase yield and productivity to cater for the increasing world population. As a consequence, limited adoption of crop rotations under intensive cropping systems for vegetable farming in both open field and control environment has built up a negative influence on yield and yield quality as a result of biotic or abiotic factors. These negative effects have led to a consequent increase in use of agricultural input such as pesticides, inorganic fertilizer and water use, which in turn has a detrimental effect on health and environment. The absence of effective chemicals for controlling diseases, coupled with sexual incompatibility between wild and elite cultivars, indicates that a short-term approach to the problem of cultivating susceptible eggplant is the grafting of elite cultivars on rootstocks with multiple pathogen resistance [4].

Grafting is a significant integrated approach to the management of biotic and abiotic limitations, fruit yield and quality of solanaceous and cucurbitaceous crops. Management of biotic and abiotic stress conditions is one of the major reasons for the development and use of vegetable grafting technique [2]. This technique has been successfully used in the management of diseases such as *Verticillium, Fusarium, Ralstonia*, root-knot nematodes and several soil-borne pathogens [2]. Hence, vegetable grafting onto resistant rootstocks is a tool for the improvement of susceptible high-yielding scion against biotic and abiotic stresses [3]. Despite the comparatively high cost of grafted plants resulting from the higher labour and production input compared to normal (un-grafted) plants, grafting has developed into a specific cultural practice that helps to increased yield and production output and improves economic viability in sustainable vegetable production under both open field and glasshouse conditions [1].

Research has shown that positive relationships exist between plants grafted on resistant and vigorous rootstocks, and improvements in production and fruit quality parameters (e.g., phytochemical content, such as phenolic compound [2]. Today, the resistance and fight against soil-borne diseases remain the major objective for applying the grafting technique. It is an effective and fast method for selecting desirable crop characteristics than species hybridization, selection processes or natural screening [5]. Eggplant scions grafted with tomato rootstock have significant increase in fruit yields as a result of increased fruit size and number of fruits in comparison self-rooted plants and self-grafted eggplants [5]. Grafting eggplant onto interspecific eggplant hybrids such as SI×SM hybrid has proved advantageous for eggplant production, as the high vigour and good compatibility of the rootstock with the scion results in improved early and total yield without negative effects on apparent fruit quality or composition [2]. Studies on grafting between *S. melongena* and *S. incanum*, *S. aethiopicum* and *S. habrochaites* germplasm, as well as intraspecific and interspecific eggplant hybrids, have shown promising results for developing new rootstocks for eggplant production [2]. As reported by [2–4], grafting can influence yield and fruit quality in eggplant. Generally, grafting increases precocity and plant health, providing more vigorous and stronger root systems compared to un-grafted plants, and the major advantage of this is improved nutrient absorption and water use efficiency. Furthermore, grafted plants provide fruit that is considerably larger than un-grafted plants, while other fruit traits such as soluble solids, skin colour, flesh texture, and shape are negatively influenced by the rootstock despite being under genetic control [2]. Therefore, the objective of this research is to validate the assumption that the use of wild relative rootstocks may be a good approach to enhance eggplant performance by exploiting the advantage of grafting to improve eggplants rooting systems and vigour through the affinity between rootstock and scion.

2. Results

2.1. Observation of Seedlings at the Nursery Stage and Grafting Success (%)

The results indicate that there were highly significant ($p < 0.01$) differences between the different varieties in terms of seed germination, germination percentage (%) and days for the seedling to reached the grafting stage (Figure 1). Among the varieties studied, the earliest germination (3 days) was observed in NCV, followed by MCV2 (3.67 days after sowing), whereas BWR (wild type) germinated later, 24 days after sowing. However, the cultivated genotypes germinated from 3 to 7 days after sowing, while the wild relatives MWR, BWR and TWR germinated between 16 to 24 days after sowing. All the genotypes exhibited a high percentage of germination (>90%) except for the wild relatives (Figure 1). The highest germination percentage (98.47%) was recorded in NCV, followed by CCV1 with 97.83%, whereas a significantly lower ($p < 0.05$) germination percentage (65.83%) was recorded in MWR. Grafting with the cleft technique was relatively more successful with over 89% success in all grafting combinations including self-grafted plants compared to the splice method with over 65% success in grafting combinations (Figure 2). Using the cleft technique, the success recorded among graft combinations ranges between 89% to 100% (Figure 2). The highest graft success was observed in MCV2/MWR, TCV/MWR, MCV1/BWR, MCV2/BWR, TCV/BWR, MCV2/TWR, TCV/TWR and TCV/TCV, whereas NCV/BWR has relatively lower graft success. In the splicing technique, graft success was comparably higher in MCV1/BWR than in TCV/BWR, TCV/TWR, CCV2/TWR, NCV/TWR and MCV1/TWR (Figure 2). The lowest graft success was observed in CCV2/BWR and NCV/BWR, respectively.

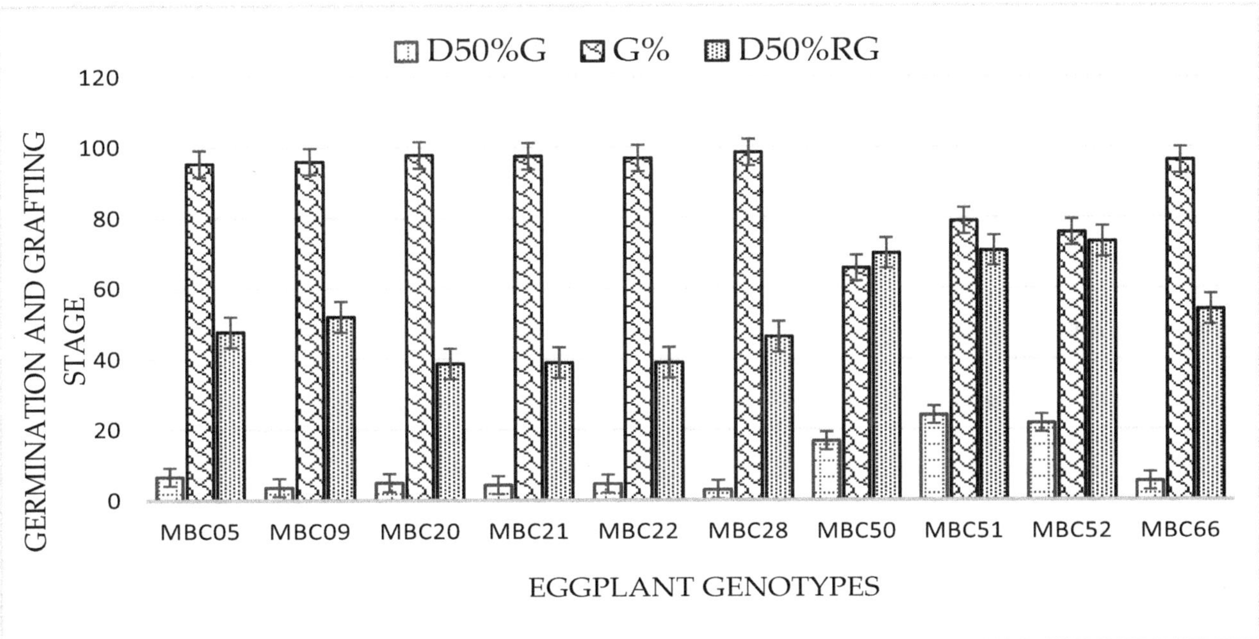

Figure 1. Day to 50% germination (D50%G), germination percentage (G%) and days to 50% reach the grafting (D50%RG) stage, with significance level $p = 0.05$.

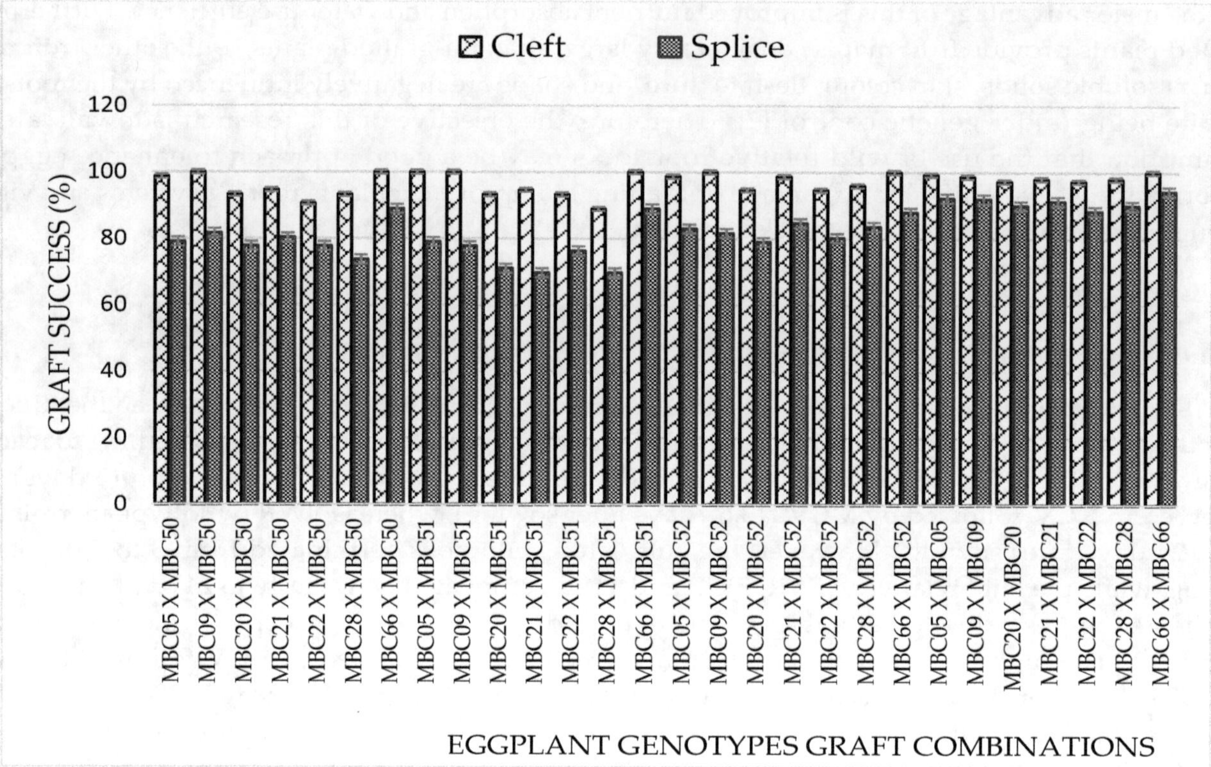

Figure 2. Percentage of graft success (%) with cleft and splice techniques, with significance level $p = 0.05$.

2.2. Analysis of Variance for Vegetative, Yield and Yield Component Traits across the Cropping Systems

The combined analysis of variance for vegetative, yield and yield component traits is presented in Table 1. Highly significant differences ($p \leq 0.01$) were observed among the genotypes for earliness and growth traits. On the other hand, highly significant differences ($p \leq 0.01$) were observed in the interaction between cropping systems with the genotype for all growth and earliness traits except in stem diameter at 90 days after transplant. Additionally, the results indicated that there was no significant difference among replications for growth and earliness traits except for the number of secondary branches at 45 days after transplant, where a significant difference ($p \leq 0.05$) was observed. Highly significant differences ($p \leq 0.01$) were observed among the genotypes, cropping system, and interaction between cropping systems with the genotype (G × Cs) in chlorophyll content, photosynthesis rate, stomatal conductance and transpiration rate. Furthermore, the result showed that there were highly significant differences ($p \leq 0.01$) among replications for traits such as chlorophyll content and transpiration rate, while significant differences ($p \leq 0.05$) were recorded in photosynthesis rate. No significant differences were observed in stomatal conductance. Highly significant differences ($p \leq 0.01$) were observed among the genotypes for yield and yield-related traits. The results showed that a highly significant difference ($p \leq 0.01$) exists in the interaction between the cropping system and genotype for all yield and yield-related traits. In cropping systems, highly significant differences ($p \leq 0.01$) were recorded for all yield and yield-related traits except for fruit diameter where a significant difference ($p \leq 0.05$) was observed. The results showed that there were no significant differences in blocks within each cropping systems for all yield and yield-related traits.

Table 1. Analysis of variance for earliness and growth characters of the grafted, self-grafted and non-grafted eggplants genotypes studied across the cropping systems.

Source of Variation	df	D50%F	FH	NSB45DAT	NSB90DAT	PH45DAT	PH90DAT
Cropping systems (Cs)	1	262.98 **	123.43 **	993.65 **	3554.70 **	30391 **	93773 **
Blocks within Cs	4	0.86 ns	5.02 ns	0.40 **	0.82 ns	5.5487 ns	9.5181 ns
Genotypes (G)	34	546.39 **	530.70 **	9.72 **	42.72 **	158.50 **	1332.80 **
G × Cs	34	45.03 **	51.68 **	2.93 **	13.81 **	252.88 **	6555.00 **
Error	136	0.40	1.59	0.10	0.82	7.76	8.44

Source of Variation	df	SD45DAT	SD90DAT	PR (μmol m^{-2} s^{-1})	SC (mmol m^{-2} s^{-1})	TR (mL cm d^{-1})	CC (ml/cm^2)
Cropping systems (Cs)	1	111.76 **	148.75 **	1000.60 **	0.20 **	34.49 **	44.44 **
Blocks within Cs	4	3.08 *	1.60 ns	0.80 *	0.01 ns	0.07 **	0.20 **
Genotypes (G)	34	10.37 **	28.27 **	49.57 **	0.06 **	4.66 **	3.01 **
G × Cs	34	2.26 **	1.81 ns	1.93 **	0.01 **	0.32 **	0.13 **
Error	136	0.89	2.41	0.31	0.01	0.01	0.01

Source of Variation	df	FL (cm)	FD (mm)	AFW (g)	NFPP	MFY (kg)
Cropping systems (Cs)	1	145.97 **	37.04 *	1433.57 **	8510.96 **	53.56 **
Blocks within Cs	4	1.29 ns	16.28 ns	67.06 ns	58.51 ns	0.01 ns
Genotypes (G)	34	257.95 **	2048.90 **	18,502.06 **	130,551.80 **	4.01 **
G × Cs	34	4.95 **	97.61 **	2548.87 **	661.94 **	4.36 **
Error	136	1.04	9.62	31.16	34.32	0.01

* Significant at $p \leq 0.05$; ** Highly significant at $p \leq 0.01$; ns, not significant at $p > 0.05$; df, degrees of freedom. D50%F, days to 50% flowering; FH, first harvest; NSB45DAT, number of secondary branches at 45 days after transplanting; NSB90DAT, secondary branches at 90 days after transplanting; PH45DAT, plant height at 45 days after transplanting; PH90DAT, plant height at 90 days after transplanting; SD45DAT, stem diameter at 45 days after transplanting; SD90DAT, stem diameter at 90 days after transplanting; PR, photosynthesis rate; SC, stomata conductance; TR, transpiration rate; CC, chlorophyll content; Fl, fruit length; FD, fruit diameter; AFW, average fruit weight; NFPP, number of fruit per plant; MFY, marketable fruit yield.

2.3. Analysis of Variance of Interaction between Rootstock and Scion for Vegetative, Yield and Yield Component Across the Cropping Systems

The combined analysis of variance for interaction between rootstock and scion on vegetative, yield and yield component traits is presented in Table 2. Highly significant differences ($p \leq 0.01$) were observed in cropping systems, rootstock, scion and interaction between rootstock and scion (R × S) on earliness and growth traits in this study except in rootstock where significant differences ($p \leq 0.05$) was recorded in plant height. No significant difference was observed in rootstock effects on transpiration rate. Additionally, the results implied that there was no significant difference among replications for earliness and growth traits. Highly significant differences ($p \leq 0.01$) were observed among the cropping system for all yield traits except for fruit diameter which showed no significant difference. For rootstock, highly significant differences ($p \leq 0.01$) were recorded for all yield traits except yield per plant and yield in ton per hectare, where no significant difference was observed. On the other hand, a highly significant difference ($p \leq 0.01$) was recorded in the interaction between rootstock and scion for all yield traits except for yield per plant and yield in ton per hectare, where significant interaction ($p \leq 0.05$) was observed.

Table 2. Analysis of variance for interaction between rootstock and scion across cropping systems.

Source of Variation	df	D50%F	FH	NSB45DAT	NSB90DAT	PH45DAT (cm)	PH90DAT (cm)
Cropping system (Cs)	1	613.37 **	360.07 **	574.72 **	2407.96 **	18,089.32 **	46,712.82 **
Blocks within Cs	4	0.39 ns	2.60 ns	0.15 ns	0.45 ns	12.56 ns	7.84 ns
Rootstocks (R)	2	3045.10 **	3112.45 **	34.94 **	241.92 **	172.20 *	3256.27 **
Scions (S)	6	152.20 **	169.61 **	7.33 **	15.05 **	196.37 **	727.81 **
R × S	12	297.37 **	280.65 **	9.59 **	34.42 **	161.28 **	1443.39 **
Error	100	5.33	6.69	0.71	3.73	49.61	133.70

Source of Variation	df	SD45DAT (mm)	SD90DAT (mm)	PR (μmol m^{-2} s^{-1})	SC (mmol m^{-2} s^{-1})	TR (mL cm d^{-1})	CC (ml/cm^2)
Cropping system (Cs)	1	61.66 **	82.85 **	401.43 **	0.13 **	18.86 **	26.64 **
Block within Cs	4	1.32 ns	2.03 ns	0.46 ns	0.01 ns	0.03 ns	0.13 ns
Rootstocks (R)	2	25.46 **	126.54 **	48.89 **	0.01 **	0.15 ns	13.62 **
Scions (S)	6	5.47 **	29.07 **	6.34 **	0.02 **	0.53 **	3.18 **
R × S	12	10.40 **	20.94 **	26.10 **	0.05 **	2.81 **	1.88 **
Error	100	1.02	2.45	0.32	0.01	0.11	0.05

Source of Variation	df	FL (cm)	FD (mm)	AFW (g)	NFPP	MFY (kg)
Cropping system (Cs)	1	104.38 **	1.46ns	5893.89*	1922.71 **	35.51 **
Block within Cs	4	1.29 ns	2.80ns	101.93ns	25.89ns	0.00 ns
Rootstocks (R)	2	2167.44 **	1195.31 **	99,079.51 **	109,510.59 **	8.87 **
Scions (S)	6	59.67 **	1274.15 **	8753.32 **	4338.70 **	1.37 ns
R × S	12	44.21 **	3114.52 **	14,525.13 **	4645.01 **	3.44 **
Error	100	1.82	22.24	618.25	131.35	1.18

* Significant at $p \leq 0.05$; ** Highly significant at $p \leq 0.01$; ns, not significant at $p > 0.05$; SOV, source of variation; df, degrees of freedom. D50%F, days to 50% flowering; FH, first harvest; NSB45DAT, secondary branches at 45 days after transplanting; NSB90DAT, secondary branches at 90 days after transplanting; PH45DAT, plant height at 45 days after transplanting; PH90DAT, plant height at 90 days after transplanting; SD45DAT, stem diameter at 45 days after transplanting; SD90DAT, stem diameter at 90 days after transplanting; PR, photosynthesis rate; SC, stomata conductance; TR, transpiration rate; CC, chlorophyll content; Fl, fruit length; FD, fruit diameter; AFW, average fruit weight; NFPP, number of fruit per plant; MFY, marketable fruit yield.

2.4. *Effect of Grafting on Early Flowering and Harvesting across Two Cropping Systems*

The mean comparison of vegetative traits is presented in Table 3. The genotypes began flowering at 52 to 85 days from the sowing date. Non-grafted CCV1 produced flower earlier at 52.33 days after sowing (DAS), whereas late flowering was observed in self-grafted MCV2 at 85.17 days after sowing (DAS). For harvesting, the fruits of non-grafted CCV1 were harvested earlier at 63.67 days after sowing (DAS), whereas fruits of self-grafted MCV2 were picked later (95.33 DAS).

In this study, the number of secondary branches in grafted plants was significantly higher than non-grafted and self-grafted eggplants at 45 and 90 DAT (Table 3). A significantly higher number of branches was observed in TCV grafted onto TWR at 45 and 90 DAT (10.00 and 19.28, respectively), whereas the lowest value for this trait was recorded in non-grafted and self-grafted eggplants. In terms of height, non-grafted NCV (72.80 cm) had relatively taller plants compared to eggplants of NCV grafted onto rootstocks of MWR, BWR and TWR, respectively, at 45 DAT. At 90 DAT, NCV grafted onto TWR (147.00 cm) recorded significantly higher plants compared to non-grafted and self-grafted plants (Table 3). Grafting with wild rootstock recorded significantly higher stem diameter compared to non-grafted and self-grafted eggplants at 45 and 90 DAT (Table 3). The highest stem diameter was observed in NCV grafted onto TWR (14.22 mm at 45 DAT and 21.80 mm at 90 DAT, respectively), whereas the lowest value for stem diameter was observed in non-grafted CCV1 (9.02 mm at 45 DAT) and CCV3 grafted onto MWR (12.28 at 90 DAT).

Table 3. Means of growth characters of the grafted, self-grafted and non-grafted eggplant genotypes studied across the cropping systems.

Scion/Rootstock	D50%F	FH	SB45	SB90	PH45 (cm)	PH90 (cm)	SD45 (mm)	SD90 (mm)
MCV1	73.83 ^{i}	84.00 ^{hi}	5.57 $^{p-s}$	15.83 ^{cd}	56.03 $^{k-p}$	85.55 ^{u}	9.57 ^{lm}	13.85 $^{h-0}$
MCV2	75.50 ^{h}	88.33 $^{d-f}$	5.65 $^{o-r}$	15.00 $^{d-h}$	58.85 $^{i-l}$	105.58 $^{g-k}$	10.97 $^{f-i}$	14.17 $^{g-n}$
CCV1	52.33 ^{t}	63.67 ^{q}	4.58 ^{xy}	9.88 ^{op}	53.40 ^{pq}	88.88 ^{t}	9.02 ^{m}	14.40 $^{f-m}$
CCV2	57.17 ^{rs}	68.83 ^{no}	4.97 $^{u-w}$	11.17 $^{l-n}$	62.12 $^{f-h}$	96.38 $^{n-p}$	10.05 $^{h-m}$	13.05 $^{m-o}$
CCV3	58.33 ^{pq}	70.00 $^{l-n}$	4.45 ^{y}	13.47 ^{j}	55.93 $^{l-p}$	101.05 ^{lm}	9.95 $^{i-m}$	13.43 $^{j-o}$
NCV	72.50 ^{j}	83.17 ^{i}	5.70 $^{n-q}$	15.30 $^{d-h}$	72.80 ^{a}	123.95 ^{d}	12.93 ^{bc}	18.67 ^{b}
TCV	72.33 ^{j}	83.67 ^{hi}	6.12 $^{i-l}$	17.00 ^{b}	58.70 $^{i-l}$	89.18 st	10.32 $^{g-l}$	13.45 $^{i-o}$
MCV1/MCV1	78.33 ^{ef}	89.67 ^{cd}	6.22 ^{hij}	15.00 $^{d-h}$	58.62 $^{i-l}$	89.97 ^{rst}	11.17 $^{e-g}$	14.47 $^{i-o}$
MCV2/MCV2	85.17 ^{a}	95.33 ^{a}	5.70 $^{n-q}$	13.47 ^{j}	57.98 $^{i-m}$	107.58 $^{f-i}$	11.68 $^{d-f}$	14.03 $^{g-o}$
CCV1/CCV1	63.67 ^{k}	76.67 ^{j}	5.27 ^{stu}	14.13 $^{h-j}$	56.97 $^{j-n}$	94.98 $^{o-p}$	10.28 $^{g-l}$	13.27 $^{k-o}$
CCV2/CCV2	60.50 ^{n}	70.50 ^{lm}	4.67 $^{w-y}$	9.58 ^{p}	58.70 $^{i-l}$	108.08 $^{f-h}$	9.72 ^{lm}	12.57 ^{no}
CCV3/CCV3	58.33 ^{pq}	70.67 ^{l}	4.75 $^{v-y}$	10.83 $^{m-o}$	52.17 ^{q}	91.85 $^{q-t}$	9.88 $^{i-m}$	15.00 $^{d-k}$
NCV/NCV	76.33 ^{g}	85.83 ^{g}	5.57 $^{p-s}$	14.70 $^{e-h}$	60.12 ^{gh}	128.52 ^{c}	12.32 $^{c-e}$	16.37 ^{cd}
TCV/TCV	78.17 ^{f}	89.00 $^{d-f}$	8.22 ^{d}	16.17 ^{bc}	55.10 $^{m-q}$	92.45 $^{q-t}$	9.88 $^{i-m}$	13.55 $^{i-o}$
MCV1/MWR	76.67 ^{g}	87.83 ^{f}	5.77 $^{l-o}$	18.78 ^{a}	52.13 ^{q}	94.75 $^{o-q}$	12.63 ^{cd}	15.23 $^{c-i}$
MCV2/MWR	77.83 ^{f}	88.83 $^{d-f}$	6.05 $^{i-n}$	14.70 $^{e-h}$	70.10 ^{ab}	118.00 ^{e}	12.28 $^{c-e}$	16.93 $^{g-o}$
CCV1/MWR	56.50 ^{s}	66.83 ^{p}	5.88 $^{k-p}$	12.18 ^{kl}	59.47 $^{g-j}$	104.97 $^{h-k}$	11.03 $^{f-h}$	13.13 $^{l-o}$
CCV2/MWR	57.67 ^{o}	67.83 ^{op}	6.08 $^{i-m}$	10.65 ^{no}	57.78 $^{i-n}$	104.20 $^{j-l}$	10.82 $^{f-k}$	14.88 $^{d-j}$
CCV3/MWR	59.67 ^{o}	71.00 $^{l-n}$	4.83 $^{v-x}$	9.37 ^{p}	57.55 $^{i-n}$	102.82 ^{kl}	9.58 ^{lm}	12.28 ^{o}
NCV/MWR	76.83 ^{g}	88.17 ^{ef}	5.37 $^{q-t}$	13.50 ^{j}	59.27 $^{h-j}$	135.08 ^{b}	13.32 $^{a-c}$	20.74 ^{a}
TCV/MWR	76.83 ^{g}	87.83 ^{f}	8.58 ^{c}	18.70 ^{a}	53.71 $^{o-q}$	106.95 $^{f-j}$	10.37 $^{g-l}$	15.63 $^{c-f}$
MCV1/BWR	78.00 ^{f}	90.83 ^{bc}	5.67 $^{o-q}$	14.58 $^{i-e}$	58.72 $^{i-l}$	93.20 $^{p-r}$	11.38 $^{e-g}$	16.17 $^{c-e}$
MCV2/BWR	80.33 ^{b}	91.50 ^{b}	6.58 $^{f-h}$	15.55 $^{c-e}$	66.82 ^{cd}	122.00 ^{d}	12.35 $^{c-e}$	15.98 $^{c-g}$
CCV1/BWR	61.33 ^{m}	71.33 ^{l}	5.95 $^{j-o}$	13.58 ^{ij}	62.60 $^{e-g}$	108.80 ^{fg}	12.32 $^{c-e}$	16.03 $^{c-f}$
CCV2/BWR	59.83 ^{no}	70.67 ^{l}	5.75 $^{m-p}$	9.87 ^{op}	59.60 $^{g-j}$	93.95 $^{o-q}$	10.90 $^{i-m}$	14.45 $^{c-h}$
CCV3/BWR	62.50 ^{l}	73.50 ^{k}	4.97 $^{u-w}$	11.53 $^{k-n}$	53.47 ^{pq}	94.17 $^{o-q}$	10.85 $^{f-j}$	12.88 $^{m-o}$
NCV/BWR	74.50 ^{i}	84.83 ^{gh}	6.62 $^{f-h}$	14.20 $^{g-j}$	65.38 ^{de}	135.93 ^{b}	13.82 ^{a}	18.35 ^{b}
TCV/BWR	76.67 ^{g}	87.83 ^{f}	9.03 ^{b}	15.22 $^{c-g}$	53.57 $^{o-q}$	96.97 ^{o}	11.02 $^{f-h}$	14.97 $^{d-k}$
MCV1/TWR	79.50 ^{cd}	89.66 ^{cd}	6.98 ^{e}	16.25 ^{bc}	56.70 $^{j-o}$	91.85 $^{q-t}$	9.88 $^{i-m}$	13.93 $^{g-o}$
MCV2/TWR	79.00 ^{de}	88.33 $^{d-f}$	6.77 ^{ef}	14.42 $^{f-j}$	65.15 $^{d-f}$	115.33 ^{e}	12.43 $^{c-e}$	16.35 ^{cd}
CCV1/TWR	58.33 ^{pq}	69.17 $^{m-o}$	5.30 $^{r-u}$	10.87 $^{m-o}$	54.73 $^{n-q}$	104.60 $^{i-k}$	10.37 $^{g-l}$	12.75 $^{m-o}$
CCV2/TWR	62.17 ^{l}	72.83 ^{k}	5.05 $^{t-v}$	11.70 $^{k-m}$	55.00 $^{m-q}$	104.28 $^{i-l}$	9.73 $^{k-m}$	15.18 $^{c-j}$
CCV3/TWR	58.83 ^{p}	70.83 ^{l}	4.83 $^{v-x}$	11.33 $^{k-n}$	55.35 $^{m-p}$	99.10 ^{mn}	10.57 $^{g-l}$	15.58 $^{c-g}$
NCV/TWR	77.83 ^{f}	89.00 $^{d-f}$	6.25 $^{h-j}$	14.62 $^{e-i}$	69.52 ^{bc}	147.00 ^{a}	14.22 ^{a}	21.80 ^{a}
TCV/TWR	79.83 ^{bc}	89.17 $^{d-f}$	10.0 ^{a}	19.28 ^{a}	59.12 ^{k}	109.52 ^{f}	11.82 $^{d-f}$	15.48 $^{c-h}$
Mean	69.52	80.48	6.01	13.75	58.96	105.35	11.13	15.11
SEM	0.68	0.68	0.18	0.36	1.02	1.92	0.12	0.19
LSD (p = 0.05)	0.72	1.44	0.35	1.04	3.18	3.32	1.08	1.77

SEM, standard error of mean; LSD, least significant difference; D50%F, days to 50% flowering; FH, first harvest; SB, secondary branches; PH45DAT, plant height at 45 days after transplanting; PH90DAT, plant height at 90 days after transplanting; SD45DAT, stem diameter at 45; SD90DAT, stem diameter at 90. Means with the same letter in the same column are not significantly different at p = 0.05 with LSD test.

2.5. *Effect of Grafting on Physiological, Yield and Yield Component Traits across Two Cropping System*

The physiological traits measured include chlorophyll content, photosynthesis rate, stomatal conductance and transpiration rate. There was no influence of rootstock on chlorophyll content in this study (Table 4). The highest chlorophyll content was recorded in non-grafted MCV2 (6.13 mL/cm^2) while NCV grafted onto BWR showed the lowest value (3.80 mL/cm^2). Significant influence of rootstock on photosynthetic rate was observed (Table 4). The highest photosynthetic rate recorded was observed in MCV2 scion grafted onto MWR rootstock (32.18 $\mu mol\ m^{-2}\ s^{-1}$), whereas non-grafted MCV2 recorded the lowest value for this trait (Table 4). For stomatal conductance, CCV3 scion grafted onto MWR rootstock recorded the highest conductance (0.75 $mmol\ m^{-2}\ s^{-1}$), whereas the lowest value was observed in non-grafted MCV1 (0.20 $mmol\ m^{-2}\ s^{-1}$) (Table 4). There was a remarkable influence of grafting on the transpiration rate, MCV1 grafted onto MWR had the highest transpiration rate (6.03 $mL\ cm\ d^{-1}$) among the graft combinations. The lowest transpiration rate was recorded in self-grafted MCV1 (2.75 $mL\ cm\ d^{-1}$) (Table 4).

Table 4. Means of physiological, yield and yield component traits of the grafted, self-grafted and non-grafted eggplants genotypes studied across the cropping systems.

Scion/Rootstock	CC	P R	S C	T R	FL (cm)	FD (mm)	AFW (g)	NFPP	MFY (kg)
MCV1	4.72^{i}	22.10^{l}	0.20^{k}	3.43^{no}	18.9^{j}	32.6^{pq}	90.2^{ij}	54.8^{l-n}	2.5^{tu}
MCV2	6.13^{a}	18.63^{m}	0.43^{ef}	3.37^{no}	20.1^{h-j}	34.1^{n-g}	142.6^{e}	53.8^{k-m}	4.0^{fg}
CCV1	4.02^{p}	23.32^{j}	0.30^{j}	3.02^{q}	15.9^{kl}	33.3^{o-q}	95.7^{i}	51.3^{m-o}	2.6^{st}
CCV2	3.98^{p}	25.35^{h}	0.33^{i}	3.37	21.2^{g}	36.4^{m-o}	90.5^{ij}	54.1^{l-n}	2.3^{v}
CCV3	4.45^{k}	25.32^{h}	0.35^{hi}	3.30^{op}	15.7^{l}	72.2^{c}	154.1^{d}	46.0^{op}	2.9^{r}
NCV	3.93^{pq}	25.37^{h}	0.28^{j}	3.32^{n-p}	11.3^{op}	27.3^{r}	33.9^{k-m}	128.2^{f}	3.2^{n}
TCV	4.37^{kl}	24.02^{i}	0.43^{ef}	3.48^{m}	6.53^{r}	19.7^{s}	27.8^{m-p}	115.3^{g}	2.1^{w}
MCV1/MCV1	4.37^{l}	23.32^{j}	0.35^{hi}	2.75^{r}	22.6^{f}	39.4^{j-m}	114.4^{g}	50.2^{no}	2.8^{r}
MCV2/MCV2	5.57^{d}	24.40^{j}	0.45^{de}	3.63^{l}	25.0^{de}	38.1^{k-m}	116.1^{g}	62.7^{h-j}	4.3^{e}
CCV1/CCV1	4.82^{h}	22.12^{l}	0.30^{j}	3.22^{p}	10.5^{pq}	37.0^{l-n}	95.0^{i}	66.6^{h}	3.8^{ij}
CCV2/CCV2	4.55^{j}	22.53^{kl}	0.35^{hi}	3.73^{kl}	25.5^{de}	30.8^{q}	85.5^{j}	55.0^{l-n}	2.2^{v}
CCV3/CCV3	4.82^{h}	22.82^{jk}	0.42^{fg}	3.80^{k}	17.0^{k}	66.4^{d}	178.2^{c}	41.9^{p}	2.9^{q}
NCV/NCV	3.85^{rs}	26.80^{g}	0.45^{de}	4.17^{ij}	13.8^{m}	52.6^{f}	25.7^{op}	177.3^{b}	3.5^{lm}
TCV/TCV	4.45^{k}	25.25^{h}	0.40^{g}	3.80^{k}	8.68^{q}	27.4^{r}	28.6^{m-p}	135.6^{e}	2.5^{tu}
MCV1/MWR	6.05^{b}	29.35^{c}	0.50^{c}	6.03^{a}	29.4^{b}	43.4^{hi}	134.2^{f}	54.4^{l-n}	3.8^{i}
MCV2/MWR	5.73^{c}	32.18^{a}	0.55_{b}	5.03^{e}	31.4^{a}	47.9^{g}	195.3^{a}	52.2^{m-o}	5.3^{a}
CCV1/MWR	5.23^{g}	25.68^{h}	0.55^{b}	4.90^{f}	25.5^{de}	38.4^{k-m}	103.2^{h}	65.6^{hi}	4.0^{fg}
CCV2/MWR	4.18^{mn}	27.97^{de}	0.50^{c}	5.20^{d}	27.3^{c}	36.0^{m-p}	116.1^{g}	55.1^{k-n}	3.0^{pq}
CCV3/MWR	4.22^{m}	28.43^{d}	0.75^{a}	5.80^{b}	20.4^{g-i}	90.3^{b}	186.7^{b}	46.0^{op}	3.5^{lm}
NCV/MWR	5.58^{J}	31.37^{b}	0.55^{b}	5.47^{c}	13.9^{m}	45.8^{gh}	38.9^{k}	130.8^{ef}	3.8^{hi}
TCV/MWR	4.57^{j}	29.41^{c}	0.55^{b}	5.02^{e}	11.3^{op}	39.4^{j-m}	25.2^{p}	157.6^{c}	2.6^{s}
MCV1/BWR	5.37^{f}	27.95^{de}	0.43^{ef}	5.40^{c}	25.8^{d}	45.9^{gh}	136.5^{ef}	52.2^{m-o}	3.6^{jkl}
MCV2/BWR	5.15^{g}	31.02^{b}	0.50^{c}	4.55^{h}	30.0^{b}	40.7^{i-k}	139.9^{ef}	59.9^{i-l}	4.9^{c}
CCV1/BWR	5.23^{m}	25.78^{h}	0.55^{b}	4.60^{h}	23.0^{f}	42.9^{h-j}	115.9^{h}	58.9^{i-l}	3.8^{hi}
CCV2/BWR	5.17^{g}	26.80^{g}	0.50^{c}	5.07^{e}	25.8^{d}	40.1^{i-l}	103.3^{h}	61.8^{h-k}	3.2^{n}
CCV3/BWR	4.07^{o}	28.10^{de}	0.55^{b}	5.47^{c}	21.1^{gh}	33.2^{o-q}	192.3^{ab}	49.3^{no}	3.7^{ij}
NCV/BWR	3.80^{h}	27.80^{d-f}	0.55^{b}	5.35^{c}	13.8^{m}	58.7^{e}	31.9^{l-o}	181.4^{b}	4.6^{d}
TCV/BWR	4.33^{l}	27.62^{ef}	0.50^{c}	4.57^{h}	10.9^{p}	37.1^{l-n}	33.4^{k-n}	148.5^{d}	3.2^{n}
MCV1/TWR	6.07^{b}	27.53^{ef}	0.45^{de}	4.63^{h}	24.7^{de}	40.7^{i-k}	136.8^{ef}	48.9^{no}	3.1^{op}
MCV2/TWR	5.50^{e}	27.58^{ef}	0.40^{g}	4.27i	27.2^{c}	42.5^{h-j}	172.6^{c}	54.7^{l-n}	5.1^{b}
CCV1/TWR	4.82^{h}	24.55^{i}	0.37^{h}	4.08^{k}	24.5^{e}	39.4^{j-m}	117.6^{g}	57.5^{j-m}	3.7^{ij}
CCV2/TWR	4.20^{m}	25.30^{h}	0.40^{g}	5.83^{b}	22.5^{f}	28.7^{r}	86.5^{j}	57.5^{j-m}	2.4^{u}
CCV3/TWR	4.13^{no}	27.53^{ef}	0.50^{c}	4.80^{fg}	21.2^{g}	99.5^{a}	192.8^{ab}	49.2^{no}	4.1^{f}
NCV/TWR	4.13^{no}	27.17^{fg}	0.47^{d}	4.67^{gh}	12.9^{mn}	57.6^{e}	36.2^{kl}	144.8^{d}	3.9^{gh}
TCV/TWR	3.88^{qr}	27.12^{fg}	0.45^{de}	4.27^{i}	12.2^{op}	34.4^{n-p}	27.4^{n-p}	193.8^{a}	3.7^{ij}
Mean	4.68	26.22	0.44	4.35	19.91	45.14	102.89	82.09	3.44
SEM	0.06	0.25	0.01	0.07	0.46	1.30	4.06	3.31	0.09
LSD (p = 0.05)	0.08	0.64	0.02	0.13	1.17	3.54	6.37	6.69	0.11

SEM, standard error of mean; LSD, least significant difference; CC, chlorophyll content (mL/cm^2); PR, photosynthetic rate ($\mu molm^{-2} s^{-1}$); SC, stomata conductance ($mmolm^{-2} s^{-1}$); TR, transpiration rate ($mL cm d^{-1}$); FL, fruit length; FD, fruit diameter; AFW, average fruit weight; NFPP, number of fruit number; MFY, marketable fruit yield. Means with the same letter in the same column are not significantly different at $p = 0.05$ with LSD test.

There was a notable influence of grafting with wild rootstock on fruit length (Table 4). MCV2 scion grafted onto MWR wild rootstock (31.37 cm) recorded significantly higher fruit length than the non-grafted (20.05 cm) and self-grafted (25.02 cm) MCV2. In contrast, self-grafting of MCV2 resulted in a significant increase in fruit length compared to the non-grafted plant (MCV2). On average, eggplants grafting on suitable rootstocks had a positive effect on fruit diameter (Table 4). Commercial MCV2 scion grafted onto BWR wild rootstock recorded comparably higher fruit diameter (99.55 mm) compared to the self-grafted (66.40 mm) and non-grafted (36.35 mm) plants. However, within the same genotypes, self-grafted eggplant had higher fruit diameter compared to their non-grafted counterparts.

Results on fruit yield components indicated that grafting with wild rootstocks significantly increased the average fruit weight (Table 4). Average fruit weight in eggplants from MCV2 scion grafted onto MWR rootstock (195.27 g) was significantly higher than controls, while TCV grafted onto MWR rootstock recorded the lowest average fruit weight (25.22 g).

Significant differences were observed among treatments in the number of fruits per plant (Table 4). TCV scion grafted with TWR rootstocks recorded a higher fruit number (193.80) compared to the non-grafted and self-grafted plants. Across cropping systems, self-grafted TCV eggplants (135.55) outperformed their non-grafted (115.33) counterparts. The average marketable yields in the two cropping systems were significantly higher in grafted plants with the MWR rootstocks compared to the control plant MCV2 (Table 4).

On the average, grafting of MCV2 onto MWR rootstock recorded a higher marketable fruit yield (5.30 kg) in comparison to self-grafted (4.28 kg) and non-grafted (4.03 kg) MCV2. In contrast, marketable fruit yield was higher in self-grafted eggplant than in their non-grafted counterparts, while the least value for this trait was observed in TCV (Table 4).

3. Discussion

Plant growth and development is influenced by physiological processes (e.g., photosynthesis) that depend on environmental factors to proceed optimally [6,7]. Research has shown that grafting is an effective method of increasing yield, disease resistance and quality of several vegetable crops [8,9]. It is believed that rootstocks enhance the yield and quality of fruits. This is attainable through the use of rootstocks that is resistant to soil diseases or pests, tolerant to stress conditions, improves nutrients uptake, and give a high vigour to the scion [5,9]. In this study, the effect of grafting eggplant varieties of MCV1, MCV2, CCV1, CCV2, CCV3, NCV and TCV onto the rootstock of different wild relatives' eggplants, i.e., MWR, BWR and TWR, were tested. The results of the experiment show that grafting can be used to improve the production of eggplant. The advantages of rootstock grafts such as synchronization and high germination rates of the rootstock and scion, as well as remarkable graft success and stand establishment post-transplant both in field and glasshouse, can adequately compensate for the difficulties encountered in the process of producing good grafted plants.

It is essential to take note of seed germination when materials of wild species are used as rootstocks. Seeds of some wild Solanum species usually have slow germination, requiring up to 24 days to germinate. Aside from this, research has shown that germination percentage often varies between 15–50% in *S. insanum* L. *S. torvum*, *S. integrifolium*, *S. surattense* Burm., *S. khasianum* C.B. Clarke, *S. sanitwongsei* Craib and in hybrids of *S. melongena* × *S. integrifolium* [10]. *S. torvum*, the most commonly grafted Solanum eggplant relative undergo a longer germination period and often has a low germination percentage [10]. Consequently, this has limited the use of *S. torvum* rootstock due to complexity in achieving speedy and homogeneous germination [11]. In this research, it was observed that soft-seeded natural NCV and MCV2 germinate earlier than the others, at 3–4 days, while their wild relatives, i.e., MWR, BWR and TWR, took longer (17 to 24 days) to germinate. High germination percentage (≥90%) was recorded using seed for MCV1, MCV2, CCV1, CCV2, CCV3, NCV and TCV scions. For MWR, BWR and TWR, the rates recorded (≤80%) were slightly lower, despite both the scion and rootstock being germinated under GA_3 treatment conditions. From our trial, germinating the seeds of this species in the commercial substrate is attainable; however, despite the optimal conditions, their germination can best be described as irregular and somewhat erratic. In this regard, it is pertinent to know that the wild species *S. incanum* has characteristic low and irregular germination.

Grafting success is dependent on the combination of the grafts and compatibility of the rootstock and scion [11]. Eggplant grafting is mostly performed using the cleft or splice grafting approach [9]. In this research, the cleft and splice grafting techniques were both studied. The result obtained indicated that grafting using the cleft technique was very successful with ≥85% success in all grafting combinations including self-grafted eggplants, compared with the splice method with a success rate of ≥65%. This result shows the cleft technique to be a very efficient method of combining the scion and rootstock of eggplants. The lower success rate recorded using the splice grafting method may be suggestive that the method is not suitable for use in eggplants, but rather for tomato, which is phylogenetically related to eggplant [9].

There are several reports of success in the grafting of eggplant varieties using the wild S. torvum which is the phylogenetically most distant of the rootstocks under study [2,3,5]. Using the cleft technique, a 100% survival rate was recorded in plants grafted with TCV scion including self-grafted eggplants and treatments combinations such as MCV1/BWR and MCV2 grafted onto all rootstock. The lower success recorded using splice technique for grafting wild rootstocks may be due to graft union. The interaction between rootstock and scion has been frequently observed in many crops, and in this study, it was noted that rootstock source can have a major influence on eggplant vigour, earliness, yield and fruit quality properties.

Generally, in grafted plants, the formation of graft union between scion and rootstock is necessary for the translocation of food material from the root system to apical parts of the plant and leads to delayed flowering in grafted plants. In this study, greater earliness was observed in most non-grafted plants. Comparable effect of delayed flowering in grafted plants of eggplant and tomato was reported by Moncada et al. [12] and Ibrahim et al. [10]. The concluding phenomenon is attributable to the stress experienced by these plants following the grafting operation [13]. However, in this study, contrasting results were obtained where grafted plants showed early flowering compared to the non-grafted plants. For example, MCV1 grafted onto MWR produced flower earlier than its corresponding non-grafted MCV1. This phenomenon was related to the absence of graft incompatibility and environmental stress. Gisbert et al. [11], explained that when there is no constraint due to incompatibility and abiotic stress, grafted plants may develop faster, thereby resulting in earliness of the grafted plant. There are reports on earliness in eggplant grafted onto two S. Lycopersicum hybrids [13] and melon plants grafted onto Cucurbita rootstocks [14]. Early harvest may be more important for the farmer to capture higher market prices [9]. Generally, non-grafted plants CCV1 produced fruits faster due to flowering earlier than their grafted counterpart. This may be attributed to the delay resulting from the formation of graft union between scion and rootstock and where necessary to translocate food material from the root system to apical parts of the plant, thereby leading to delayed flowering and fruiting in grafted plants. Additionally, delayed harvesting in the grafted plant may be due to the healing crisis and shock during transplanting. The flowering date is an important aspect of vegetable production as it affects fruit harvesting time, which can, in turn, have a direct bearing on fruit quality [9]. In this study, contrasting results were obtained where some grafted plants bear fruits earlier than the non-grafted plants. For example, MCV1 grafted onto MWR produced fruits earlier than the corresponding non-grafted MCV1. Khah [15] reported a similar observation where grafted tomato fruit earlier than the non-grafted ones.

Grafting with vigorous rootstocks (MWR, BWR and TWR) led to improved growth of scion which results in a higher number of branches compared to controls at 45 and 90 DAT. The effect of rootstock on the mineral content in the aerial portion of the plant may be related to the physical properties of the root system, such as lateral and vertical development, which may lead to improved uptake of water and minerals, thereby resulting in a greater number of branches in grafted plants. These findings are in line with the results reported by [11,15].

Plant height, which has been widely regarded as an indicator of plant vigour was highest in plants with MWR, BWR and TWR rootstock at both 45 and 90 days after transplant, while the lowest height was observed in plants with self-grafted, non-grafted and BWR rootstock grafts, thus showing that the vigour of the rootstock is essential in conferring scion vigour. The wild rootstock MWR, BWR and TWR had good root systems which ensured better plant height and vigorous growth through the absorption of an optimal level of water and nutrients. As reported by Gisbert et al. [14], the vigorous root system of the rootstock enhances the ability to absorb water and nutrients compared to the non-grafted plants while serving as a better supplier of endogenous plant hormones. Observations similar to this were reported by Passam et al. [16], showing that the impact of grafting on plant height differ according to the rootstock/eggplant's varieties combined.

The increase in stem diameter was due to the vigour of the wild rootstock. The effect of rootstock on the level of minerals in the aerial portion of the plant may be related to the physical properties of the root system, including lateral and vertical development which may function in improved uptake of nutrients. The vigorous root system of the rootstock enhanced better growth of scion which resulted in larger stem diameter. Buller et al. [17] found that the stem diameter of Cherokee Purple was larger in plants grafted onto the rootstock Maxifort in a soil infested with verticillium wilt. Similarly, Buller et al. [17] noted that stem diameter was larger in scions of an eggplant variety grafted onto *S. torvum*. In contrast, Leonardi and Giuffrida [18] in their study reported that self-grafted eggplants had similar sized or lower stem diameter compared to eggplant grafted onto different tomato rootstock's.

Chlorophyll content, which has a direct influence on the photosynthetic efficiency of leaves, is a major index in photosynthetic capacity and may be considered as a source of food and energy [19]. Chlorophyll content was higher in grafted MCV1/TWR under the open field condition, whereas non-grafted had the highest chlorophyll content at the glasshouse conditions. In this study, the lowest value for chlorophyll was recorded in the plants of NCV/BWR under open field cropping system and self-grafted NCV/NCV under the glasshouse environment. Trinchera et al. [19] reported that grafted plants had a relatively higher concentration of chlorophyll pigments compared to controls. Kumar et al. [20] stated that the reduction of water content often leads to increased chlorophyll content. In the incompatible graft combinations, when the synthesis of vascular bundles in the graft union decrease, water and nutrient transport will be inhibited [19]. This may, however, be responsible for higher chlorophyll content at the graft union. Liu et al. [21], in an experiment on tomatoes, found that chlorophyll content was significantly higher in the grafted tomatoes than in the non-grafted ones.

The photosynthetic rate may be considered as the amount of food manufactured and energy released to other parts of the plant. The rate of photosynthesis was higher in MCV2 grafted onto MBC5 under both cropping systems, whereas the lowest value for this physiological trait was recorded in non-grafted MCV2. A study by Kumar et al. [20] reported that better uptake of nutrients in grafted seedlings leads to improved photosynthesis, which is evident especially when plants are subjected to adverse condition. Stomata conductance, measuring the degree of stomatal opening and serves as an indicator of plant water status, was higher in CCV3 grafted onto MBC5 at both cropping systems, whereas the lowest stomatal conductance was observed in non-grafted MCV1. Higher stomatal conductance in a grafted plants may be due to their vigorous root systems, which are often capable of absorbing water, nutrients and enhancing the efficient accumulation of dry matter compared to non-grafted plants.

The transpiration rate, considered as the rate of water loss by plant depending on the nature of stomata, was higher in MCV1 grafted onto both MWR and BWR at both cropping systems, respectively, whereas the lowest value for transpiration rate was reported in self-grafted MCV1. The rootstock's vigorous root system enhances the absorption of water, nutrients and efficient accumulation of dry matter compared to the non-grafted plants. Grafting with wild rootstock can enhance transpiration rate. This claim has been supported by several researches such as report by [19–21], who stated that grafting with inter-specific rootstocks (*Cucurbita maxima* × *C. moschata*) improves transpiration rate in grafted plants.

Fruit quality is vital for fruit marketability, and grafting can have an impact on the traits associated with fruit quality [2]. In our study, it was observed that plants grafted with MWR rootstocks had higher fruit length than the non-grafted plants and self-grafted controls. This can be related to changes in the concentration of growth regulators induced by the eggplant rootstock which leads to increased length of the fruit. Johnson et al. [22] recorded an increase in the fruit lengths of eggplant Epic grafted onto Beaufort, while Bekhradi et al. [23] observed longer fruits in melon plant Charleston Grey grafted onto Ferro rootstocks. In terms of fruit diameter, it was observed that eggplants grafted with TWR rootstocks had higher fruit diameter compared to the non-grafted plants and self-grafted plants treatments, respectively. This was explained to be due to changes in the concentration of growth regulators induced by the wild brinjal rootstock which leads to increased diameter of the fruit. Wide fruits have been reported for eggplant Epic grafted on to Beaufort [22], while cucumber Hady scion grafted onto Ferro rootstocks gave the highest value for this trait [23]. Fruit shape in eggplant has high heritability [1], and rootstocks influence fruit length and fruit diameter, possibly due to changes in the concentration of growth regulators induced by the rootstock.

In this study, average fruit weight was also found to be higher in eggplants grafted with MWR and BWR rootstock. This corresponds with results from other studies that showed positive rootstock–scion interactions on the average fruit weight. Bletsos et al. [24] reported that the fruit weight of grafted plants (Tsakoniki grafted onto *Solanum torvum* Swartz) were observed to be higher than in non-grafted plants in two planting seasons; in this respect, grafting eggplant plants onto wild relative eggplant rootstock resulted in higher average fruit and fruit weight when compared with self-grafted and non-grafted eggplants.

The findings from this study indicated that the use of all wild rootstock significantly increased fruit number per plant in MCV2, CCV3 and TCV. However, MCV1 grafted onto MWR, BWR and TWR gave low fruit number compared to their non-grafted counterparts. This result was supported by Yetisir and Sari [25], who stated that higher number of fruits (2.4 fruits/plant) was obtained when cultivar of Crimson Tide was grafted onto the rootstock of Skopje (*Lagenaria* hybrid), whereas fewer fruits were recorded (1.0 fruits/plant) when the cultivar was grafted onto the rootstock of *Cucurbita moschata* (CMO).

The results from this study show that grafting with resistant and vigorous rootstocks, i.e., MWR, BWR and TWR, can enhance the marketable yield of eggplant plants better than non-grafted and self-grafted control. The increase in marketable yields in grafted treatments combinations in this study has also been documented in other research studies. Khah et al. [13] highlighted that grafting can increase the marketable yield of tomato by up to 54%. The enhanced marketable fruit yield in eggplants from MCV2 and CCV3 grafted onto rootstock MWR, BWR and TWR compared to controls MCV2 and CCV3 under the open field and glasshouse environments, respectively, can be attributed to increase in the production of endogenous hormones leading to improved water and nutrient uptake. Observations similar to this were reported by Gisbert et al. [11] and Khah [13] who explained that higher marketable yield in grafted plants is mainly due to improved water and nutrient uptake by vigorous rootstock.

Generally, the yield improvement in eggplant which occur due to grafting with rootstocks of MWR, BWR and TWR as observed in this study was consistent with previous reports of grafted plant production. Schwarz [26] reported that the use of Maxifort, which is a vigorous rootstock, tends to be highly successful and gives a higher yield of tomato. Louws et al. [5] observed that there was comparatively higher yield increase on grafting eggplant onto wild solanum rootstock than in self-grafted controls. According to Khah et al. [15], a remarkable increase in yield recorded in eggplant grafted onto tomato rootstock was due to increased fruit size and fruit number which was significantly higher than in non-grafted controls and those with eggplant rootstock. Leonardi and Giuffrida [18] and Kumar et al. [20] also reported that increased fruit yield observed in grafted tomato was mainly due to a larger fruit size of the plant.

4. Materials and Methods

4.1. Planting Materials and Agronomic Practices

Seven high-yielding eggplant cultivars of *Solanum melongena* were used as scions, among which three were collected from China (CCV1, CCV2 and CCV3), two from Malaysia (MCV1 and MCV2), one was collected from Nigeria (NCV) and Thailand (TCV), respectively. For rootstock, three wild eggplant cultivars of *Solanum torvum* from Malaysia (MWR), Bangladesh (BWR) and Thailand TWR) were used (Table 5). The scions and rootstocks were used to develop 21 interspecific grafted combinations, while the control (seven self-grafted and seven non-grafted scion cultivars) was used as comparison. The 35 grafted, self-grated and non-grafted eggplant seedlings were evaluated under two cropping systems from June to December 2019 for the open field experiment and November 2019 to May 2020 for glasshouse condition. The experiment was laid out in a Randomized Complete Block Design (RCBD) in three replications with a planting distance of 50 cm within plant and 70 cm between the row in each growing system in Field 15, Universiti Putra Malaysia (UPM), Serdang, Selangor,

Malaysia, which is geographically located between 3°02′ N latitude and 101°42′ East longitude, at 31 m above sea level altitude. Average minimum and maximum temperature were about 24 °C and 33 °C and 24 °C and 38 °C, respectively, for open field and glasshouse conditions. The mean values of relative humidity were 89 and 78 for the open field and glasshouse, respectively. The mean rainfall during the growing period was 934.7 mm and 743.4 mm, with a monthly mean of 233.7 and 195.9, respectively. The eggplant materials were transplanted to the open field and glasshouse after 15 days of hardening after grafting procedures, with ten plants for each treatment. Agronomic practices and plant maintenance such as fertilizer application, pest and disease management, and weeding were carried out as recommended by the Department of Agriculture, Malaysia (http://jpn.penang.gov.my/index.php/perkhidmatan/teknologi-tanaman/sayur-sayuran/78-terung-sp-3424) for the open field trial, while the Fertigation System Manual [9,10] was used for the glasshouse trial. Plants were examined for insect, pest and disease incidence. Pesticides were regularly applied at the appropriate growth stages of the plant. Daily visit to the study site was conducted for crop maintainance, to observe crop performance and irrigate them when necessary using a rain port sprinkler system. The open field cropping system was fertilized using compound fertilizer NPK 15-15-15+2S and NPK 12-12-17-2+8S+TE following the recommended dose. The first and second doses were applied at two and six weeks after the transplant. The fertigation system in the second study environment (glasshouse) was laid out following the Fertigation System Manual (first edition) recommended by MARDI. Plants were given a daily dosage of copper standard formulation fertilizer comprising of N 14.29, P 1.94, K 7.67, Ca 4.24, Mg 2.06, Fe 0.22, Mn 0.042, B 0.14, Zn 0.0015, Cu 0.0016 and Mo 0.0021, (mmol L^{-1}) with electrical conductivity (EC) reading increasing progressively following.

Table 5. Accessions (*Solanum melongena*) and species used for this study.

No	Code	Original Code	Type of Material	Fruit Type	Origin
		Scion			
1	MCV1	Green world white eggplant 330	Commercial variety	Long	Malaysia
2	MCV2	Green world purple dream eggplant 302	Commercial variety	Long	Malaysia
3	CCV1	China vegetable seed technology 62146129	Commercial variety	Long	China
4	CCV2	China vegetable seed technology 62119631	Commercial variety	Long	China
5	CCV3	China vegetable seed technology 62119631	Commercial variety	Round	China
6	NCV	Yalo garden egg eggplant	Commercial variety	Round	Nigeria
7	TCV	Round eggplant (Chao-phaya)	Commercial variety	Round	Thailand
		Rootstock			
8	MWR	MFS eggplant	Wild relatives	Round	Malaysia
9	BWR	Plate brush eggplant	Wild relatives	Round	Bangladesh
10	TWR	Pea eggplant	Wild relatives	Round	Thailand

4.2. Seed Germination and Grafting Methods

The seeds of wild relatives and commercial varieties were soaked in GA_3 for 24 h before sowing to facilitate early, good and uniform germination. Seeds of rootstock (MWR, BWR and TWR) were sown 2 and 4 weeks before sowing the scion seeds. The scions were sown at three different stages at 2, 4 and 6 weeks. The rootstock seeds were sown in seed trays with 108 holes containing peat moss and later transplanted into 24 celled pro-trays at the two-leaf stage. Two grafting techniques (splices and cleft techniques) were used for this research on 35 days old seedlings (Figure 3), as described by [9]. The experiment was laid out in a Completely Randomized Design (CRD) in three replications. For the splice grafting technique, the rootstock stem was cut at 45° below the cotyledons, a clip was placed over the rootstock stem, and the scion was also cut at 45° above the cotyledons. The scion was placed over the rootstock and the graft was held together using a grafting clip. For the cleft grafting technique, the rootstock stem was cut horizontally below the cotyledons and the top was discarded. The cut was a 0.5-cm-long vertical incision into the centre of the rootstock stem. The scion stem was cut in the same diameter as the rootstock stem diameter at 0.5 cm long wedge shape cut, then the scion stem was inserted into the vertical incision of the rootstock. A plastic clip was placed around the graft union to hold the plant firmly together.

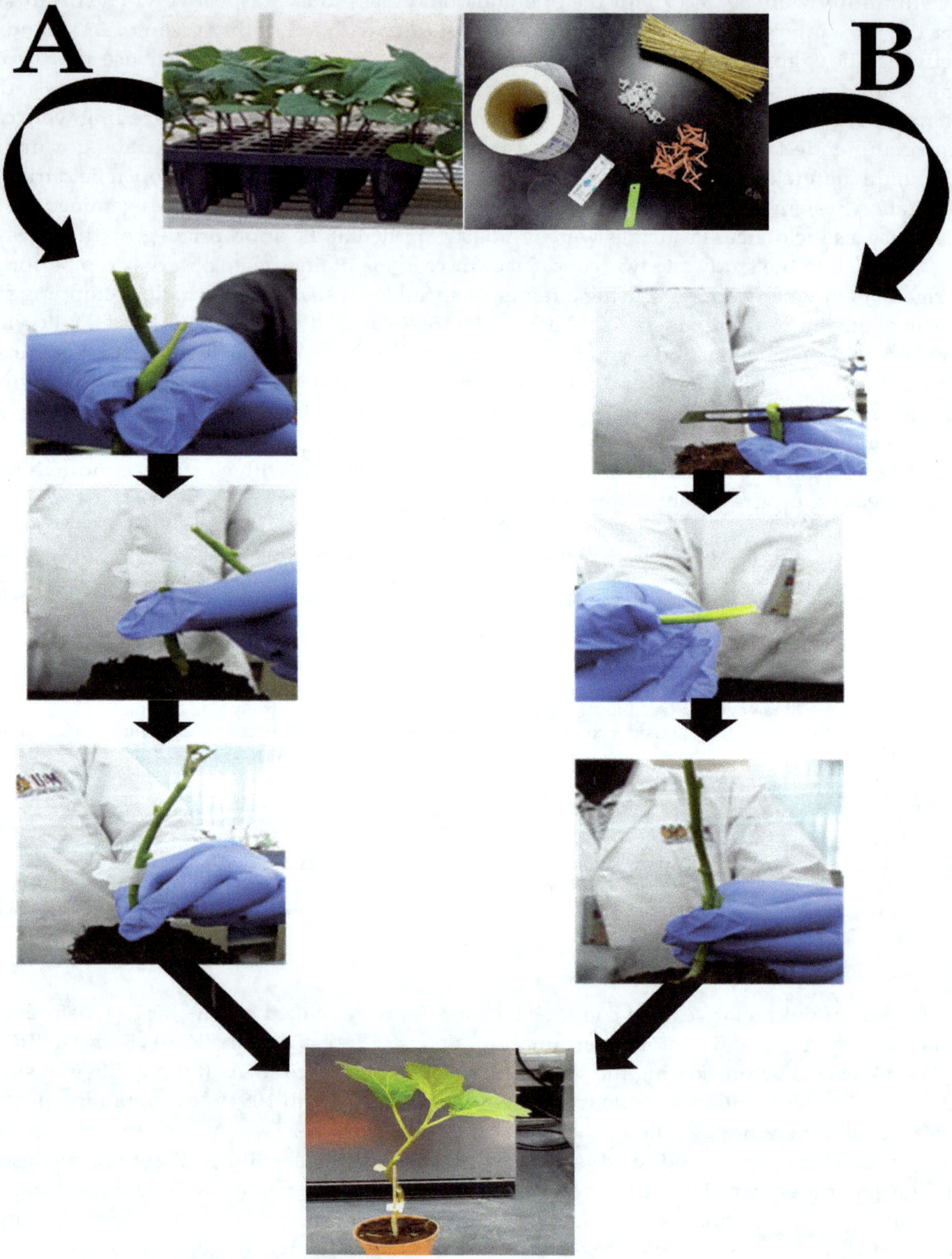

Figure 3. Grafting procedure for splice grafting (**A**) and cleft grafting (**B**) techniques.

4.3. Data Collection

From each grafted, self-grafted and non-grafted genotype, five plants were randomly sampled from each replication for data collection. The data on morphological, physiological and yield characteristics were collected after transplanting (Table 6). The measurement of total chlorophyll content was done following Coombs et al. (1987) at the fruiting stages with the formula chlorophyll-a = [(13.19 × A664) − (2.57 × A647)], chlorophyll b = [(22.1 × A647) − (5.26 × A664)], Total chlorophyll = chlorophyll a + chlorophyll b, and Actual chlorophyll = (3.5 × total chlorophyll)/1. The content was measured in actively growing leaf from the third or fourth fully expanded leaves from the tip, and was determined after extraction in 80% (*v*/*v*) acetone/water. Photosynthetic rate, stomatal conductance, and transpiration rate for the 35 treatments were measured at the fruiting stages using an infrared gas analyzer model Li-6400XT (Li-cor Inc., Lincoln, NE, USA). The measurement was conducted on young fully expanded and exposed leaves (third and fourth leaf from the tip) of the tagged plants from each treatment between 9:00 am to 11:00 am when stomata opening was at the optimal point.

Table 6. Data and description in the nursery, and morphological traits.

Parameters	Denotation	Data Collection
Days for 50% seeds germination	D50%SG	Daily visual observations were recorded until the reaching of 50% of emerged seedlings
Germination %	G%	The observation was recorded at every day by visual observation until all germinated
Days taken to 50% of grafting stage plants	G50%RGS	Daily visual observations were recorded until 50% of plants reached grafting stage
Graft success (%)	GS	A graft success was recorded at 15 days after grafting based on wilting of the grafts at healing region.
Days taken for 50% flowering	D50%F	Number of days from sowing to 50% flowering.
First harvest	FH	Days to first fruit picking was measured by counting the number of days after transplanting to the day of the first picking at the marketable stage.
Plant height (cm)	PH	Plant height 45 and 90 days after transplanting.
Number of secondary branches	NSB	Numbers of secondary branches 45 and 90 days after transplanting.
Stem diameter	SD	The diameter of the stem was measured 5 cm above the ground level at 45 and 90 days after transplanting with an electronic digital caliper.
Fruit length (cm)	FL	The length of one matured fruit per plant from calyx to the apex of the fruit
Fruit diameter (cm)	FD	The diameter of one matured fruit (0.3 cm below the calyx) using venire calipers
Fruit weight (g)	FW	The weight of one matured fruit per plant.
Number of fruits per plant	NFPP	Total number of fruits from the first harvest until 90 days after transplanting
Average fruit weight (g)	AFW	By weighing of individual fruits.
Marketable fruit yield (Kg)	MFY	Total weight of fruit from the first harvest until 90 days after transplanting

4.4. Statistical Analysis

The analysis of variance (ANOVA) was carried out on 35 accessions consisting of interspecific grafted (21), self-grafted (7) and non-grafted (7) scions evaluated in two cropping systems, namely, open field and glasshouse, using SAS software version 9.4 (SAS Institute, Inc., Cary, N.C., USA). The accession means comparison was determined using the Least Significant Difference (LSD) at $p = 0.05$ significant level.

5. Conclusions

Grafting technology has the potential to promote the production of eggplant in marginal and fragile agro-ecosystem and has proven to be a rapid alternative tool to confer resistance to bacterial wild disease, promote plant vigour, increase fruit yield and fruit quality in eggplant production. Since grafting gives increased vigour to crops, it will be valuable in sustainable low input horticulture of the future. In Malaysia, where the eggplant production is still carried out mostly through conventional methods and modern production techniques are limited, the grafting technique could help in solving many problems related to eggplant production. Grafting with the cleft technique exhibited high graft success and eggplant scions of MCV1, MCV2 and TCV grafted onto all rootstock displayed good vigour as well as excellent graft compatibility. Overall, the fruit character (fruit length and diameter) recorded was highest in scions of MCV2 and CCV3 grafted on MWR and TWR, respectively. Higher yields were obtained in eggplants with scions of MCV2 and CCV3 grafted onto rootstocks of MWR and TWR respectively, whereas the highest number of fruits were recorded in TCV grafted onto TWR rootstock. The effect of three rootstocks on the agronomic features of the seven eggplant varieties (scions) allow for considerations of the feasibility of this technique on the varieties tested. The grafted plants produced higher yield with better economic value than their self-rooted counterpart. Our results demonstrated that the use of wild relative rootstocks of eggplant species offers a valuable approach for improving eggplant production.

Author Contributions: Methodology, software, validation, formal analysis, and investigation, I.M. (Ibrahim Musa), M.Y.R., K.A., S.I.R., M.A.M.H.; writing—original draft preparation, writing—review and editing I.M. (Ibrahim Musa), M.Y.R., Y.O., I.M. (Isma'ila, Muhammad), S.C.C., N.N.M.S., A.F.A., J.H.; supervision, M.Y.R., K.A., S.I.R., M.A.M.H. All authors have read and agreed to the published version of the manuscript.

References

1. Sulaiman, N.N.M.; Rafii, M.Y.; Duangjit, J.; Ramlee, S.I.; Phumichai, C.; Oladosu, Y.; Datta, D.R.; Musa, I. Genetic variability of eggplant germplasm evaluated under open field and glasshouse cropping conditions. *Agronomy* **2020**, *10*, 436. [CrossRef]
2. Sabatino, L.; Iapichino, G.; D'Anna, F.; Palazzolo, E.; Mennella, G.; Rotino, G.L. Hybrids and allied species as potential rootstocks for eggplant: Effect of grafting on vigour, yield and overall fruit quality traits. *Sci. Hortic.* **2018**, *228*, 81–90. [CrossRef]
3. Yang, R.Y. Application of Antioxidant Activity Analytical Methods for Studies on Antioxidant Activities of Vegetables. Ph.D. Thesis, Institute of Tropical Agriculture and International Cooperation of National Ping-Tung University of Science and Technology, Neipu, Taiwan, 2006.
4. Miceli, A.; Sabatino, L.; Moncada, A.; Vetrano, F.; D'Anna, F. Nursery and field evaluation of eggplant grafted onto unrooted cuttings of *Solanum torvum* Sw. *Sci. Hortic.* **2014**, *178*, 203–210. [CrossRef]
5. Louws, F.J.; Rivard, C.L.; Kubota, C. Grafting fruiting vegetables to manage soilborne pathogens, foliar pathogens, arthropods and weeds. *Sci. Hortic.* **2010**, *127*, 127–146. [CrossRef]
6. Oladosu, Y.; Rafii, M.Y.; Magaji, U.; Abdullah, N.; Miah, G.; Chukwu, S.C.; Hussin, G.; Ramli, A.; Kareem, I. Genotypic and phenotypic relationship among yield components in rice under tropical conditions. *Biomed Res. Int.* **2018**, *2018*. [CrossRef]
7. Usman, M.G.; Rafii, M.Y.; Martini, M.Y.; Oladosu, Y.; Kashiani, P. Genotypic character relationship and phenotypic path coefficient analysis in chili pepper genotypes grown under tropical condition. *J. Sci. Food Agric.* **2017**, *97*, 1164–1171. [CrossRef]
8. Kacjan Maršić, N.; Mikulič-Petkovšek, M.; Stampar, F. Grafting influences phenolic profile and carpometric traits of fruits of greenhouse-grown eggplant (*Solanum melongena* L.). *J. Agric. Food Chem.* **2014**, *62*, 10504–10514. [CrossRef]
9. Lee, J.M.; Kubota, C.; Tsao, S.J.; Bie, Z.; Echevarria, P.H.; Morra, L.; Oda, M. Current status of vegetable grafting: Diffusion, grafting techniques, automation. *Sci. Hortic.* **2010**, *127*, 93–105. [CrossRef]

10. Ibrahim, M.; Munira, M.K.; Kabir, M.S.; Islam, A.K.M.S.; Miah, M.M.U. Seed germination and graft compatibility of wild solanum as rootstock of tomato. *J. Biol. Sci.* **2001**, *1*, 701–703.
11. Gisbert, C.; Prohens, J.; Raigón, M.D.; Stommel, J.R.; Nuez, F. Eggplant relatives as sources of variation for developing new rootstocks: Effects of grafting on eggplant yield and fruit apparent quality and composition. *Sci. Hortic.* **2011**, *128*, 14–22. [CrossRef]
12. Moncada, A.; Miceli, A.; Vetrano, F.; Mineo, V.; Planeta, D.; D'Anna, F. Effect of Grafting on yield and quality of eggplant (*Solanum melongena* L.). *Sci. Hortic.* **2013**, *149*, 108–114. [CrossRef]
13. Khah, E.M.; Kakava, E.; Mavromatis, A.; Chachalis, D.; Goulas, C. Effect of grafting on growth and yield of tomato (*Lycopersicon esculentum* Mill.) in greenhouse and open-field. *J. Appl. Hortic.* **2006**, *8*, 3–7. [CrossRef]
14. Fita, A.; Picó, B.; Roig, C.; Nuez, F. Performance of Cucumis melo spp. agrestis as a rootstock for melon. *J. Hortic. Sci. Biotechnol.* **2004**, *82*, 184–190. [CrossRef]
15. Khah, E.M. Effect of grafting on growth, performance and yield of aubergine (*Solanum melongena* L.) in greenhouse and open-field. *Int. J. Plant. Prod.* **2011**, *5*, 1735–8043.
16. Passam, H.C.; Stylianoy, M.; Kotsiras, A. Performance of eggplant grafted on tomato and eggplant rootstocks. *Eur. J. Hortic. Sci.* **2005**, *70*, 130–134.
17. Buller, S.; Inglis, D.; Miles, C. Plant growth, fruit yield and quality, and tolerance to verticillium wilt of grafted watermelon and tomato in field production in the Pacific Northwest. *HortScience* **2013**, *48*, 1003–1009. [CrossRef]
18. Leonardi, C.; Giuffrida, F. Variation of plant growth and macronutrient uptake in grafted tomatoes and eggplants on three different rootstocks. *Eur. J. Hortic. Sci.* **2006**, *71*, 97–101.
19. Trinchera, A.; Pandozy, G.; Rinaldi, S.; Crinò, P.; Temperini, O.; Rea, E. Graft union formation in artichoke grafting onto wild and cultivated cardoon: An anatomical study. *J. Plant Physiol.* **2013**, *170*, 1569–1578. [CrossRef]
20. Kumar, P.; Edelstein, M.; Cardarelli, M.; Ferri, E.; Colla, G. Grafting affects growth, yield, nutrient uptake, and partitioning under cadmium stress in tomato. *HortScience* **2015**, *50*, 1654–1661. [CrossRef]
21. Liu, Y.F.; Qi, H.Y.; Bai, C.M.; Qi, M.F.; Xu, C.Q.; Hao, J.H.; Li, Y.; Li, T.L. Grafting helps improve photosynthesis and carbohydrate metabolism in leaves of muskmelon. *Int. J. Biol. Sci.* **2011**, *7*, 1161–1170. [CrossRef]
22. Johnson, S.; Inglis, D.; Miles, C. Grafting effects on eggplant growth, yield and *Verticillium* wilt incidence. *Int. J. Veg. Sci.* **2014**, *20*, 3–20. [CrossRef]
23. Bekhradi, F.; Kashi, A.; Delshad, M. Effect of three cucurbits rootstocks on vegetative and yield of 'Charleston Gray' watermelon. *Int. J. Plant Prod.* **2011**, *5*, 105–110.
24. Bletsos, F.; Thanassoulpoulos, C.; Roupakias, D. Effect of grafting on growth, yield and *Verticillium* wilt of eggplant. *HortScience* **2003**, *38*, 183–186. [CrossRef]
25. Yetisir, H.; Sari, N. Effect of different rootstock on plant growth, yield and quality of watermelon. *Aust. J. Exp. Agric.* **2003**, *43*, 1269–1274. [CrossRef]
26. Schwarz, D.; Rouphael, Y.; Colla, G.; Vanema, J.K. Grafting as tool to improve tolerance of fruit vegetable to abiotic stresses: Thermal stress, water stress, and organic pollutants. *Sci. Hort.* **2010**, *127*, 162–171. [CrossRef]

3

Augmenting the Sustainability of Vegetable Cropping Systems by Configuring Rootstock-Dependent Rhizomicrobiomes that Support Plant Protection

Mariateresa Cardarelli [1], Youssef Rouphael [2], Marios C. Kyriacou [3], Giuseppe Colla [4] and Catello Pane [1,*]

1 Consiglio per la Ricerca in Agricoltura e L'analisi Dell'economia Agraria, Centro di Ricerca Orticoltura e Florovivaismo, 84098 Pontecagnano Faiano, Italy; mteresa.cardarelli@crea.gov.it
2 Department of Agricultural Sciences, University of Naples Federico II, 80055 Portici, Italy; Youssef.rouphael@unina.it
3 Department of Vegetable Crops, Agricultural Research Institute, Nicosia 1516, Cyprus; m.kyriacou@ari.gov.cy
4 Department of Agriculture and Forest Sciences, University of Tuscia, 01100 Viterbo, Italy; giucolla@unitus.it
* Correspondence: catello.pane@crea.gov.it

Abstract: Herbaceous grafting is a propagation method largely used in solanaceous and cucurbit crops for enhancing their agronomic performances especially under (a)biotic stress conditions. Besides these grafting-mediated benefits, recent advances about microbial networking in the soil/root interface, indicated further grafting potentialities to act as soil environment conditioner by modulating microbial communities in the rhizosphere. By selecting a suitable rootstock, grafting can modify the way of interacting root system with the soil environment regulating the plant ecological functions able to moderate soilborne pathogen populations and to decrease the risk of diseases. Genetic resistance(s) to soilborne pathogen(s), root-mediate recruiting of microbial antagonists and exudation of antifungal molecules in the rhizosphere are some defense mechanisms that grafted plants may upgrade, making the cultivation less prone to the use of synthetic fungicides and therefore more sustainable. In the current review, new perspectives offered by the available literature concerning the potential benefits of grafting, in enhancing soilborne disease resistance through modulation of indigenous suppressive microbial communities are presented and discussed.

Keywords: solanaceae; cucurbitaceae; defense mechanisms; soilborne pathogen; genetic resistance; microbial communities; grafting; soil/root interface

---

1. Introduction

Modern agriculture needs innovative strategies inspired by the principles of agroecology, aimed to guarantee soil conservation and fertility, to face the adversities that affect crop productivity and, generally, to increase the sustainability of intensive systems. One of the most important challenges for sustainability concerns the control of soil-borne diseases being considered a major limitation to crop production. The massive use of chemicals against plant pathogens is no longer a viable practice for environmental risks and human health care. The most commonly applied eco-friendly approaches include solarisation, biofumigation, crop rotation, tillage management practices, residue management and organic amendments [1]; additional strategies such as applications of plant growth promoting rhizobacteria (PGPRs), endo- and ectomycorrhizal fungi, cyanobacteria and other organisms can also improve plant resistance to soilborne pathogens. Several studies reported that plants can recruit a specific beneficial rhizosphere microflora which can contribute to reduce the activity of plant pathogens and to make plants more resistant to environmental stressors [2,3]. Investigations on rhizosphere microbiome carried out by next-generation sequencing technologies allowed to identify and quantify microorganisms associated with the root apparatus of vegetable crops highlighting an evolutionistic mechanism of microbial recruiting adopted by plants.

Vegetable grafting is a propagation method largely used in solanaceous and cucurbit crops to increase plant resistance to soilborne pathogens as well as other environmental stresses, and to enhance crop productivity and fruit quality [4]. In soils infected by highly destructive plant pathogens, the use of resistant grafted plants represents the main biological-based method that allows cultivation of the highly susceptible cultivars [5]. In the latest years, increasing studies on the ecological role of the grafting revealed new interesting opportunities of this technology to contrast and limit the activity of soilborne pathogens and their damages on vegetable crops, by modifying the presence of beneficial root-associated microbes. For instance, Poudel et al. [6] used a grafted tomato system to study the effect of rootstock genotypes and grafting on soil bacteria communities and their results highlighted an effect of rootstock genotype on bacterial diversity and composition in the rhizosphere. Moreover, Duan et al. [7] reported that some pepper grafting combinations significantly increased the populations of fungi and actinomycetes in the rhizosphere enhancing the activities of peroxidase, catalase, phosphatase, invertase, urease, and nitrate reductase in root rhizosphere soil. Similar results were also reported in grafted eggplants [8]. Besides plant species, additional environmental factors such as soil type can also affect the microbial communities in the rhizosphere indicating the need to evaluate the rootstock-mediated effects on rhizosphere microbiome under different environmental conditions [9,10]. The above findings indicate the potential of using plant rootstocks as a mean to recruit specific soil beneficial microorganisms for biotic stress management.

This review offers a novel perspective on the potentialities of vegetable grafting as sustainable practices to enhance crop resistance to environmental stresses through the modulation of microbial community structure in the rhizosphere.

2. Soil Microorganisms for Biological Control of Plant Diseases

Rhizosphere and bulk soil around growing plants constitute the habitat for a large number of microbial species with different lifestyles, interacting with one another, with the soil and plants. The culturable bacteria and fungi associated with rhizosphere can contain up to 10^{11} microbial cells per gram of roots [11]. Some microorganims have a neutral effect on the plant, but many known microorganisms are beneficial to the plants for nutrient and carbon cycling, soil organic matter formation and stabilization, so influencing agricultural productivity. They are symbiotic nitrogen-fixing bacteria (*Rhizobium leguminosarum*), endo- and ectomycorrhizal fungi, plant growth-promoting rhizobacteria (PGPR) and other fungi [12]. Many of these microorganisms are even investigated for their ability to prevent or limit soil-borne plant pathogens (fungi and oomycetes are the most important) through mechanisms as hyperparasitism, antibiosis and competition for ecological niches and nutrients [13].

Microbial hyperparasites have a predatory behavior: they enter host cells of fungi and oomycetes helped by secreting cell-wall lytic enzymes (as chitinases, cellulases and proteases) and feed on the pathogen as long as it dies. Parasitic activity is quite common, and well documented for *Trichoderma* and *Gliocladium* against fungal pathogens as *Rhizoctonia*, *Sclerotinia*, *Verticillium* and *Gaeumannomyces* [14–16]. Even *Coniothyrium minitans* [17] and *Sporidesmium sclerotivorum* [18] are effective in controlling diseases caused by sclerotia-forming fungi. Chitinolytic activity of *Pseudomonas* spp. is responsible of antagonistic activity towards *R. solani* [12].

Antibiosis results from the release of low-molecular weight compounds produced by microorganisms in the surrounding environment which are deleterious to the metabolism or growth of plant pathogens [19,20]. Fluorescent *Pseudomonas* spp., *Bacillus* spp., *Streptomyces* spp. and *Trichoderma* spp. produce antibiotic molecules (phenazines, 2,4-diacetylphloroglucinol, pyoluteorin, and pyrrolnitrin) affecting the electron transport chain, metalloenzymes, membrane integrity, or cell membrane and zoospores [20–22].

With the competition for space, pathogens are prevented from accessing root surface and plant tissue whereas the competition for nutrients, especially for carbon released as root exudates, affect the saprophytic phase of pathogens, and may cause lacking spore germination and microbiostasis [11]. Siderophore-producing *Pseudomonas* spp. are involved in the competition for iron, a micronutrient essential for growth and activity of the pathogens and are able to reduce disease incidence or severity of pathogenic fungi [23].

Literature survey reveals many pivotal examples of rhizosphere-competent microbial biocontrol agents of plant pathogens. Fungi belonging to *Trichoderma* genus, for example, have antagonistic properties towards a plethora of plant pathogens relying on all of the antagonist modes [24]. *T. harzianum* and *T. asperellum* showed antagonistic effects against *Fusarium oxysporum* of tomato [25–27] and melon [28] under field conditions. Gava et al. [28] tested different *Trichoderma* species (*T. harzianum*, *T. viride*, *T. koningii*, and *T. polysporum*) in a naturally infested soil and obtained the highest control of melon wilt using *T. polysporum*.

Malolepsza [29] observed a significant stimulation of systemic defenses (activation of antioxidant enzymes and enhancement of phenols) by *T. virens* inoculation of tomato plants, leading to lowered *Rhizoctonia solani* infection. Some authors verified the involvement of the proteins Sm1 and Epl1 in the systemic protection of tomato plants mediated by *Trichoderma* spp. [30]: in presence of these proteins it was observed an increase in disease resistance against *Alternaria solani*, *Botrytis cinerea*, and *Pseudomonas syringae* due to the increased expression of peroxidase and α-dioxygenase encoding genes [30], the elicitation of the salicylic acid pathway [31,32] and the ethylene pathway [31]. Even the accumulation of phenolic acids, flavonoids and *de novo* synthesis of catechins, enhanced by *T. atroviride* inoculation, is supposed to contribute to cucumber protection against *R. solani* [33]. In cucumber seedlings inoculated with *T. harzianum* inoculations, Chen [34] found alterations in nuclear DNA content and cell cycle-related genes expression that might maintain a lower ROS accumulation and higher root cell viability counteracting *Fusarium* disease in open field. Quantitative proteomics studies on black pepper plants primed with *T. harzianum* confirmed the plant defense response against the pathogen *Phytophthora capsicum* through an increase of ethylene synthesis and activating both the isoflavanoid pathway and lignin synthesis [35].

Coating of tomato seeds with *T. asperellum* and *Bacillus subtilis* decreased the susceptibility of plants to *Pythium aphanidermatum* [36]. Frequently the combination of different antagonists in microbial consortia improves their effectiveness for the involvement of diverse biocontrol mechanisms [37]. In watermelon, a systemic acquired resistance against *F. oxysporum* f.sp. *niveum* was related to the inoculation with a consortium of *T. harzianum*, *Paenibacillus polymyxa* and other antagonistic microorganisms that activated defense-related enzymes [38]. *T. harzianum* and *Bacillus amyloliquefaciens* together inhibited the growth and production of mycelia and sclerotia protecting over 80% of tomato, squash and eggplant seedlings against *Sclerotinia sclerotiorum* [39]. Similarly, to *T. asperellum*, even the arbuscular mycorrhizal fungus *Rhizophagus irregularis* (previously known as *Glomus intraradices*) lowered disease incidence of *Fusarium* wilt in tomato plants [40]. Both these agents increased plant height, chlorophyll content and Ca, Mg, S, Mn, B and Si uptake. *T. harzianum* combined with *Glomus intraradices* induced a plant basal resistance attenuating the hormone (ethylene and abscisic acid) disruption induced by *F. oxysporum* in melon [41] whereas associated with *Glomus mossae* is useful in cucumber against *Phytophthora melonis* increasing the transcription level of defensive genes as phenylalanine ammonialyase, cucumber pathogen-induced 4, lipoxygenase and galactinol synthase [42]. In field experiment the consortium mycorrhizae, *Trichoderma* and plant growth-promoting bacteria enhanced the pepper yield and modulated the activities of defense enzymes as polyphenol oxidase, peroxidase, superoxide dismutase, and catalase [43]. Even arbuscular mycorrhizal fungi alone can control plant soil-borne diseases [44,45] activating a systemic plant immune response. Mycorrhizal fungi reduced the disease severity index of eggplant Verticillium wilt [46]; the disease tolerance was related to a lower proline content and relative electrical conductivity in leaves, and to higher activity of browning related enzymes (phenylalanine ammonia-lyase, polyphenol oxidase and peroxidise). Panda et al. [47]

found that pre-colonization of tomato roots with mycorrhizal fungus *Piriformospora indica* systemically induced resistance against early blight by *Alternaria solani*; after pathogen attack *P. indica* induced a rapid activation of jasmonic acid/ethylene-mediated basal defenses against pathogen infection by altering the expression of JA/ET related genes. *Glomus mossae* suppresses *F. oxysporum* development in the roots and rhizosphere of watermelon and modulates the composition of root exudates. Bacterial species of the *Bacillus* and *Pseudomonas* genus control plant disease by producing antibiotics or stimulating the host resistance [48]. Tomato plants inoculated with *Bacillus amyloliquefaciens* exhibited significantly low *F. oxysporum* f. sp. *lycopersici* infection compared with the control [49]. *Clonostachys rosea* counteracted gray mold disease not only by suppressing development and sporulation of *Botrytis cinerea* but also by inducing the resistance of tomato plants against *B. cinerea* [50].

3. How Plants Recruit Antagonist Microbes to Prevent the Infection by Pathogens

The rhizosphere microbiome has been recognized as the second genome of the plants [11], and it may reveal the greater ability of the plant in adapting to the external environment, including the protection degree against pathogenic agents. Plants through their roots are in dynamic communication with the surrounding bulking soil microbial communities [51]. Roots exudates are a main food source for microorganisms and a driving force for their assembling and activities. Carbon and nitrogen are exudates as simple molecules (sugars, organic acids, secondary metabolites) or complex polymers (mucilage) but their composition varies with the plant genotype, developmental stage, and the presence of (a)biotic stresses [52]. It has been seen that microbial communities in the rhizosphere of different plant species growing on the same soil are often different and vice versa [53,54] thus demonstrating that plants modulate exudates profile (alteration of biosynthesis and transport of molecules) and immune system activities to recruit specific beneficial microbes. Moreover, plants can detect communication signals between bacteria in the rhizosphere (quorum sensing signals) and produce molecules that stimulate or deactivate these signals so influencing the outcome of microbe-microbe and/or plant-microbe interactions [55]. L-malic acid is the small signalling molecule exudated from tomato plant roots that is responsible for biofilm formation and root colonization of the antagonistic *Bacillus subtilis* [56]. Similarly, Tan et al. [57] confirmed the important role of malic and citric acids in tomato root surface colonization of *Bacillus amyloliquefaciens*. Moreover, different organic acids in watermelon and cucumber roots exudates stimulate the beneficial *Paenibacillus polymyxa* [58] and *Trichoderma harzianum* [59], respectively. Glucose, succinic acid, *p*-hydroxybenzoic acid, *p*-coumaric acid and glutamic acid in cucumber root exudates recruit *Trichoderma* with the most relevant effect from glucose [60].

Recent advances reveal that multiple signals operate in the establishment and the maintenance of arbuscular mycorrhizal symbiosis including calcium spiking, reactive oxygen species and phytohormones [61,62]. The plant hormones strigolactones are actively exuded into rhizosphere as ex-planta signalling molecules that attract arbuscular mycorrhizal fungi affecting both pre-symbiotic and symbiotic phases [63]. Other important hormones involved in the onset of the AM symbiosis are jasmonates [64], gibberellins [65], ethylene [66], and auxins [67]. A possible crosstalk between auxin and strigolactones is also postulated in tomato plants [68]. Martínez-Medina et al. [69] studied the interaction between tomato roots and the AM fungus *Rhizophagus irregularis* showing a regulatory role of nitric oxide mediated by a specific phytoglobin during the symbiosis.

4. Defense Mechanisms in Grafted Plants against Soilborne Pathogens

Disease control-related mechanisms of grafting may rely on the genetic traits of the rootstock-type. Interspecific grafting, for example, is often secured with the non-host resistance of the rootstock against species-specific soil-borne pathogens, such as the wilting causal agents belonging to *F. oxysporum* group that harbour differentially host-compatible *formae specialis* [70,71]. Limitation in spread of tomato wilting-associated bacteria in resistant cultivars could avoid the generalization of infection to the entire vascular tissues of susceptible scion [72]. Graft-transmissible resistance to airborne disease has been observed for sweet pepper cultivar grafted on a resistant cherry pepper rootstock [73]. In this situation, the effects of rootstock on the scion is the major determinant of resistance. Disease resistance of grafted plants may also result from the enhanced vigour for which grafted plants grow more fortified and more efficient in resource utilization, as well as developing vigorous root apparatus that allow them to better withstand pathogenic attacks caused by the parenchymatous parasites [74]. For example, reduction of *Verticillium* wilting by effective eggplant grafting on *S. lycopersicum* × *S. habrochaites* rootstock is associated to promotion of plant vigour [75]. Another mechanisms of disease resistance promoted by grafting is the release in the rhizosphere of exudates having antifungal activity on pathogen propagules. Liu et al. [76], for example, found root exudates from grafted eggplants, suppressive against the *Verticillium dahliae* mycelial development, contrarily to those released by the non-grafted ones. Grafting can affect root ability to harbour rhizosphere-competent microbes and to regulate beneficial antagonistic functions by modifying the root-architecture and the exudate profiling [77,78]. Rootstock/scion interaction can modulate regulations of transcripts and metabolites affecting disease susceptibility to fungal pathogens [79]. For instance, rootstock can influence disease resistance of scion by modifying secondary metabolites in the sap flow moving through the vascular system [80]. Shibuya et al. [81] attributed the transient reduction of disease symptoms on cucumber scion to changes in morphology or physiology in response to the modified water relations immediately after grafting onto squash. Defense mechanisms in grafted plants against soil-borne pathogens are summarized in Figure 1.

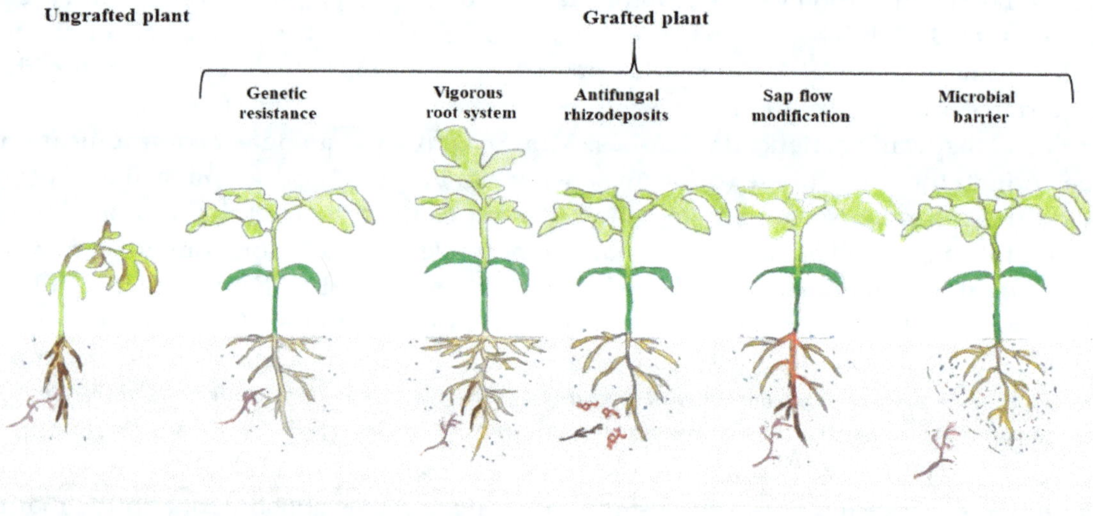

Figure 1. Disease-resistance mechanisms of grafted plants in soil infested by pathogens.

Experimental evidences suggested that the above listed mechanisms act synergistically in the regulation of the disease resistance of grafted roots. Greenhouse trials carried out by Song et al. [77] allowed to define as changes in root exudates from watermelon grafted on *Lagenaria siceraria* and *Cucurbita pepo* rootstocks, have promoted the non-host holobiont resistance to *F. oxysporum* f. sp. *niveum* infections through additional pathogen exclusion effects. In particular, niche chemical shifts around watermelon grafted roots affected the sheltered microbial diversity, increasing, for example, biocontrol-associated bacteria populations, such as *Sphingobacteriia* and *Bacillus* [77]. Biochemical investigations on watermelon/bottle gourd rhizodeposits have allowed to detect the *ex-novo* formation of bioactive proteins associated with plant resistance to biotic and abiotic stresses [82] and the release of chlorogenic and caffeic acids involved in the microbiostasis against propagules of the wilting causal agent [83]. Findings suggesting that the ungrafted watermelon delivery of molecular signals that are normally used as stimuli by pathogens harbouring the rhizosphere [84], with the grafting, likely, it is interrupted. In this light, combining grafting with environmentally friendly soil treatments (e.g., soil solarization) could slow down the recovery of pathogen recrudescence, often observed in conductive soils with the continuous monoculture cropping [84]. Structural variations of a root-associated microbiome after grafting have been also reported in open field tomato production, where Poudel et al. [6] showed rootstock-specific filtering effects driven by vigorous rooting, on endosphere and rhizosphere microbiomes enriched by representatives of *Firmicutes*, *Verrucomicrobia*, *Planctomycetes*, *Proteobacteria* and *Acidobacteria*. Under greenhouse, the use of *Solanum torvum* as rootstook for eggplant, induced an incremental shift in *Bacteria* and *Actinomycetes* on the root system, and in the responsive rhizosphere soil enzymes were found both associated to the pathogen population lowering and *Verticillium* wilting decrease [85].

Plant domestication privileged the selection of agronomic traits strictly related to the edibility, quantity and quality of the yields, leaving on the way all those alleles that gave to wild relatives the ability to profitably deal with soil microorganisms, reducing the potential of modern cultivars to lead rhizosphere microbiome assembly and functions [86]. In order, to bridge this gap, only recently some breeding programs are including, among the genetic improvement objectives, the enhanced ability of crops in recruiting beneficial host-specific root microbiota too [87]. In the meantime, horticultural experiences indicated that, at the moment, grafting better than others available technologies allows to reach this phenotype more quickly depending on the specific influence of rootstock genotype [84,88]. The metabolic structure of rootstock-recruited microorganisms, for example, has been implemented for managing tree resistance to the apple replant disease complex in replanted orchards [89]. Thereby, also vegetable productions may alleviate detrimental risks hidden into the intensive cropping systems following a holistic approach, in which grafting improves telluric environment adaptation of cultivar through the reinforcement of partnerships with beneficial and biodiverse microbiota and the enhancement of the efficient resource utilization. A summary of the possible directional interactions occurring in the soil environments shared by grafted roots and microbiome aiming at reducing pathogens pressure is reported in Table 1.

Table 1. Possible directional interactions occurring in the soil between grafted roots and microbiome aiming at reducing pathogens pressure.

To	Grafted Roots	Microbiota
Grafted roots	Allelopathy Competition for space, water and nutrients	Selection-filtering effects C-food providing Activation of the endophytism
Microbiota	Nutrients availability Growth promotion Induction of resistance to (a)biotic stresses	Antibiosis Hyperparassitism Niche occupying Essential food depletion

Relationships-mediated by rootstock between the vegetable scion and the infected soil open new opportunities for disease risks management in a more environmentally friendly way (Table 2).

Table 2. Soilborne disease risk management using ungrafted and grafted plants.

Management Goal	Ungrafted Plants	Grafted Plants
Reduction of disease risks	Cultivation of susceptible cultivars on healthy soils	Grafting of susceptible cultivars on disease resistant rootstocks
Genetic innovation regarding plant protection objectives	New cultivars resistant/tolerant to pathogens	New rootstock genotypes resistant/tolerant to pathogens New grafting combinations with enhanced root ability to recruit protective rhizobiome
Engineering soil microbiome for enhancing its antagonistic structure	Soil conditioning techniques mitigating soil sickness phenomena	Actions for maintenance of ecological bio-barriers proliferating on the rootstock roots
Increased effectiveness of the chemical control means	Wide-spectrum preventive and curative applications of fungicides	Targeted use of active molecules both in time and in dosage
Increased effectiveness of the biological control agents	Mass applications of the bio-based formulates	Coordinating application of the external microbial antagonists, with the naturally present rhizobiome Targeted use of specific microbial strains characterized by grafted-rhizosphere competence

Moreover, the integration of grafting technology with other agronomic, chemical and/or biological control means can allow to further reduce the risks of soilborne diseases especially in a long term. However, the use of chemical inputs must be carefully evaluated to avoid depression of beneficial indigenous soil microflora in the rootstock rizhosphere. Under these boosted agroecological conditions monitoring of plant and soil health status over time is crucial.

5. Conclusion and Future Prospects

Grafting represents a relatively recent innovation in the vegetable production systems of Western Countries; this advanced technology allowed to reduce the negative impact of soilborne pathogens in solanaceous and cucurbit crops depending on the disease resistance level of rootstock. Grafting can enhance plant disease resistance through several multiple defense mechanisms in plant. Recent studies demonstrated that rootstock-mediated effects on rhizosphere microbiome can contribute to reduce the soilborne diseases by stimulating indigenous microflora able to compete for food and space with plant pathogens and to reduce their activity through antibiosis and hyperparasitism. Moreover, plant-beneficial microbes recruited by rootstock can enhance soilborne disease resistance indirectly through the increase of plant nutrient availability, the stimulation of plant growth and the induction of resistance to biotic stresses. Despite experimental evidences demonstrated a significant role of rhizosphere microbiome in enhancing soilborne disease resistance in grafted plants, more studies are necessary to better understand the scion-rootstock-rhizosphere microbiome interaction under different environmental conditions. Advanced technologies like metagenomics can help to identify and characterize the microbial strains in the rhizosphere of grafting combinations in order to link the changes in rhizosphere microbial community to enhanced plant resistance to specific soilborne pathogens. This knowledge could allow to develop new disease control strategies based on the combined application of selected microbial inoculants and specific grafting combinations.

Author Contributions: Conceptualization, M.C. and C.P.; writing—original draft preparation, M.C. and C.P.; writing—review and editing, M.C., Y.R., M.C.K., G.C. and C.P. All authors have read and agreed to the published version of the manuscript.

Acknowledgments: This work was partially supported by MIUR (Minister for education, University and Research), Law 232/2016, 'Department of Excellence'.

References

1. Raaijmakers, J.M.; Paulitz, T.C.; Steinberg, C.; Alabouvette, C.; Moënne-Loccoz, Y. The rhizosphere: A playground and battlefield for soilborne pathogens and beneficial microorganisms. *Plant Soil* **2009**, *321*, 341–361. [CrossRef]
2. Tena, G. Recruiting microbial bodyguards. *Nat. Plants* **2018**, *4*, 857. [CrossRef] [PubMed]
3. Compant, S.; Samad, A.; Faist, H.; Sessitsch, A. A review on the plant microbiome: Ecology, functions, and emerging trends in microbial application. *J. Adv. Res.* **2018**, *19*, 29–37. [CrossRef]
4. Colla, G.; Rouphael, Y.; Cardarelli, M.; Salerno, A.; Rea, E. The effectiveness of grafting to improve alkalinity tolerance in watermelon. *Environ. Exp. Bot.* **2010**, *68*, 283–291. [CrossRef]
5. Louws, F.J.; Rivard, C.L.; Kubota, C. Grafting fruiting vegetables to manage soilborne pathogens, foliar pathogens, arthropods and weeds. *Sci. Hortic.* **2010**, *127*, 127–146. [CrossRef]
6. Poudel, R.; Jumpponen, A.; Kennelly, M.M.; Rivard, C.L.; Gomez-Montano, L.; Garrett, K.A. Rootstocks shape the rhizobiome: Rhizosphere and endosphere bacterial communities in the grafted tomato system. *Appl. Environ. Microbiol.* **2018**, *85*. [CrossRef]
7. Duan, X.; Bi, H.G.; Wei, Y.Y.; Li, T.; Wang, H.T.; Ai, X.Z. Effect of grafting on rhizosphere soil environment and its relationship with disease resistance and yield of pepper. *Chin. J. Appl. Ecol.* **2016**, *27*, 3539–3547.
8. Zhou, B.L.; Xu, Y.; Yin, Y.L.; Ye, X.L. Effects of different years continuous cropping and grafting on the biological activities of eggplant soil. *Chin. J. Appl. Ecol.* **2010**, *29*, 290–294.

9. Garbeva, P.; van Elsas, J.D.; Veen, J.A. Rhizosphere microbial community and its response to plant species and soil history. *Plant Soil* **2008**, *302*, 19–32. [CrossRef]
10. Viebahn, M.; Doornbos, R.; Wernars, K.; van Loon, L.C.; Smit, E.; Bakker, P.A.H.M. Ascomycete communities in the rhizosphere of fieldgrown wheat are not affected by introductions of genetically modified *Pseudomonas putida* WCS358r. *Environ. Microbiol.* **2005**, *7*, 1775–1785. [CrossRef]
11. Berendsen, R.L.; Pieterse, C.M.J.; Bakker, P.A.H.M. The rhizosphere microbiome and plant health. *Trends Plant Sci.* **2012**, *17*, 478–486. [CrossRef] [PubMed]
12. Garbeva, P.; van Veen, J.A.; van Elsas, J.D. Assessment of the diversity and antagonism toward *Rhizoctonia solani* AG3 of *Pseudomonas* species in soil from different agricultural regimes. *FEMS Microbiol. Ecol.* **2004**, *47*, 51–64. [CrossRef]
13. McSpadden Gardener, B.B. Diversity and ecology of biocontrol *Pseudomonas* spp. in agricultural systems. *Phytopathology* **2007**, *97*, 221–226. [CrossRef] [PubMed]
14. Reithner, B.; Ibarra-Laclette, E.; Mach, R.L.; Herrera-Estrella, A. Identification of mycoparasitism-related genes in *Trichoderma atroviride*. *Appl. Environ. Microbiol.* **2011**, *73*, 4361–4370. [CrossRef]
15. Pinto, Z.V.; Cipriano, M.A.P.; dos Santos, A.S.; Pfenning, L.H.; Patrício, F.R.A. Control of lettuce bottom rot by isolates of *Trichoderma* spp. *Summa Phytopathol.* **2014**, *40*, 141–146. [CrossRef]
16. Salamone, A.L.; Gundersen, B.; Inglis, D.A. *Clonostachys rosea*, a potential biological control agent for *Rhizoctonia solani* AG-3 causing black scurf on potato. *Biocontrol Sci. Technol.* **2018**, *28*, 895–900. [CrossRef]
17. Jones, E.E.; Clarkson, J.P.; Mead, A.; Whipps, J.M. Effect of inoculum type and timing of application of *Coniothyrium minitans* on *Sclerotinia sclerotiorum*: Control of *Sclerotinia* disease in glasshouse lettuce. *Plant Pathol.* **2004**, *53*, 621–623. [CrossRef]
18. Mischke, S. Mycoparasitism of selected sclerotia-forming fungi by *Sporidesmium sclerotivorum*. *Can. J. Bot.* **1998**, *76*, 460–466. [CrossRef]
19. Raaijmakers, J.M.; Vlami, M.; De Souza, T.J. Antibiotic production by bacterial biocontrol agents. *Antonie Leeuwenhoek* **2002**, *81*, 537–547. [CrossRef]
20. Haas, D.; Défago, G. Biological control of soil-borne pathogens by fluorescent pseudomonads. *Nat. Rev. Microbiol.* **2005**, *3*, 307–319. [CrossRef]
21. Raaijmakers, J.M.; de Bruijn, I.; de Kock, M.J.D. Cyclic lipopeptide production by plant-associated *Pseudomonas* species: Diversity, activity, biosynthesis and regulation. *Mol. Plant-Microbe Interact.* **2006**, *19*, 699–710. [CrossRef] [PubMed]
22. Bae, S.J.; Kumar, T.; Chung, J.Y.; Ryu, M.; Park, G.; Shim, S.; Hong, S.B.; Seo, H.; Bae, D.W.; Bae, I.; et al. *Trichoderma* metabolites as biological control agents against *Phytophthora* pathogens. *Biol. Control* **2016**, *92*, 128–138. [CrossRef]
23. Alabouvette, C.; Olivain, C.; Steinberg, C. Biological control of plant diseases: The European situation. *Eur. J. Plant Pathol.* **2006**, *114*, 329–341. [CrossRef]
24. Kumar, G.; Maharshi, A.; Patel, J.; Mukherjee, A.; Singh, H.B.; Sarma, B.K. *Trichoderma*: A potential fungal antagonist to control plant diseases. *SATSA Mukhapatra* **2017**, *21*, 206–218.
25. Fitrianingsih, A.; Martanto, E.A.; Abbas, B. The effectiveness of fungi *Gliocladium fimbriatum* and *Trichoderma viride* to control *Fusarium* wilt disease of tomatoes (*Lycopersicum esculentum*). *Indian J. Agric. Res.* **2019**, *53*, 57–61.
26. Li, Z.B.; Zhou, R.J.; Fu, J.F. Allelopathic effects of phenolic acids from ginseng rhizosphere soil on *Cylindrocarpon destructans* (Zinss) scholten. *Allelopath. J.* **2018**, *43*, 53–64. [CrossRef]
27. Mwangi, M.W.; Muiru, W.M.; Narla, R.D.; Kimenju, J.W.; Kariuki, G.M. Management of *Fusarium oxysporum* f. sp. *lycopersici* and root-knot nematode disease complex in tomato by use of antagonistic fungi, plant resistance and neem. *Biocontrol Sci. Technol.* **2019**, *29*, 207–216. [CrossRef]
28. Gava, C.A.T.; Pinto, J.M. Biocontrol of melon wilt caused by *Fusarium oxysporum* Schlect f. sp. *melonis* using seed treatment with *Trichoderma* spp. and liquid compost. *Biol. Control* **2016**, *97*, 13–20. [CrossRef]
29. Małolepsza, U.; Nawrocka, J.; Szczech, M. *Trichoderma virens* 106 inoculation stimulates defence enzyme activities and enhances phenolic levels in tomato plants leading to lowered *Rhizoctonia solani* infection. *Biocontrol Sci. Technol.* **2017**, *27*, 180–199. [CrossRef]

30. Salas-Marina, M.A.; Isordia-Jasso, M.I.; Islas-Osuna, M.A.; Delgado-Sánchez, P.; Jiménez-Bremont, J.F.; Rodríguez-Kessler, M.; Rosales-Saavedra, M.T.; Herrera-Estrella, A.; Casas-Flores, S. The Epl1 and Sm1 proteins from *Trichoderma atroviride* and *Trichoderma virens* differentially modulate systemic disease resistance against different life style pathogens in *Solanum lycopersicum*. *Front. Plant Sci.* **2015**, *6*. [CrossRef]
31. Martinez, C.; Blanc, F.; Le Claire, E.; Besnard, O.; Nicole, M.; Baccou, J.C. Salicylic acid and ethylene pathways are differentially activated in melon cotyledons by active or heat-denatured cellulase from *Trichoderma longibrachiatum*. *Plant Physiol.* **2001**, *127*, 334–344. [CrossRef] [PubMed]
32. Gomes, E.V.; Ulhoa, U.J.; Cardoza, R.E.; Silva, R.N.; Gutiérrez, S. Involvement of *Trichoderma harzianum* Epl-1 protein in the regulation of *Botrytis* virulence- and tomato defense-related genes. *Front. Plant Sci.* **2017**, *8*, 1–11. [CrossRef]
33. Nawrocka, J.; Szczech, M.; Małolepsza, U. *Trichoderma atroviride* enhances phenolic synthesis and cucumber protection against *Rhizoctonia solani*. *Plant Prot. Sci.* **2018**, *54*, 17–23. [CrossRef]
34. Chen, S.; Zhao, H.; Zou, C.; Li, Y.; Chen, Y.; Wang, Z.; Jiang, Y.; Liu, A.; Zhao, P.; Wang, M.; et al. Combined inoculation with multiple arbuscular mycorrhizal fungi improves growth, nutrient uptake and photosynthesis in cucumber seedlings. *Front. Microbiol.* **2017**, *8*. [CrossRef] [PubMed]
35. Umadevi, P.; Anandaraj, M. Proteomic analysis of the tripartite interaction between black pepper, *Trichoderma harzianum* and *Phytophthora capsici* provides insights into induced systemic resistance mediated by *Trichoderma* spp. *Eur. J. Plant Pathol.* **2019**, *154*, 607–620. [CrossRef]
36. Kipngeno, P.; Losenge, T.; Maina, N.; Kahangi, E.; Juma, P. Efficacy of *Bacillus subtilis* and *Trichoderma asperellum* against *Pythium aphanidermatum* in tomatoes. *Biol. Control* **2015**, *90*, 92–95. [CrossRef]
37. Akköprü, A.; Demir, S. Biological control of *Fusarium* wilt in tomato caused by *Fusarium oxysporum* f. sp. lycopersici by AMF *Glomus intraradices* and some *Rhizobacteria*. *J. Phytopathol.* **2005**, *153*, 544–550. [CrossRef]
38. Wu, F.; Liu, B.; Zhou, X. Effects of root exudates of watermelon cultivars differing in resistance to *Fusarium* wilt on the growth and development of *Fusarium oxysporum* f.sp. *niveum*. *Allelopath. J.* **2010**, *25*, 403–414.
39. Abdullah, M.T.; Ali, N.Y.; Suleman, P. Biological control of *Sclerotinia sclerotiorum* (Lib.) de Bary with *Trichoderma harzianum* and *Bacillus amyloliquefaciens*. *Crop Prot.* **2008**, *27*, 1354–1359. [CrossRef]
40. Bidellaoui, B.; Segarra, G.; Hakkou, A.; Trillas, M.I. Beneficial effects of *Rhizophagus irregularis* and *Trichoderma asperellum* strain T34 on growth and Fusarium wilt in tomato plants. *J. Plant Pathol.* **2019**, *101*, 121–127. [CrossRef]
41. Martínez-Medina, A.; Pascual, J.A.; Pérez-Alfocea, F.; lbacete, A.; Roldán, A. *Trichoderma harzianum* and *Glomus intraradices* modify the hormone disruption in duced by *Fusarium oxysporum* infection in melon plants. *Phytopathology* **2010**, *100*, 682–688. [CrossRef] [PubMed]
42. Sabbagh, S.K.; Roudini, M.; Panjehkeh, N. Systemic resistance induced by *Trichoderma harzianum* and *Glomus mossea* on cucumber damping-off disease caused by *Phytophthora melonis*. *Arch. Phytopathol. Plant Prot.* **2007**, *50*, 375–388. [CrossRef]
43. Duc, N.H.; Mayer, Z.; Pék, Z.; Helyes, L.; Posta, K. Combined inoculation of arbuscular mycorrhizal fungi, *Pseudomonas Fluorescens* and *Trichoderma* spp. For enhancing defense enzymes and yield of three pepper cultivars. *Appl. Ecol. Environ. Res.* **2017**, *15*, 1815–1829. [CrossRef]
44. Azcón-Aguilar, C.; Palenzuela, J.; Roldán, A.; Bautista, S.; Vallejo, R.; Barea, J.M. Analysis of the mycorrhizal potential in the rhizosphere of representative plant species from desertification-threatened Mediterranean shrublands. *Appl. Soil Ecol.* **2003**, *22*, 29–37. [CrossRef]
45. Xavier, L.J.C.; Boyetchko, S.M. Arbuscular mycorrhizal fungi as biostimulants and bioprotectants of crops. *Appl. Mycol. Biotechnol.* **2002**, *2*, 311–340. [CrossRef]
46. Zhou, B.L.; Zheng, J.D.; Bi, X.H.; Cai, L.L.; Guo, W.W. Effects of mycorrhizal fungi on eggplant *Verticillium* wilt and eggplant growth. *Chin. J. Ecol.* **2015**, *34*, 1026–1030.
47. Panda, S.; Busatto, N.; Hussain, K.; Kamble, A. *Piriformospora indica*-primed transcriptional reprogramming induces defense response against early blight in tomato. *Sci. Hortic.* **2019**, *255*, 209–219. [CrossRef]
48. Gao, P.; Qin, J.; Li, D.; Zhou, S. Inhibitory effect and possible mechanism of a *Pseudomonas* strain QBA5 against gray mold on tomato leaves and fruits caused by *Botrytis cinerea*. *PLoS ONE* **2018**, *10*, e0190932. [CrossRef]
49. Patakioutas, G.; Dimou, D.; Yfanti, P.; Karras, G.; Ntatsi, G.; Savvas, D. Root inoculation with beneficial micro-organisms as a means to control *Fusarium oxysporum* f. sp. *lycopersici* in two Greek landraces of tomato grown on perlite. *Acta Hortic.* **2017**, *1168*, 277–286. [CrossRef]

50. Zheng, N.; Feng, K.; Jingyan, J.; Xuefei, W.; Li, C.W.; Jiahui, L.; Yu, Z.; Zi, Y.; Bin, Z. Growth improvement and salt tolerance mechanisms of tomato seedlings mediated by plant growth-promoting rhizobacteria from contaminated soils. *Chin. J. Appl. Environ. Biol.* **2018**, *24*, 47–52.
51. Lareen, A.; Burton, F.; Schäfer, P. Plant root-microbe communication in shaping root microbiomes. *Plant Mol. Biol.* **2016**, *90*, 575–587. [CrossRef] [PubMed]
52. Sasse, J.; Martinoia, E.; Northen, T. Feed your friends: Do plant exudates shape the root microbiome? *Trends Plant Sci.* **2018**, *23*, 25–41. [CrossRef] [PubMed]
53. Berg, G.; Opelt, K.; Zachow, C.; Lottmann, J.; Gotz, M.; Costa, R.; Smalla, K. The rhizosphere effect on bacteria antagonistic towards the pathogenic fungus *Verticillium* differs depending on plant species and site. *FEMS Microbiol. Ecol.* **2006**, *56*, 250–261. [CrossRef] [PubMed]
54. Miethling, R.; Wieland, G.; Backhaus, H.; Tebbe, C.C. Variation of microbial rhizosphere communities in response to crop species, soil origin, and inoculation with *Sinorhizobium meliloti* L33. *Microb. Ecol.* **2000**, *40*, 43–56. [CrossRef]
55. Faure, D.; Dessaux, Y. Quorum sensing as a target for developing control strategies for the plant pathogen *Pectobacterium*. In *New Perspectives and Approaches in Plant Growth-Promoting Rhizobacteria Research*; Bakker, P.A.H.M., Raaijmakers, J.M., Bloemberg, G., Höfte, M., Lemanceau, P., Cooke, B.M., Eds.; Springer: Dordrecht, The Netherlands, 2007; pp. 353–365. [CrossRef]
56. Chen, Y.; Cao, S.; Chai, Y.; Clardy, J.; Kolter, R.; Guo, J.H.; Losick, R. A *Bacillus subtilis* sensor kinase involved in triggering biofilm formation on the roots of tomato plants. *Mol. Microbiol.* **2012**, *85*, 418–430. [CrossRef]
57. Tan, S.; Yang, C.; Mei, X.; Shen, S.; Raza, W.; Shen, Q.; Xu, Y. The effect of organic acids from tomato root exudates on rhizosphere colonization of *Bacillus amyloliquefaciens* T-5. *Appl. Soil Ecol.* **2013**, *64*, 15–22. [CrossRef]
58. Ling, N.; Huang, Q.; Guo, S.; Shen, Q. *Paenibacillus polymyxa* SQR-21systemically affects root exudates of watermelon to decrease the conidial germination of *Fusarium oxysporum* f.sp. *niveum*. *Plant Soil* **2011**, *341*, 485–493. [CrossRef]
59. Zhang, F.; Zhu, Z.; Yang, X.; Ran, W.; Shen, Q. *Trichoderma harzianum* T-E5 significantly affects cucumber root exudates and fungal community in the cucumber rhizosphere. *Appl. Soil Ecol.* **2013**, *72*, 41–48. [CrossRef]
60. Zhou, X.G.; Wang, J.; Jin, X.; Li, D.L.; Shi, Y.J.; Wu, F.Z. Effects of selected cucumber root exudates components on soil *Trichoderma* spp. communities. *Allelopath. J.* **2019**, *47*, 257–265. [CrossRef]
61. Pozo, M.J.; Lopez-Raez, J.A.; Azcon-Aguilar, C.; Garcia-Garrido, J.M. Phytohormones as integrators of environmental signals in the regulation of mycorrhizal symbioses. *New Phytol.* **2015**, *205*, 1431–1436. [CrossRef]
62. Muller, L.M.; Harrison, M.J. Phytohormones, miRNAs, and peptide signals integrate plant phosphorus status with arbuscular mycorrhizal symbiosis. *Curr. Opin. Plant Biol.* **2019**, *50*, 132–139. [CrossRef] [PubMed]
63. López-Ráez, J.A.; Shirasu, K.; Foo, E. Strigolactones in plant interactions with beneficial and detrimental organisms. The Yin and Yang. *Trends Plant Sci.* **2017**, *22*, 527–537. [CrossRef]
64. Hause, B. Jasmonates in arbuscular mycorrhizal interactions. *Phytochemistry* **2007**, *68*, 101–110. [CrossRef]
65. Foo, E.; Ross, J.J.; Jones, W.T.; Reid, J.B. Plant hormones in arbuscular mycorrhizal symbioses: An emerging role for gibberellins. *Ann. Bot.* **2013**, *111*, 769–779. [CrossRef]
66. De Los Santos, R.T.; Vierheilig, H.; Ocampo, J.A.; Garrido, J.M.G. Altered pattern of arbuscular mycorrhizal formation in tomato ethylene mutants. *Plant Signal. Behav.* **2011**, *6*, 755–758. [CrossRef]
67. Etemadi, M.; Gutjahr, C.; Couzigou, J.M.; Zouine, M.; Lauressergues, D.; Timmers, A.; Audran, C.; Bouzayen, M.; Becard, G.; Combier, J.P. Auxin perception is required for arbuscule development in arbuscular mycorrhizal symbiosis. *Plant Physiol.* **2014**, *166*, 281–292. [CrossRef] [PubMed]
68. Guillotin, B.; Etemadi, M.; Audran, C.; Bouzayen, M.; Becard, G.; Combier, J.P. Sl-IAA27 regulates strigolactone biosynthesis and mycorrhization in tomato (var. MicroTom). *New Phytol.* **2017**, *213*, 1124–1132. [CrossRef] [PubMed]
69. Martínez-Medina, A.; Pescador, L.; Fernandez, I.; Rodrıguez-Serrano, M.; Garcıa, J.M.; Romero-Puertas, M.C.; Pozo, M.J. Nitric oxide and phytoglobin PHYTOGB1 are regulatory elements in the *Solanum lycopersicum–Rhizophagus irregularis* mycorrhizal symbiosis. *New Phytol.* **2019**, *223*, 1560–1574. [CrossRef] [PubMed]
70. King, S.R.; Davis, A.R.; Liu, W.; Levi, A. Grafting for disease resistance. *HortScience* **2008**, *43*, 1673–1676. [CrossRef]

71. Yetışır, H.; Sari, N.; Yücel, S. Rootstock resistance to *Fusarium* wilt and effect on watermelon fruit yield and quality. *Phytoparasitica* **2003**, *31*, 163. [CrossRef]
72. Grimault, V.; Prior, P. Grafting tomato cultivars resistant or susceptible to bacterial wilt: Analysis of resistance mechanisms. *J. Phytopathol.* **1994**, *141*, 330–334. [CrossRef]
73. Albert, R.; Künstler, A.; Lantos, F.; Ádám, A.L.; Király, L. Graft-transmissible resistance of cherry pepper (*Capsicum annuum* var. *cerasiforme*) to powdery mildew (*Leveillula taurica*) is associated with elevated superoxide accumulation, NADPH oxidase activity and pathogenesis-related gene expression. *Acta Physiol. Plant* **2017**, *39*, 53. [CrossRef]
74. Lee, J.M.; Kubota, C.; Tsao, S.J.; Bie, Z.; Echevarria, P.H.; Morra, L.; Oda, M. Current status of vegetable grafting: Diffusion, grafting techniques, automation. *Sci. Hortic.* **2010**, *127*, 93–105. [CrossRef]
75. Johnson, S.; Inglis, D.; Miles, C. Grafting effects on eggplant growth, yield, and *Verticillium* wilt incidence. *Int. J. Veg. Sci.* **2014**, *20*, 3–20. [CrossRef]
76. Liu, N.; Zhou, B.; Zhao, X.; Lu, B.; Li, Y.; Hao, J. Grafting eggplant onto tomato rootstock to suppress *Verticillium dahliae* infection: The effect of root exudates. *HortScience* **2009**, *44*, 2058–2062. [CrossRef]
77. Song, Y.; Zhu, C.; Raza, W.; Wang, D.; Huang, Q.; Guo, S.; Ling, N.; Shen, Q. Coupling of the chemical niche and microbiome in the rhizosphere: Implications from watermelon grafting. *Front. Agric. Sci. Eng.* **2016**, *3*, 249–262. [CrossRef]
78. Cichy, K.A.; Snapp, S.S.; Kirk, W.W. *Fusarium* root rot incidence and root system architecture in grafted common bean lines. *Plant Soil* **2007**, *300*, 233–244. [CrossRef]
79. Chitarra, W.; Perrone, I.; Avanzato, C.G.; Minio, A.; Boccacci, P.; Santini, D.; Gilardi, G.; Siciliano, I.; Gullino, M.L.; Delledonne, M.; et al. Grapevine grafting: Scion transcript profiling and defense-related metabolites induced by rootstocks. *Front. Plant Sci.* **2017**, *8*, 654. [CrossRef]
80. Wallis, C.M.; Wallingford, A.K.; Chen, J. Grapevine rootstock effects on scion sap phenolic levels, resistance to *Xylella fastidiosa* infection, and progression of Pierce's disease. *Front. Plant Sci.* **2013**, *4*, 502. [CrossRef]
81. Shibuya, T.; Itagaki, K.; Wang, Y.; Endo, R. Grafting transiently suppresses development of powdery mildew colonies, probably through a quantitative change in water relations of the host cucumber scions during graft healing. *Sci. Hortic.* **2015**, *192*, 197–199. [CrossRef]
82. Song, Y.; Ling, N.; Ma, J.; Wang, J.; Zhu, C.; Raza, W.; Shen, Y.; Huang, Q.; Shen, Q. Grafting resulted in a distinct proteomic profile of watermelon root exudates relative to the un-grafted watermelon and the rootstock plant. *J. Plant Growth Regul.* **2016**, *35*, 778–791. [CrossRef]
83. Ling, N.; Zhang, W.; Wang, D.; Mao, J.; Huang, Q.; Guo, S.; Shen, Q. Root exudates from grafted-root watermelon showed a certain contribution in inhibiting *Fusarium oxysporum* f. sp. *niveum*. *PLoS ONE* **2013**, *8*, e63383. [CrossRef] [PubMed]
84. Liu, L.; Chen, S.; Zhao, J.; Zhou, X.; Wang, B.; Li, Y.; Zheng, G.; Zhang, J.; Cai, Z.; Huang, X. Watermelon planting is capable to restructure the soil microbiome that regulated by reductive soil disinfestation. *Appl. Soil Ecol.* **2018**, *129*, 52–60. [CrossRef]
85. Yin, Y.; Luo, S.; Li, Y.; Zhou, J.; Tang, Y.; Liu, Y. Grafting shaping the microbial community structure to suppress *Verticillium dahliae* in the rhizosphere of eggplants. *Am. J. Agric. For.* **2018**, *6*, 132–140. [CrossRef]
86. Pérez-Jaramillo, J.E.; Mendes, R.; Raaijmakers, J.M. Impact of plant domestication on rhizosphere microbiome assembly and functions. *Plant Mol. Biol.* **2016**, *90*, 635–644. [CrossRef]
87. Wei, Z.; Jousset, A. Plant breeding goes Microbial. *Trends Plant Sci.* **2017**, *22*, 555–558. [CrossRef]
88. Marasco, R.; Rolli, E.; Fusi, M.; Michoud, G.; Daffonchio, D. Grapevine rootstocks shape underground bacterial microbiome and networking but not potential functionality. *Microbiome* **2018**, *6*, 3. [CrossRef]
89. Nicola, L.; Insam, H.; Pertot, I.; Stres, B. Reanalysis of microbiomes in soils affected by apple replant disease (ARD): Old foes and novel suspects lead to the proposal of extended model of disease development. *Appl. Soil Ecol.* **2018**, *129*, 24–33. [CrossRef]

4

Grafting Tomato as a Tool to Improve Salt Tolerance

Hira Singh [1], Pradeep Kumar [2,*], Ashwani Kumar [3], Marios C. Kyriacou [4], Giuseppe Colla [5] and Youssef Rouphael [6,*]

1 Department of Vegetable Science, Punjab Agricultural University, Ludhiana 141004, India; hira@pau.edu
2 ICAR-Central Arid Zone Research Institute, Jodhpur 342003, India
3 ICAR-Central Soil Salinity Research Institute, Karnal 132001, India; Ashwani.Kumar1@icar.gov.in
4 Department of Vegetable Crops, Agricultural Research Institute, 1516 Nicosia, Cyprus; m.kyriacou@ari.gov.cy
5 Department of Agriculture and Forest Sciences, University of Tuscia, 01100 Viterbo, Italy; giucolla@unitus.it
6 Department of Agricultural Sciences, University of Naples Federico II, 80055 Portici, Italy
* Correspondence: pradeephort@gmail.com (P.K.); youssef.rouphael@unina.it (Y.R.)

Abstract: Salinity in soil or water is a serious threat to global agriculture; the expected acreage affected by salinity is about 20% of the global irrigated lands. Improving salt tolerance of plants through breeding is a complex undertaking due to the number of traits involved. Grafting, a surgical mean of joining a scion and rootstock of two different genotypes with the desired traits, offers an alternative to breeding and biotechnological approaches to salt tolerance. Grafting can also be used to circumvent other biotic and abiotic stresses. Increasing salinity tolerance in tomato (*Solanum lycopresicum* L.), a highly nutritious and economical vegetable, will have greater impact on the vegetable industry, especially in (semi) arid regions where salinity in soil and water are more prevalent. Besides, plants also experience salt stress when water in hydroponic system is recycled for tomato production. Grafting high yielding but salt-susceptible tomato cultivars onto salt-resistant/tolerant rootstocks is a sustainable strategy to overcome saline stress. Selection of salt-tolerant rootstocks though screening of available commercial and wild relatives of tomato under salt stress conditions is a pre-requisite for grafting. The positive response of grafting exerted by tolerant rootstocks or scion-rootstock interactions on yield and fruit characteristics of tomato under saline conditions is attributed to several physiological and biochemical changes. In this review, the importance of tomato grafting, strategies to select appropriate rootstocks, scion-rootstock interaction for growth, yield and quality characteristics, as well as the tolerance mechanisms that (grafted) plants deploy to circumvent or minimize the effects of salt stress in root zones are discussed. The future challenges of grafting tomato are also highlighted.

Keywords: Tomato grafting; salinity tolerance; rootstock; physio-biochemical mechanisms; *Solanum lycopresicum* L.

---

1. Introduction

Salinity in soil or water is a serious threat to plant growth that prevents plants in achieving their genetic potential. Salinity annually damages about 20% of the world's crops grown under irrigation [1]. The high salt content of productive arable lands could render half of this land unusable for agriculture by 2050 [1]. In the world's arid and semiarid areas, less rain, high evaporation, saline irrigation water and inefficient management of water leads to salinity problems. Approximately 20% of the total cultivated lands and 33% of the irrigated agricultural lands worldwide are afflicted by high salinity and such areas are increasing at a rate of 10% per year [2]. It is estimated that about 3 ha of productive land per minute is lost due to salinity [3]. Thus, it is imperative to take appropriate measures timely to improve the salt tolerance of crops [4], especially tomato (*Solanum lycopersicum*

L.), which enjoys the prime position worldwide among different vegetables owing to having high nutritional and economical significance.

Tomato is a member of the nightshade family and cultivated for fresh and processing purposes as an annual crop [5]. Additionally, this crop is used as a model plant to study physiology to molecular genetics and genomics in the angiosperms [6,7]. Salinity represents a substantive threat to tomato production [8]; it causes considerable reductions in tomato growth and yield [2]. Most commercial cultivars of tomato are considered to be moderately sensitive to salt stress, which affects seed germination and the vegetative and reproductive stages of growth [5,9,10]. Ongoing efforts to improve salt tolerance in tomato using plant breeding, biotechnological approaches and other management practices have met with limited success due to the genetic and physiological complexity of the traits involved in salt tolerance [4,7,10,11]. According to Rao et al. [12], improvement of salt tolerance using plant breeding is an intricate task because of the high number of traits involved, their quantitative nature, epistatic gene action, low to moderate heritability and high sensitivity to the environment. Genetic engineering has been used to increase salt tolerance in plants and has claimed some success [11,12], but its commercial success is still to be witnessed.

An environmentally friendly, sustainable and effective method is grafting that enables to exploit the benefit of resistant genotypes (as rootstocks) to improve the performance of commercial cultivars (as scion) that are susceptible to (a)biotic stresses [13]. Grafting offers an alternative to breeding and biotechnological approaches to rapidly enhance salt tolerance in vegetable plants [8,11,14–16]. This technique in woody perennial fruit plants has been a routine practice in Asia for more than 2000 years [14]. In herbaceous vegetables, grafting initially was practiced in 1920s in watermelon to increase resistance to soil-borne diseases [17]. After the first scientific publication about grafting [14,17], use of the technique spread to other cucurbitaceous and solanaceous vegetable crops to address soil-borne diseases and other environmental stresses [18,19]. Commercial tomato grafting became popular in the 1960s in East Asia, Europe and later in North America [20,21]. Application of vegetable grafting has spread in many countries by the end of the 1990s, but it got momentum only after the banning of methyl bromide (MB) in the Montreal protocol in 2005, particularly in developed countries. Although, the protocol was authorized until 2015 in developing countries, with the efforts of United Nations Industrial Development Organization (UNIDO), United Nations Development Program (UNDP), World Bank and lateral agencies from Europe, many developing countries (some of Latin America, Africa, Middle East and Asia) were able to control the use of MB well before 2015, and started using grafting as an alternative to MB [18]. MB was the most commonly used soil fumigant against deadly soil borne pathogens prevailing in intensive vegetables production in protected cultivation [22]. Besides, being a sustainable practice, grafting is a useful component of organic vegetable production, particularly of tomato. In recent pasts, the horizon of grafting use expanded to abiotic stresses too and among which, salinity got the most priority. For instance, watermelon produced in countries like Japan, Korea, Turkey, Greece and parts of Spain and Italy with almost 100% grafted seedlings, and the use of grafted plants is also increasing immensely in other vegetables like tomato, eggplant, pepper, cucumber and melons across the world [23]. In major countries where grafting is a popular technique, out of the total tomato cultivation proportion share of grafted tomato seedlings use is around 1% in China, though it also ranks first in the number of grafted seedlings used, 25% in Korea, 33% in Vietnam, 40% in Japan, 50% in France and 75% in Netherlands. In other countries like the USA, Italy, Morocco and Spain the uses of grafted seedlings were available in numbers, i.e., 18, 15.1, 44, 72.8 and millions [24,25]. In this review, we present an outline of the potential of a grafting tool to enhance salt tolerance in a tomato based on recent researches done across the world. We also propose a strategy for future research as well as adoption for its better exploitation for the growth of the agriculture sector.

2. Effect of Salinity on Tomato Plants

Salinity is the most serious of all environmental stresses [26] and poses a great threat to agricultural sustainability [27]. It occurs when there is an excessive accumulation of salts (especially high Na^+, Cl^- and SO_4^-) in the soil [28] or irrigation water [11,16]. The elevated level of salts generally causes a reduction of water potential in the root medium, thereby leading to a water deficit within plants [5,29], besides their excess level can cause ion-toxicity and nutrient imbalance, especially of K^+, Ca^{2+} and Mg^{2+} by disturbing their uptake and/or transport to the shoots [30].

In general, the main factor of inhibiting growth of salt stressed plants is elevated levels of Na^+ and Cl^-; with roots remaining the primary sites for stress perception [31] and the subsequent responses at the cell, organ or whole plant levels [32]. According to the two-phase model of salt-induced growth reduction, plants suffer initially due to osmotic stress impairing their ability to absorb sufficient water, and then from salt specific injuries ascribed to, in general, toxic levels of Na^+ and Cl^- interfering with key cellular processes and causing damage to the cell membranes and organelles; altering nutrient ratios, endogenous growth regulator concentrations and enzymatic activities and suppressing photosynthetic assimilation and causing plant death in the extreme cases [33]. In fact, the second phase of salinity stress, i.e., ion-specific toxicity, is a long-term process and depends on the intracellular salt ion levels, which mostly tend to increase with an increase in the magnitude and duration of salinity stress.

Salinity has been reported to disturb plants physiological and biochemical processes and induce changes in morphological characteristics that finally lead to losses in yield [2,27,28,34]. Salt stress causes decrease in plant height, shoot and root biomass and root length in tomato plants [35]. According to Najla et al. [36] salinity adversely affected plant height and leaf area, and the overall development process of tomato plants. However, the salinity induced decrease in plant height was more related to reduced internodal length than the number of nodes, and lower plant growth rate was associated with a decrease in leaflet growth and the number of leaflets per leaf in salt stressed plants [36]. The adverse effects of salinity on shoot and root morphology were the result of an alteration of plant physiology that includes altered absorption of water and nutrients, hormonal production and disrupted root to shoot signals [37]. Salt induced inhibition in photosynthesis and oxidative stress has been widely documented. Salt stress can also alter leaf metabolites concentrations as have been reported by Khavari-Nejad and Mostofi [38]. They observed a notable decrease in the level of chlorophyll and β-carotene contents, along with an increase of soluble sugars and total saccharides in the leaves of tomato plants treated with 100 mM NaCl. In addition, aggregated chloroplasts and distorted and wrinkled cell membranes were also noticed in tomato leaves by these researchers. In spite of inhibiting plant height and shoots and roots dry matter contents, excess salt concentration in water reduced water use efficiency (WUE) and amount of K^+ and K/Na ratio in all studied tomato cultivars [39]. The accumulation of monovalent and bivalent Na^+, K^+ and Ca^{2+} ions in foliage and roots under saline conditions was, though, genotype-dependent [40]. Further, salinity affects almost all the plant growth stages, but the severity depends on the growth stage, salinity level and cultivar [35]. However, despite the inhibitory effects of salinity on growth and yield, the enhancement of some fruit quality characteristics (i.e., higher levels of sugars and organic acids) in tomato have been reported [41]. An increase of up to 40% carotenoid content in fruits of tomato plants exposed to moderate salinity has also been observed (4.4 dS m^{-1}) [42].

3. Grafting to Improve Salt Tolerance in Tomato

3.1. Growth and Yield

The persuasive response of grafting on plant growth and yield characteristics under saline conditions can vary; this can be the result of intrinsic characteristics of scion, rootstock and their functional interactions, and severity of saline stress. A wide range of studies suggests that the adverse effects of salt stress on vegetable plants can be mitigated by grafting. Improved plant growth and yield performance of susceptible tomato cultivars under salt stress is the manifestation of positive response of grafting; these ascribed to the right choice of scion-rootstocks combination [11]. Concerning the response of scion-rootstock combination on tomato growth and yield, grafting 'Cuore di Bue' scion onto 'Arnold' rootstock was found superior to either non-grafted or those grafted onto other rootstocks ('Maxifort' and 'Armstrong') under a moderate salinity level (20 mM NaCl), while the response of rootstock grafting was not evident at higher NaCl concentration (40 mM) [43]. The response of different rootstocks to salt susceptible scion 'Moneymaker' varied for yield and fruit parameters under saline conditions; graft combination involving 'Beaufort' and 'He-man' rootstocks was more productive under a mild saline condition. However, between two, only 'He-man' rootstock could provide sustenance to susceptible scion 'Moneymaker' for fruit yield under a high level of salinity. Contrarily, 'HPG' and 'Energy' rootstocks grafted tomatoes showed negative effect and produced lower yields [15].

Salt stress decreased both vegetative growth (i.e., stem diameter, plant height and shoot fresh weight) and fruit yield of both grafted and non-grafted plants. At this salinity level (3.76 dS m^{-1}), no significant decrease in fruit yield was noticed in grafted plants ('Faridah' scion onto 'Unifort' rootstock) [10]. Concerning scion-rootstock combinations, a significant variation in plant growth (e.g., stem growth rate) was observed under saline conditions by Balliu et al. [44], who revealed that 'Charlotte' grafted onto 'Cyndia' exhibited higher mean stem growth rate than non-grafted plants. However, no difference in this parameter was noted between grafted and non-grafted tomato plants under normal (non-saline) condition. These authors, further stated that 'Bona' plants grafted onto 'Energy' rootstock produced higher mean yield compared to non-grafted plants, while this combination displayed lowest stem growth rate. The other scion, 'Charlotte', gave the maximum yield with rootstock 'Prospero' under saline conditions. The performance of salt sensitive cultivar 'Moneymaker' improved in terms of both plant growth and fruit yield under salt stress (50 mM NaCl), when it was grafted onto the salt tolerant rootstock 'Pera' [45]. Santa-Cruz et al. [46] reported that the 'UC-82B'/'Kyndia' (scion/rootstock) combination showed a tolerance response to excess salt (100 mM NaCl) by producing higher shoot growth and fruit yield than the rest of the other graft combinations. Among yield-contributing traits, fruit weight was found to be the single factor to determine yield, while the number of fruits affected was not significantly affected by salt stress; as also was obvious in this study where graft combination 'UC-82B'/'Kyndia' exhibited higher fruit weight. Al-Harbi et al. [10] concluded that tomato could be grown successfully with a satisfactory yield by grafting onto suitable rootstock under salt stress (EC 3.76 dS m^{-1}). While comparing the response of tomato grafting onto different rootstocks under saline conditions in indoor or outdoor grown plants, Voutsela et al. [47] revealed that tomato fruit yield was significantly higher in grafted plants than non-grafted plants in both indoor and outdoor grown plants under salt stress (6.0 mS/cm). The benefit of grafting for increased fruit yield over non-grafted plants under salt stress was more pronounced in indoor condition (from 208% to 259%) than outdoor condition (from 0% to 149%). These workers have also reported significantly higher fruit yield in salt stressed self-grafted plants than non-grafted plants regardless of growing conditions. However, Iseri et al. [48] stated that enhanced salt tolerance and adaptive response of tomato scions was rootstock dependent rather than graft-induced changes per se. These reports indicate that selection of rootstocks and scion cultivars to be made reasonably for harnessing the benefit of grafting in tomato under saline conditions.

3.2. Fruit Quality

Despite the reduction of the fruit yield, enhancement of fruit quality traits is a general response of mild stress (i.e., water and salt); these are due to the accumulation of more metabolite contents under stress conditions. The interaction between grafting and salinity for fruit quality traits may be positive, negative or even neutral under stress conditions. The response of grafting on growth and fruit yield was positive when tomato cultivar 'Cuore di Bue' grafted onto 'Arnold' rootstock, whereas no obvious effect of grafting was observed in this graft combination on fruit quality traits (i.e., total soluble solids, fruit dry matter percentage, titratable acidity and TA) at any of the levels of NaCl (i.e., 0, 20 or 40 mM) added medium [43]. Similarly, grafting 'Durinta' onto 'He-man' rootstock showed promising response for yield traits, especially under high saline medium (8.8 dS m^{-1}), but the fruits of this graft combination had low titratable acidity content [15]. Turhan et al. [49] have also reported a reduction in the tomato fruit quality in grafted plants than non-grafted plants under salt stress. Contrary to these, findings of Balliu et al. [44] reported positive response of both grafting and salinity level; the fruit quality characteristics namely fruit dry matter percentage, vitamin C and total soluble solids contents increased in grafted plants with the increase of NaCl concentration from 0 to 5 mM. Flores et al. [50] noticed that grafting 'Moneymaker' onto 'UC82B' did not increase fruit yield either under optimal or saline condition, fruit yield rather decreased in grafted plants, but fruit quality parameters (i.e., soluble solid content and titratable TA) were higher in grafted plants, more conspicuously under salt stress condition (50 mM). However, grafting 'Moneymaker' onto 'Radja' rootstock was found as promising as both fruit yield and quality were increased in this graft combination under a saline condition (25 and 50mM NaCl) [50]. The inference drawn from these is that the combination of scion and rootstock should be selected carefully to get the maximum benefit of the grafting technique.

4. Mechanisms of Salt Tolerance in Grafted Plants

In order to overcome the harmful effects of salinity, grafted tomato plants employ certain adaptive strategies such as salt exclusion or retention, osmotic adjustment, activation of antioxidant defense system, nutrient homeostasis, plant hormonal balances and a gene expression led favorable response. Roots being the primary plant organ have to face any soil related stress (e.g., salinity); their intrinsic characteristics would determine overall plants performance.

4.1. Morpho-Physiological Traits

4.1.1. Root System Architecture and Salt Tolerance

Performance of grafted plants compared to non-grafted or self-grafted plants under a stressful condition is often dependent on the rootstock's root system characteristics; a vigorous root system could be the most important criterion for increasing salt tolerance [44]. The rootstock's root systems architecture specified by root length and density, root hairs and root surface area plays a critical role in ion and water uptake, thus determining salt tolerance of grafted plants [11,51]. A vigorous root system, for instance, produced more cytokinins and transported water to the shoot system by xylem sap, which positively affected plant growth and crop yield [44,52]. Furthermore, hydraulic conductivity of the roots may control plant growth by manipulation of the water supply to epigeous plant parts [53]; however, this process is still not clear. Roots being the primary organ exposed to salt stress, salt induced inhibition of root growth is quite obvious. For instance, salt stress (100 and 150 mM NaCl) invariably reduced root growth of both control and grafted tomato plants, but the rate of root dry mass reduction was lesser in grafted plants (on 'ZhezhenNo.1' rootstock) than non-grafted plants. However, no effect of grafting was observed on either epigeous or hypogeous dry mass under non-saline (0 mM NaCl) conditions [51]. Martínez-Rodríguez et al. [54] studied tomato rootstocks 'Radja' and 'Pera' for salt tolerance and they found that 'Radja' had a unique root genotype for salt tolerance with the ability to decrease salt ion transportation to the shoot system, as indicated by a lower ion concentration in the leaves. 'Pera' has the ability to improve salt tolerance in plants cultivated at any salt stress level. Tomato scion (cv. 'Faridah') grafted onto rootstock 'Unifort' was less affected by excess salt than the non-grafted scion; reflecting that 'Unifort' roots had a better capacity to limiting the transport of Na^+ and Cl^- to above-ground parts [10]. Similar results were also recorded in cucumber grafted onto a figleaf gourd differing in their ability to regulate water and salt absorption and subsequent translocation to the scion [55].

4.1.2. Plant Water Relations

Plant water relations can be measured through root hydraulic conductance by estimating the nutrient and water uptake in grafted plants [56]. Rootstocks confer to grafted plants a stronger and larger root system, which provide an increase in minerals and water uptake in comparison to non-grafted plants [57]. Rootstocks mainly set roots physical characteristics, such as its enhanced uptake capacity or vertical and lateral expansion, rendering water uptake more facilitated. Water and nutrients uptake have been found more efficiently in a vigorous rootstock than scion roots [58]. Root hydraulic conductance was higher in grafted plants due to higher resistance of the graft union to water-flow [59]. Shalhevet et al. [60], Munns and Passioura [61] and Evlagon et al. [62] reported that hyperosmotic stress and ionic imbalance caused by the high apoplastic concentrations of Na+ and Cl^-. The decrease in root hydraulic conductance could be closely related to the decrease in the activity or concentration of aquaporins in the root plasma membrane due to increased salinity [63]. An experiment conducted by Estañ et al. [64] on tomato illustrated that, in order to exclude salts, various tomato genotypes' roots were used as rootstocks for a commercial tomato hybrid (cv. 'Jaguar'). Estañ and his collaborators noted that especially at 50 and 75 mM NaCl, the 'Jaguar'/'Jaguar' combination plants demonstrated equal or paradoxically higher leaf water contents than the other grafting combinations, indicating that the 'Jaguar' rootstock was capable in salinity conditions to maintain water uptake. The rootstock–scion vascular connection played an important role in determining water and nutrient translocation, which mainly depends on the rootstock's vigorous root system [58,65], enhanced production of endogenous-hormones [66] and enhancement of scion vigor [67].

4.1.3. Ion Uptake

Studies have shown that rootstocks protect the scion shoots from salt damage mainly by reducing the ionic stress, and to some extent, by increasing the translocation of K^+, Ca^{2+} and Mg^{2+} to the shoots and leaves; whereas they have little role in reducing the osmotic stress [68,69]. Reduced translocation of Na^+ and/or Cl^- to the shoot system is achieved either by exclusion or restricted absorption by the roots [54,64]. Pérez-Alfocea et al. [70] opined that depending on the salt source plants employ its intrinsic potential to exclude Na^+ and/or Cl^- from shoots by maintaining energy consuming the root toxic ion efflux. The positive effects of grafting on salt tolerance in a tomato was widely attributed to restricted entry of Na^+ and Cl^- ion in epigeous biomass [10,64,71], though grafted plants were able to accumulate more nutrient elements (e.g., Ca^{2+} and K^+) in the leaves than normal plants under saline stress [10]. Similarly, tomato cultivars 'Fanny' and 'Goldmar' grafted onto rootstock 'AR-9704' showed differential accumulation of Na^+ and Cl^-, with Cl^- and Na^+ concentrations being significantly higher in non-grafted than in grafted plants in both the cultivars and 'Fanny', respectively [72]. Semiz and Suarez [73] reported improved performance of tomato scion cv. 'Big Dena' grafted onto 'Maxifort' rootstock across five tested salinity levels; the positive response of grafting was attributed to a lesser concentration of Na^+ in the leaves of a grafted tomato than non-grafted plants. However, these researchers could not find any relation between Cl^- in irrigation water and yield in a tomato. Apart from osmotic balance and root exclusion or restricted root-to-shoot translocation of the toxic ion (i.e., Na^+), the third strategy that plants employ is ion accumulation and subsequent partitioning among the plant organs or compartmentalization in cellular organs such as vacuoles that eases toxic effects of salts. For instance, tomato scions grafted onto rootstock 'Arnold' tended to partition a bulk of shoot Na^+ into older leaves and thus reduced Na^+ levels in the actively growing younger leaves. This allowed the grafted plants to maintain favorable K^+/Na^+, Ca^{2+}/Na^+ and Mg^{2+}/Na^+ ratios in actively growing leaves [43].

The response of salt stress varies widely among *Solanum* species [37], besides salt concentration and exposure period also exert a severity response [64,74] In contrast to cultivated tomatoes (*Solanum lycopersicum* L.) that are moderately sensitive to salt stress, certain wild species are reportedly salt resistant. Unlike the cultivated tomato that generally excludes the salt ions [2,54,75,76], most of the wild types (e.g., *S. peruvianum* [77], *S. cheesmaniae* [78], *Lycopersicon pimpinellifolium*, *L. hirsutum* and *L. pennellii* [79]) accumulate higher concentrations of Na^+ and Cl^- in the leaves. However, there are some salt tolerant tomato ecotypes (e.g., 'Edkawy' and 'Pera') [75,80,81] that exhibit a pronounced salt inclusion capacity similar to the salt-loving wild species. This implies that tolerance or sensitivity to salt would depend on the ability of a genotype (excluder or includer) to maintain benign ion levels through mechanisms like root exclusion, phloem recirculation and dilution such that physiological homeostasis is not disturbed [82]. If ion levels exceed the critical threshold, salt tolerance capacity diminishes, adversely affecting growth and fruiting. Use of salt tolerant types as rootstocks can protect the sensitive types from deleterious effects of salinity [54,64].

He et al. [51] found that with the enhancement of NaCl concentration, the Na^+ levels in leaves and roots increased considerably. They estimated that non-grafted, self-grafted and plants of scion-rootstock combinations elicited the same levels of Na^+ in leaves and roots under similar salt concentrations, whereas K^+ concentration in both leaves and roots decreased considerably with higher NaCl levels. Nevertheless, the roots of scion-rootstock grafted plants showed more K^+/Na^+ ratio and water use efficiency (WUE) at 150 mM NaCl concentration than non-grafted and self-grafted plants.

In another study, Santa-Cruz et al. [46] selected tomato cultivars 'Moneymaker' (indeterminate growth habit, shows excluder character under salt stress) and 'UC-82B' (determinate growth habit and shows includer behavior) as scions, and tomato hybrid 'Kyndia' as a rootstock for the study of grafting behavior under 50 and 100 mM NaCl salt stress. Grafting had a positive effect on salt tolerance of scion 'UC-82B' grafted onto 'Kyndia' rootstock in terms of the shoot biomass at a 100 mM NaCl concentration. The same combination was noted for higher water content in leaves under salt stress compared to self-grafted 'UC-82B' plants. Accumulation of Na^+ and Cl^- ions in leaves varied with the graft combination. Lower Na^+ and similar Cl^- concentration was found in the 'Moneymaker' and 'Kyndia' combination as compared to self-grafted 'Moneymaker' plants. However, this response was different for the 'UC-82B' and 'Kyndia' combination, which elicited lower accumulation of both Na^+ and Cl^- ions. Na^+/K^+ ratio in leaves of this combination decreased three times compared to self-grafted plants of 'UC-82B'. The toxic effect of ions was mitigated by the 'UC-82B' and 'Kyndia' combination.

The improvement in salt tolerance of tomato plants was related to a high leaf K^+ content in many studies [51,83,84], though the direct relationship between leaf K^+ homeostasis and salinity tolerance of grafted plants has not been yet adequately worked out [11]. Studying the response of tomato cv. 'Faridah' onto 'Unifort' rootstock under saline conditions, fruit yield, Ca^{2+} and K^+ ions were noticeably higher in grafted plants compared to non-grafted plants. Besides, accumulation of Na^+ and Cl^- was lower than in non-grafted tomato plants, under salt stress condition [10].

4.2. Gas Exchange Attributes

One of the early responses of glycophytes to salt stress is manifested as a reduction in leaf growth that seems to be caused by a decrease in stomatal conductance in tomato [85]. Stomatal factors decrease the photosynthetic rate in salt treated grafted and non-grafted plants by adversely affecting CO_2 diffusion into leaves as a result of impaired stomatal and mesophyll conductance [86]. Salt stress blocks photosynthesis by suppressing the electron transport chain, which leads to the photoinhibition of photosynthesis. The inhibition of photosynthesis causes a reduction in the plant growth. In previous reports, it is established that grafting of salt tolerant rootstocks can enhance the photosynthetic rate by protecting the structure of chloroplast and alleviating the oxidative damage, which ultimately leads to the delay in the rate of photoinhibition [87,88].

The rate of net photosynthesis under salt stress was higher in tomato plants (cv. 'Durinta') grafted onto 'Energy', 'He-man' and 'Resistar' rootstocks, though no significant effect of grafting on epigeous dry matter was noticed [15]. High photosynthetic rates as well as WUE were reported by He et al. [51] in tomato plants (cv. 'Hezuo903') grafted onto 'Zhezhen No.1' rootstock under severe salt stress. However, comparing the response of four tomato cultivars ('Jenin1,' 'Hebron,' 'Ramallah' and 'Maramand') under four salinity levels (0, 50, 100 and 150 mM NaCl), Sholi [5] reported that the higher salt tolerance (up to 100 mM NaCl) in cultivar 'Ramallah' was attributed to the highest photosynthesis rate along with K^+/Na^+ ratio than the others.

Two commercial hybrid tomato cultivars, 'Belle' and 'Gardel' scions grafted on two interspecific hybrids, 'Beaufort' and 'Maxifort' rootstock showed no significant effect on net photosynthesis and stomatal conductance when exposed to NaCl compared to the non-grafted plants [89]. Feng et al. [90] also found that salinity treated tomato scion grafted on wolfberry (*Lycium chinense*) rootstock displayed higher net photosynthesis, transpiration, stomatal conductance, Fv/Fm and electron transport rate in comparison with non-grafted tomato plants.

4.3. Molecular Responses

Salt stress sets in motion a chain of events—reduced stomatal conductance, decreased photosynthetic electron transport and increased production of harmful reactive oxygen species (ROS)—that lead eventually to plant damage and death. ROS (e.g., hydrogen peroxide and superoxide radicals), by causing oxidative damage to the proteins, nucleic acid and lipids, disturbs the plant metabolism. Both enzymatic (i.e., ascorbate peroxidase (APX), catalase (CAT), superoxide dismutase (SOD), monodehydro ascorbate reductase, dehydro ascorbate reductase and glutathione reductase (GR)) and non-enzymatic (i.e., reduced glutathione, reduced ascorbate, carotenoids and tocopherols) antioxidant systems are activated by the plants to minimize ROS-induced oxidative damage. In grafted plants, salt-induced production of ROS generally remains quite low than in non-grafted plants as evidenced by a significant increase in the antioxidant activity in salt-treated (grafted) tomatoes. While SOD and CAT activities tended to increase with an increase in salinity in grafted plants, the reverse was true for non-grafted and self-grafted plants [51]. Besides enzymatic antioxidants, a non-enzymatic antioxidant system (ascorbate peroxidase, monodehydro ascorbate reductase, reduced glutathione, etc.) is also known to remove the excess ROS [91]. Higher expression of CAT mRNA, Cu/Zn-SOD, Mn-SOD and higher activities of CAT, Cu/Zn-SOD, Mn-SOD and SOD have been observed in salt stressed grafted plants [92]. Grafted eggplant produced lower O_2-, H_2O_2 and MDA compared to non-grafted seedlings when subjected to $Ca(NO_3)_2$ stress. Higher activities of SOD, APX and GR were also seen in grafted than in non-grafted seedlings [93]. Ruiz et al. [94] explored a salt tolerance ability in another solanaceous species: tobacco (*Nicotiana* spp.) and a tomato–tobacco (scion–rootstock) combination (named as 'Tomacco' by Yasinok et al. [95]). Tobacco roots, in fact, were able to restrain the concentration of Na^+ and Cl^-, and lipid peroxidation in their leaves, besides inducing an accumulation of proline and sucrose concentrations under saline conditions [94]. The ability of tobacco roots in conferring better adaptive responses to salt stress in comparison to tomato was supported by the upregulation of proline and antioxidant enzymes (APOX and CAT) levels in grafted plants [48].

The commercial cultivars of tomato show more or less sensitivity to salt stress [96] and undergo drastic losses due to salts [97]. The genetics of this trait specific in tomato have also been discussed nicely by Cuartero et al. [34]. In tomato, a universal stress protein gene (SpUSP) cloned from *S. pennellii*, which is a wild relative of tomato and functionally characterized in commercial tomato genotype revealed enhanced expression under stress conditions such as dehydration, salt, oxidative and abscisic acid (ABA) treatment. About 15 primers were synthesized using fifteen salinity responsive candidate genes. The results from these primers displayed that the genotype UC82B showed the maximum vegetative growth parameters and yield, so, exploitation of this identified genotype could be as used for the salt tolerance [97]. Most recently in 2020, Yveline et al. [98] noticed that the salinity tolerant accessions collected from the Islands of Galapagos use different mechanisms to survive under salt stress conditions, which exhibited natural diversity among the Galapagos tomatoes for high salinity tolerance. They found some accessions namely LA0317, LA1449 and LA1403, which showed significantly higher tolerance under salt stress at the seedling stage. Their tolerance mechanism at the genetic and molecular level need to be studied intensively and comprehensively so that it could be helpful to develop salt tolerant commercial tomato cultivars to sustain tomato production under changing climate scenario across the globe. Such identified genotypes can be used as rootstocks in tomato grafting to raise the salinity tolerance of commercial scion cultivars after investigating their graft compatibility and subsequent effects on fruit characteristics.

Asins et al. [99] reported the genetic dissection of tomato rootstock effects on scion traits under moderate salinity. They documented that the rootstock HKT1 genotype affects the Na^+ level in fruits, which was found higher when the rootstock genotype was homozygote for SpHKT1. Along with this, they found the 37 QTLs, which controls the important rootstock-mediated scion traits such as the leaf concentration of nutrients, fruit yield, soluble-solids content of fruits and harvest time under moderate salinity.

In the salt-tolerance mechanism, it is now well established that calcium has an important role. The exogenous application of calcium can protect the plant from the negative effect of salinity [100]. The high amount of Na^+ can induce a calcium signaling cascade, which leads to the activation of the salt-overly-sensitive (SOS) pathway and stimulation of Na^+/H^+ exchange [101]. An ER chaperone protein, Calreticulin, is considered as a regulating protein for the Ca^{2+} homeostasis by binding to Ca^{2+}. Salt stress causes the calreticulin to bind with the Ca^{2+} ions, which later accumulated in the cell. The accumulated calcium ions act as a secondary messenger for ABA resulting into the ameliorations of salt stress [102]. Shaterian et al. [103] reported the effect of grafting and ABA on calreticulin gene expression in a potato. They observed an increase in the expression of calreticulin in salt stressed grafted plants and showed a positive association in the presence of abscisic acid.

4.4. Growth Regulators

During the past few decades, the role of phytohormones in modulating the union between shoot (scion) and root (rootstock) components in grafted plants has become a focal point of research: in addition to regulating key metabolic processes at the graft union, endogenous plant bio-regulators (e.g., auxins, gibberellins, cytokinins and ethylene) are also involved in signal transduction across the graft union [104]. Functions of some important phytohormones in grafted plants are briefly summarized in the following sections:

Auxins: While other phytohormones play minor roles, auxins, both alone and in interaction with other phytohormones, remain central to vascular tissue regeneration and connection, and thus to the growth and development of vascular tissues [105]. Two main protein families (PIN-FORMED auxin transport proteins (PINs) and proteins of ATP-binding cassette subfamily B (ABCB)) regulate a process called 'polar auxin transport' (PAT) that ensures the maintenance of adequate auxin levels for the proper development of xylem tissues [106]. For example, increased expression of CcPIN1b and CcLAX3 genes encoding the efflux carriers for PAT was found to induce graft union development in *Carya cathayensis* [107]. Similarly, Sauer et al. [108] found higher PIN1 expression at damaged stem portions to activate the process of xylem cell differentiation in pea (*Pisum sativum*). Transcriptome analysis of grafted *Torreya grandis* plants indicated that auxins modulate the key MAPK signaling pathway during graft development [109]. Auxins are also believed to enhance graft union by regulating metabolic pathways linked to phenylpropanoids, cytochrome P450 and carbohydrates. Available evidence suggests that besides promoting the formation of lateral roots, auxins also regulate xylem development and cambium growth, processes critical to the success of grafting in plants [104].

Cytokinins: The significant role of cytokinins (CKs) in callus proliferation at the graft union can be ascribed to their increased biosynthesis during the process of wound healing and to the elevated zeatin riboside levels at the graft unions [105]. CKs are known to stimulate the regeneration of vessels and sieve tubes. Moreover, CKs interact with the auxins to promote the vascular differentiation and increase the phloem/xylem ratio [110]. Increased levels of CKs in the rootstock xylem tend to induce auxin translocation from the aerial parts that in turn promotes the development of graft union [111]. Exogenously applied CKs accelerate the graft union and growth by enhancing the callus formation, phloem regeneration and soil nutrient supply to the scion [112,113]. While CKs alone are often less effective in modulating the development of vascular tissues, they act synergistically with other phytohormones (e.g., auxins) to enhance the cell division, xylem fiber development, cambium activity and regeneration of phloem/xylem tissues at the site of injury [114].

Since salinity stress decreased plant CK status, increasing root-to-shoot CK transport by using a rootstock overexpressing the CK biosynthesis genes such as isopentenyltransferase improved tomato salt tolerance by increasing vegetative and fruit growth and also delaying leaf senescence and maintaining stomatal conductance and PSII efficiency, thereby avoiding or delaying the accumulation of toxic ions [115].

Gibberellins: Gibberellins (GAs) are known to regulate the processes like cambium activity, xylem fiber differentiation and expansion, and secondary growth. Frequent movement of GAs across the graft union is believed to contribute to the union of shoot and root components, and to the normal development of vascular tissues; as evidenced by their involvement in the reunion of cortex in cucumber and tomato [116,117]. Since they hasten xylogenesis, GAs might also trigger vascular bundles formation at the graft union [118]. Experiments with mutants lacking GA-biosynthetic enzymes and wild type pea plants pointed to the role of GAs in regulating the normal growth of vascular tissues, induce the processes of wood formation, xylem expansion and cambial activity [119]. Development of a particular vascular tissue also depends on the ratio of auxins and GAs; while a higher IAA:GA ratio hastens xylem formation, low IAA:GA ratio promotes phloem formation [120]. These two hormones also act synergistically to regulate the cell division and secondary growth of the vascular tissues. For instance, GAs stimulate PAT by up-regulating the key auxin transporter PIN1 in the cambial region [119]. Up-regulation of the expression of GA20OX, a gene involved in GA-biosynthesis, points to increased biosynthesis of GAs at the site of graft union. Up-regulation of Cla015407, a gene encoding enzyme gibberellin 3-beta-hydroxylase involved in the conversion of GA20 to GA1, in grafted watermelons also supports the positive role of GAs in the growth and development of composite plants [121].

Abscisic acid: There is little evidence for any direct role of ABA in the growth and development of grafted plants; however, ABA seems to be indirectly involved in the processes of wound formation and vascular differentiation: abiotic stresses in general including wounding tend to hasten ABA biosynthesis [122]. In contrast to ABA-deficient mutants, wild type plants display increased expression of wound-activated genes in response to ABA application [122,123]. Salinity enhances ABA concentration in shoot limiting water loss through transpiration (by inducing stomatal closure) and maintaining leaf water relations in plants grown under saline conditions. Chen et al. [124] reported that the scion genotype, and its ABA level, played the major role in the growth of grafted plants, regardless of the rootstock genotype and the salinity of the growth medium. Although ABA has been suggested to restrict the synthesis of the potential growth inhibitor ethylene, thus maintaining growth, in other circumstances such as citrus leaf abscission under salinity, ABA apparently stimulates 1-aminocyclopropane-1-carboxylic acid (ACC) synthesis [85]; these authors also reported that ratios between CKs and the ethylene precursor ACC were most closely correlated with both leaf biomass and PSII efficiency (Fv/Fm) of tomato grown under saline conditions whereas the ratio ACC/ABA was negatively correlated with leaf biomass.

Ethylene: Although ethylene signaling is implicated in tissue reunion in the wounded stems, it is hardly of any significance in the grafted hypocotyls, suggesting that differential roles of ethylene might be due to the differences in tissue age and type. Transcriptomic analyses of grafted *Arabidopsis* hypocotyls showed that ethylene biosynthesis genes are activated at the point of graft union [125]. However, mutating ANAC071 reduced the formation of vascular tissues at the graft junction [126]. Under salinity, ethylene production is quickly stimulated in plant tissues [127]. Ungrafted tomato genotype cv. Boludo F1 had 40% higher ACC concentration in the xylem sap in comparison with plants grafted onto some low and high vigor rootstocks (derived from the recombined inbred line (RIL) population of *Solanum lycopersicum* × *Solanum cheesmaniae*) after growing for 50 days under moderate salinity (75 mM NaCl). Accumulation of the ethylene precursor ACC indicated the onset of salt-induced oxidative damage in the leaves, preceding leaf senescence induction and leaf growth impairment under salinity [85].

A summary of the main agronomic effects and physiological and molecular mechanisms of tomato grafted plants in comparison to non-grafted or self-grafted plants grown under non-saline and saline conditions is reported in Table 1 and Figure 1.

Table 1. Selected examples of grafting combinations to improve salt stress tolerance in tomato.

Tomato (*S. lycopersicum* L.) Scion Cultivar and Commercial source	Rootstock Species/Cultivar and Commercial Source	Salt Treatments	Growing Conditions	Effects of Grafting	Key Mechanisms of Grafted Plant Tolerance	Reference
'Faridah' (Golden Valley Seed)	'Unifort' (*S. lycopersicum* × *S. habrochaetes*) (DeRuiter Seed)	Fresh water (0.52 dSm^{-1}) and brackish water (3.76 $dS\ m^{-1}$)	Soil medium under greenhouse	High growth and yield	Na^{+} and Cl^{-} exclusion; high Ca^{+2} and K^{+}	Al-Harbi et al. [10]
'Boludo' (Seminis Vegetable Seeds)	RILs derived from cross *S. lycopersicum* × *S. cheesmaniae*	75 mM NaCl (8.6 dSm^{-1}) designated as moderate saline	Nutrient solution culture using rockwool substrate, in greenhouse (long-term, 120 d)	High yield, leaf fresh weight, *Fv/Fm*,	High K^{+} concentration and K+/Na+ ratio, and cytokinins	Albacete et al. [84]
'Boludo' (Seminis Vegetable Seeds]	RILs derived from *Solanum lycopersicum* × *S. pimpinellifolium*	75 mM NaCl (8.94 $dS\ m^{-1}$)	Open soilless system in coco fiber substrate, in greenhouse (long-term, 234 d)	High yield, fruit quality and delay fruit maturity		Asins et al. [99]
'Charlotte' (NA), 'Bona' (Clause Seeds)	'Cyndia' (*Lycopersicon* spp.) (NA)	0.0, 2.5 and 5.0 mM NaCl	Natural soil and Klasmann grow bags as substrate under greenhouse	High growth and fruit size but not yield. Soluble solids and ascorbic acids also increased.	Vigorous root system, high growth.	Balliu et al. [44]
'Cuore di Bue' (Vita Sementi Italian Seeds)	'Arnold' (*S. lycopersicum* × *S. habrochaites*) (Syngenta Seeds)	NaCl concentrations 0, 20 and 40 mM	Nutrient solution culture in pots (perlite–peat substrate; 3:1, v/v), in greenhouse (long-term; 130 d)	High marketable yield (23–33%)	Shoot Na^{+} partitioning to older leaves (maintenance of higher K^{+}/Na^{+}, Ca^{2+}/Na^{+}, and Mg^{2+}/Na^{+} ratios)	Di Gioia et al. [43]
'Jaguar' (Ramiro Arnedo S.A.)	'Radja', 'Pera', 'Volgogradskij' (*S. lycopersicum* L.) (NA)	Control, 25, 50 and 75 mM NaCl corresponding to EC of 2.3, 4.8, 4.8, 7.3 and 9.1 $dS\ m^{-1}$, respectively	Nutrient solution culture in pots (siliceous sand), in greenhouse	High yield	Na^{+} and Cl^{-} exclusion	Estañ et al. [64]

Table 1. *Cont.*

Tomato (*S. lycopersicum* L.) Scion Cultivar and Commercial source	Rootstock Species/Cultivar and Commercial Source	Salt Treatments	Growing Conditions	Effects of Grafting	Key Mechanisms of Grafted Plant Tolerance	Reference
'Boludo' (Seminis Vegetable Seeds)	RIL's of cross *S. lycopersicum* × *S. pimpinellifolium*	75 and 125 mM NaCl, corresponding EC value of 8.6 and 13.7 dS m^{-1}, respectively	Nutrient solution culture in fiber coconut coast, in greenhouse (long-term, 120 d)	High fruit yield	Useful QTL's favoring salt tolerance to rootstocks	Estañ et al. [6]
'Fanny', 'Goldmar' (NA)	'AR-9704' (*S. lycopersicum*) (NA)	0 and 75 mM NaCl	Hydroponic system, in controlled condition (short duration, 10 d salt treatment)	Similar hydraulic conductance	Na^+ and Cl^- exclusion	Fernández-García et al. [59]
'Fanny' (NA)	'AR-9704' (*S. lycopersicum*) (NA)	0, 30 and 60 mM NaCl corresponding EC of 2.3, 5.3 and 8.3 dS m^{-1}, respectively	Nutrient solution culture- in perlite sacs, in greenhouse	High growth and stomatal conductance	Na^+ and Cl^- exclusion by shoot	Fernández-García et al. [72]
'Kyndia', 'Moneymaker' (HL), 'Boludo' (Seminis Vegetable Seeds)	'UC832B', Pera, Radja (*S. lycopersicum*) and RILs of cross *S. lycopersicum* × *S. cheesmaniae* (NA)	0, 20, 50 and 75 mM NaCl corresponding EC 2.2, 4.5, 7.0 and 9.5 dSm^{-1}, respectively	Nutrient solution culture in pots (coconut fiber substrate), in greenhouse	Fruit yield and especially the fruit quality (soluble solids and titratable acidity)	-	Flores et al. [50]
'Hezuo903' (Jiangsu Seed)	'Zhezhen No. 1' (*S. lycopersicum*) (NA)	0, 50, 100 and 150 mM NaCl concentration	Nutrient solution culture in pots, in controlled condition (short duration; 2-weeks)	High plant growth	High photosynthetic and antioxidant activities; but similar Na^+ and Cl^-	He et al. [51]

Table 1. *Cont.*

Tomato (*S. lycopersicum* L.) Scion Cultivar and Commercial source	Rootstock Species/Cultivar and Commercial Source	Salt Treatments	Growing Conditions	Effects of Grafting	Key Mechanisms of Grafted Plant Tolerance	Reference
'Elazig' (LC)	'Hasankeyf' (*Nicotiana tabacum*); 'Samsun' *N. rustica*)	0.2, 0.4 and 0.6 M NaCl	In pots, in control condition (short duration; 10 d salt treatment)	High shoot growth (stem elongation and leafing)	High proline content and antioxidant activity (ascorbate peroxidase (APX) and catalase (CAT))	Iseri et al. [48]
'Moneymaker' (HL)	'Radja', 'Pera' (*S. lycopersicum*) (NA)	0, 25 and 50 mM (I expt.) and 75, 150 and 225 mM (II expt.) NaCl	Nutrient solution culture in pots under greenhouse	High fruit yield	Na+ and Cl- exclusion	Martínez-Rodríguez et al. [54]
'Durinta' (Western Seeds)	'He-man' (Syngenta seeds)	2.8 (control) and 8.8 dS m^{-1} by adding NaCl	Open soilless system, in pots (sand substrate), in greenhouse	High fruit yield	High net photosynthesis	Martorana et al. [15]
'Gokce (191)' (Zeraim Gedera)	'Vigomax' (De Ruiter seed) ' Yedi' (*L. esculentum* × *L. hirsutum*) (Rijk Zwaan)	50, 100, 150, 200, 250 and 300 mM	Nutrient solution culture on Styrofoam, in greenhouse (15 d salt treatment)	Least leaf damage and high leaf number, stem thickness	High antioxidant activity (CAT, POX)	Oztekin and Tuzel [52]
'Moneymaker' (HL)	'Pera' (*S. lycopersicum*) (NA)	50 mM NaCl (EC, 6.5 dSm^{-1}), without control	In soil with fertigation system under greenhouse (long term)	High fruit yield and fruit number	High Na^+ and Cl^-, but similar osmotic potential and high K^+	Santa-Cruz et al. [45]
'UC-82B' (Tecni seeds/Starke Ayres seeds)	'Kyndia' (*S. lycopersicum*) (NA)	0, 50 and 100 mM NaCl	Nutrient solution culture without (short duration, 35 d) or with substrate (siliceous sand, long term, 90 d) in greenhouse	Fruit yield and fruit weight, shoot biomass	Na+ and Cl- exclusion; low Na^+/K^+ ratio	Santa-Cruz et al. [46]

Table 1. *Cont.*

Tomato (*S. lycopersicum* L.) Scion Cultivar and Commercial source	Rootstock Species/Cultivar and Commercial Source	Salt Treatments	Growing Conditions	Effects of Grafting	Key Mechanisms of Grafted Plant Tolerance	Reference
'Belladona' (Hazera seeds)	'He-man' (*S. lycopersicum* × *S. habrochaites*), (Syngenta seeds) 'Resistar' (Hazera Seeds)	0.3 (control), 22 and 45 mM NaCl, corresponding to EC of 2.5, 5.0 and 7.5 dS m^{-1}, respectively	Recirculating nutrient solution in channel, in greenhouse (long term, 6 moths)	High yield, fruit number and least leaf chlorosis	Low Na^+	Savvas et al. [128]
'Big Dena' (Syngenta seeds)	'Maxifort' (*S. lycopersicum* L. × *S. habrochaites*) (De Ruiter Seed)	Salinity levels: 0.03, –0.15, –0.30, –0.45 and –0.60 MPa osmotic pressure that were correspond to EC 1.2, 4.0, 8.5, 12 and 15.8 dS m^{-1}, respectively	Sand tank fed by nutrient solution of different EC	High absolute yield	Na+ exclusion	Semiz and Suarez [73]
'Farida' (Golden Valley Seed)	'Unifort' (*S. lycopersicum* L. × *S. habrochaites*) (De Ruiter Seed)	Non-saline (EC, 1.2 dSm^{-1}) and saline water (EC, 4.5 dSm^{-1}); these were applied at 100%, 75%, 50% etc.	In soil plots (area, 8 m^2) under greenhouse	High vegetative growth, yield and WUE	Na+ and Cl- exclusion	Wahb-Allah [129]

HL: Heirloom; LC: Local; NA: Not available in literature

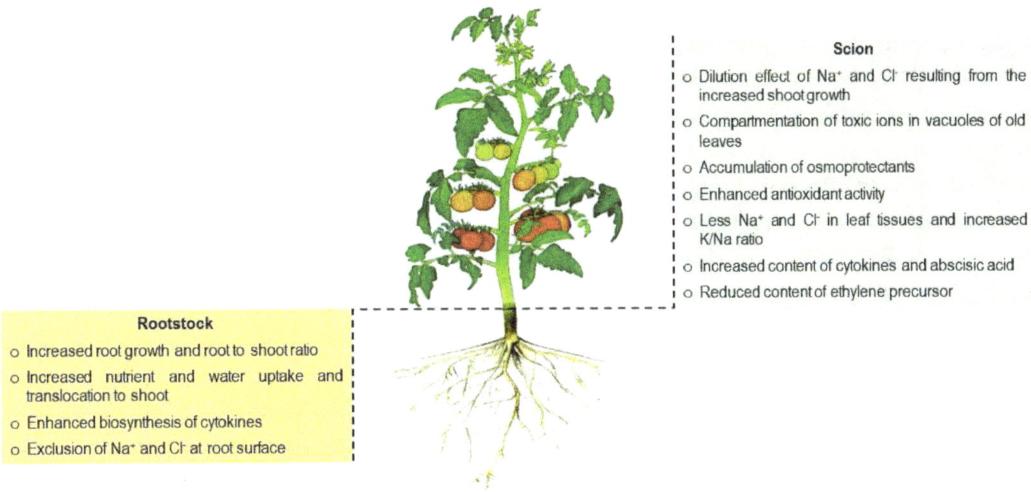

Figure 1. Physiological adaptive traits to mitigate the effect of salt stress in grafted plants.

5. Potential Use of Wild Tomato Species as Rootstocks for Improving Salt Tolerance in Tomato

Selection of salt-tolerant rootstocks is an important strategy to enhance salt tolerance. Testing and screening of available commercial and wild relatives under salt stress conditions is a prerequisite for grafting. Natural genetic variation with specific root characters can be utilized effectively through grafting [14]. Various wild tomato/*Solanum* species including *S. cheesmaniae*, *S. chmielewskii*, *S. habrochaites*, *S. pennellii*, *S. pimpinellifolium* and *S. peruvianum* demonstrate salt tolerance [12,85]. Due to incompatibility in crossing and linkage drag, it is cumbersome to transfer salt tolerant genes from wild to cultivated tomato cultivars [12]. However, these wild species could be exploited as salt-resistant rootstocks for grafting susceptible but high yielding commercial tomato cultivars. Compared to commercially grown tomato (*S. lycopersicum*) that usually excludes toxic ions [47], several wild relatives possess an ionic inclusion mechanism. These genotypes have the capability to accumulate higher Na^+ and Cl^- concentrations in the leaves [85]. Breeding new rootstocks with tolerance or resistance to a spectrum of biotic and abiotic stresses is a focus of the private [19] as well as public sectors [11].

Use of interspecific hybrids as rootstocks for salt tolerance is well documented [6,64,85]. In this technique, recombinant inbred lines (RILs) created from interspecific crosses are directly used as rootstocks to enhance salinity tolerance. Evaluation of RILs as rootstocks derived from interspecific crosses under salt stress was demonstrated by Albacete et al. [85]. A commercial tomato hybrid cv. 'Boludo' was used to graft onto rootstocks (RILs derived from the cross of *S. lycopersicum* and *S. cheesmaniae*) and grown on soil with 75 mM NaCl salt concentration. These RILs showed good performance under salt stress by affecting growth, and target ionic and hormonal factors in the leaves. They displayed the correlation of rootstock-altered leaf area with 20% more fruit yield compared to self-grafted plants of the scion hybrid under salt pressure, but not under the control [7,75], indicating that canopy development only limited yield under salt stress [14]. For high salt tolerance, rootstock plants should capable of decreasing the salt load in the foliage cells by excluding the salt ions or accumulating salt in the roots, resulting in a reduction of salt transportation to the shoot system of the plant [21].

While intraspecific grafting is a common practice, efforts also have been made to use interspecific grafting to alleviate wide-spectrum pathogen and environmental pressure without affecting scion fruit yield and quality. Accessions of wild tomato species *S. pimpinellifolium* were screened for salt tolerance by Rao et al. [12], who suggested the species could be a potential salt tolerance source for tomato breeding and at the same time be used for interspecific grafting to enhance salt tolerance in commercial cultivars. Their results indicated that shoot dry weight and the ratio of K^+ and Na^+ are the two most critical factors for survival, while fruit number is critical for yield per plant under saline conditions. Two genotypes exhibited high survival rates and produced more fruit under salt stress, and these accessions showed tolerance to 40 dS m^{-1} EC level.

6. Conclusions and Challenges Ahead

Climate change, water scarcity and human activities are increasing soil salinity and depleting arable land. High salinity in soil or water is a serious threat to plant growth and prevents plants from reaching their genetic potential, which reduces yields and threatens food security. Grafting can reduce the negative effects of salinity on tomato scions and thus provides an alternative way to enhance salt tolerance and maintain fruit yield and quality under salt stress. The selection of salt-tolerant rootstocks is an important strategy for enhancing salt tolerance. Testing and screening of available commercial cultivars, landraces and wild relatives, and inbred lines or hybrids involving a resistant parent under salt stress conditions is a prerequisite for successful grafting. This technique can cause both positive and negative effects on yield and fruit characteristics when tomatoes are grown under saline conditions. Weather, salinity level, soil type, cultural practices and consumer preferences also vary in different locations or countries, so the combination of scion and rootstock should be selected carefully at plants morphometric, physio-biochemical and molecular planes to get the maximum benefit from grafting.

Author Contributions: H.S. Wrote the initial draft of the review including yield and quality of grafted tomatoes under salinity; A.K. First draft of physio-biochemical and molecular aspects. P.K. Improved the first draft, contributed to the morphological, physio-biochemical mechanism, and prepared the graph. M.C.K., G.C. and Y.R. critically revised and improved the review paper. All authors have read and agreed to the published version of the manuscript.

Acknowledgments: The authors acknowledge Punjab Agricultural University, Ludhiana and ICAR-Central Arid Zone Research Institute, Jodhpur, India for providing supports.

References

1. Roychoudhury, A.; Paul, S.; Basu, S. Cross-talk between abscisic acid-dependent and abscisic acid-independent pathways during abiotic stress. *Plant Cell Rep.* **2013**, *32*, 985–1006. [CrossRef] [PubMed]
2. Foolad, M.R. Recent advances in genetics of salt tolerance in tomato. *Plant Cell Tissue Organ. Cult.* **2004**, *76*, 101–119. [CrossRef]
3. Arzani, A. Improving salinity tolerance in crop plants: A biotechnological view. *In Vitro Cell. Dev. Biol. Plant* **2008**, *44*, 373–383. [CrossRef]
4. Zheng, Q.; Liu, J.; Liu, R.; Wu, H.; Jiang, C.; Wang, C.; Guan, Y. Temporal and spatial distributions of sodium and polyamines regulated by brassinosteroids in enhancing tomato salt resistance. *Plant Soil* **2016**, *400*, 147–164. [CrossRef]
5. Sholi, N.J. Effect of salt stress on seed germination, plant growth, photosynthesis and ion accumulation of four tomato cultivars. *Am. J. Plant Physiol.* **2012**, *7*, 269–275. [CrossRef]
6. Estañ, M.T.; Villalta, I.; Bolarin, M.C.; Carbonell, E.A.; Asins, M.J. Identification of fruit yield loci controlling the salt tolerance conferred by *Solanum* rootstocks. *Theor. Appl. Genet.* **2009**, *118*, 305–312. [CrossRef]
7. Lim, M.Y.; Jeong, B.R.; Jung, M.; Harn, C.H. Transgenic tomato plants expressing strawberry d-galacturonic acid reductase gene display enhanced tolerance to abiotic stresses. *Plant Biotechnol. Rep.* **2016**, *10*, 105–116. [CrossRef]
8. Keatinge, J.D.H.; Lin, L.J.; Ebert, A.W.; Chen, W.Y.; Hughes, J.A.; Luther, G.C.; Wang, J.F.; Ravishankar, M. Overcoming biotic and abiotic stresses in the *Solanaceae* through grafting: Current status and future perspectives. *Biol. Agric. Hortic.* **2014**, *30*, 272–287. [CrossRef]
9. Dehyer, R.; Gordon, I. Irrigation water quality. I-salinity and soil structure stability. *Nat. Res. Sci.* **2005**, *55*, 55–60.
10. Al-Harbi, A.; Hejazi, A.; Al-Omran, A. Responses of grafted tomato (*Solanum lycopersiocon* L.) to abiotic stresses. *Saudi, J. Biol. Sci.* **2016**, *24*, 1274–1280. [CrossRef]
11. Colla, G.; Rouphael, Y.; Leonardi, C.; Bie, Z. Role of grafting in vegetable crops grown under saline conditions. *Sci. Hortic.* **2010**, *127*, 147–155. [CrossRef]

12. Rao, E.S.; Kadirvel, P.; Symonds, R.C.; Ebert, A.W. Relationship between survival and yield related traits in *Solanum pimpinellifolium* under salt stress. *Euphytica* **2013**, *190*, 215–228. [CrossRef]
13. Kumar, P.; Rouphael, Y.; Cardarelli, M.; Colla, G. Vegetable Grafting as a Tool to Improve Drought Resistance and Water Use Efficiency. *Front. Plant Sci.* **2017**, *8*, 1130. [CrossRef] [PubMed]
14. Albacete, A.; Martínez-Andújar, C.; Martínez-Pérez, A.; Thompson, A.J.; Dodd, I.C.; Pérez-Alfocea, F.R. Unravelling rootstock-scion interactions to improve food security. *J. Exp. Bot.* **2015**, *66*, 2211–2226. [CrossRef]
15. Martorana, M.; Giuffrida, F.; Leonardi, C.; Kaya, S. Influence of rootstock on tomato response to salinity. *Acta Hortic.* **2007**, *747*, 555–561. [CrossRef]
16. Colla, G.; Rouphael, Y.; Jawad, R.; Kumar, P.; Rea, E.; Cardarelli, M. The effectiveness of grafting to improve NaCl and $CaCl_2$ tolerance in cucumber. *Sci. Hortic.* **2013**, *164*, 380–391. [CrossRef]
17. Tateishi, K. Grafting watermelon onto pumpkin. *J. Jpn. Hortic.* **1927**, *39*, 5–8.
18. Besri, M. Current situation of tomato grafting as alternative to methyl bromide for tomato production in Morocco. In Proceedings of the Annual International Research Conference on Methyl Bromide Alternatives and Emission Reductions, San Diego, CA, USA, 28–31 October 2007; pp. 62.1–62.5.
19. King, S.R.; Davis, A.R.; Zhang, X.; Crosby, K. Genetics, breeding and selection of rootstocks for *Solanaceae* and *Cucurbitaceae*. *Sci. Hortic.* **2010**, *127*, 106–111. [CrossRef]
20. Kubota, C.; McClure, M.A.; Kokalis-Burelle, N.; Bausher, M.G.; Rosskopf, E.N. Vegetable grafting: History, use, and current technology status in North America. *Sci. Hortic.* **2008**, *43*, 1644–1669. [CrossRef]
21. Edelstein, M.; Cohen, R.; Baumkoler, F.; Ben-Hur, M. Using grafted vegetables to increase tolerance to salt and toxic elements. *Isr. J. Plant Sci.* **2016**, 1–18. [CrossRef]
22. Cohen, R.; Omari, N.; Porat, A.; Edelstein, M. Management of *Macrophomina* wilt in melons using grafting or fungicide soil application: Pathological, horticultural and economical aspects. *Crop Prot.* **2012**, *35*, 58–63. [CrossRef]
23. Kubota, C. *How to Produce Grafted Vegetable Plants*; Kubota, C., Miles, C., Zhao, X., Eds.; Grafting manual: Washington, DC, USA, 2016; pp. 1–5.
24. Besri, M. Economic aspects of grafting tomato in some Mediterranean countries. In Proceedings of the Annual International Research Conference on Methyl Bromide Alternatives and Emissions Reductions proceedings, San Diego, CA, USA, 29 October–1 November 2007; p. 62-1-5.
25. Singh, H.; Kumar, P.; Chaudhari, S.; Edelstein, M. Tomato grafting: A Global perspective. *HortScience* **2017**, *52*, 1328–1336. [CrossRef]
26. Sardoo, M.; Fatemeh, S. Evaluation of salt tolerance of tomato (*Lycopersicon esculentum*). *Agric. Adv.* **2016**, *5*, 221–226.
27. Abbasi, H.; Jamil, M.; Haq, A.; Ali, S.; Ahmad, R.; Malik, Z. Salt stress manifestation on plants, mechanism of salt tolerance and potassium role in alleviating it: A review. *Zemdirbyste Agric.* **2016**, *103*, 229–238. [CrossRef]
28. Parida, AK.; Das, A.B. Salt tolerance and salinity effects on plants. *Ecotox. Environ. Saf.* **2005**, *60*, 324–349. [CrossRef]
29. Rivero, R.M.; Ruiz, J.M.; Romero, L. Role of grafting in horticultural plants under stress conditions. *J. Food Agric. Environ.* **2003**, *1*, 70–74.
30. Colla, G.; Kumar, P.; Cardarelli, M.; Rouphael, Y. Grafting an effective tool for abiotic stress alleviation in vegetables. In *Horticulture for Food and Environment Security*; Chadha, K.L., Singh, A.K., Singh, S.K., Dhillon, W.S., Eds.; Westville Publishing House: New Delhi, India, 2012; pp. 15–28.
31. Rajaei, S.M.; Niknam, V.; Seyedi, S.M.; Ebrahimzadeh, H.; Razavi, K. Contractile roots are the most sensitive organ in *Crocus sativus* to salt stress. *Biol. Plant.* **2009**, *53*, 523. [CrossRef]
32. Chaves, M.M.; Costa, J.M.; Saibo, N.J.M. Recent advances in photosynthesis under drought and salinity. *Adv. Bot. Res.* **2011**, *57*, 49–104.
33. Mahajan, S.; Tuteja, N. Cold, salinity and drought stresses: An overview. *Arch. Biochem. Biophys.* **2005**, *444*, 139–158. [CrossRef]
34. Cuartero, J.; Bolarin, M.C.; Asins, M.J.; Moreno, V. Increasing salt tolerance in the tomato. *J. Exp. Biol.* **2006**, *57*, 1045–1058. [CrossRef]
35. Hajer, A.S.; Malibari, A.A.; Al-Zahrani, H.S.; Almaghrabi, O.A. Responses of three tomato cultivars to sea water salinity 1. Effect of salinity on the seedling growth. *Afr. J. Biotechnol.* **2006**, *5*, 855–861.

36. Najla, S.; Vercambre, G.; Page's, L.; Grasselly, D.; Gautier, H.; Ge'nard, M. Tomato plant architecture as affected by salinity: Descriptive analysis and integration in a 3-D simulation model. *Botany* **2009**, *87*, 893–904. [CrossRef]
37. Cuartero, J.; Fernández-Muñoz, R. Tomato and salinity. *Sci. Hortic.* **1999**, *78*, 83–125. [CrossRef]
38. Khavari-Nejad, R.A.; Mostofi, Y. Effects of NaCl on photosynthetic pigments, saccharides, and chloroplast ultrastructure in leaves of tomato cultivars. *Photosynthetica* **1998**, *35*, 151–154. [CrossRef]
39. Al-Karaki, G.N. Growth, water use efficiency, and sodium and potassium acquisition by tomato cultivars grown under salt stress. *J. Plant. Nutr.* **2000**, *23*, 1–8. [CrossRef]
40. Turhan, A.; Seniz, V.; Kusçu, H. Genotypic variation in the response of tomato to salinity. *Afr. J. Biotechnol.* **2009**, *8*, 1062–1068.
41. Gao, Z.; Sagi, M.; Lips, S.H. Carbohydrate metabolism in leaves and assimilate partitioning in fruits of tomato (*Lycopersicon esculentum* L.) as affected by salinity. *Plant Sci.* **1998**, *135*, 149–159. [CrossRef]
42. De Pascale, S.; Maggio, A.; Fogliano, V.; Ambrosino, P.; Ritieni, A. Irrigation with saline water improves carotenoids content and antioxidant activity of tomato. *J. Hortic. Sci. Biotechnol.* **2001**, *76*, 447–453. [CrossRef]
43. Di Gioia, F.; Signore, A.; Serio, F.; Santamaria, P. Grafting improves tomato salinity tolerance through sodium partitioning within the shoot. *HortScience* **2013**, *48*, 855–862. [CrossRef]
44. Balliu, A.; Vuksani, G.; Nasto, T.; Haxhinasto, L.; Kaçiu, S. Grafting effects on tomato growth rate, yield and fruit quality under saline irrigation water. *Acta Hortic.* **2007**, *801*, 1161–1166. [CrossRef]
45. Santa-Cruz, A.; Martínez-Rodríguez, M.M.; Cuartero, J.; Bolarin, M.C. Response of plant yield and ion contents to salinity in grafted tomato plants. *Acta Hortic.* **2001**, *559*, 413–417. [CrossRef]
46. Santa-Cruz, A.; Martínez-Rodríguez, M.M.; Pérez-Alfocea, F.; Romero-Aranda, R.; Bolarin, M.C. The rootstock effect on the tomato salinity response depends on the shoot genotype. *Plant Sci.* **2002**, *162*, 825–831. [CrossRef]
47. Voutsela, S.; Yarsi, G.; Petropoulos, S.A.; Khan, E.A. The effect of grafting of five different rootstocks on plant growth and yield of tomato plants cultivated outdoors and indoors under salinity stress. *Afr. J. Agric. Res.* **2012**, *7*, 5553–5557. [CrossRef]
48. Iseri, O.D.; Körpe, D.A.; Sahin, F.I.; Haberal, M. High salt induced oxidative damage and antioxidant response in tomato grafted on tobacco. *Chil. J. Agric. Res.* **2015**, *75*, 192–201.
49. Turhan, A.; Ozmen, N.; Serbeci, M.S.; Seniz, V. Effects of grafting on different rootstocks on tomato fruit yield and quality. *Sci. Hortic.* **2011**, *38*, 142–149. [CrossRef]
50. Flores, F.B.; Sanchez-Bel, P.; Estañ, M.T.; Martinez-Rodriguez, M.M.; Moyano, E.; Morales, B.; Campos, J.F.; Garcia-Abellán, J.O.; Egea, M.I.; Fernández-García, N.; et al. The effectiveness of grafting to improve tomato fruit quality. *Sci. Hortic.* **2010**, *125*, 211–217. [CrossRef]
51. He, Y.; Zhu, Z.; Yang, J. Grafting increases the salt tolerance of tomato by improvement of photosynthesis and enhancement of antioxidant enzymes activity. *Environ. Exp. Bot.* **2009**, *66*, 270–278. [CrossRef]
52. Oztekin, G.B.; Tuzel, Y. Salinity response of some tomato rootstocks at seedling stage. *Afr. J. Agric. Res.* **2011**, *6*, 4726–4735.
53. Gregory, P.J.; Atkinson, C.J.; Bengough, A.G.; Else, M.A.; Fernández-Fernández, F.; Harrison, R.J.; Schmidt, S. Contributions of roots and rootstocks to sustainable, intensified crop production. *J. Exp. Bot.* **2013**, *64*, 1209–1222. [CrossRef]
54. Martínez-Rodríguez, M.M.; Estañ, M.T.; Moyano, E.; Garcia-Abellan, J.O.; Flores, F.B.; Campos, J.F.; Al-Azzawi, M.J.; Flower, T.J.; Bolarin, M.C. The effectiveness of grafting to improve salt tolerance in tomato when an 'excluder' genotype is used as scion. *Environ. Exp. Bot.* **2008**, *63*, 392–401. [CrossRef]
55. Zhen, A.; Bie, Z.; Huang, Y.; Liu, Z.; Li, Q. Effects of scion and rootstock genotypes on the anti-oxidant defense systems of grafted cucumber seedlings under NaCl stress. *Soil Sci. Plant Nutr.* **2010**, *56*, 263–271. [CrossRef]
56. Turquois, N.; Malone, M. Non-destructive assessment of developing hydraulic connections in the graft union of tomato. *J. Exp. Bot.* **1996**, *298*, 701–707. [CrossRef]
57. Huang, Y.; Zhao, L.Q.; Kong, Q.S.; Cheng, F.; Niu, M.L.; Xie, J.J.; Nawaz, M.A.; Bie, Z.L. Comprehensive mineral nutrition analysis of watermelon grafted onto two different rootstocks. *Hort. Plant J.* **2016**, *2*, 105–113. [CrossRef]
58. Ruiz, D.; Martínez, V.; Cerda, A. Citrus response to salinity, growth and nutrient uptake. *Tree Physiol.* **1997**, *17*, 141–150. [CrossRef] [PubMed]

59. Fernández-García, N.; Martinez, V.; Cerdá, A.; Carvajal, M. Water and nutrient uptake of grafted tomato plants grown under saline conditions. *J. Plant Physiol.* **2002**, *159*, 899–905. [CrossRef]
60. Shalhevet, J.; Maas, E.V.; Hoffman, G.J.; Ogata, G. Salinity and the hydraulic conductance of roots. *Physiol. Plant.* **1976**, *38*, 224–232. [CrossRef]
61. Munns, R.; Passioura, J.B. Hydraulic resistance of plants. III. Effects of NaCl in barley and lupin. *Aust. J. Plant Physiol.* **1984**, *11*, 351–359. [CrossRef]
62. Evlagon, D.; Ravina, Y.; Neumann, P.M. Interactive effects of salinity and calcium on hydraulic conductivity, osmotic adjustment and growth in primary roots of maize seedlings. *Isr. J. Bot.* **1990**, *39*, 239–247.
63. Carvajal, M.; Martínez, V.; Alcaraz, C.F. Physiological function of water-channels, as affected by salinity in roots of paprika pepper. *Physiol. Plant.* **1999**, *105*, 95–101. [CrossRef]
64. Estañ, M.T.; Martinez-Rodriguez, M.M.; Perez-Alfocea, F.; Flowers, T.J.; Bolarin, M.C. Grafting raises the salt tolerance of tomato through limiting the transport of sodium and chloride to the shoot. *J. Exp. Bot.* **2005**, *56*, 703–712. [CrossRef]
65. Lee, J.M. Cultivation of grafted vegetables I. Current status, grafting methods, and benefits. *HortScience* **1994**, *29*, 235–239. [CrossRef]
66. Zijlstra, S.; Groot, S.P.C.; Jansen, J. Genotypic variation of rootstocks for growth and production in cucumber; possibilities for improving the root system by plant breeding. *Sci. Hortic.* **1994**, *56*, 185–196. [CrossRef]
67. Leoni, S.; Grudina, R.; Cadinu, M.; Madeddu, B.; Carletti, M.G. The influence of four rootstocks on some melon hybrids and a cultivar in greenhouse. *II Inter. Sym. Protec. Culti. Veg. Mild Winter Clim.* **1990**, *287*, 127–134. [CrossRef]
68. Moya, J.L.; Primo-Millo, E.; Talón, M. Morphological factors determining salt tolerance in citrus seedlings: The shoot to root ratio modulates passive root uptake of chloride ions and their accumulation in leaves. *Plant Cell. Environ.* **1999**, *22*, 1425–1433. [CrossRef]
69. Zhu, J.; Bie, Z.; Huang, Y.; Han, X. Effect of grafting on the growth and ion concentrations of cucumber seedlings under NaCl stress. *Soil Sci. Plant Nutr.* **2008**, *54*, 895–902. [CrossRef]
70. Pérez-Alfocea, F.; Albacete, A.; Ghanem, M.E.; Dodd, I.C. Hormonal regulation of source–sink relations to maintain crop productivity under salinity: A case study of root-to-shoot signalling in tomato. *Funct. Plant Biol.* **2010**, *37*, 592–603. [CrossRef]
71. Martinez-Rodriguez, M.M.; Santa-Cruz, A.; Estan, M.T.; Caro, M.; Bolarin, M.C. Influence of rootstock in the tomato response to salinity. *Acta Hortic.* **2002**, *573*, 455–460. [CrossRef]
72. Fernández-García, N.; Martínez, V.; Carvajal, M. Effect of salinity on growth, mineral composition, and water relations of grafted tomato plants. *J. Plant Nutr. Soil Sci.* **2004**, *167*, 616–622. [CrossRef]
73. Semiz, G.D.; Suarez, D.L. Tomato salt tolerance: Impact of grafting and water composition on yield and ion relations. *Turk. J. Agric. For.* **2015**, *39*, 876–886. [CrossRef]
74. Munns, R. Genes and salt tolerance: Bringing them together. *New Phytol.* **2005**, *167*, 645–663. [CrossRef]
75. Pérez-Alfocea, F.; Estañ, M.T.; Caro, M.; Bolarín, M.C. Response of tomato cultivars to salinity. *Plant Soil.* **1993**, *150*, 203–211. [CrossRef]
76. Pérez-Alfocea, F.; Balibrea, M.E.; Santa Cruz, A.; Estañ, M.T. Agronomical and physiological characterization of salinity tolerance in a commercial tomato hybrid. *Plant Soil.* **1996**, *180*, 251–257. [CrossRef]
77. Tal, M. Salt tolerance in the wild relatives of the cultivated tomato: Responses of *Lycopersicon esculentum*, *L. peruvianum*, and *L. esculentum* minor to sodium chloride solution. *Aust. J. Agric. Res.* **1971**, *22*, 631–638. [CrossRef]
78. Rush, D.W.; Epstein, E. Comparative studies on the sodium, potassium, and chloride relations of a wild halophytic and a domestic salt-sensitive tomato species. *Plant Physiol.* **1981**, *68*, 1308–1313. [CrossRef]
79. Bolarin, M.C.; Fernandez, F.G.; Cruz, V.; Cuartero, J. Salinity tolerance in four wild tomato species using vegetative yield-salinity response curves. *J. Am. Soc. Hort. Sci.* **1991**, *116*, 286–290. [CrossRef]
80. Cuartero, J.; Yeo, A.R.; Flowers, T.J. Selection of donors for salt-tolerance in tomato using physiological traits. *New Phytol.* **1992**, *121*, 63–69. [CrossRef]
81. Shannon, M.C. Adaptation of plants to salinity. In *Advances in Agronomy*; Academic Press: Cambridge, MA, USA, 1997; Volume 60, pp. 75–120.
82. Sacher, R.F.; Staples, R.C.; Robinson, R.W. Saline tolerance in hybrids of *Lycopersicon esculentum* X *Solanum penellii* and selected breeding lines. *Environ. Sci. Res.* **1982**, 325–336.

83. Huang, Y.; Zhu, J.; Zhen, A.; Chen, L.; Bie, Z.L. Organic and inorganic solutes accumulation in the leaves and roots of grafted and ungrafted cucumber plants in response to NaCl stress. *J. Food Agric. Environ.* **2009**, *7*, 703–708.
84. Albacete, A.; Martínez-Andújar, C.R.; Ghanem, M.E.; Acosta, M.; Sánchez-bravo, J.O.; Asins, M.J.; Cuartero, J.; Lutts, S.; Dodd, I.C.; Pérez-Alfocea, F.R. Rootstock-mediated changes in xylem ionic and hormonal status are correlated with delayed leaf senescence, and increased leaf area and crop productivity in salinized tomato. *Plant Cell Environ.* **2009**, *32*, 928–938. [CrossRef]
85. Xu, H.L.; Gauthier, L.; Gosselin, A. Photosynthetic responses of greenhouse tomato plants to high solution electrical conductivity and low soil water content. *J. Hortic. Sci.* **1994**, *69*, 821–832. [CrossRef]
86. Flexas, J.; Bota, J.; Loreto, F.; Cornic, G.; Sharkey, T.D. Diffusive and metabolic limitations to photosynthesis under drought and salinity in C_3 plants. *Plant Biol.* **2004**, *6*, 269–279. [CrossRef] [PubMed]
87. Zhen, A.; Bie, Z.; Huang, Y. Effects of salt-tolerant rootstock grafting on ultrastructure, photosynthetic capacity, and H_2O_2-scavenging system in chloroplasts of cucumber seedlings under NaCl stress. *Acta Physiol. Plant.* **2011**, *33*, 2311. [CrossRef]
88. Liu, Z.; Bie, Z.; Huang, Y. Rootstocks improve cucumber photosynthesis through nitrogen metabolism regulation under salt stress. *Acta Physiol. Plant.* **2013**, *35*, 2259–2267. [CrossRef]
89. Marsic, N.K.; Vodnik, D.; Mikulic-Petkovsek, M.; Veberic, R.; Sircelj, H. Photosynthetic Traits of Plants and the Biochemical Profile of Tomato Fruits Are Influenced by Grafting, Salinity Stress, and Growing Season. *J. Agric. Food Chem.* **2018**, *66*, 5439–5450. [CrossRef]
90. Feng, X.; Guo, K.; Yang, C.; Li, J.; Chen, H.; Liu, X. Growth and fruit production of tomato grafted onto wolfberry (*Lycium chinense*) rootstock in saline soil. *Sci. Hortic.* **2019**, *255*, 298–305.
91. Bartoli, C.G.; Buet, A.; Grozeff, G.G. Ascorbate-glutathione cycle and abiotic stress tolerance in plants. In *Ascorbic Acid in Plant. Growth, Development and Stress Tolerance*; Springer: Cham, Germany, 2017; pp. 177–200.
92. Gao, J.J.; Zhang, L.; Qin, A.G.; Yu, X.C. Relative expression of SOD and CAT mRNA and activities of SOD and CAT in grafted cucumber leaves under NaCl stress. Ying yong sheng tai xue bao. *J. Appl. Ecol.* **2008**, *19*, 1754–1758.
93. Wei, G.P.; Yang, L.F.; Zhu, Y.L.; Chen, G. Changes in oxidative damage, antioxidant enzyme activities and polyamine contents in leaves of grafted and non-grafted eggplant seedlings under stress by excess of calcium nitrate. *Sci. Hortic.* **2009**, *120*, 443–451. [CrossRef]
94. Ruiz, J.M.; Ríos, J.J.; Rosales, M.A.; Rivero, R.M.; Romero, L. Grafting between tobacco plants to enhance salinity tolerance. *J. Plant Physiol.* **2006**, *163*, 1229–1237. [CrossRef]
95. Yasinok, A.E.; Feride, S.I.; Eyidogan, F.; Kuru, M.; Haberal, M. Grafting tomato plant on tobacco plant and its effect on tomato plant yield and nicotine content. *J. Sci. Food Agric.* **2009**, *89*, 1122–1128. [CrossRef]
96. Singh, J.; Sastry, E.V.D.; Singh, V. Effect of salinity on tomato (*Lycopersicon esculentum* Mill.) during seed germination stage. *J. Plant Physiol. Molecular Biol.* **2012**, *18*, 45–50. [CrossRef] [PubMed]
97. Ja'afar, U.; Aliero, A.A.; Shehu, K.; Abubakar, L. Genetic diversity in tomato genotypes (*Solanum lycopersicum*) based on salinity responsive candidate gene using simple sequence repeats. *International Lett. Nat. Sci.* **2018**, *72*, 37–46. [CrossRef]
98. Yveline, P.; Awlia, M.; Julkowska, M.; Passone, L.; Zemmouri, K.; Negrão, S.; Schmöckel, S.M.; Tester, M. Diverse traits contribute to salinity tolerance of wild tomato seedlings from the Galapagos Islands. *Plant. Physiol.* **2020**, *182*, 534–546.
99. Asins, M.J.; Raga, V.; Roca, D.; Belver, A.; Carbonell, E.A. Genetic dissection of tomato rootstock effects on scion traits under moderate salinity. *Theor. Appl Genet.* **2015**, *128*, 667–679. [CrossRef]
100. Nedjimi, B. Calcium application enhances plant salt tolerance: A review. *Essent. Plant Nutr. Springer* **2017**, 367–377.
101. Zhao, N.; Wang, S.; Ma, X. Extracellular ATP mediates cellular K+/Na+ homeostasis in two contrasting poplar species under NaCl stress. *Trees.* **2016**, *30*, 825–837. [CrossRef]
102. Mahajan, S.; Pandey, G.K.; Tuteja, N. Calcium-and salt-stress signaling in plants: Shedding light on SOS pathway. *Arch. Biochem. Bioph.* **2008**, *471*, 146–158. [CrossRef]
103. Shaterian, J.; Georges, F.; Hussain, A. Root to shoot communication and abscisic acid in calreticulin (CR) gene expression and salt-stress tolerance in grafted diploid potato clones. *Environ. Exp. Bot.* **2005**, *53*, 323–332. [CrossRef]

104. Sharma, A.; Zheng, B. Molecular responses during plant grafting and its regulation by auxins, cytokinins, and gibberellins. *Biomolecules* **2019**, *9*, 397. [CrossRef]
105. Nanda, A.K.; Melnyk, C.W. The role of plant hormones during grafting. *J. Plant Res.* **2018**, *131*, 49–58. [CrossRef]
106. Ruzicka, K.; Ursache, R.; Hejatko, J.; Helariutta, Y. Xylem development-from the cradle to the grave. *New Phytol.* **2015**, *207*, 519–535. [CrossRef] [PubMed]
107. Kumar, R.M.S.; Ji, G.; Guo, H.; Zhao, L.; Zheng, B. Over-expression of a grafting-responsive gene from hickory increases abiotic stress tolerance in Arabidopsis. *Plant Cell Rep.* **2018**, *37*, 541–552. [CrossRef] [PubMed]
108. Sauer, M.; Balla, J.; Luschnig, C.; Wisniewska, J.; Reinohl, V.; Friml, J.; Benkova, E. Canalization of auxin flow by Aux/IAA-ARF-dependent feedback regulation of PIN polarity. *Genes Dev.* **2006**, *20*, 2902–2911. [CrossRef]
109. Yuan, H.; Zhao, L.; Qiu, L.; Xu, D.; Tong, Y.; Guo, W.; Yang, X.; Shen, C.; Yan, D.; Zheng, B. Transcriptome and hormonal analysis of grafting process by investigating the homeostasis of a series of metabolic pathways in *Torreya grandis* cv. *Merrillii. Ind. Crops Prod.* **2017**, *108*, 814–823. [CrossRef]
110. Aloni, R. The role of cytokinin in organised differentiation of vascular tissues. *Funct. Plant Biol.* **1993**, *20*, 601–608. [CrossRef]
111. Shehata, S.; El-Shraiy, A.M. Regulating cucumber grafting by interactions of cytokinins in xylem exudates of rootstock and basipetol polar auxin transport of scion at graft union. *Aus. J. Basic Appl. Sci.* **2010**, *4*, 6179–6184.
112. Aloni, B.; Cohen, R.; Karni, L.; Aktas, H.; Edelstein, M. Hormonal signaling in rootstock–scion interactions. *Sci. Hortic.* **2010**, *127*, 119–126. [CrossRef]
113. Ghanem, M.E.I.; Hichri, A.C.; Smigocki, A.; Albacete, M.L.; Fauconnier, E.; Diatloff, C.; Martı'nez-Andu´jar, M.; Acosta, J.; Sanchez-Bravo, S.; Lutts, I.C.; et al. Root-targeted biotechnology to mediate hormonal signaling and improve crop stress tolerance. *Plant Cell Rpt.* **2011**, *30*, 807–823. [CrossRef]
114. Aloni, R.; Aloni, E.; Langhans, M.; Ullrich, C.I. Role of cytokinin and auxin in shaping root architecture: Regulating vascular differentiation, lateral root initiation, root apical dominance and root gravitropism. *Ann. Bot.* **2006**, *97*, 883–893. [CrossRef]
115. Ghanem, M.E.; Albacete, A.; Smigocki, A.C.; Frébort, I.; Pospíšilová, H.; Martínez-Andújar, C.; Acosta, M.; Sánchez-Bravo, J.; Lutts, S.; Dodd, I.C.; et al. Root-synthesized cytokinins improve shoot growth and fruit yield in salinized tomato (*Solanum lycopersicum* L.) plants. *J. Exp. Bot.* **2011**, *62*, 125–140. [CrossRef]
116. Asahina, M.; Gocho, Y.; Kamada, H.; Satoh, S. Involvement of inorganic elements in tissue reunion in the hypocotyl cortex of *Cucumis sativus. J. Plant Res.* **2006**, *119*, 337–342. [CrossRef] [PubMed]
117. Asahina, M.; Yamauchi, Y.; Hanada, A.; Kamiya, Y.; Kamada, H.; Satoh, S.; Yamaguchi, S. Effects of the removal of cotyledons on endogenous gibberellin levels in hypocotyls of young cucumber and tomato seedlings. *Plant Biotechnol.* **2007**, *24*, 99–106. [CrossRef]
118. Mo, Z.; Feng, G.; Su, W.; Liu, Z.; Peng, F. Transcriptomic Analysis Provides Insights into Grafting Union Development in Pecan (*Carya illinoinensis*). *Genes* **2018**, *9*, 71. [CrossRef] [PubMed]
119. Bjorklund, S.; Antti, H.; Uddestrand, I.; Moritz, T.; Sundberg, B. Cross-talk between gibberellin and auxin in development of *Populus* wood: Gibberellin stimulates polar auxin transport and has a common transcriptome with auxin. *Plant J.* **2007**, *52*, 499–511. [CrossRef] [PubMed]
120. Dayan, J.; Voronin, N.; Gong, F.; Sun, T.P.; Hedden, P.; Fromm, H.; Aloni, R. Leaf-induced gibberellin signaling is essential for internode elongation, cambial activity, and fiber differentiation in tobacco stems. *Plant Cell.* **2012**, *24*, 66–79. [CrossRef]
121. Liu, N.; Yang, J.; Fu, X.; Zhang, L.; Tang, K.; Guy, K.M.; Hu, Z.; Guo, S.; Xu, Y.; Zhang, M. Genome-wide identification and comparative analysis of grafting-responsive mRNA in watermelon grafted onto bottle gourd and squash rootstocks by high-throughput sequencing. *Mol. Genet. Genom.* **2016**, *291*, 621–633. [CrossRef]
122. Peña-Cortés, H.; Sánchez-Serrano, J.J.; Mertens, R. Abscisic acid is involved in the wound-induced expression of the proteinase inhibitor II gene in potato and tomato. *Proc. Natl. Acad. Sci. USA* **1989**, *86*, 9851–9855. [CrossRef]
123. Peña-Cortés, H.; Fisahn, J.; Willmitzer, L. Signals involved in wound-induced proteinase inhibitor II gene expression in tomato and potato plants. *Proc. Natl. Acad. Sci. USA* **1995**, *92*, 4106–4113. [CrossRef]

124. Chen, G.; Fu, X.; Lips, S.H.; Sagi, M. Control of plant growth resides in the shoot, and not in the root, in reciprocal grafts of flacca and wild-type tomato (*Lysopersicon esculentum*), in the presence and absence of salinity stress. *Plant Soil* **2003**, *256*, 205–215. [CrossRef]
125. Yin, H.; Yan, B.; Sun, J.; Jia, P.; Zhang, Z.; Yan, X.; Chai, J.; Ren, Z.; Zheng, G.; Liu, H. Graft-union development: A delicate process that involves cell-cell communication between scion and stock for local auxin accumulation. *J. Exp. Bot.* **2012**, *63*, 4219–4232. [CrossRef]
126. Matsuoka, K.; Sugawara, E.; Aoki, R.; Takuma, K.; Terao-Morita, M.; Satoh, S.; Asahina, M. Differential Cellular Control by Cotyledon-Derived Phytohormones Involved in Graft Reunion of Arabidopsis Hypocotyls. *Plant Cell Physiol.* **2016**, *57*, 2620–2631. [CrossRef] [PubMed]
127. Tao, J.J.; Chen, H.W.; Ma, B.; Zhang, W.K.; Chen, S.Y.; Zhang, J.S. The role of ethylene in plants under salinity stress. *Front. Plant Sci.* **2015**, *6*, 1059. [CrossRef] [PubMed]
128. Savvas, D.; Savva, A.; Ntatsi, G.; Ropokis, A.; Karapanos, I.; Krumbein, A.; Olympios, C. Effects of three commercial rootstocks on mineral nutrition, fruit yield, and quality of salinized tomato. *J. Plant Nutr. Soil Sci.* **2011**, *174*, 154–162. [CrossRef]
129. Wahb-Allah, M.A. Effectiveness of grafting for the improvement of salinity and drought tolerance in tomato (*Solanum lycopersicum* L.). *Asian J. Crop. Sci.* **2014**, *6*, 112–122.

5

Effect of Supplementary Light Intensity on Quality of Grafted Tomato Seedlings and Expression of Two Photosynthetic Genes and Proteins

Hao Wei [1,†], Jin Zhao [1,†], Jiangtao Hu [1] and Byoung Ryong Jeong [1,2,3,*]

1 Department of Horticulture, Division of Applied Life Science (BK21 Plus Program), Graduate School of Gyeongsang National University, Jinju 52828, Korea; oahiew@gmail.com (H.W.); jzhao9006@gmail.com (J.Z.); jiangtaoh@yahoo.com (J.H.)

2 Institute of Agriculture & Life Science, Gyeongsang National University, Jinju 52828, Korea

3 Research Institute of Life Science, Gyeongsang National University, Jinju 52828, Korea

* Correspondence: brjeong@gmail.com

† These authors contributed equally to this work.

Abstract: Lower quality and longer production periods of grafted seedlings, especially grafted plug seedlings of fruit vegetables, may result from insufficient amounts of light, particularly in rainy seasons and winter. Supplemental artificial lighting may be a feasible solution to such problems. This study was conducted to evaluate light intensity's influence on the quality of grafted tomato seedlings, 'Super Sunload' and 'Super Dotaerang' were grafted onto the 'B-Blocking' rootstock. To improve their quality, grafted seedlings were moved to a glasshouse and grown for 10 days. The glasshouse had a combination of natural lighting from the sun and supplemental lighting from LEDs ($W_1R_2B_2$) for 16 h/day. Light intensity of natural lighting was 490 $\mu mol \cdot m^{-2} \cdot s^{-1}$ photosynthetic photon flux density (PPFD) and that of supplemental lighting was 50, 100, or 150 $\mu mol \cdot m^{-2} \cdot s^{-1}$ PPFD. The culture environment had 30/25 °C day/night temperatures, 70% ± 5% relative humidity (RH), and a natural photoperiod of 14 h as well. Compared with quality of seedlings in supplemental lighting of 50 $\mu mol \cdot m^{-2} \cdot s^{-1}$ PPFD, that of seedlings in supplement lighting of 100 or 150 $\mu mol \cdot m^{-2} \cdot s^{-1}$ PPFD improved significantly. With increasing light intensity, diameter, fresh weight, and dry weight, which were used to measure shoot growth, greatly improved. Leaf area, leaf thickness, and root biomass were also greater. However, for quality of seedlings, no significant differences were discovered between supplement lighting of 100 $\mu mol \cdot m^{-2} \cdot s^{-1}$ PPFD and supplement lighting of 150 $\mu mol \cdot m^{-2} \cdot s^{-1}$ PPFD. Expressions of *PsaA* and *PsbA* (two photosynthetic genes) as well as the corresponding proteins increased significantly in supplement lightning of 100 and 150 $\mu mol \cdot m^{-2} \cdot s^{-1}$ PPFD, especially in 100 $\mu mol \cdot m^{-2} \cdot s^{-1}$ PPFD. Overall, considering quality and expressions of two photosynthetic genes and proteins, supplemental light of 100 $\mu mol \cdot m^{-2} \cdot s^{-1}$ PPFD ($W_1R_2B_1$) would be the best choice to cultivate grafted tomato seedlings.

Keywords: LED; PPFD; *PsaA*; *PsbA*; Western Blot

1. Introduction

Light, temperature, humidity, concentration of CO_2, and nutrition supply are vital factors that influence photosynthesis of plants. Light intensity, quality, photoperiod, and direction influence plant growth, development, photomorphogenesis, and anatomical structure as energy sources and regulatory signals [1–3]. Previous studies have proven that light intensity can adjust formation of the chloroplast protein complex, electron transport, and quantum yield between photosystems I (PSI) and II (PSII) [4,5].

Grafting as an efficient technique can be applied to overcome the obstacles of continuous cropping and improve yield, quality, and stress resistance in various plant species including fruit trees, vegetables, and ornamental flowers [6]. In addition to rootstock, affinity of the scion, and grafted methods, the surviving rates, formation of new vascular bundles, and biological properties of grafted seedlings are also related to environmental conditions and management in healing and acclimatization processes [7]. Light-emitting diodes (LEDs) are widely used in facility horticulture as an artificial light source for production of high-quality seedlings. LEDs have shown promising foreground applications because of their advantages of being light-weight, having a small volume, having a specific wavelength, they are easy to integrate, and they have low heat dissipation. Previous studies have shown that the LEDs play a vital role in toughening grafted seedlings in healing and acclimatization processes [8,9].

Generally, stocky plug seedlings are preferred by crop growers for the convenience of transportation and to insure a higher survival ratio and enhanced development after transplanting. Each country has different standards for evaluating the quality of seedlings. In South Korea, it is generally agreed that high-quality seedlings are those without overgrowth, with short and not lodging stems and leaves, and with well-developed roots that hold the medium in the plug tray. Light intensity has been proven to affect plant height and biomass growth [10,11]. Leaf area also changes with light intensity as plants use leaves to absorb light energy [12,13]. Moreover, higher light intensities may result in more root biomass so the absorption of water and mineral elements could be promoted [14]. Therefore, seedlings with good quality could be obtained by supplementing them with optimal light intensity.

This study was carried out aimed at finding an optimal light intensity to improve the quality of two kinds of grafted tomato seedlings. After grafting, well-healed grafted seedlings were moved to and cultivated for an additional 10 days in a glasshouse, which had natural lighting and supplemental lighting. The supplementary light intensities were set at 50, 100, or 150 $\mu mol \cdot m^{-2} \cdot s^{-1}$ photosynthetic photon flux density (PPFD) supplied from mixed LEDs ($W_1R_2B_1$) to further improve the quality of seedlings. After treatment, expressions of two photosynthesis-related genes (*PsaA* and *PsbA*) and the corresponding proteins were analyzed to evaluate the capacity of photosynthesis.

2. Materials and Methods

2.1. Plant Materials

Two cultivars of tomato (*Solanum lycopersicum* L.) 'Super Sunload (SS)' and 'Super Dotaerang (SD)' were selected as the scions and tomato 'B-Blocking' as a common rootstock. Cultivars 'SS' and 'SD', widely used for grafting, were purchased from Sakata Seed Korea Co., Ltd. (Seoul, Korea) and Koregon Co., Ltd. (Anseong, Korea), respectively. The 'SS' has high fruit firmness with an average fruit weight of 220–240 g. The primary advantages of 'SS' are a low ratio of fruit deformity and resistance to high temperatures. The 'SD' is a cultivar with disease resistance and has a high sugar content, firm flesh, and a low deformity ratio. The seeds of these cultivars were sown into 40 square cell plug trays at the same time they were filled with commercial growing medium (Super Mix, NongKyung Co., Jincheon, Korea). The scion and rootstock were cleft-grafted at 20 d after sowing when both scion cultivars had two same sizes of compound leaves. After grafting, the grafted seedlings were immediately transported to a healing chamber, which had a temperature of 23 °C and relative humidity of 95% to 100% throughout the healing period. Five days later, the well-healed grafted seedlings were moved from the healing chamber to the greenhouse for light treatment when the average plant height was about 8 ± 0.7 cm and each seedling had two and a half true leaves.

2.2. Light Treatments

Mixed LEDs ($W_1R_2B_1$) (Custom made, Sung Kwang LED Co., Ltd., Incheon, Korea), which were selected as the best artificial light source in precedent research [15], were used as the sole light source. The grafted seedlings were cultivated for an additional 10 d in a glasshouse with a daily light integral (DLI) of 11.59 $mol \cdot m^{-2} \cdot d^{-1}$ from the sun and supplemental lighting for 16 h/d. The supplementary light intensities were set at 50, 100, or 150 $\mu mol \cdot m^{-2} \cdot s^{-1}$ PPFD (henceforth shortened as *50*, *100*, and *150*). The DLIs of the three supplementary light treatments were 2.88, 5.76, and 8.64 $mol \cdot m^{-2} \cdot d^{-1}$, respectively. The distance between the LEDs and seedlings was 50 cm. Light intensity was measured at the horizontal position at the seedling stem. The culture environment had 30/25 °C day/night temperatures, 70% ± 5% relative humidity (RH), with a natural photoperiod of 14 h. A portable photo/radiometer (Type specification: HD-2102.2, Delta, OHM, Padova, Italy) was used to measure the photon flux density. Different light intensities were controlled by adjustable electric currents.

2.3. Stomatal Conductance

Mature and expanded leaves were selected for the measurement of stomatal conductance. Measurements were carried out during the day (9:00–10:00) on a porometer (Type specification: Sc-1 Porometer, Decagon Devices Inc., Pullman, WA, USA).

2.4. Quantitative Real-Time PCR Analysis

Tomato leaves were ground under RNase-free conditions after being frozen in liquid nitrogen. An Easy-Spin total RNA extraction kit (Type Specification: iNtRON Biotechnology, Seoul, Republic of Korea) and a GoScript Reverse Transcription System (Specification: Promega, Madison, WI, USA) were used for RNA extraction and cDNA synthesis, respectively, following the manufacturer's instructions. The Rotor-Gene Q detection system (Qiagen, Hilden, Germany) was applied in the examination of gene expression. Reaction volumes (20 μL) contained 1 μL cDNA, 0.5 μL primer (10 μM), 10 μL of 2× AMPIGENE qPCR Green Mix Lo-ROX (Enzo life Sciences Inc., Farmingdale, NY, USA), and 8 μL ddH_2O. The $2^{-\Delta\Delta Ct}$ method was applied to determine the relative gene expressions, and the *18S* gene was used as housekeeping gene. All primers used are shown in Supplementary Table S1.

2.5. Protein Extraction and Western Blotting

Protein extraction was conducted following the method described elsewhere [16], with slight modifications. An extraction buffer was used to homogenize approximately 500 mg tomato leaves. The buffer contained 100 mM Tris-HCl (pH 8.0), 0.5 mM $EDTA \cdot Na_2$, 1% (v/v) polyvinylpyrrolidone, 10 mM β-mercaptoethanol, and 200 mM sucrose. At 4 °C, the mixture was centrifuged at 12,000 rpm for 10 min. Subsequent to determination of protein concentrations, the supernatant (10 μL) was mixed together with protein loading buffer and then separated on 12% SDS-PAGE gel. After that, proteins were moved to a polyvinylidene fluoride (PVDF) blotting membrane (Type specification: Millipore, Bedford, MA, USA) using an electrotransfer instrument (Type specification: Bio-Rad, Hercules, CA, USA). Following electrotransfer, the PVDF blotting membrane was transferred into 5% nonfat skimmed milk, dissolved in tris buffered saline tween (TBST) overnight at 4 °C to prevent nonspecific adsorption, washed with TBST, and incubated by using the following polyclonal primary antibodies: 1:3000 dilution of anti-PsaA (Specification: Agrisera AS06 172, Vännäs, Sweden) for PsaA, 1:10,000 dilution of anti-PsbA (Specification: Agrisera AS05 084, Vännäs, Sweden) for PsbA, and 1:10,000 dilution of anti-RbcL (Specification: Agrisera AS03 037, Vännäs, Sweden) for loading control overnight at 4 °C, respectively. After that, PVDF blotting membranes were incubated in 1:10,000 dilution of HRP (horseradish peroxidase)-linked anti-rabbit 1gG (Bethyl A120 101P, Montgomery, TX, USA) for 1 h at ambient temperature as a secondary antibody before visualization of immunoreactive proteins using Clarity™ Western ECL Substrate (#6883, Bio-Rad, Hercules, CA, USA) on an iBright™ Imaging Systems (Type specification: Thermo Fisher Scientific, Waltham, MA, USA).

2.6. Data Collection and Analysis

After supplemental lighting treatments for 10 d, this study measured the following parameters: lengths of scion and root, diameters of scions, number of leaves, leaf length, leaf width, leaf area and thickness, chlorophyll (SPAD), fresh weight of root, dry weight of root, specific weight of leaf, and dry weight to height ratio of scion.

The experiment was carried out in three replicates with a design of randomized complete block, using 9 seedlings in each replication. Locations of replications were randomly mixed to eliminate position effect within a controlled environment. One-way analysis of variance ANOVA by SAS program Statistical Analysis System (V.9.1, Cary, NC, USA), was applied to statistically analyze the collected data. Duncan's multiple range test was adopted to test significant differences between different means.

3. Results

3.1. Growth and Development of Tomato Seedlings as Affected by Supplemental Lighting Treatment in the Glasshouse

After cultivation in a glasshouse for 10 d, growth and development parameters of the grafted tomato seedlings were measured. Results of the measurements showed that the characteristics of the seedlings' morphologies were significantly influenced by light intensities. The two cultivars of grafted tomato cultivated in *100* and those in *150* showed the greatest scion length and root fresh weight. No significant differences were discovered in scion stem diameter and leaf length between supplementary lighting of *50*, *100*, or *150*. Leaf area and width, number of leaves, and root length of 'SS' were significantly greater in supplementary lighting of *100* and *150* than in supplementary lighting *50*. The greatest root length was observed in light intensity of *150* in 'SD'. In general, there were little differences between supplementary lighting of *100* and *150*. Compared with supplementary lighting of *50*, both of these two treatments significantly improved quality of the two types of grafted tomato seedlings (Table 1).

Dry weight of shoot and dry weight of root are shown in Figure 1. Considering these two weights, there were no significant differences between light intensity of *100* and light intensity of *150*. However, compared to the treatment of light intensity of *50*, these two treatments significantly improved dry weight of the two types of tomato cultivars.

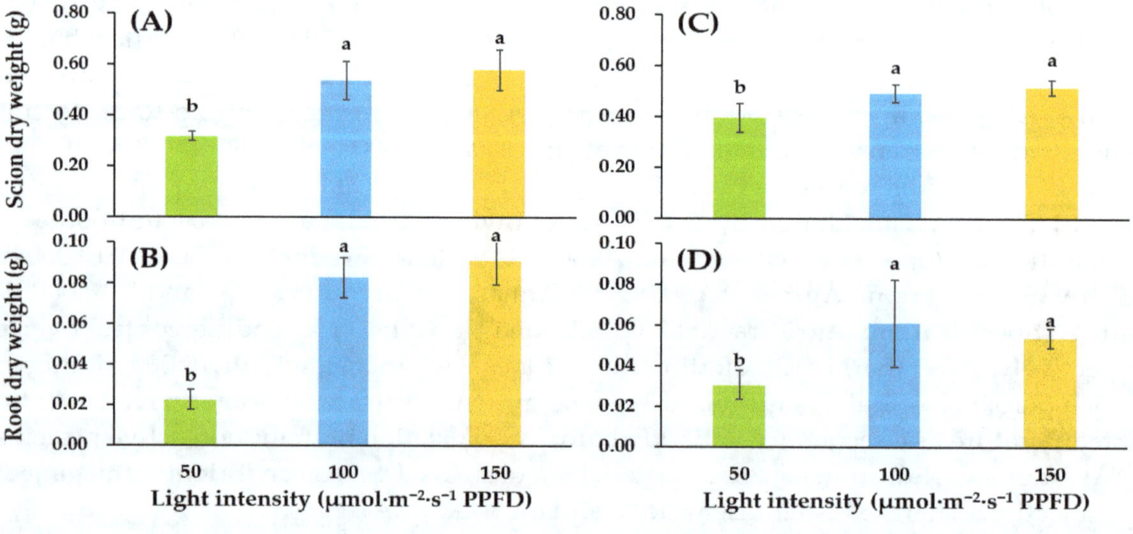

Figure 1. Influence of supplementary lighting intensity on dry weight of shoots and dry weight of roots of two cultivars of tomato 'Super Sunload' ('SS') (**A**,**B**) and 'Super Dotaerang' ('SD') (**C**,**D**) grafted onto the 'B-Blocking' rootstock. Error bars represent SEs of nine biological replicates ($n = 9$). According to the Duncan test, lowercase letters at $p \leq 0.05$ indicate significant differences between different treatments.

Table 1. Influence of supplemental light intensity on growth and development of grafted tomato seedlings grown for 10 d.

Cultivar (C)	Light Intensity (LI) ($\mu mol \cdot m^{-2} \cdot s^{-1}$ PPFD)	Scion		Leaf						Root	
		Length (cm)	Stem Diameter(mm)	No.	Length(cm)	Width (cm)	Area (cm^2)	Thickness (mm)	Chlorophyll (SPAD)	Length (cm)	Fresh Weight (g)
'SS'	50	13.2 b [z]	5.13 a	7.8 b	14.3	8.7 b	61.9 b	0.35 b	44.1 b	8.6 b	0.4 b
	100	19.1 a	5.31 a	8.8 a	15.0	9.0 ab	73.6 ab	0.39 b	47.6 ab	17.2 a	1.1 ab
	150	20.2 a	5.69 a	9.3 a	15.8	9.9 a	84.5 a	0.47 a	50.3 a	18.6 a	1.9 a
'SD'	50	16.3 b	5.52 a	8.7 a	15.3	8.8 a	70.0 a	0.36 a	39.5 c	9.9 c	0.4 b
	100	20.1 a	5.63 a	9.2 a	15.7	9.3 a	69.9 a	0.39 a	48.2 a	14.6 b	1.0 a
	150	19.5 a	5.96 a	9.8 a	16.0	8.7 a	68.1 a	0.39 a	42.6 ab	16.9 a	1.0 a
F-test											
LI		*** [y]	*	***	NS	*	*	**	*	***	***
C		NS	*	*	NS	NS	NS	NS	NS	NS	NS
LI * C		NS	NS	NS	NS	NS	**	NS	NS	NS	NS

[z] Means followed by same letter (s) within a column are not significantly different ($p \leq 0.05$). [y] NS, not significant. *, **, and ***, significant at $p = 0.05$, 0.01, or 0.001, respectively.

Growing compact and sturdy seedlings is an objective of high-quality seedling production. Dry weight to height ratio of scion (WHR) was conducted to indicate the quality of seedlings. For 'SS', supplementary lighting of *100* and *150* significantly improved WHR as compared to supplementary lighting of *50*. There were no significant differences in WHRs between supplementary lighting of *100* and *150*. For 'SD', the greatest WHRs showed in supplementary light of *150*. For both tomato cultivars, the specific leaf weight (a leaf fresh weight per unit leaf area) was significantly improved in light intensities of *100* and *150* as compared to light intensity of *50*. There were no significant differences in the specific leaf weight between light intensity of *100* and 150 (Figure 2).

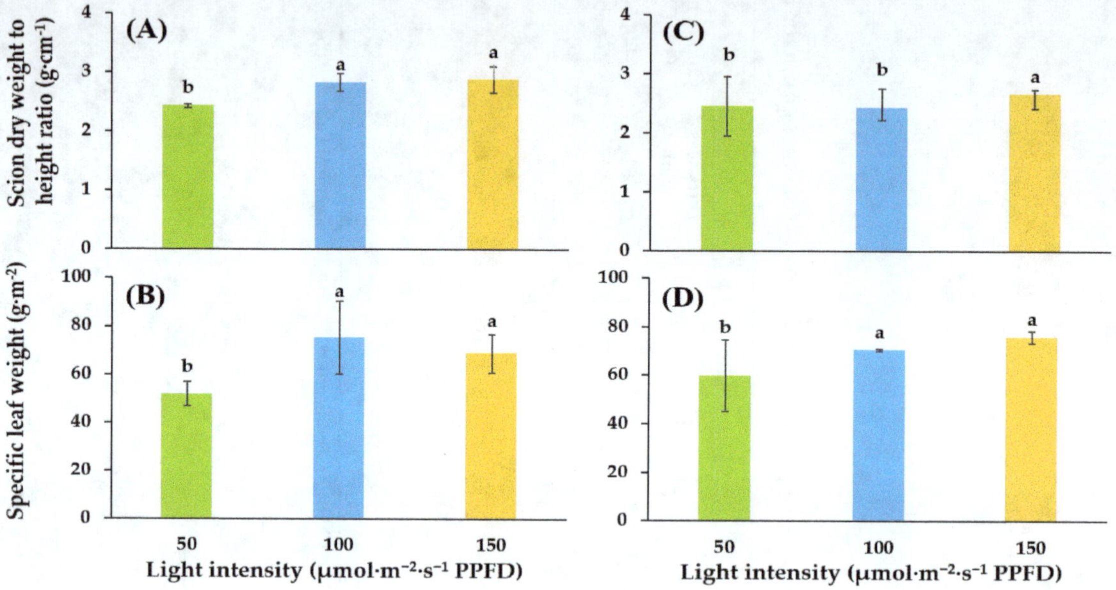

Figure 2. Influence of supplementary lighting intensity on dry weight to height ratio of scion and specific leaf weight of two cultivars of tomato 'SS' (**A**,**B**) and 'SD' (**C**,**D**) grafted onto 'B-Blocking' rootstock. The error bars represent SEs of biological replicates ($n = 9$). According to the Duncan test, lowercase letters at $p \leq 0.05$ indicate significant differences between different treatments.

Moreover, we determined the effect of lighting intensity on stomatal conductance of the two tomato cultivars. Results showed, compared with light intensity of *50*, light intensity of *100* increased stomatal conductance of the two cultivars. Interestingly, stomatal conductance in light intensity of *150* was similar to that in light intensity of *50* treatment (Figure 3).

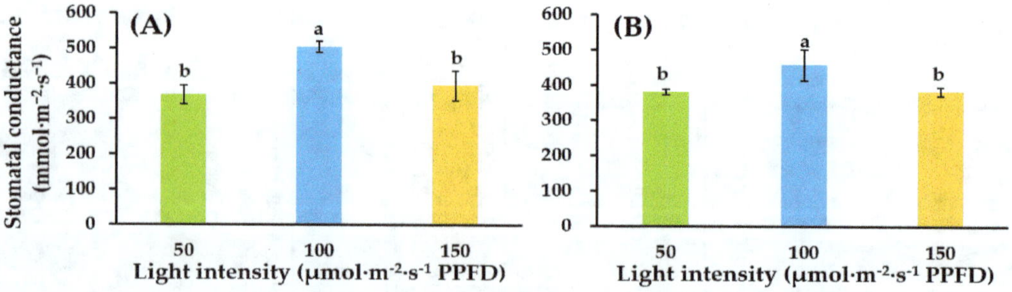

Figure 3. Influence of supplementary lighting intensity on stomatal conductance of two cultivars of tomato 'SS' (**A**) and 'SD' (**B**) grafted onto the 'B-Blocking' rootstock. The error bars represent SEs of three biological replicates ($n = 3$). According to the Duncan test, lowercase letters at $p \leq 0.05$ indicate significant differences between treatments.

Figure 4 illustrated seedling morphologies of 'SS' and 'SD' after treating for 10 d by using supplementary lighting with a different light intensity. It showed that tomato seedlings grown in supplementary lighting of *100* and *150* had more biomass and were sturdier.

Figure 4. Influence of supplementary light intensity on morphology of 'SS' (**A**) and 'SD' (**B**) after 10 d of cultivation.

3.2. Expressions of Photosynthesis-Related Genes

Light intensity significantly influenced *PsaA* and *PsbA* genes (Figure 5). Light intensity of *100* enhanced expressions of *PsaA* and *PsbA*. Relative expressions of *PsaA* in 'SS' and 'SD' were 1.37- and 2.27-fold, while that of *PsbA* was 4.01- and 10.29-fold, as compared to that in the light intensity of *50*. However, a dramatic decrease in expression was observed in light intensity of *150* in 'SS' and 'SD'. Particularly, the relative expression of *PsaA* in 'SS' was only 0.44-fold compared to the expression level in *50*.

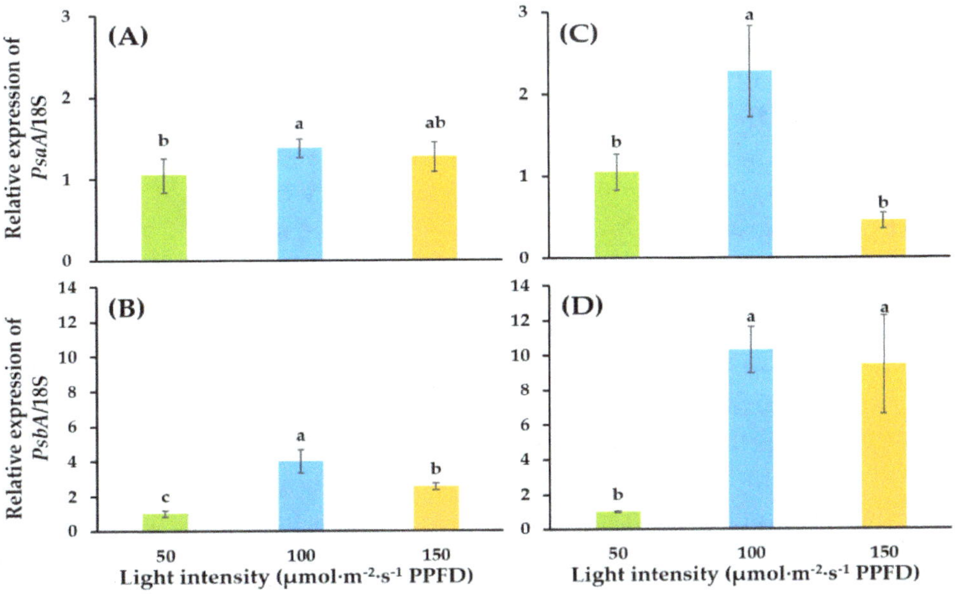

Figure 5. Relative expressions of *PsaA* and *PsbA* genes in 'SS' (**A**,**B**) and 'SD' (**C**,**D**) grafted onto the 'B-Blocking' rootstock. The bars among treatments represent the mean ± SEs of technological replicates ($n = 3$). According to the Duncan test, lowercase letters at $p \leq 0.05$ indicate significant differences between different treatments.

3.3. PsaA and PsbA Immunoblots

The *PsaA* gene is located in chloroplast genome (cpDNA). It encodes a core protein of a protein-pigment complex in PSI [17] and encodes D1 protein, which is a core component of PSII [18]. Figure 6 shows the protein expressions of PsaA and PsbA in 'SD' and 'SS' cultivated in supplementary lighting with different light intensities (*50*, *100*, or *150*). For both tomato cultivars, expression levels of PsaA and PsbA in light intensities of *100* and *150* increased as compared to those in light intensity of *50*. The highest expression levels of the PsaA and PsbA proteins appeared in light intensity of *100*.

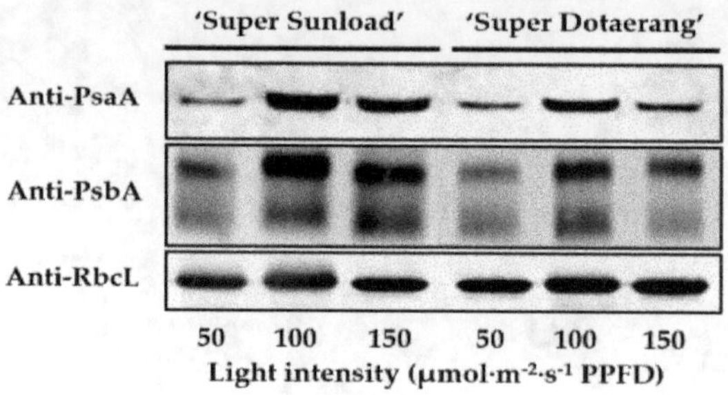

Figure 6. Expressions of PsaA and PsbA protein in 'SD' and 'SS' grafted onto the 'B-Blocking' rootstock. Equal loading was verified by the Rubisco large subunit (RbcL) level.

4. Discussion

Among regulatory factors for light condition, light intensity as an essential element is involved in regulation of growth, yield, and nutritional quality of vegetables [19]. Previous studies have proven that increased light intensity within a certain range may promote growth in lettuce [20,21]. Light condition is the foundation and guarantees healthy growth and yield of tomatoes [22]. Recent studies have shown that tomato seedlings under low-light conditions showed elongated internodes and petioles as well as a larger leaf area, which were caused by a shade-avoidance response [23,24]. Low-light conditions also affect the differentiation time and quality of flower buds in tomatoes [25]. For tomato cultivars 'Matador' and 'Blizzard', plant height, leaf width, number of leaves, and dry weight of aerial plant parts significantly improved due to increased levels of supplementary light (30, 60, and 90 μmol·m^{-2}·s^{-1} PPFD) [26]. However, these findings do not mean the higher light intensities are always better. Tomato seedlings treated with 150 μmol·m^{-2}·s^{-1} PPFD using white and blue LEDs showed continuous light-induced injuries, and the degree of the injury was more serious under white and blue LEDs than under orange and red LEDs of the same intensity [27]. This study showed that, compared with stem diameter of grafted tomato seedlings in supplementary lighting of *50*, that of grafted tomato seedlings increased in supplementary lighting of *100* and *150*. Leaf width and leaf thickness were also greater with treatments of light intensity of *100* and *150* (Table 1). Higher over-ground biomass and activeness of antioxidant enzymes were discovered in treatments with increased light intensity (100, 200, and 400 μmol·m^{-2}·s^{-1} PPFD) in lettuce [28]. For *Nicotiana tabacum*, seedling biomass, soluble sugar concentrations, and primary root growth increased dramatically with light intensity of 300 μmol·m^{-2}·s^{-1} PPFD as compared to 60 μmol·m^{-2}·s^{-1} PPFD [14]. It had been proven that low light intensity decreased above- and below- ground biomass in radish [29]. The results also showed, compared with light intensity of *100* and *150*, dry weight of scion and dry weight of root were much lower in light intensity of *50* (Figure 1). Additionally, light intensity influences the development of stomata as well. According to the study of Lee [30], the stomatal conductance and the number of stomata per unit area of leaf increased with light intensity, increasing up to 60 μmol·m^{-2}·s^{-1} PPFD in *Withania somnifera*. However, width and length of stomata declined in light intensity of 90 μmol·m^{-2}·s^{-1} PPFD. These findings were consistent with our results in Figure 3. The *100* promoted stomatal conductance of both tomato cultivars, while *150* affected it adversely.

The aim of supplementary lighting in greenhouses are as follows: to increase the light intensity in greenhouse and enable crops to receive proper light, to provide crops with light with specific wavelengths, to alter the time of flowering and fruiting, and to increase time for photosynthesis via changing the lighted period [31,32]. Advantages of light-emitting diodes (LEDs) include their small volume, longer service life, energy savings, and use of a single wavelength, which are efficient to encourage plant growth by producing the type of light with the specific wavelength needed by plants [33]. Studies have shown that red and blue lights absorbed by plants account for 90% of total light absorption [34], which are involved in regulating gene expression in photomorphogenesis [35]. Our previous study showed that compared with high pressure sodium (HPS), far-red (FR), metal halide (MH), and white LEDs (W), mixed LEDs ($W_1R_2B_1$) as a supplementary light source had the best effect to improve the quality of grafted tomato seedlings [15]. The application of monochromatic light may cause physiological defects of plants such as disturbance of photosynthetic mechanisms [36], damage to the granule and thylakoid [37], and restrained synthesis of chlorophyll [38], which can be overcome or relieved by the mixture of red and blue light.

Photosynthesis is one of the supreme chemical reactions in plants. Photoreaction takes place in the chloroplast thylakoid membrane. Two light energy-driven systems, PSI and PSII, synergistically function in primary energy conversion reactions. Driven by light, plants and other photoautotrophs are able to synthesize carbon compounds directly from CO_2 and H_2O and release O_2. Processes of consumption and formation in photosynthesis are related to photosynthetic phosphorylation coupled with electron transport. Electrons generated from splitting of H_2O by the oxygen-evolving complex (OEC) are used to reduce $NADP^+$ to NADPH. At the same time, synthesis of ATP is driven by a transmembrane electrochemical proton gradient [39,40]. The PSII, a multi-subunit protein-pigment complex, is associated with the photolysis of water and synthesis of ATP [41]. Within the core of the complex, PsbC and PsbB are bond to the photosynthetic pigments and transfer the excitation energy to reaction center proteins D1 (Qb, PsbA) and D2 (Qa, PsbD) [42]. D1 dimerize with D2 as a heterodimer. Studies have suggested that synthesis of D1 is coordinated with linear photosynthetic electron transport as well as the relative activities of PSI and PSII [43]. Light has an important role in the translation of Dl protein [44]. The PSI complex contains reaction center P700, as well as five electron transfer centers (A0, A1, FX, FA, and FB) bound to PsaA and PsaB proteins [45]; it mediates electron transfer and is involved in light-driven conversion of $NADP^+$ to NADPH [46]. The two largest polypeptides, PsaA and PsaB, carrying P700 are believed to form the center Fx through dimerization [47]. In higher plants, the genes *PsaA* and *PsaB* encode the two polypeptides respectively and are located adjacent to each other in the plastid genome [48].

A previous study proved that expression of the *PsaA* gene product during rice plastid development was light-regulated on the translational or post-translational level [49]. In *Synechococcus*, blue light photoreceptor regulates expressions of *PsbA* genes [50]. Moreover, in *Synechocystis* the transcription levels of promoters of *Psba* gene changed in response to dark and light conditions [51].

Expression levels of *PsaA* and *PsbA* genes in light intensities of *100* and *150* increased as compared to those in light intensity of *50* in both cultivars. Compared with those in light intensity of *100*, lower expression levels of genes appeared in light intensity of *150* (Figure 5). Western blotting showed the same tendency of variation in protein expression as shown in Figure 6. Previous studies pointed that high light intensities caused degradation and phosphorylation of PSII core protein and damaged the photosynthetic machinery [52]. It could be speculated that the photosynthetic apparatuses suffered slight damage in light intensity of *150*.

Although light is the basic energy for photosynthesis, light that is too strong will cause light stress for plants, which has side effects for photosynthesis, especially under low temperatures, drought, or other unfavorable conditions [53]. In 1956, Kok firstly proposed the concept of photoinhibition and defined it as 'the photochemical inactivation of complete pigment complexes or photosynthetic units' [54]. Powles (1984) pointed that excessive light energy absorbed by the photosynthetic apparatus caused photoinhibition with the characteristic of reduced photosynthetic capacity [55]. PSII is very sensitive to strong light, and it is the main occurrence site of photoinhibition [40]. There are two mechanisms of photoinhibition in PSII, acceptor-side photoinhibition and donor-side photoinhibition [56]. In the former situation, the hindered assimilation of CO_2 will cause the reduction of the plastoquinone pool. Consequently, the double-reduced form of QA^- accumulates and promotes the formation of the triplet state of the primary donor ($^3P680$). The $^1O_2$ generated from the reaction between $^3P680$ and O_2 will cause damage to proteins and pigments, especially the amino acid residue sites in the D1 protein. In another mechanism, electron transport is impaired due to the hindered oxidation of water and prolonged half-life period of $P680^+$ resulting in high oxidation potential on the donor side. Adjacent proteins and pigments are damaged as a consequence [57].

In 1994, Terashima firstly reported photoinhibition in PSI [58]. High-light intensity [59,60], low temperatures [61], and fluctuating light conditions [62,63] will cause PSI photoinhibition. Studies have shown that changes of reducing state and production of ROS (reactive oxygen species) and hydroxyl peroxide on the acceptor side are the main causes of PSI photoinhibition [64,65].

Plants have developed various mechanisms to prevent photoinhibition under high-light intensities because of their long-term evolution and development. The main mechanisms of photoprotection are as follows: to reduce the light absorption via movements of leaf and chloroplast [66]; to increase photosynthetic efficiency via increased levels of the photosynthetic electron transport carriers and contents of the photosynthetic key enzymes [67,68]; to protect the photosynthetic apparatus by photochemical and non-photochemical pathway such as the xanthophyll cycle, photorespiration, and the Mehler reaction [69–71]; to strengthen the active oxygen eliminating enzyme system [72,73]; and to distribute excitation energy and heat dissipation equally via state transition between PSII and PSI [74,75].

Usually, sun and shade plants possess different morphological, physiological, and biochemical characteristics to adapt to different light conditions. Chloroplasts with distinctive characteristics will contribute to levels of photosynthesis. As compared to sun plants, shade plants have higher thylakoid stack areas, fewer grana stacks, and smaller widths of grana stacks to inhibit degradation of the D1 protein [76]. The levels of photosynthetic electron transport carriers and activities of photosynthesis-related enzymes, especially Rubisco, are higher in sun plants [77,78]. When light intensity is changed, the activities of RuBP and Rubisco do respond to the fluctuating environment [79]. Studies in mulberry showed that the leaves, which were at the daily maximum photosynthetic state, showed higher reversibility of photosynthetic quantum conversion, confronting the environment with changing light intensity [80]. After being transferred to an intercropping shaded environment, the lodging-resistant soybean genotypes showed elongated main stems and internodes as well as decreased biomass to the length of the stem ratio. The reasons may be due to downregulated gene expressions related to biosynthesis of structural carbohydrates [81]. Additionally, shade conditions influenced the photosynthetic rate by altering the chlorophyll fluorescence performance in soybean [82] and *Athyrium pachyphlebium* [83]. Shade treatment changed concentrations of chlorophyll and chlorophyll a/b ratios in soybeans [84] and wheat [85]. These studies suggested that light capture and utilization, PSII activity, and the transfer of electron and energy between PSII and PSI may involve the regulation of photosynthetic capacity in shade and light conditions. Phenotypic plasticity of plants refers to the ability to conform their morphology and physiology to the environment [86]. Even though studies have shown that the phenotypic plasticity of some shade plant species is lower [87,88], the mechanism of optimizing light energy capture should be lucubrated. To illustrate that, the mechanism of photosynthesis in natural growth condition with fluctuating light can help plants to adapt to sun and shade conditions and increase yields as a consequence [86]. Additionally, more attention should be paid to the relationship between genotype plasticity and productivity as well as the intraspecific variation in plasticity to fluctuating light [89].

Light required by plants is related to different species, cultivars, growth stages, developmental stages, and environmental factors as well as target of production [19]. Although light has a vital function in photosynthesis, excess light may depress photosynthetic efficiency, which is referred to as photoinhibition [55]. Therefore, to maximize economic benefits of grafted seedlings with high quality and quantity, detailed studies on optimizing light intensity are in urgent need.

5. Conclusions

When moved to and cultivated in a glasshouse having supplementary lighting, light intensities of *100* and *150* had significantly improved qualities of grafted tomato seedlings as compared to light intensity of *50*. However, considering improvement of quality of grafted seedlings, there were no significant differences between light intensities of *100* and *150*. Owing to enhanced photosynthesis, the upregulated gene expression levels of *PsaA* and *PsbA* and the corresponding protein expression levels indicated that light intensity of *100* might be the most suitable supplementary lighting to improve the quality of grafted tomato seedlings. That is, supplementary lighting with a light intensity of *100* would be the best choice to improve the quality of grafted vegetable seedlings when both power consumption and economic benefits are taken into account. This study provides new thoughts on supplemental lighting strategies that might be applied in greenhouses and lays the foundation for further studies on the utilization of light energy during early development stages of tomato seedlings.

Author Contributions: Conceptualization, B.R.J.; Methodology, B.R.J. and H.W.; Formal Analysis, H.W., J.Z., and J.H.; Resources, B.R.J.; Data Curation, H.W. and J.Z.; Writing—Original Draft Preparation, H.W.; Writing—Review & Editing, B.R.J.; Project Administration, B.R.J.; Funding Acquisition, B.R.J.

Acknowledgments: This study was carried out with support from the Korea Institute of Planning and Evaluation for Technology in Food, Agriculture, and Forestry (Project No.319008-01). H.W., J.Z., and J.H. were supported by a scholarship from the BK21 Plus Program, Ministry of Education, Republic of Korea.

References

1. Muneer, S.; Kim, E.J.; Park, J.S.; Lee, J.H. Influence of green, red and blue light emitting diodes on multiprotein complex proteins and photosynthetic activity under different light intensities in lettuce leaves (*Lactuca sativa* L.). *Int. J. Mol. Sci.* **2014**, *15*, 4657–4670. [CrossRef] [PubMed]
2. Berta, M.; Giovannelli, A.; Sebastiani, F.; Camussi, A.; Racchi, M.L. Transcriptome changes in the cambial region of poplar (*Populus alba* L.) in response to water deficit. *Plant Biol.* **2010**, *12*, 341–354. [CrossRef] [PubMed]
3. Baba, K.; Karlberg, A.; Schmidt, J.; Schrader, J.; Hvidsten, T.R.; Bako, L.; Bhalerao, R.P. Activity-dormancy transition in the cambial meristem involves stage-specific modulation of auxin response in hybrid aspen. *Proc. Natl. Acad. Sci. USA* **2011**, *108*, 3418–3423. [CrossRef] [PubMed]
4. Sorce, C.; Giovannelli, A.; Sebastiani, L.; Anfodillo, T. Hormonal signals involved in the regulation of cambial activity, xylogenesis and vessel patterning in trees. *Plant Cell Rep.* **2013**, *32*, 885–898. [CrossRef] [PubMed]
5. Lee, J.M.; Kubota, C.; Tsao, S.; Bie, Z.; Echevarria, P.H.; Morra, L.; Oda, M. Current status of vegetable grafting: Diffusion, grafting techniques, automation. *Sci. Hortic.* **2010**, *127*, 93–105. [CrossRef]
6. Vu, N.T.; Kim, Y.S.; Kang, H.M.; Kim, I.S. Effect of red LEDs during healing and acclimatization process on the survival rate and quality of grafted tomato seedlings. *Prot. Hortic. Plant Fact.* **2014**, *23*, 43–49. [CrossRef]
7. Yorio, N.C.; Goins, G.D.; Kagie, H.R.; Wheeler, R.M.; Sager, J.C. Improving spinach, radish, and lettuce growth under red light-emitting diodes (LEDs) with blue light supplementation. *HortScience* **2001**, *36*, 380–383. [CrossRef]
8. Yano, A.; Fujiwara, K. Plant lighting system with five wavelength-band light-emitting diodes providing photon flux density and mixing ratio control. *Plant Methods* **2012**, *8*, 46. [CrossRef]
9. Kim, K.; Kook, H.S.; Jang, Y.J.; Lee, W.H.; Kamala-Kannan, S.; Chae, J.C.; Lee, K.J. The effect of blue-light-emitting diodes on antioxidant properties and resistance to *Botrytis cinerea* in tomato. *J. Plant Pathol. Microbiol.* **2013**, *4*, 203. [CrossRef]
10. Beon, M.S.; Bartsch, N. Early seedling growth of pine (*Pinus densiflora*) and oaks (*Quercus serrata, Q. mongolica, Q. variabilis*) in response to light intensity and soil moisture. *Plant Ecol.* **2003**, *167*, 97–105. [CrossRef]
11. Du, Y.X.; Xin, J.; Zhang, J.; Li, J.Z.; Sun, H.Z.; Zhao, Q.Z. Research progress on the impacts of low light intensity on rice growth and development. *Chin. J. Eco-Agric.* **2013**, *11*, 1307–1317. (In Chinese) [CrossRef]
12. Perrin, P.M.; Mitchell, F.J. Effects of shade on growth, biomass allocation and leaf morphology in European yew (*Taxus baccata* L.). *Eur. J. For. Res.* **2013**, *132*, 211–218. [CrossRef]
13. Shao, Q.; Wang, H.; Guo, H.; Zhou, A.; Huang, Y.; Sun, Y.; Li, M. Effects of shade treatments on photosynthetic characteristics, chloroplast ultrastructure, and physiology of *Anoectochilus roxburghii*. *PLoS ONE* **2014**, *9*, e85996. [CrossRef] [PubMed]

14. Nagel, K.A.; Schurr, U.; Walter, A. Dynamics of root growth stimulation in *Nicotiana tabacum* in increasing light intensity. *Plant Cell Environ.* **2006**, *29*, 1936–1945. [CrossRef] [PubMed]
15. Wei, H.; Hu, J.T.; Liu, C.; Wang, M.Z.; Zhao, J.; Kang, D.I.; Jeong, B.R. Effect of supplementary light source on quality of grafted tomato seedlings and expression of two photosynthetic genes. *Agronomy* **2018**, *8*, 207. [CrossRef]
16. Li, Y.Y.; Mao, K.; Zhao, C.; Zhao, X.Y.; Zhang, H.L.; Shu, H.R.; Hao, Y.J. MdCOP1 ubiquitin E3 ligases interact with MdMYB1 to regulate light-induced anthocyanin biosynthesis and red fruit coloration in apple. *Plant Physiol.* **2012**, *160*, 1011–1022. [CrossRef]
17. Choquet, Y.; Goldschmidt-Clermont, M.; Girard-Bascou, J.; Kück, U.; Bennoun, P.; Rochaix, J.D. Mutant phenotypes support a trans-splicing mechanism for the expression of the tripartite psaA gene in the C. reinhardtii chloroplast. *Cell* **1988**, *52*, 903–913. [CrossRef]
18. Singh, M. Turnover of D1 protein encoded by *PsbA* gene in higher plants and cyanobacteria sustains photosynthetic efficiency to maintain plant productivity under photoinhibitory irradiance. *Photosynthetica* **2000**, *38*, 161–169. [CrossRef]
19. Kang, J.H.; Krishna, K.S.; Atulba, S.L.S.; Jeong, B.R.; Hwang, S.J. Light intensity and photoperiod influence the growth and development of hydroponically grown leaf lettuce in a closed-type plant factory system. *Hortic. Environ. Biotechnol.* **2013**, *54*, 501–509. [CrossRef]
20. Hunter, D.C.; Burritt, D.J. Light quality influences adventitious shoot production from cotyledon explants of lettuce (*Lactuca sativa* L.). *In Vitro Cell Dev. Biol.* **2004**, *40*, 215–220. [CrossRef]
21. Li, Q.; Kubota, C. Effects of supplemental light quality on growth and phytochemicals of baby leaf lettuce. *Environ. Exp. Bot.* **2009**, *67*, 59–64. [CrossRef]
22. Franklin, K.A.; Larner, V.S.; Whitelam, G.C. The signal transducing photoreceptors of plants. *ITLE Int. J. Dev. Biol.* **2005**, *49*, 653–664. [CrossRef] [PubMed]
23. Fan, X.X.; Xu, Z.G.; Liu, X.Y.; Tang, C.M.; Wang, L.W.; Han, X.L. Effects of light intensity on the growth and leaf development of young tomato plants grown under a combination of red and blue light. *Sci. Hortic.* **2013**, *153*, 50–55. [CrossRef]
24. Chitwood, D.H.; Kumar, R.; Ranjan, A.; Pelletier, J.M.; Townsley, B.T.; Ichihashi, Y.; Martinez, C.C.; Zumstein, K.; Harada, J.J.; Maloof, J.N.; et al. Light-Induced Indeterminacy Alters Shade-Avoiding Tomato Leaf Morphology. *Plant Physiol.* **2015**, *169*, 2030–2047. [CrossRef] [PubMed]
25. Samach, A.; Lotan, H. The transition to flowering in tomato. *Plant Biotechnol.* **2007**, *24*, 71–82. [CrossRef]
26. McCall, D. Effect of supplementary light on tomato transplant growth, and the after-effects on yield. *Sci. Hortic.* **1992**, *51*, 65–70. [CrossRef]
27. Matsuda, R.; Yamano, T.; Murakami, K.; Fujiwara, K. Effects of spectral distribution and photosynthetic photon flux density for overnight LED light irradiation on tomato seedling growth and leaf injury. *Sci. Hortic.* **2016**, *198*, 363–369. [CrossRef]
28. Fu, W.G.; Li, P.P.; Wu, Y.Y.; Tang, J. Effects of different light intensities on anti-oxidative enzyme activity, quality and biomass in lettuce. *Sci. Hortic.* **2012**, *39*, 129–134.
29. Hole, C.C.; Dearman, J. The effect of photon flux density on distribution of assimilate between shoot and storage root of carrot, red beet and radish. *Sci. Hortic.* **1993**, *55*, 213–225. [CrossRef]
30. Lee, S.H.; Tewari, R.K.; Hahn, E.J.; Paek, K.Y. Photon flux density and light quality induce changes in growth, stomatal development, photosynthesis and transpiration of *Withania somnifera* (L.) Dunal. plantlets. *Plant Cell Tissue Organ Cult.* **2007**, *90*, 141–151. [CrossRef]
31. Seginer, I.; Albright, L.D.; Ioslovich, I. Improved strategy for a constant daily light integral in greenhouses. *Biosyst. Eng.* **2006**, *93*, 69–80. [CrossRef]
32. Warner, R.M.; Erwin, J.E. Effect of photoperiod and daily light integral on flowering of five *Hibiscus* sp. *Sci. Hortic.* **2003**, *97*, 341–351. [CrossRef]
33. Brown, C.S.; Schuerger, A.C.; Sager, J.C. Growth and photomorphogenesis of pepper plants under red light-emitting diodes with supplemental blue or far-red lighting. *J. Am. Soc. Hortic. Sci.* **1995**, *120*, 808–813. [CrossRef] [PubMed]
34. Terashima, I.; Fujita, T.; Inoue, T.; Chow, W.S.; Oguchi, R. Green light drivesleaf photosynthesis more efficiently than red light in strong white light: Revisiting the enigmatic question of why leaves are green. *Plant Cell Physiol.* **2009**, *50*, 684–697. [CrossRef] [PubMed]

35. Barnes, C.; Bugbee, B. Morphological responses of wheat to changes in phytochrome photoequilibrium. *Plant Physiol.* **1991**, *97*, 359–365. [CrossRef] [PubMed]
36. Hogewoning, S.W.; Douwstra, P.; Trouwborst, G.; van Ieperen, W.; Harbinson, J. An artificial solar spectrum substantially alters plant development compared with usual climate room irradiance spectra. *Exp. Bot.* **2010**, *61*, 1267–1276. [CrossRef] [PubMed]
37. Wang, X.Y.; Xu, X.M.; Cui, J. The importance of blue light for leaf area expansion, development of photosynthetic apparatus, and chloroplast ultrastructure of *Cucumis sativus* grown under weak light. *Photosynthetica* **2015**, *53*, 213–222. [CrossRef]
38. Bach, A.; Krol, A. Effect of light quality on somatic embryogenesis in *Hyacinthus orientalis* L. 'Delft's Blue'. *Biol. Bull. Pozn.* **2001**, *38*, 103–107.
39. Lander, G.C.; Estrin, E.; Matyskiela, M.E.; Bashore, C.; Nogales, E.; Martin, A. Complete subunit architecture of the proteasome regulatory particle. *Nature* **2012**, *482*, 186–191. [CrossRef]
40. Huang, W.; Zhang, S.B.; Liu, T. Moderate photoinhibition of photosystem II significantly affects linear electron flow in the shade-demanding plant *Panax notoginseng*. *Front. Plant Sci.* **2018**, *9*, 250–256. [CrossRef]
41. Raymond, J.; Blankenship, R.E. The evolutionary development of the protein complement of photosystem II. *Biochim. Biophys. Acta Bioenerg.* **2004**, *1655*, 133–139. [CrossRef] [PubMed]
42. Grotjohann, I.; Jolley, C.; Fromme, P. Evolution of photosynthesis and oxygen evolution: Implications from the structural comparison of Photosystems I and II. *Phys. Chem. Chem. Phys.* **2004**, *6*, 4743–4753. [CrossRef]
43. Trebitsh, T.; Danon, A. Translation of chloroplast *psbA* mRNA is regulated by signals initiated by both photosystems II and I. *Proc. Natl. Acad. Sci. USA* **2001**, *98*, 12289–12294. [CrossRef] [PubMed]
44. Gruissem, W. Chloroplast gene expression: How plants turn their plastids on. *Cell* **1989**, *56*, 161–170. [CrossRef]
45. Chitnis, P.R.; Xu, Q.; Chitnis, V.P.; Nechushtai, R. Function and organization of photosystem I polypeptides. *Photosynth. Res.* **1995**, *44*, 23–40. [CrossRef] [PubMed]
46. Altman, A.; Cohen, B.N.; Weissbach, H.; Brot, N. Transcriptional activity of isolated maize chloroplasts. *Arch. Biochem. Biophys.* **1984**, *235*, 26–33. [CrossRef]
47. Webber, A.N.; Malkin, R. Photosystem I reaction-centre proteins contain leucine zipper motifs: A proposed role in dimer formation. *FEBS Lett.* **1990**, *264*, 1–4. [CrossRef]
48. Hiratsuka, J.; Shimada, H.; Whittier, R.; Ishibashi, T.; Sakamoto, M.; Mori, M.; Kondo, C.; Honji, Y.; Sun, C.R.; Meng, B.Y.; et al. The complete sequence of the rice (*Oryza sativa*) chloroplast genome: Intermolecular recombination between distinct tRNA genes accounts for a major plastid DNA inversion during the evolution of the cereals. *Mol. Gen. Genet.* **1989**, *217*, 185–194. [CrossRef]
49. Chen, S.G.; Cheng, M.C.; Chung, K.R.; Yu, N.J.; Chen, M.C. Expression of the rice chloroplast *psaA-psaB-rps14* gene cluster. *Plant Sci.* **1992**, *81*, 93–102. [CrossRef]
50. Tsinoremas, N.F.; Schaefer, M.R.; Golden, S.S. Blue and red light reversibly control *psbA* expression in the cyanobacterium *Synechococcus* sp. strain PCC 7942. *J. Biol. Chem.* **1994**, *269*, 16143–16147.
51. Albers, S.C.; Peebles, C.A. Evaluating light-induced promoters for the control of heterologous gene expression in *Synechocystis sp.* PCC 6803. *Biotechnol. Prog.* **2017**, *33*, 45–53. [CrossRef] [PubMed]
52. Chen, L.; Jia, H.; Tian, Q.; Du, L.; Gao, Y.; Miao, X.; Liu, Y. Protecting effect of phosphorylation on oxidative damage of D1 protein by down-regulating the production of superoxide anion in photosystem II membranes under high light. *Photosynth. Res.* **2012**, *112*, 141–148. [CrossRef] [PubMed]
53. Antal, T.K.; Kovalenko, I.B.; Rubin, A.B.; Tyystjärvi, E. Photosynthesis-related quantities for education and modeling. *Photosynth. Res.* **2013**, *117*, 1–30. [CrossRef] [PubMed]
54. Kok, B. On the inhibition of photosynthesis by intense light. *Biochim. Biophys. Acta.* **1956**, *21*, 234–244. [CrossRef]
55. Powles, S.B. Photoinhibition of photosynthesis induced by visible light. *Annu. Rev. Plant Physiol.* **1984**, *35*, 15–44. [CrossRef]
56. Aro, E.M.; Virgin, I.; Andersson, B. Photoinhibition of photosystem II. Inactivation, protein damage and turnover. *Biochim. Biophys. Acta* **1993**, *1143*, 113–134. [CrossRef]
57. Henmi, T.; Yamasaki, H.; Sakuma, S.; Tomokawa, Y.; Tamura, N.; Shen, J.R.; Yamamoto, Y. Dynamic Interaction between the D1 protein, CP43 and OEC33 at the lumenal side of photosystem II in spinach chloroplasts: Evidence from light-induced cross-Linking of the proteins in the donor-side photoinhibition. *Plant Cell Physiol.* **2003**, *44*, 451–456. [CrossRef]

58. Terashima, I.; Funayama, S.; Sonoike, K. The site of photoinhibition in leaves of *Cucumis sativus* L. at low temperatures is photosystem I, not photosystem II. *Planta* **1994**, *193*, 300–306. [CrossRef]
59. Huang, W.; Yang, Y.J.; Zhang, J.L.; Hu, H.; Zhang, S.B. PSI photoinhibition is more related to electron transfer from PSII to PSI rather than PSI redox state in *Psychotria rubra*. *Photosynth. Res.* **2016**, *129*, 85–92. [CrossRef]
60. Gollan, P.J.; Lima-Melo, Y.; Tiwari, A.; Tikkanen, M.; Aro, E.M. Interaction between photosynthetic electron transport and chloroplast sinks triggers protection and signalling important for plant productivity. *Philos. Trans. R. Soc. B* **2017**, *372*, 20160390. [CrossRef]
61. Tjus, S.E.; Moller, B.L.; Scheller, H.V. Photosystem I is an early target of photoinhibition in barley illuminated at chilling temperatures. *Plant Physiol.* **1998**, *116*, 755–764. [CrossRef] [PubMed]
62. Suorsa, M.; Järvi, S.; Grieco, M.; Nurmi, M.; Pietrzykowska, M.; Rantala, M.; Kangasjärvi, S.; Paakkarinen, V.; Tikkanen, M.; Jansson, S.; et al. PROTON GRADIENT REGULATION5 is essential for proper acclimation of *Arabidopsis* photosystem I to naturally and artificially fluctuating light conditions. *Plant Cell* **2012**, *24*, 2934–2948. [CrossRef]
63. Tikkanen, M.; Grebe, S. Switching off photoprotection of photosystem I—A novel tool for gradual PSI photoinhibition. *Physiol. Plant.* **2018**, *162*, 156–161. [CrossRef] [PubMed]
64. Munekage, Y.; Hojo, M.; Meurer, J.; Endo, T.; Tasaka, M.; Shikanai, T. *PGR5* is involved in cyclic electron flow around photosystem I and is essential for photoprotection in *Arabidopsis*. *Cell* **2002**, *110*, 361–371. [CrossRef]
65. Tikkanen, M.; Mekala, N.R.; Aro, E.M. Photosystem II photoinhibition-repair cycle protects Photosystem I from irreversible damage. *Biochim. Biophys. Acta* **2014**, *1837*, 210–215. [CrossRef]
66. Kagawa, T.; Sakai, T.; Suetsugu, N.; Oikawa, K.; Ishiguro, S.; Kato, T.; Tabata, S.; Okada, K.; Wada, M. Arabidopsis NPL1: A phototropin homolog controlling the chloroplast high-light avoidance response. *Science* **2001**, *291*, 2138–2141. [CrossRef] [PubMed]
67. Anderson, J.M.; Osmond, C.B. Shade-sun responses: Compromises between acclimation and photoinhibition. In *Photoinhibition*; Kyle, D.J., Osmond, C.B., Arntzen, C.J., Eds.; Elsevier Science BV: Amsterdam, The Netherlands, 1987; Volume 96, pp. 2–38.
68. Demmig-Adams, B.; Adams Iii, W.W. Photoprotection and other responses of plants to high light stress. *Annu. Rev. Plant Biol.* **1992**, *43*, 599–626. [CrossRef]
69. Sacharz, J.; Giovagnetti, V.; Ungerer, P.; Mastroianni, G.; Ruban, A.V. The xanthophyll cycle affects reversible interactions between PsbS and light-harvesting complex II to control non-photochemical quenching. *Nat. Plants* **2017**, *3*, 16225. [CrossRef]
70. Bräutigam, A.; Gowik, U. Photorespiration connects C3 and C4 photosynthesis. *J. Exp. Bot.* **2016**, *67*, 2953–2962. [CrossRef] [PubMed]
71. Badger, M.R.; Caemmerer, S.; Ruuska, S.; Nakano, H. Electron flow to oxygen in higher plants and algae: Rates and control of direct photoreduction (Mehler reaction) and rubisco oxygenase. *Philos. Trans. R. Soc. B* **2000**, *355*, 1433–1446. [CrossRef]
72. Guo, Y.Y.; Tian, S.S.; Liu, S.S.; Wang, W.Q.; Sui, N. Energy dissipation and antioxidant enzyme system protect photosystem II of sweet sorghum under drought stress. *Photosynthetica* **2018**, *56*, 861–872. [CrossRef]
73. Li, J.; Cang, Z.; Jiao, F.; Bai, X.; Zhang, D.; Zhai, R. Influence of drought stress on photosynthetic characteristics and protective enzymes of potato at seedling stage. *J. Saud. Soc. Agric. Sci.* **2017**, *16*, 82–88. [CrossRef]
74. Takahashi, H.; Iwai, M.; Takahashi, Y.; Minagawa, J. Identification of the mobile light-harvesting complex II polypeptides for state transitions in Chlamydomonas reinhardtii. *Proc. Natl. Acad. Sci. USA* **2006**, *103*, 477–482. [CrossRef] [PubMed]
75. Falkowski, P.G.; Chen, Y.B. Photoacclimation of light harvesting systems in eukaryotic algae. In *Light Harvesting Antennas in Photosynthesis*; Green, B.R., Parson, W.W., Eds.; Springer: Dordrecht, The Netherlands, 2003; Volume 13, pp. 423–447.
76. Lichtenthaler, H.K.; Meier, D.; Buschmann, C. Development of chloroplasts at high and low light quanta fluence rates. *Isr. J. Bot.* **1984**, *33*, 185–194.
77. Pastenes, C.; Horton, P. Effect of high temperature on pho to synthesis in beans. *Plant Physiol.* **1996**, *11*, 1253–1260. [CrossRef] [PubMed]
78. Mathur, S.; Jain, L.; Jajoo, A. Photosynthetic efficiency in sun and shade plants. *Photosynthetica* **2018**, *56*, 354–365. [CrossRef]

79. Pearcy, R.W.; Osteryoung, K.; Calkin, H.W. Photosynthetic responses to dynamic light environments by Hawaiian trees: Time course of CO_2 uptake and carbon gain during sunflecks. *Plant Physiol.* **1985**, *79*, 896–902. [CrossRef] [PubMed]
80. Hu, Y.B.; Sun, G.Y.; Wang, X.C. Induction characteristics and response of photosynthetic quantum conversion to changes in irradiance in mulberry plants. *J. Plant Physiol.* **2007**, *164*, 959–968. [CrossRef] [PubMed]
81. Hussain, S.; Iqbal, N.; Rahman, T.; Ting, L.; Brestic, M.; Safdar, M.E.; Asghar, M.A.; Farooq, M.U.; Shafiq, I.; Ali, A.; et al. Shade effect on carbohydrates dynamics and stem strength of soybean genotypes. *Environ. Exp. Bot.* **2019**, *162*, 374–382. [CrossRef]
82. Hussain, S.; Iqbal, N.; Brestic, M.; Raza, M.A.; Pang, T.; Langham, D.R.; Safdar, M.E.; Ahmed, S.; Wen, B.; Gao, Y.; et al. Changes in morphology, chlorophyll fluorescence performance and Rubisco activity of soybean in response to foliar application of ionic titanium under normal light and shade environment. *Sci. Total Environ.* **2019**, *658*, 626–637. [CrossRef]
83. Huang, D.; Wu, L.; Chen, J.R.; Dong, L. Morphological plasticity, photosynthesis and chlorophyll fluorescence of *Athyrium pachyphlebium* at different shade levels. *Photosynthetica* **2011**, *49*, 611–618. [CrossRef]
84. Yao, X.; Li, C.; Li, S.; Zhu, Q.; Zhang, H.; Wang, H.; Yu, C.; Martin, S.K.S.; Xie, F. Effect of shade on leaf photosynthetic capacity, light-intercepting, electron transfer and energy distribution of soybeans. In *Plant Growth Regulation*; Springer: Berlin/Heidelberg, Germany, 2017; Volume 83, pp. 1–8.
85. Kunderlikova, K.; Brestic, M.; Zivcak, M.; Kusniarova, P. Photosynthetic responses of sun-and shade-grown chlorophyll b deficient mutant of wheat. *J. Cent. Eur. Agric.* **2016**, *17*, 950–956. [CrossRef]
86. Valladares, F.; Niinemets, Ü. Shade tolerance, a key plant feature of complex nature and consequences. *Annu. Rev. Ecol. Syst.* **2008**, *39*, 237–257. [CrossRef]
87. Portsmuth, A.; Niinemets, U. Structural and physiological plasticity in response to light and nutrients in five temperate deciduous woody species of contrasting shade tolerance. *Funct. Ecol.* **2007**, *21*, 61–77. [CrossRef]
88. Valladares, F.; Wright, S.J.; Lasso, E.; Kitajima, K.; Pearcy, R.W. Plastic phenotypic response to light of 16 congeneric shrubs from a Panamanian rainforest. *Ecology* **2000**, *81*, 1925–1936. [CrossRef]
89. Aspinwall, M.J.; Loik, M.E.; Resco de Dios, V.; Tjoelker, M.G.; Payton, P.R.; Tissue, D.T. Utilizing intraspecific variation in phenotypic plasticity to bolster agricultural and forest productivity under climate change. *Plant Cell Environ.* **2015**, *38*, 1752–1764. [CrossRef]

6

Combined Influence of Cutting Angle and Diameter Differences between Seedlings on the Grafting Success of Tomato using the Splicing Technique

José-Luis Pardo-Alonso [1], Ángel Carreño-Ortega [1,*], Carolina-Clara Martínez-Gaitán [2] and Ángel-Jesús Callejón-Ferre [1]

[1] Department of Engineering, University of Almería, Agrifood Campus of International Excellence (CeiA3), La Cañada de San Urbano, Almería 04120, Spain; jolupa@ual.es (J.-L.P.-A.); acallejo@ual.es (A.-J.C.-F.)
[2] Tenova, Technological Center: Foundation for Auxiliary Technologies for Agriculture; Parque Tecnológico de Almería, Avda. de la Innovación, 23, Almería 04131, Spain; cmartinez@fundaciontecnova.com
* Correspondence: acarre@ual.es

Abstract: Herbaceous crop yield intensification creates favourable conditions for the development of pests that intensify the attack of soil pathogens traditionally controlled by disinfectant, which are mostly prohibited and unlisted because of their toxicity. The use of grafted plants solves this problem and assists in addressing abiotic stress conditions. Within Solanaceae, specifically tomato crops (*Solanum lycopersicum*), the use of the splicing technique (simple and easily automated) is of special interest. This experiment attempts to present the combined influence of cutting angle and different random diameters on grafting success with the objective of detecting an optimum working range that will be applicable to automated and robotic grafting systems. An increase in the grafting angle is associated with a higher survival of grafted plants despite variations in diameter. Moreover, a threshold cutting angle is observed from which the success rate no longer increases but decreases drastically. Therefore, for a given working range with a significant cutting angle, whether the seedlings of origin are similar in diameter is not important, and this factor is more influential outside the optimal cutting angle range.

Keywords: tomato grafting; splice grafting technique; graft angle; random diameter

---

1. Introduction

The objective of the experimental study was to determine the combined importance of random rootstock and scion diameters at different cutting angles on splice grafting success. The proposed working hypothesis suggests that both parameters have a statistically significant relationship with grafting success and an optimum working range can be defined to achieve successful grafts.

The experiment was developed as part of a larger study to optimize working conditions for the automation of grafting via the splicing technique. The study is autonomous and independent and presents sufficient and consistent results for the definition and specification of the splice grafting conditions that provide the most optimal results, whether performed manually or automated.

Grafting can be defined as a natural or deliberate fusion of plant parts by which vascular continuity is established [1], so that the resulting organisms function as a single plant [2].

The portion of the upper tissue or crown of a plant, which is also known as the stem or scion, adheres to another portion of the plant, specifically its root and lower part, which is commonly called rootstock, under stock or stock, and both parts come in contact and join with each other so that the resulting composite plant grows and develops as a single organism (graft). The callus corresponds to the mass of parenchyma cells that develop from the plant tissue of the scion and the rootstock around the wound and where the development of vascular connections of the resulting graft union occurs [3].

Reducing the impact of pathogens is a challenge in all agricultural production systems [4], and monocultures are even more vulnerable than more diversified agricultural production systems [5]. Thus, grafting has become a tool of enormous potential to quickly enhance the efficiency of modern vegetable cultivars to promote wider adaptability or resistance to different stress situations [6].

The sequence of structural and biological events that occur in the development of a compatible graft between plants has been described in many studies, and the following development pattern is observed in which three fundamental phases can be distinguished: fusing of the rootstock and scion; proliferation of the callus around the union; and vascular re-differentiation through the interface establishing continuity between rootstock and scion [1,3,7,8]:

1. The meristematic tissues of the stem are placed in direct contact with the tissues of the rootstock. Once both components of the graft are in intimate contact, cambium cells from the rootstock and the scion produce parenchyma cells that fuse forming a callus tissue [9]. This first phase of cohesion that forms the callus is a reaction similar to wound healing, and it does not require recognition between rootstock and scion, occurring in both compatible and incompatible grafts.
2. If the graft is compatible, a differentiation of certain vessels and sieve tube elements of the phloem is observed in the callus, and they are not derived from the cambium and constitute the first transitional and continuous union between the rootstock and the scion.
3. In the last part of the grafting process, the cambium layer newly formed in the bridge of the callus begins its own meristematic tissue activity, thus forming new vascular tissue. Production of these new vascular elements that join xylem and phloem allows establishing a symplastic communication between rootstock and scion [1].

The success of the graft performed with a variety of compatible seedlings is determined by the three events previously described, assuming an important role of the plant hormones related to growth, such as auxin, in the grafting process [10]. Thus, the graft may originate from a combined mechanism of wound healing and conductive vessels formation [11]. Therefore, vascular connection is the last event in a successful healing process and represents the most important event because once such a vascular connection is established, water and solute transport begins from the stock to the scion, and the mechanical strength at the graft union increases [7,12]. One difficulty is to understand when the grafting process is completed [13], since a simple technique for continuous evaluation of graft development is not available [14]. Nonetheless, the assessment can be based on various techniques, including destructive and non-destructive techniques as follows:

(1) Visual estimation of the constituent seedlings and the appearance of graft growth [15]; (2) thermal camera imagery because the temperature of the leaves is 2 to 3 °C lower than that in a failed graft due to the transpiration of a successful graft [16]; (3) vertical cut performed on the graft surface and observation of the curvature of the vascular system formed at the union [17]; (4) measurement of the electrical resistance transferred from the scion to the rootstock through the surface area of its connection, which undergoes variations associated with histological changes during the union of the rootstock and the scion [14]; (5) assay performed to evaluate the tensile strength of the graft and analyse the strength of the graft union between rootstock and scion until breaking [18,19]; (6) displacement transducers used to perform a continuous and non-disruptive evaluation of the functional hydraulic connections within the plant [20]; and (7) NMR-based method (Nuclear magnetic resonance), that reveals water flux vectors inside individual vessels of intact plants [21].

The tomato is one of the world's most important herbaceous crops [22], and grafting of tomato plants is a widespread practice. Grafting methods among seedlings vary greatly and considerably depending upon the type of crops, farmer's experience and preferences, availability of facilities and machines, the number of grafts to perform and even the purpose and destination of the grafts, i.e., whether they are for the farmer's own uses or for sale and commercial distribution [23]. Grafting is a common practice among herbaceous crops, and its use for Solanaceae is highlighted as follows: cleft or split grafting [21,24], splice or tube grafting [25,26], plug-in grafting [27], double-stem grafting [28,29], or pin grafting [15,30], among others.

Splice grafting, also known as tube grafting, top grafting or slant cut grafting, is the most popular [31] and widely used technique for tomato as well as eggplant [32]. The rootstock is cut below the cotyledons, thus eliminating the need to continually eliminate the sprouting of the stock over the plant's life [33]. The scion is also slant cut on a complementary angle above the cotyledons. Both parts are then placed in contact and secured by means of a tube or elastic tube-shaped clip with side slit. This method has the disadvantage of being highly demanding in terms of post-grafting microclimatic conditions, which require meticulous timing and delicate handling after the cut until healing and the maintenance of optimal temperature and humidity conditions to stimulate rooting [34]. As an advantage, the splicing method allows grafting with smaller plants, which reduces the pre-graft cultivation time and takes up less space in cultivation chambers and nurseries [35].

Velasco [12] and Villasana [36] affirmed that successful grafting is contingent upon similar stem diameters and the alignment of the vascular cambium area. However, such claims are dependent on the effect that the seedling diameter variable has on grafting success and do not consider the interrelated influence of other parameters and constraints of the grafting process.

For studying the success or failure of splice grafting, Yamada [37] established three factors of importance for its execution: (a) area of the cut surface; (b) gripping force of the union clip; and (c) smoothness of the cut surface and sharpness of the blade. For the first factor, allusive to the coincidence rate between seedlings, from coincidences of 50% graft successes of over 85% were achieved, and up to 95% success rates were observed for alignments over 80%; all this referred exclusively to a test angle of 30°, so these results were obtained regardless of the effect of the cutting angle on the success of the process.

Furthermore, although the number of vascular bundles does not affect the grafting success, differences in diameters between the rootstock and the scion [38], that mark the alignment between them do have an effect. Thus, when a graft is performed, it is important to increase the chances that the vascular bundles from the rootstock and the scion come in direct contact, maximizing the area of the cut surfaces that are joined by pressing them together [39].

For manual and automated grafting, the alignment of diameter of both seedlings must be identified, classified, and visually paired. This is an expensive and arduous task that applies a series of calibrations and visual comparison criteria based on morphological characteristics, which may be subjective and susceptible to human error [40,41]. The present study analyses the importance of these pre-grafting tasks based on grafting success and whether an optimal working zone can be established that guarantees adequate grafting success without having to pre-sort the seedlings.

2. Materials and Methods

2.1. *Definition of Operating Conditions*

The experimental study was carried out at the Tenova Technological Center: Foundation for Auxiliary Technologies for Agriculture (Centro Tecnológico Tecnova: Fundación para las Tecnologías Auxiliares de la Agricultura) in Almería between March and June 2017. Almería's surroundings correspond to a model of agricultural exploitation of high technical and economic performance of greenhouse herbaceous fruit crops, especially for the tomato crop (Figure 1).

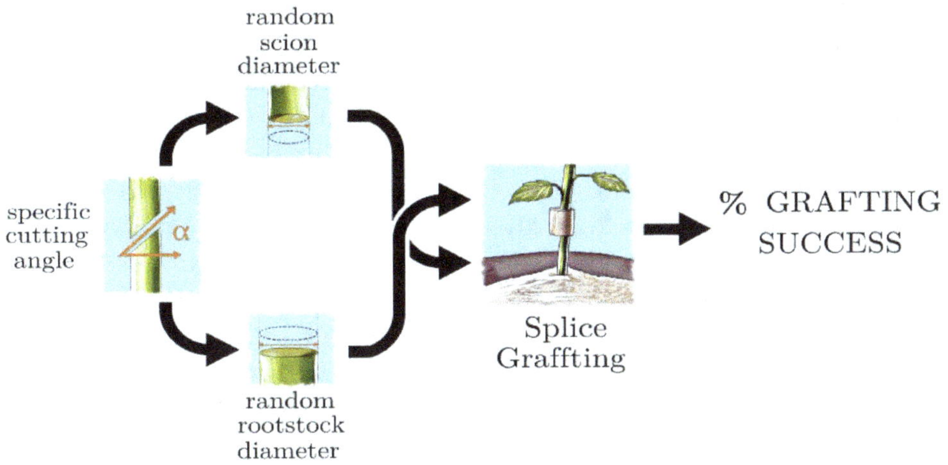

Figure 1. Working procedure of the experimental study to determine the combined importance of random rootstock and scion diameters at different cutting angles on splice grafting success.

For the study, rootstocks of the interspecific hybrid KNVF (*L. esculentum x L. hirsutum*) were used since it is the most used stock for tomato grafting [42–44]. The commercial rootstock Maxifort was used because of its strong roots and vigour and good performance at lower temperatures and in high salinity conditions. It presents high resistance (HR): ToMV: 0–2/Fol: 0,1/For/PI/Va: 0/Vd: 0; intermediate resistance (IR): Ma/Mi/Mj [45]. Likewise, the Ventero variety has been used as a grafting scion as an indeterminate tomato hybrid for truss harvesting, and it presents medium vigour, with good foliar coverage, very uniform fruits, slightly flattened of good red colour and deep shine, very good cracking and micro-cracking tolerance, and compact and well-formed clusters. It presents high genetic resistance (HR): ToMV: 0–2/Ff: B, D/Fol: 0,1/Va: 0/Vd: 0; and intermediate resistance (IR): TYLCV/Ma/Mi/Mj [39]. Both varieties are commonly used for manual grafting using the "tomato on tomato" (ToT) splicing technique, which demonstrated their compatibility prior to the experiment.

The working environment during the study was maintained under stable and controlled environmental conditions throughout all grafting experiments, with temperatures oscillating between 20° and 25°, relative humidity conditions occasionally forced between 75% and 90% and stable brightness conditions of natural in-direct light. Oda [39] indicated that grafting must be performed in the shade in an area protected from the wind and the sun to avoid wilting of grafted seedlings. Grafting was performed at the lowest period of plant transpiration during morning hours [46], between 8 h and 12 h to maintain transpiration similar among the experiments and at the time period when the transport of water from the roots to the leaves is slowest, which makes the graft less susceptible to water stress and therefore to water loss. Other parameters with possible influence, such as atmospheric CO2 and other air contaminants, were not been controlled.

In the nursery, the plants were cultivated and attended to from sowing until 25 to 35 days after, and the scions were sown 2 to 5 days before the rootstock seeds. This variability of days is determined by the growth rate since different plants require different germination periods [47], and such periods are directly related to climatic conditions of the month of growing.

For the experiment, the plants were considered mature and ready for grafting when they had 2 to 4 well-defined true compound leaves [32], preferably with little foliage, thus decreasing the transpiration demand and post-grafting stress. The peat root ball remained wet but not soggy at all times during the grafting process, thus ensuring proper root respiration. The substrate used was 80% black peat with 10% perlite and 10% mulch. The experiments were always conducted with seedlings whose stems were still green and tender (herbaceous and non-woody).

For the splicing method, the seedling stems diameters should be at least 1.5 mm [48], and not too thick but with some natural and random variation. In the study, the diameters in the area close to the cut have varied from 1.5 to 2.5 mm for the scion and 2 to 3 mm for the rootstock.

The matching of the rootstock and scion samples was established randomly among plants that were healthy, had an acceptable anatomy and growth and presented diameters between the established reference limits. Seedlings with anomalous growth and diameters outside the established range were discarded. Diameters were measured using a digital calliper with a resolution of 1 dmm (0.1 mm) and repeatability of 1 dmm in the areas close to where the cut was performed both for the rootstocks and for the scions. The cut in the rootstock was always performed below the cotyledons, whereas the cut in the scion was performed above the cotyledons.

The complete experiment consisted of 10 individual events of 150 grafts each distributed over 4 months. Each graft consisted of 10 series of 15 grafts per tested angle. Therefore, 10 representative angles of the possible cutting range were selected: 0°, 10°, 20°, 30°, 40°, 50°, 60°, 70°, 80° and 85°. Once each of the experiments was performed, grafts were grouped for healing on a single grafting tray of 15 × 10 cells, placing each group of grafts of a given angle in each of the 10 rows of the tray. Each of the grafted plants was rotated within their row for every tested angle eliminating in this way the position factor and its possible effects (ventilation and luminosity among others).

2.2. Device Description

For the experiment, two seedling-cutting machines were implemented to ensure the accuracy of the cutting angle required for each test and the integrity of the dissected seedling. The machines were similar and complementary to each other, where one was designated for cutting the rootstock and the other for cutting the scion. Each of the machines had a double acting dual rod cylinder, model CXSM15-15 by SMC, which operated at 3.5 bars and used dry-pressed air to produce a clean bisection of the seedling via a stainless steel cutting blade (type BA-160-9 mm from NT Cutter) that can be attached as a tip and is interchangeable with an adjustable inclination angle, thus providing optimal sharpness (Figure 2). To ensure a clean cut, the machine has a fitting notch adapted to accommodate the seedling, ensuring the verticality of the stem ahead of the blade and another notch fitted for the blade at each cutting angle. The blade was replaced prior to each experiment (150 grafts per blade), and it was below the limit of 5000 grafts per blade established by Yamada [35], who determined that as the number of cuts per blade increases, the roughness of the cutting area becomes notable, thus reducing the grafting success. Before each use, the blade was cleaned and disinfected as Bumgarner suggests by soaking in alcohol [45], exposure to flame and air drying.

Figure 2. Cutting device. As can be seen, a fitting device is used to guarantee or ensure the verticality of seedling in the cutting process, beside a groove for insertion of the cutting blade. A specific fitting device for each cutting angle was developed. Once inserted, the seedling is disectioned in two by a sharp blow of the sharp blade.

The union of the seedlings cut by the machines was executed manually using the traditional splice grafting technique described by Oda [49] and expanded on by DeMiguel [33]. For this, plastic grafting clips of different lengths were used according to the tested angles. Clips were cut from a continuous flexible transparent plastic roll and outfitted with lateral wings for opening and placement to allow for easy observation of the success or failure of the graft. The clip diameter was less than 3 mm, and the shape was slightly oval, which guaranteed a better grip when the rootstock and scion diameters were unequal. Grafting clips that were too long inhibited the attached graft union, and clips that were too short exerted too much pressure and deformed the graft union [50].

Manual handling of the seedlings was always performed after thorough washing with antibacterial soap. Direct contact with the wounds was constantly avoided. Once grafted, the plants were placed on the tray and introduced into a healing chamber, similar to a small acclimatization tunnel as described by Oda [51]. This chamber was itself placed inside a larger growth chamber that allows for the appropriate basic environmental conditions as follows: 14 h photoperiod and a daily light integral (DLI) of $\approx$100 $\mu mol \cdot m^{-2} \cdot s^{-1}$ PAR, (~3000 Lux) of indirect diffuse light during the callus development stage [52,53], temperature of 26 °C and relative humidity of 95%.

After the trays were transferred to the chamber, the plants were not manipulated or moved for 3 full days so that the natural healing process was not disturbed. From the fourth day of the graft, the first individual plant by plant visual inspection was conducted inside the chamber.

This inspection was routinely repeated during the following days from the 4° DAG (day after the graft) until the 9° DAG to assess changes and the healing process in each plant and therefore the success or failure of the graft. During this period and from the 6° DAG, the environmental conditions of the chamber were gradually relaxed, acclimatising to external environment, and the inside chamber was opened to decrease the humidity and temperature, acording to outside. Between the 10° DAG and the 14° DAG, the plants were eventually removed from the growth chamber and allowed to develop without being transplanted.

While the rootstock and scion establish a vascular connection during the first days [54], at least 14 days are needed for the graft union to heal completely and be considered functional. After 14 DAG of performing daily observations for each experiment, the experiment was ended. Grafts that did not survive the healing process within the stipulated period were considered failures.

The success or failure of the graft has been evaluated by daily visual estimations and observations that evaluated the development of the graft and analysed other external symptoms and evidence, such as physiological abnormalities or signs of wound healing and scarring. Symptoms of internal failure generally precede those of external failure [55]. If the graft is successful, evident progress is generally seen from the wilt stage to greater vigour in the aerial part of the graft, which is reflected in a palpable recovery and associated with a gradual disappearance of signs of dehydration, which implies that adequate vascular continuity has been generated among the elements of the xylem. In addition, this factor is accompanied with the occurrence of axillary buds in the aerial part, thus indicating that the graft is successful and the resulting plant is functional.

These factors are used to determine whether the graft is successful. Regardless, the behaviour and subsequent evolution of the grafts continue to be evaluated until 14 DAG to corroborate and validate the evolution of the natural healing process of the graft (Figure 3).

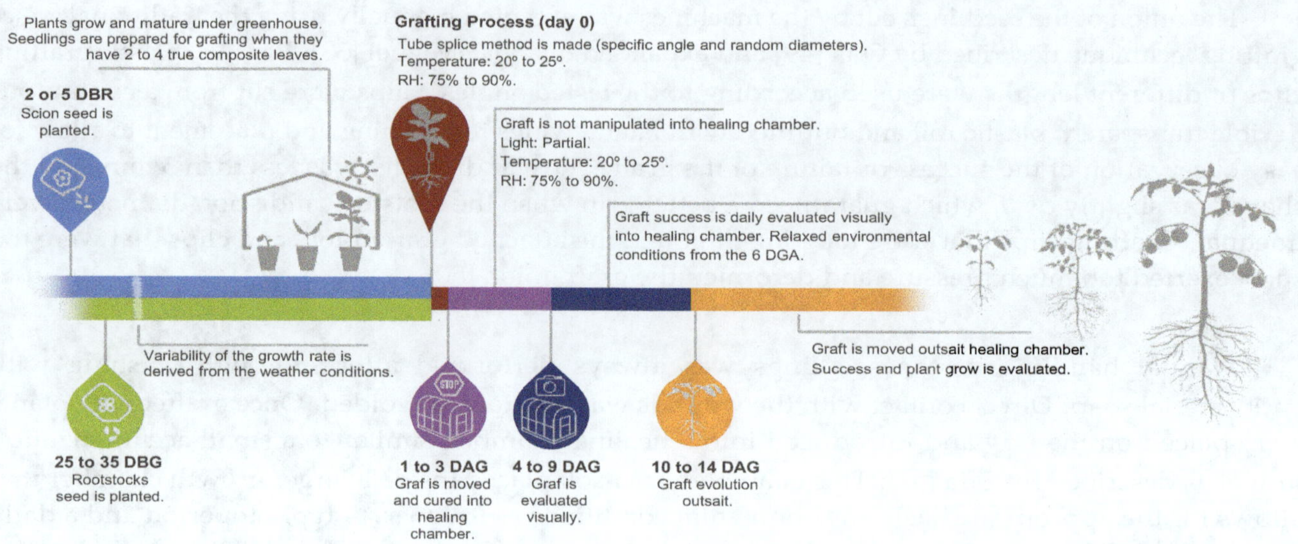

Figure 3. Timeline of the grafting process developed. DBR (days before rootstock has been planted). DBG (days before grafting process). DAG (days after grafting process).

3. Results and Discussion

The results of the 1500 individual experiences were collected and grouped polytomously in 10 sections of equal height to compare the cutting angle and differences in diameter of rootstock and scion and to evaluate the grafted plant survival for each case.

The data analysis process included two stages: the first phase consisted of conducting a descriptive analysis of the data distribution and their correlation through the application of One-way ANOVA for Randomized Complete Block Design (RCBD ANOVA), and the second phase consisted of a two-way analysis of variance in which only one sample per group was run, and the results were then assessed by a post-hoc comparison test, such as Student's *t*-test.

3.1. Data Analysis: Descriptive Statistics

An analysis of the experimental results showed that the effect of variations in diameter on the grafting success decreases as the grafting angle increases, and the differences are nearly negligible in the range between 50° and 80° and between 60° and 80°, where percentage changes between the grafted seedling and the successful graft were maintained at an overall success rate of greater than 90% or even greater than 95%, respectively. This finding confirms that for greater angles, the success probability depends less on the diameter of the seedlings. From 80° onward, successful execution of the graft began to be materially more complicated due to two factors: physical limitations related to the technology used for the cutting and subsequent union of the seedlings; and the exponential increase in the sectioned surface that was directly related to the tangent of the cutting angle, which determined both the exposed surface and the rigidity and firmness of the structure of the dissected and subsequently joined seedlings.

Grouped data confirm that independent of section variation among the seedlings of origin, a working zone between 50° and 80° offers good results in terms of graft success (Figure 4).

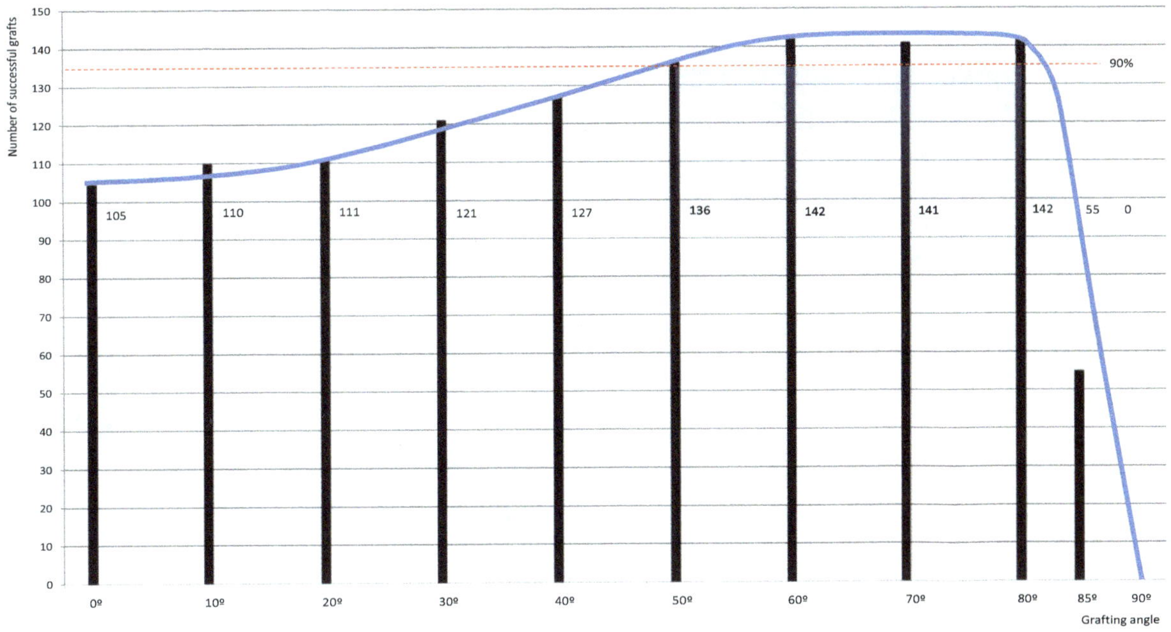

Figure 4. Distribution of successful grafts according to different cutting angles. It can be seen that the zone between 50° and 80° (blue zone), has a success rate higher than 90%, so we can consider it an optimal work zone. This graph represents the absolute number of successes for each cutting angle, without considering the variable of difference between diameters.

Having studied the diametric differences with respect to grafting angles, it is apparent that at small cutting angles and with highly variable diameters, the failure probability is high, and success was not observed, while at larger cutting angles, success associated with high diametric differences was recorded. A slightly greater range between quartiles is observed at angles between 50° and 70°, which indicates a greater tolerance to variable diameters during the grafting process (Figure 5).

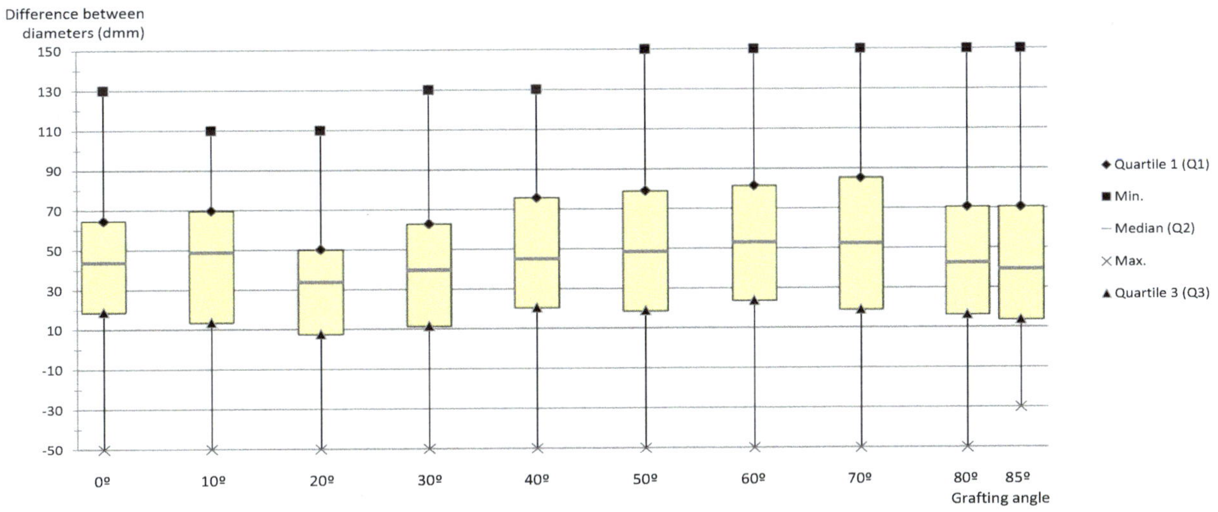

Figure 5. Distribution of successful grafts according to the differences between diameters of plants for each angle of union. This graph represents the density function of successes for each cutting angle (1 dmm is a tenths of a millimeter, 10^{-4} m). A slightly greater amplitude between quartiles is appreciable at angles between 50° and 70° degrees, which indicates a greater tolerance in this range to the disparity of diameters.

The combined representation of cutting angle and diameter differences between plants versus the success of the graft provide evidence of the combined effect that both factors have on the successful execution of the graft. This representation has been developed through the use of the software Surfer12 and the Local Polynomial gridding method for the interpolation of points of the spatial matrix, which only uses points within the defined neighbourhood and adjusts the matrix to a first-order polynomial to the power of two. Polynomial interpolation allows us to create a uniform surface and identify long-range trends in the data set (Figure 6).

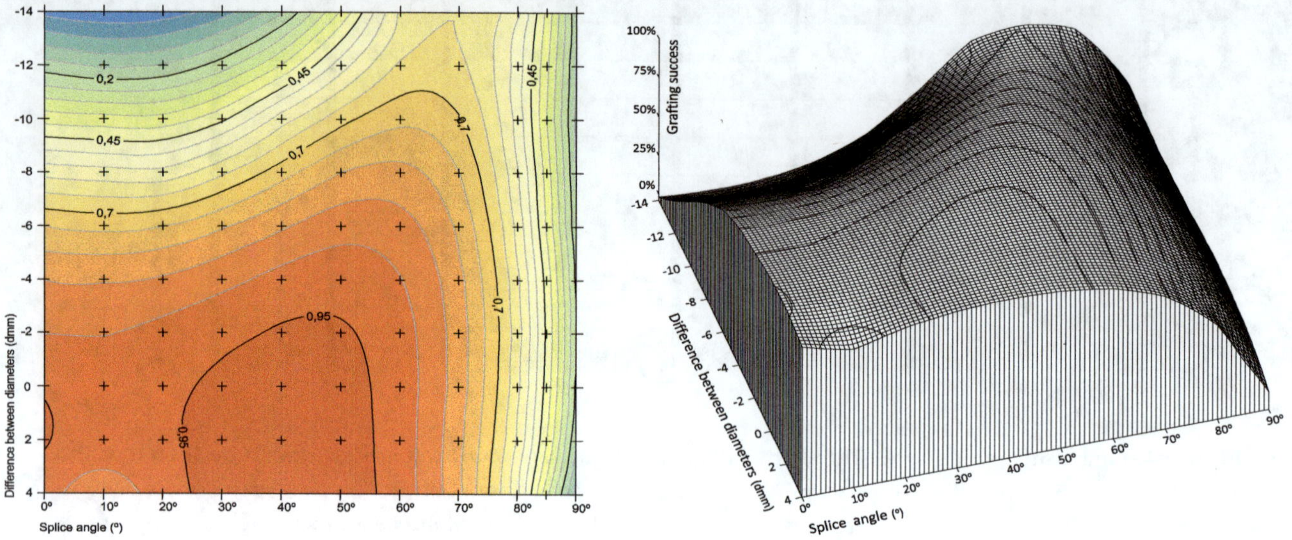

(a) (b)

Figure 6. Graphical representation of the combined influence of cutting angle and diameter differences between seedlings on the grafting success of tomato using the splice grafting thecnique. These graphs represent the successes for each concrete difference of diameters, associated to each angle of union (1 dmm is a tenths of a millimeter, 10^{-4} m). Results represented by the Local Polynomial Gridding Method (Polynomial Order 1, Power 2). (**a**) Coloured contour diagram of successful grafts (%), as a function of cutting angle (°) and difference between diameters of grafted (dmm); (**b**) 3D wireframe of successful grafts (%), as function of cutting angle (°) and difference between diameters of grafted (dmm).

At small differences in diameter, the success rate is high for cutting angles between 30° and 60°, whereas when this difference between diameters increases, the maximum values move to values between 50° and 70°, and the percentage of success gradually decreases.

One of the possible causes that justify grafting success within this range of working angles is the exponential increase of the contact surface, which increases in equal measure the possibility of matching between vascular bundles arranged in a circle around the stem [34]. Effective contact depends on the surface and arrangement of the bundles in the two plants that are grafted; therefore, at a larger cutting surface with an appropriate arrangement and matching of the seedlings, a greater area of contact is observed. Thus, this range of graft angles between 50° and 70° is associated with a decrease in failure and substantially less importance and influence of uniformity in stem diameter between both plants on grafting success.

By increasing the difference between the sections at the point of union and at greater graft angles, the area of contact surface increases exponentially, thus increasing the probability of vascular correlation. Moreover, for uncovered surfaces, the area remaining outside the contact area is exposed to pathogens, such as bacteria and fungi, which cause the graft to fail [56]. In addition, greater stress associated with scarring is evident based on the proliferation of a larger callus in response to the wound. As the cutting angle approaches 90°, the successful execution of the graft begins to become more mechanically complicated due to the technology used in the process and the firmness of the dissected seedling themselves. In addition, the uncovered or unmatched surface between the seedlings increases excessively.

Therefore, when significant differences in diameter occur between the rootstock and the scion, the probability of failure is higher at small cutting angles close to horizontal, while the probability of success is higher at similar diameters. This correlation reflects the farmer's own practice and experience and represents a frequently observed factor that is directly related to the success of the graft as observed in the first known publication referring to seedling grafting for herbaceous crops, which indicated that seedlings with similar diameters should be selected [57]. However, subsequent studies have corroborated the direct relationship between the cutting angle and the success of the graft [19], which supports the premise that seedlings should have similar diameters in the cutting zone. Zhao [58] stated that expanding the area of contact between the rootstock and scion is the key to graft survival.

3.2. Data Analysis: ANOVA

The experimental results for grafting success were tested via two-way ANOVA of the cutting angle and diameter difference, where each of these factors has been grouped into 10 blocks, with a single sample or repetition per group (ANOVA). The randomized complete block design (RCBD ANOVA) analysis technique used as the usual standard for agriculture was used, where similar experimental units were grouped into blocks. We consider $\alpha = 0.05$ (95% confidence level). The statistical package Real Statistics Resource Pack 5.8 in Microsoft Excel 2010 was used for the study. Analysis of variance was performed to answer the following general research question (RQ): Are statistically significant differences observed between the means of grafting success for different cutting angles and different diameters between seedlings? (Tables 1 and 2).

Table 1. Analysis of variance of two factors without replication. Factor 1: the difference between the rootstock and scion diameters, where the positive values represent a larger diameter of the rootstock and the negative ones a larger diameter of the scion. Factor 2: cutting angles of the seedlings.

Analysis of Variance of Two Factors without Replication				
Summary	**Count**	**Sum**	**Mean**	**Variance**
15 to 13 dmm	10	3.3333	0.3333	0.1806
13 to 11 dmm	10	3.7242	0.3724	0.1253
11 to 9 dmm	10	5.8789	0.5879	0.1009
9 to 7 dmm	10	7.1258	0.7126	0.0589
7 to 5 dmm	10	8.4487	0.8449	0.0545
5 to 3 dmm	10	9.0306	0.9031	0.0202
3 to 1 dmm	10	9.1009	0.9101	0.0245
1 to (−1) dmm	10	9.2110	0.9211	0.0284
(−1) to (−3) dmm	10	9.2333	0.9233	0.0239
(−3) to (−5) dmm	10	8.6000	0.8600	0.1071
Summary	**Count**	**Sum**	**Mean**	**Variance**
0°	10	6.7621	0.6762	0.1398
10°	10	6.3766	0.6377	0.1391
20°	10	6.0410	0.6041	0.1725
30°	10	6.9445	0.6945	0.1644
40°	10	7.5000	0.7500	0.1405
50°	10	8.5650	0.8565	0.0645
60°	10	9.5815	0.9582	0.0031
70°	10	9.1847	0.9185	0.0117
80°	10	9.4521	0.9452	0.0022
85°	10	3.2792	0.3279	0.0326

Table 2. Combined influence of the cutting angle and the difference betwen diameters in graft success. All statistical analyzes were done using a significance factor of 95% ($p \leq 0.05$). ANOVA summary tables (1 dmm is a tenths of a millimeter, 10^{-4} m). The result of the analysis ANOVA (two factors without replication) indicates that the statistical value of "F" is much higher than the critical value for "F" for both factors: angles and differences between diameters. Therefore, we can assure that the results of our tests are significant.

ANOVA						
Sourface of Variation	**Sum of Squares**	**Degrees of Freedom**	**Mean of the Squares**	**F**	**Probability Value**	**Critical Value for F**
Diameter difference	4.7160	9.0000	0.5240	13.6138	0.0000	1.9976
Cutting angle	3.4002	9.0000	0.3778	9.8154	0.0000	1.9976
Error	3.1178	81.0000	0.0385			
Total	11.2340	99.0000				

After running the ANOVA analysis, the null hypotheses H0 were rejected for both cases, and the alternative hypotheses Hi were accepted. Therefore, confirmable cases of significant differences between success means and cutting angles and seedling diametric differences were observed with a 95% confidence. To compare the differences, post-hoc rank tests were conducted to determine which means differ from each other. Student's *t*-type comparison tests (RCBD ANOVA and *t*-test) were performed (Table 3).

Table 3. Comparison of means differences between angles using Student's *t*-test (*t*-test). Use of contrasts to determine whether there is a significant difference ($p \leq 0.05$).

	0°	**10°**	**20°**	**30°**	**40°**	**50°**	**60°**	**70°**	**80°**	**85°**
0°		N	N	N	N	Y	Y	Y	Y	Y
10°			N	N	N	Y	Y	Y	Y	Y
20°				N	N	Y	Y	Y	Y	N
30°					N	N	N	N	N	Y
40°						N	N	N	N	Y
50°							N	N	N	Y
60°								N	N	Y
70°									N	Y
80°										Y
85°										

Significant differences in the mean grafting success values were not observed for similar angles, whereas clearly significant differences were observed when larger angles were compared, especially for angles equal to 20° or less and angles equal to 50° or greater. A cutting angle of 85° produced significant differences in the mean grafting success compared with most angles, including angles close to each other and distant, since the response of grafting success to cutting angle was random and irregular (Table 4).

Variations in the diameters of grafting seedlings greater than 90 cmm produced significant differences in the mean success with respect to the other variations in diameter, which may be due to a random and unpredictable response to the success of the graft from these diametric differences. The remaining variations in diameters below 90 cmm did not produce significant differences between their success means.

Table 4. Comparison of means differences between diameter of seedlings using Student's t-test (1 dmm is a tenths of a millimeter, 10^{-4} m). Use of contrasts to determine whether there is a significant difference ($p \leq 0.05$).

	15 to 13 dmm	13 to 11 dmm	11 to 9 dmm	9 to 7 dmm	7 to 5 dmm	5 to 3 dmm	3 to 1 dmm	1 to (−1) dmm	(−1) to (−3) dmm	(−3) to (−5) dmm
15 to 13 dmm		N	Y	Y	Y	Y	Y	Y	Y	Y
13 to 11 dmm			Y	Y	Y	Y	Y	Y	Y	Y
11 to 9 dmm				Y	Y	Y	Y	Y	Y	Y
9 to 7 dmm					N	Y	Y	Y	Y	N
7 to 5 dmm						N	N	Y	N	N
5 to 3 dmm							N	N	N	N
3 to 1 dmm								N	N	N
1 to (−1) dmm									N	N
(−1) to (−3) dmm										N
(−3) to (−5) dmm										

4. Conclusions

A review of the scientific literature suggests that the success of the splice grafting process increases as the cutting angle increases as long as the union is based on similarities between the stems of the grafting seedlings. However, until now, the success rate based on the interaction between the cutting angle and diameter of the seedling has not been reported.

The present study has concluded that the success of a graft depends to a great extent on the cutting angle and diameter of the seedlings, with disparities or similarities between the seedling sections playing an important role in the success of smaller cutting angles, although the stem diameter shows decreasing importance as the cutting angle increases, with the minimum influence of diameter observed within a cutting angle range of 50° and 70°.

Consequently, using the splicing technique with a cutting angle between 50° and 70° can lead to a substantial improvement of grafting conditions and techniques and can eliminate, to some extent, the random factor of differences between diameters as well as the pre-selection and matching requirements for sections between seedlings, thereby simplifying the demands for manual, automated and robotized grafting systems.

Author Contributions: J.-L.P.-A. conceived and designed the experiments, performed the experiments, collected, analyzed the data and interpreted the results, and developed the manuscript. A.C.-O. conceived and designed the experiments, analyzed the data, interpreted the results, and developed the manuscript. C.-C.M.-G. conceived and designed the experiments and performed the experiments. A.-J.C.-F. provided constructive suggestions on experiment analysis.

Acknowledgments: This study was supported by Tenova, Technological Center: Foundation for Auxiliary Technologies for Agriculture. The authors would like to acknowledge to all the employees involved for their contributions to experimental setting and data collection.

References

1. Pina, A.Y.; Errea, P. A review of new advances in mechanism of graft compatibility–incompatibility. *Sci. Hortic.* **2005**, *106*, 1–11. [CrossRef]
2. Mudge, K.; Janick, J.; Scoffield, S.; Goldschmidt, E.E. A history of grafting. *Hortic. Rev.* **2009**, *35*, 437–494. [CrossRef]
3. Hartmann, H.T.; Kester, D.E.; Davies, F.T.; Geneve, R.L. Principles of grafting and budding. In *Plant Propagation. Principles and Practices*, 8th ed.; Pearson: London, UK, 2010; Chapter 11; ISBN 978-0-13-501449-3.
4. Hartman, G.; Pawlowski, M.; Herman, T.; Eastburn, D. Organically Grown Soybean Production in the USA: Constraints and Management of Pathogens and Insect Pests. *Agronomy* **2016**, *6*, 16. [CrossRef]
5. Diacono, M.; Persiani, A.; Fiore, A.; Montemurro, F.; Canali, S. Agro-Ecology for Potential Adaptation of Horticultural Systems to Climate Change: Agronomic and Energetic Performance Evaluation. *Agronomy* **2017**, *7*, 35. [CrossRef]
6. Pradeep, K.; Youssef, R.; Mariateresa, C.; Giuseppe, C. Vegetable Grafting as a Tool to Improve Drought Resistance and Water Use Efficiency. *Front. Plant Sci.* **2017**, *8*, 1130.

7. Moore, R. A model for graft compatibility-incompatibility in higher plants. *Am. J. Bot.* **1984**, *71*, 751–758. [CrossRef]
8. De Miguel, A.; Cebolla, V. Terralia 53, Pages 50–60. Octubre 2005. Unión del Injerto. Available online: https://www.terralia.com/terralias/view_report?magazine_report_id=365 (accessed on 1 September 2018).
9. Acosta Muñoz, A. La Técnica del Injerto en Plantas Hortícolas. Horticom (Extra Viveros I), Extra 2005. pp. 62–65. Available online: http://bit.do/eAgQc (accessed on 15 September 2018).
10. Saravana Kumar, R.M.; Gao, L.X.; Yuan, H.W.; Xu, D.B.; Liang, Z.; Tao, S.C.; Guo, W.B.; Yan, D.L.; Zheng, B.S.; Edqvist, J. Auxin enhances grafting success in *Carya cathayensis* (Chinese hickory). *Planta* **2018**, *247*, 761–772. [CrossRef]
11. Melnyk, C.W. Plant grafting: Insights into tissue regeneration. *Regeneration* **2017**, *4*, 3–14. [CrossRef]
12. De Velasco Alvarado, M.J. Tesis: Anatomía y Manejo Agronómico de Plantas Injertadas en Jitomate. Universidad Autónoma de Chapingo (Mexico), 2013. Available online: https://chapingo.mx/horticultura/pdf/tesis/TESISMCH2013050810128186.pdf (accessed on 15 September 2018).
13. Leonardi, C.; Romano, D. Recent issues on vegetable grafting. *Acta Hortic.* **2004**, *631*, 163–174. [CrossRef]
14. Yang, S.; Xiang, G.; Zhang, S.; Lou, C. Electrical resistance as a measure of graft union. *J. Plant Physiol.* **1993**, *141*, 98–104. [CrossRef]
15. Bletsos, F.A.; Olympios, C.M. Rootstocks and Grafting of Tomatoes, Peppers and Eggplants for Soil-borne Disease Resistance, Improved Yield and Quality. 2008. Available online: http://www.globalsciencebooks.info/Online/GSBOnline/images/0812/EJPSB_2(SI1)/EJPSB_2(SI1)62-73o.pdf (accessed on 15 September 2018).
16. Torii, T.; Kasiwazaki, M.; Okamoto, T.; Kitani, O. Evaluation of graft-take using a thermal camera. *Acta Hortic.* **1992**, *319*, 631–634. [CrossRef]
17. Oda, M.; Maruyama, M.; Mori, G. Water Transfer at Graft Union of Tomato Plants Grafted onto Solanum Rootstocks. *J. Jpn. Soc. Hortic. Sci.* **2005**, *74*, 458–463. [CrossRef]
18. De Miguel, A.; Cebolla, V. Terralia 53, Pages 50–60. Octubre 2005. Unión del Injerto.
19. Bausher, M.G. Road, South Rock, Pierce, Fort Graft Angle and Its Relationship to Tomato Plant Survival. *Hort Sci.* **2013**, *48*, 34–36. Available online: http://hortsci.ashspublications.org/content/48/1/34.short (accessed on 15 September 2018).
20. Turquois, N.; Malone, M. Non-destructive assessment of developing hydraulic connections in the graft union of tomato. *J. Exp. Bot.* **1996**, *47*, 701–707. [CrossRef]
21. Xia, Y.; Sarafis, V.; Campbell, E.O.; Callaghan, P.T. Non invasive imaging of water flow in plants by NMR microscopy. *Protoplasma* **1993**, *173*, 170–176. [CrossRef]
22. Mutisya, S.; Saidi, M.; Opiyo, A.; Ngouajio, M.; Martin, T. Synergistic Effects of Agronet Covers and Companion Cropping on Reducing Whitefly Infestation and Improving Yield of Open Field-Grown Tomatoes. *Agronomy* **2016**, *6*, 42. [CrossRef]
23. Lee, J.-M.; Kubota, C.; Tsao, S.J.; Bie, Z.; Echevarria, P.H.; Morra, L.; Oda, M. Current status of vegetable grafting: Diffusion, grafting techniques, automation. *Sci. Hortic.* **2010**, *127*, 93–105. [CrossRef]
24. Tian, S.; Xu, D. Current status of grafting robot for vegetable. In Proceedings of the 2011 International Conference on Electronic & Mechanical Engineering and Information Technology, Harbin, China, 12–14 August 2011; Volume 4, pp. 1954–1957. [CrossRef]
25. Ito, T. Present state of transplant production practices in Japanese horticultural industry. In Proceedings of the International Symposium on Transplant Production Systems, Transplantant Production Systems, Yokohama, Japan, 21–26 July 1992; Kurata, K., Kozai, T., Eds.; Kluwer Academic Publishers: Amsterdam, The Netherlands, 1992. Available online: http://www.springer.com/us/book/9780792317975 (accessed on 15 September 2018).
26. Lee, J.-M.; Oda, M. Grafting of Herbaceous Vegetable and Ornamental Crops. *Hortic. Rev.* **2003**, *28*. [CrossRef]
27. Nishiura, Y.; Murase, H.; Honami, N.; Taira, T. Devlopment of Plug-in Grafting Robotic System Osaka Prefecture University. In Proceedings of the IEEE Lnternatlonal Conference on Robotics and Automatlon, Nagoya, Japan, 25–27 May 1995; pp. 2510–2517. Available online: http://ieeexplore.ieee.org/abstract/document/525636/ (accessed on 15 September 2018).
28. Oda, M. Grafting of Vegetable Crops. *Osaka Perfecture Univ.* **2002**, *54*, 49–72. Available online: http://repository.osakafu-u.ac.jp/dspace/bitstream/10466/1053/1/KJ00000052064.pdf (accessed on 15 September 2018).
29. Oda, M. Use of Grafted Seedlings for Vegetable Production in Japan. *Acta Hortic.* **2008**, *770*, 15–20. Available online: https://www.actahort.org/books/770/770_1.htm (accessed on 15 September 2018). [CrossRef]

30. Lee, J.-M.; Bang, H.J.; Ham, H.S. Grafting of Vegetables. *Jpn. Soc. Agric. Mach. Food Eng.* **1998**, *67*, 1098–1104. Available online: https://www.jstage.jst.go.jp/article/jjshs1925/67/6/67_6_1098/_pdf (accessed on 15 September 2018). [CrossRef]
31. Singh, H.; Kumar, P.; Chaudhari, S.; Edelstein, M. Tomato Grafting: A Global Perspective. *HortScience* **2017**, *52*, 1328–1336. [CrossRef]
32. Same as 49 Miles, C.; Flores, M.; Estrada, E. *Hoja de Datos de la Extensión, FS052ES. INJERTO de Verduras Berenjenas y Tomates*; Washington State University: Washington, DC, USA, 2013; pp. 1–4. Available online: http://cru.cahe.wsu.edu/CEPublications/FS052ES/FS052ES.pdf (accessed on 15 September 2018).
33. Same as 52 De Miguel Gómez, A. Serie Documentos: El Injerto de Plantas de Tomate. Postcosecha 2011, Www.poscosecha.com–Postharvest, Www.postharvest.biz. Available online: www.poscosecha.com/es/publicaciones/ (accessed on 15 September 2018).
34. Alvarez, C.; Herzog, D. La técnica del injerto (Presentation by Rijk Zwaan Ibérica). Rijk Zwaan Ibérica, 2011. Available online: https://es.slideshare.net/directorarica/presentacin-injerto-semilleros (accessed on 15 September 2018).
35. Oda, M. *Grafting of Vegetables to Improve Greenhouse Production*; Food & Fertilizer Technology Center: Kawana, Japan, 1998; pp. 1–11. Available online: http://www.agnet.org/library.php?func=view&id=20110803135029 (accessed on 15 September 2018).
36. Villasana Rojas, J.A. *Tesis: Efecto del Injerto en la Producción de Tomate (Lycopersicon esculentum Mill.) Bajo Condiciones de Invernadero en Nuevo León*; Universidad Autónoma de Nuevo León: San Nicolás de los Garza, Mexico, 2010; Available online: http://eprints.uanl.mx/5613/1/1080194762%20%281%29.PDF (accessed on 15 September 2018).
37. Yamada, H. Research for Development of the Grafting Robot for Solanaceae. *Tech. Pap. Agric. Mach. Res. Assoc.* **2003**, *65*, 142–149. Available online: https://www.jstage.jst.go.jp/article/jsam1937/65/5/65_5_142/_pdf/-char/ja (accessed on 15 September 2018).
38. Oda, M.; Tsuji, K.; Sasaki, H. Effect of Hypocotyl Morphology on Survival Rate and Growth of Cucumber Seedling Grafted on *Cucurbita* ssp. Japan Agricultural Research Quarterly. 1993. Available online: http://www.jircas.affrc.go.jp/english/publication/jarq/26-4/26-4-259-263.pdf (accessed on 15 September 2018).
39. Same as 46 Oda, M. Vegetable seedling grafting in Japan. *Acta Hortic.* **2007**, *759*, 175–180.
40. Ashraf, M.A.; Kondo, N.; Shiigi, T. Use of Machine Vision to Sort Tomato Seedlings for Grafting Robot. *Eng. Agric. Environ. Food* **2011**, *4*, 119–125. [CrossRef]
41. Tian, S.; Ashraf, M.A.; Kondo, N.; Shiigi, T.; Momin, M.A. Optimization of Machine Vision for Tomato Grafting Robot. *Sens. Lett.* **2013**, *11*, 1190–1194. [CrossRef]
42. Division, P.B.; Crops, O. *Use of Rootstocks in Solanaceous Fruit- Vegetable Production in Japan Plant Breed*; JIRCIS: Tsukuba, Japan, 1980.
43. Camacho Ferre, F.; (Coordinador Obra). *V. autores*. Técnicas de Producción de Cultivos Protegidos. Tomo II. Caja Rural Intermediterránea, Instituto Cajamar. Ediciones Agrotécnicas, S.L.; II, 776. 2003. Available online: http://www.publicacionescajamar.es/pdf/series-tematicas/agricultura/tecnicas-de-produccion-en-cultivos.pdf (accessed on 15 September 2018).
44. Gaion, L.A.; Braz, L.T.; Carvalho, R.F. Grafting in Vegetable Crops: A Great Technique for Agriculture. *Int. J. Veg. Sci.* **2017**, 1–18. [CrossRef]
45. Montsanto, De Ruiter Product guide. Rut. Prod. Guid. 2012. Available online: http://www.deruiterseedstemp.com/global/au/products/Documents/DeRuiterAUProductGuide.pdf (accessed on 15 September 2018).
46. Johnson, S.; Kreider, P.; Miles, C. Vegetable Grafting: Eggplants and Tomatoes. 2013. Available online: http://extension.wsu.edu/publications/wp-content/uploads/sites/54/publications/fs052e.pdf (accessed on 15 September 2018). Washington State Univ.; WSU Mount Vernon Northwestern Washington, Fact Sheet, 1–6.
47. Rivard, C.L.; Louws, F.J. *Grafting for Disease Resistance in Heirloom Tomatoes*; AG-675; North Carolina Cooperative Extension Service: Raleigh, NC, USA, 2007; pp. 1–8.
48. Bumgarner, N.R.; Kleinhenz, M.D. Grafting Guide: A Pictorial Guide to the Cleft and Splice Graft Methods as Applied to Tomato and Pepper. Ohio State University. Research and Development Center, 2014. Available online: http://web.extension.illinois.edu/smallfarm/downloads/50570.pdf (accessed on 15 September 2018).
49. Oda, M. New Grafting Methods for Fruit-Bearing Vegetables in Japan. *Jpn. Agric. Res. Q.* **1995**, *194*, 187–194.
50. Maurya, A.K. Role of Grafting in vegetable Production. Vegetable Grafting. GB Pant University of Agriculture and Technology, Pantnagar Uttarakhand, 2014. Available online: http://es.slideshare.net/ashish7891/vegetable-grafting (accessed on 15 September 2018).

51. Oda, M. Grafting of Vegetables to Improve Greenhouse Production, p. 1–11. Extension Bulletin—Food & Fertilizer Technology Center, 1999. Available online: https://es.scribd.com/document/220401141/Veggie-Grafting (accessed on 15 September 2018).
52. Torres, A.P.; Lopez, G.R. Medición de Luz Diaria Integrada en Invernaderos. Purdue Agriculture. Available online: https://www.extension.purdue.edu/extmedia/HO/HO-238-SW.pdf (accessed on 15 September 2018).
53. Kubota, C.; McClure, M.A.; Kokalis-Burelle, N.; Bausher, M.G.; Rosskopf, E.N. Vegetable grafting: History, use, and current technology status in North America. *Hortscience* **2008**, *43*, 1664–1669.
54. Velasco-alvarado, M.D.J.; Castro-brindis, R.; Avitia-garcía, E.; Castillo-gonzáles, A.M.; Sahagún, J. Proceso de unión del injerto de empalme en jitomate (*Solanum lycopersicum* L.). Revista Mexicana de Ciencias Agrícolas. 2017. Available online: http://www.redalyc.org/html/2631/263152411004/ (accessed on 15 September 2018).
55. Andrews, P.K.; Marquez, C. Graft incompatibility. *Hortic. Rev.* **1993**, *1*, 183–232.
56. De Miguel, Alfredo II Jornadas Sobre Semillas y Semilleros Hortícolas. Libro Editado: ISBN: 84-87564-40-2 1997, 216. Available online: http://www.juntadeandalucia.es/export/drupaljda/1337170141II_Jornadas_sobre_Semillas_y_Semilleros_Horticolas__BAJA.pdf (accessed on 15 September 2018).
57. Tateishi, K. Grafting Watermelon onto Pumpkin Suggested Potential Applications. Available online: https://pdfs.semanticscholar.org/2d91/f2d762af27589ecffd858c5eed25bd98ad93.pdf (accessed on 15 September 2018).
58. Zhao, X. Grading system of tomato grafting machine based on machine vision. In Proceedings of the 8th International Congress on Image and Signal Processing, Shenyang, China, 14–16 October 2015; pp. 604–609. Available online: http://ieeexplore.ieee.org/document/7407950/ (accessed on 15 September 2018).

7

A Novel Graft between Pac Choi (*Brassica rapa* var. *chinensis*) and Daikon Radish (*Raphanus sativus* var. *longipinnatus*)

Tian Gong, Zachary T. Ray, Kylee E. Butcher, Zachary E. Black, Xin Zhao * and Jeffrey K. Brecht

Horticultural Sciences Department, University of Florida, Gainesville, FL 32611, USA; tiangong@ufl.edu (T.G.); ray.ztyler@ufl.edu (Z.T.R.); kbutcher@ufl.edu (K.E.B.); zackblack@ufl.edu (Z.E.B.); jkbrecht@ufl.edu (J.K.B.)
* Correspondence: zxin@ufl.edu

Abstract: Vegetable grafting has primarily been used in the commercial production of high-value crops in the Solanaceae and Cucurbitaceae families. In this study, we explored the feasibility of making a novel graft between pac choi (*Brassica rapa* L. var. *chinensis*) and daikon radish (*Raphanus sativus* L. var. *longipinnatus*) to create a plant with harvestable pac choi leafy vegetable above-ground, and a daikon radish taproot below-ground. 'Mei Qing Choi' pac choi (scion) was grafted onto 'Bora King' daikon radish (rootstock). Grafted pac choi–daikon radish plants did not show a decrease in SPAD value, canopy size, leaf number, leaf area, or above-ground weight compared with self-grafted pac choi plants. However, taproot formation was reduced in grafted pac choi–daikon radish plants, as shown by decreased taproot length, diameter, fresh weight, and dry weight compared with non- and self-grafted daikon radish plants. Surprisingly, grafting with radish increased the photosynthetic rate of the pac choi. This pilot study demonstrated the potential of creating a new pac choi–daikon radish vegetable product to help save growing space and minimize waste at consumption, as pac choi roots are not eaten and radish leaves are usually discarded. The inter-generic grafting between *B. rapa* var. *chinensis* and *R. sativus* var. *longipinnatus* could also provide a unique model system to help elucidate scion-rootstock synergy and above- and below-ground sink competition in horticultural crops.

Keywords: Brassicaceae; growth; mineral content; photosynthesis; rootstock; taproot

---

1. Introduction

Grafting has become an effective practice in the production of high-value solanaceous and cucurbitaceous vegetables to help overcome biotic and abiotic stresses and improve crop productivity [1–3]. Although grafting also has been used as a tool in plant physiology studies of Arabidopsis (*Arabidopsis thaliana* L.) [4], for accelerating the breeding work of common beans (*Phaseolus vulgaris* L.) [5], and for combating Verticillium wilt of globe artichoke (*Cynara cardunculus* L. subsp. *Scolymus*) [6], grafting in other vegetable species beyond Solanaceae and Cucurbitaceae is generally not practiced commercially. Interestingly, some attempts have been made to explore the feasibility of grafting vegetable plants in *Brassicaceae*. Oda et al. [7] tested inter-varietal, inter-specific, and inter-generic grafting among cabbage (*Brassica oleracea* L. var. *capitata*), kale (*Brassica oleracea* var. *sabellica*), kohlrabi (*Brassica oleracea* var. *gongylodes*), Chinese cabbage (*Brassica rapa* L. subsp. *pekinensis*), turnip (*Brassica rapa* subsp. *rapa*), Japanese mustard (Takana) (*Brassica juncea* L. var. *integrifolia*), and Japanese radish (*Raphanus sativus* L. var. *longipinnatus*) and obtained successful grafts. Particularly, an adhesive and hardener system was developed for making grafts between Chinese cabbage (scion) and turnip (rootstock) [8]. Recently, Chen et al. [9] evaluated the survival rate of cabbage grafted onto Chinese kale (*B. oleracea* Alboglabra group) rootstocks and assessed the feasibility of using grafting to improve cabbage head quality.

The effort of cruciferous vegetable grafting has not only broadened the potential use of vegetable grafting as a management tool, but also presents the possibility of creating a novel vegetable product with added value. In the case of grafted Chinese cabbage/turnip plants, the above-ground portion of Chinese cabbage—a leafy vegetable, and the below-ground portion of turnip—a root vegetable, can be harvested from the same plant. This type of rootstock–scion combination holds promise for space saving in small-scale intensive cultivation systems. Moreover, the grafted Chinese cabbage/turnip vegetable may possess added economic value with minimal waste, since many consumers may prefer not to eat turnip leaves and could be drawn to the novelty of this new product. However, the Chinese cabbage and turnip grafting study of Oda and Nakajima [8] only reported a 50% graft survival rate and observed restricted development of the Chinese cabbage head.

In this proof of concept study, we grafted the pac choi (*B. rapa* var. *chinensis*) scion onto the daikon radish (*R. sativus* var. *longipinnatus*) rootstock to generate a vegetable plant that produced a pac choi leafy vegetable above-ground and an edible daikon radish root below-ground. Pac choi and daikon radish are among specialty vegetables increasingly grown for local markets in the U.S. Although edible, daikon radish leaves are often discarded at consumption. Recent genetic studies supported the feasibility of making successful inter-generic grafts between *B. rapa* and *R. sativus*. Yang et al. [10] sequenced the chloroplast noncoding region and found that *R. sativus* was closely related to *B. rapa/oleracea* and proposed that *Raphanus* was derived from hybridization between *B. rapa/oleracea* and *B. nigra*, the two evolutionary lineages in the genus Brassica. Furthermore, the reciprocal hybridization between *R. sativus* and *B. rapa* has been proven viable [11]. Vigorous growth was also observed for most of the successful primary hybrids between *B. rapa* and *R. sativus* [12]. On the other hand, according to Tonosaki et al. [13], when hybridized with *R. sativus*, only one particular breeding line of *B. rapa* ('Shogoin-kabu') successfully produced hybrid seeds, whereas most other lines failed due to embryo breakdown.

By grafting the pac choi scion onto the daikon radish rootstock, the objectives of this pilot experiment were to examine the feasibility of developing successful grafts for harvesting both pac choi leaves and daikon radish taproot from the same plant, and to compare the growth and development of grafted plants with self-grafted and non-grafted pac choi and daikon radish plants.

2. Materials and Methods

Two experiments were carried out in this study. The first experiment was a pilot study to test the feasibility of grafting pac choi onto daikon radish. The second experiment was intended to provide a better understanding of above-ground growth and below-ground development of this unique scion-rootstock system over an extended post-grafting period of plant establishment. In both experiments, 'Bora King' (BK), a daikon radish with purple taproots (Johnny's Selected Seeds, Winslow, ME, USA) was used as the rootstock, while 'Mei Qing Choi' (MQ) pac choi (Johnny's Selected Seeds) was used as the scion. They were selected based on our preliminary study in which these two cultivars were found to be compatible for grafting and have similar hypocotyl diameters.

2.1. Setup of the Pilot Experiment

Pac choi and daikon radish were seeded on 7 and 13 November 2016, respectively. The pac choi was seeded 6 d earlier than the daikon radish in order to match the stem diameter of the seedlings at grafting, as the daikon radish germinated and emerged much quicker than the pac choi based on a preliminary seeding test. All the seeds were sown in 72-cell Speedling trays (Speedling Inc., Ruskin, FL, USA) and filled with Fafard-2 potting mix (Sun Gro Horticulture, Agawam, MA, USA) containing a mixture of peat moss, perlite, vermiculite, and dolomite lime. Plants were grown in a greenhouse at the University of Florida campus (Gainesville, FL, USA). Water-soluble fertilizer 20N-8.7P-16.7K (Jack's Classic; Jr Peters Inc., Allentown, PA, USA) was applied on 17 and 28 November at a nitrogen (N) concentration of 200 mg L^{-1}.

Plants were grafted on 30 November 2016 (0 d after grafting (DAG)) using the splice grafting method [1]. Twenty-four plants were grafted using seedlings with the most consistent growth. With the purpose of ensuring consistent grafting quality, only a small number of plants were grafted in this pilot experiment after earlier attempts at practicing the grafting technique. The daikon radish seedlings were severed using a double edge razor blade at approximately a 45-degree angle below cotyledons to remove the shoots, with pac choi scions cut at the hypocotyl with the same angle just above the soil surface. The cut surfaces of the pac choi scion and the daikon radish seedlings with shoot removal were conjoined using a 1.5 mm silicone grafting clip (Johnny's Selected Seeds). Grafted plants were then placed in a healing chamber constructed by wrapping a metal shelving unit with thin plastic film (Uline Econo-Wrapper (0.02 mm), Uline corporation, Pleasant Prairie, WI, USA) in a temperature-controlled room with air temperature set at 23 °C and relative humidity (RH) at 99%. Light was provided by two, 54-watt T5 fluorescent lights (Philips Lighting Company, Somerset, NJ, USA) at a photosynthetic photon flux density (PPFD) of 56 μmol m^{-2} s^{-1} at seedling canopy level for 12 h each day. An additional plastic tray with a wet sheet of germination paper was placed inside the healing chamber to help maintain humidity.

From 5 DAG, the healing chamber was gradually cut open and the ambient RH setting was reduced to 60%. At 8 DAG, the plastic film was completely removed, and grafts remained exposed in the temperature-controlled room until 13 DAG. Water was applied to plants by filling the bottom of the tray for absorption. All the grafted plants were transferred to a greenhouse at 13 DAG, and graft survival rate was determined by counting the number of live and dead plants; only plants with turgid leaves were counted as living. At harvest, the number of surviving grafted plants was counted again for calculation of the final graft survival rate, as some plants severely declined following transplanting into pots.

At 16 DAG, surviving grafted plants were transplanted into 11.36 L black plastic pots (1200C; Hummert International, Earth City, MO, USA) filled with Fafard-2 soilless mix (Sun Gro Horticulture) for continued monitoring of the survival of the grafted plants. In addition, five plants of non-grafted 'Mei Qing Choi' and 'Bora King' were potted as controls. All the plants were placed on the greenhouse bench following a completely randomized design. Organic fertilizer MicroSTART60 3N-0.9P-2.5K (Perdue AgriRecycle, LLC., Seaford, DE, USA) was applied to each pot at the rate of 80 g/pot. Drip irrigation was used by placing one 1.89 L h^{-1} emitter (Woodpecker pressure compensating junior dripper; Netafim USA, Fresno, CA, USA) in each pot; plants were watered once a day for 3 min. Irrigation increased to twice per day for 2 min each time starting at 56 DAG. Insecticidal soap (Safer Brand; Woodstream Corporation, Lancaster, PA, USA) was sprayed at 56 DAG and lacewing larvae (*Chrysoperla rufilabris* (Neuroptera: Chrysopidae); Rincon-Vitova Insectaries, Ventura, CA, USA) were released at 64 DAG for aphid control. The average day and night temperatures of the greenhouse during the plant growth were 22.8 °C and 16.5 °C, respectively.

2.2. Setup of the Follow-Up Experiment

A follow-up experiment was conducted in 2019 to further explore the above-ground growth and below-ground taproot development in grafted pac choi–daikon radish plants. 'Mei Qing Choi' (MQ) pac choi was grafted onto 'Bora King' (BK) daikon radish (MQ/BK), while non-grafted pac choi (MQ) and daikon radish (BK) as well as self-grafted pac choi (MQ/MQ) and daikon radish (BK/BK) were used as controls. A randomized complete block design with four replications (blocks) and ten grafted plants per treatment per replication (block) was used in the grafting experiment. MQ and BK were seeded into 72-cell trays on 8 and 14 February 2019, respectively. Fish & seaweed organic liquid fertilizer 2N-1.3P-0.8K (Neptune's Harvest, Gloucester, MA, USA) and 0N-0P-41.5K potassium sulfate (Big K; JHBiotech, Inc., Ventura, CA, USA) were applied at concentrations of 200 mg L^{-1} N and 200 mg L^{-1} K_2O at the seedling growth stage on 18 and 26 Feb. Plants were grafted on 1 March 2019 using the aforementioned grafting method. Grafted plants were healed in an air-conditioned laboratory room with the same set up of healing chamber as in 2016. Supplemental light was provided for 10 h each day during the healing process. The air temperature and RH of the laboratory room were 23.8 ± 0.5 °C and 47.6 ± 13.1%. The plastic film was completely removed at 8 DAG. Water was gently sprayed onto the soil surface using a wash bottle when needed to avoid wetting foliage. All the grafted plants were moved into a greenhouse at 13 DAG where the average day and night temperatures were 27.1 °C and 20.7 °C, respectively. Graft survival rate was determined for each grafting treatment in each replication by counting the number of live and dead plants at 17 DAG. At 19 DAG, 24 plants from each treatment with healthy and consistent growth were chosen and randomly reassigned to four blocks with six plants in each block for further evaluation of the growth of the grafted plants in a greenhouse pot study, following a randomized complete block design. Plants were transplanted into 11.36 L black plastic pots filled with PRO-MIX premium organic vegetable and herb mix (Premier Tech Ltd., Quakertown, PA, USA) which contained 60–75% peat moss plus peat humus, compost, perlite, gypsum, limestone, organic fertilizer, and mycorrhizae. Drip irrigation was used by placing one 1.89 L h^{-1} emitter in each pot; the plants were watered twice a day for 3 min per cycle between 21 and 39 DAG and irrigation increased to 4 min per cycle thereafter. Adventitious roots developed from the graft union area were monitored and removed once a week as needed after the plants were transplanted into the pots.

2.3. *Plant Growth Measurements*

In the 2019 follow-up experiment, leaf relative chlorophyll content and canopy size were measured at 33 and 41 DAG. A SPAD 502 Plus Chlorophyll Meter (Spectrum Technologies, Aurora, IL, USA) was used to measure leaf relative chlorophyll content on three randomly chosen plants per treatment per block by averaging four readings obtained from two distal areas of the leaf blade for each of the two most recent mature leaves per plant. The canopy size was measured on 3 plants of MQ/BK, MQ/MQ, and MQ for each block using digital photographs processed with ImageJ/Fiji (version 2.0.0) [14]. A ruler held in the frame of each photograph set the scale for pixels per linear cm and enabled digital measurement of length and width of the plant canopy. The canopy size was then determined by multiplying the canopy length and width.

2.4. *Gas Exchange Measurements*

Gas-exchange was measured in the 2019 follow-up experiment at 34 and 46 DAG between 10:00 am and 3:00 pm by using an open gas exchange system (Li-6800; Li-Cor Inc., Lincoln, NE, USA) on three plants per treatment per block. Leaf transpiration rate (E, mmol H_2O m^{-2} s^{-1}), net CO_2 assimilation rate (A, mmol CO_2 m^{-2} s^{-1}), intercellular CO_2 concentration (Ci, μmol CO_2 mol^{-1} air), and stomatal conductance to water (gsw, mmol H_2O m^{-2} s^{-1}) were measured at steady-state on the third (fully expanded) leaf from the top of each plant [15]. The PPFD was set at 800 μmol m^{-2} s^{-1}, with CO_2 concentration at 400 ppm, vapor pressure deficit at 1.2 kPa, and leaf temperature at 27–29 °C. Instantaneous water use efficiency (iWUE) (μmol CO_2 $mmol^{-1}$ H_2O) was calculated as A/E [16] and stomatal conductance (Gs, mol m^{-2} s^{-1}) was calculated as gsw/1.6 [17].

2.5. *Yield Components and Biomass Accumulation at Harvest*

For the 2016 pilot study, the above-ground part (above soil line) of all the plants of MQ/BK, non-grafted MQ, and non-grafted BK were harvested at 68 DAG. The number of leaves longer than 4 cm were counted for each plant. The MQ/BK and non-grafted BK were then uprooted, and the taproots were separated and rinsed with water to remove excess potting soil from the roots. Taproot length (from the stem base to the end of the radish taproot) of each harvested plant was recorded, and the diameter of the widest part of each radish taproot was measured with a digital caliper. For the 2019 follow-up experiment, harvest and destructive sampling were carried out at 47 DAG. Five out of six plants per treatment per replication were randomly sampled. The above-ground part (above soil line) of each plant was removed from the pot, and leaves longer than 4 cm were counted and scanned with a leaf area meter (LI-3100; Li-Cor Inc., Lincoln, NE, USA). Only taproots from MQ/BK, BK, and BK/BK were harvested and cleaned. Pak choi and daikon radish leaves and taproots from the 2016 and 2019 experiments were first weighed then dried at 65 °C for 7 d (until constant weight) to determine the above-ground and below-ground fresh and dry biomass.

2.6. Mineral Nutrient Contents in Leaf and Root Tissues

In the 2019 experiment, the dried root samples from the BK, BK/BK, and MQ/BK treatments and dried leaf samples from the MQ, MQ/MQ, and MQ/BK treatments were ground using a Thomas Wiley Laboratory mill (Model 4; Arthur H. Thomas Company, Philadelphia, PA, USA) and sent to Waters Agricultural Laboratories (Camilla, GA, USA) to measure the concentrations of the macronutrients N, phosphate (P), sulfate (S), potassium (K), calcium (Ca), and magnesium (Mg) and the micronutrients boron (B), zinc (Zn), manganese (Mn), iron (Fe), and copper (Cu). Nutrient accumulation was calculated by multiplying the nutrient concentration by dry tissue biomass.

2.7. Statistical Analyses

The pilot study followed a completely randomized design with five replications and one plant per replication. In the follow-up experiment, a randomized complete block design with four replications (blocks) and ten plants per experimental unit was used before plants were transplanted to larger pots, when the number was reduced to six plants per experimental unit. Data were analyzed using a linear mixed model in the GLIMMIX procedure of SAS (SAS Version 9.4 for Windows; SAS Institute, Cary, NC, USA). Some data were transformed by taking the square root to meet the assumptions of the model (normality, homogeneity, linearity) as needed, while results were presented using the original data following statistical analysis. Fisher's least significant difference (LSD) test ($\alpha = 0.05$) was conducted for multiple comparisons of different measurements among treatments.

3. Results and Discussion

3.1. Graft Survival Rate

In the 2016 pilot study, the survival rate of MQ/BK was 95.8% at 13 DAG but decreased to 87.5% at 68 DAG (data not shown). In the 2019 experiment, there was no difference in survival rate ($p = 0.483$) between BK/BK (79%), MQ/MQ (92%), and MQ/BK (77%) at 17 DAG (data not shown). The relatively high survival rates of MQ/BK indicated good graft compatibility between 'Mei Qing Choi' pac choi and 'Bora King' daikon radish. Both the edible pac choi leafy green part of the plant and the radish taproot developed in MQ/BK (Figure 1A–C). According to Oda and Nakajima [8], 'Taibyoh 60-nichi' Chinese cabbage grafted onto 'Taibyoh hikari' turnip had a survival rate of only approximately 50%; however, it was attributed to the small size of the seedlings at grafting rather than graft incompatibility. In our studies, we also used 1.5 mm or even smaller grafting clips to graft younger and tender seedlings as the hypocotyl tissue of radish and pac choi plants tend to become more lignified as they grow. The lower survival rate observed in the 2019 experiment may have been due to an issue with properly matching the plant stem diameters at grafting since the hypocotyl of the daikon radish plant had grown thicker than expected at the time of grafting. As stem diameter and alignment of cambial tissues affect the success of grafting [18–20], matching plant stem diameters between these two species, which have thin hypocotyls at the optimum stage for grafting, is a challenge for achieving successful grafts. In this study, daikon radish was seeded 6 d earlier than pac choi to help match their stem diameters, but with less than desirable results, especially in the 2019 experiment owing to the seasonal variability of greenhouse conditions that led to unpredictable growth rate of plants. As shown by Hayashida et al. [21] and Kwack et al. [22], the hypocotyl growth of pac choi and radish seedlings can be manipulated by light quality and intensity. Employing a more controlled environment for growing seedlings until ready for grafting seems to be advisable for future work.

Figure 1. *Cont.*

(G)

Figure 1. Grafted plants with 'Mei Qing Choi' pac choi as scion and 'Bora King' daikon radish as rootstock. (**A**) A well-healed grafted pac choi–daikon radish seedling. (**B**) Formation of the taproot of daikon radish grafted with pac choi. (**C**) Longitudinal section of the grafted pac choi–daikon radish plant at harvest. (**D**) Comparison between self-grafted 'Bora King' (left), 'Mei Qing Choi' pac choi grafted onto 'Bora King' daikon radish (middle), and non-grafted 'Bora King' (right) at harvest. (**E**) Longitudinal section of the graft union area of grafted pac choi–daikon radish plant at harvest. (**F**) Longitudinal section of the graft union area of self-grafted 'Bora King' daikon radish at harvest. (**G**) Longitudinal section of the graft union area of self-grafted 'Mei Qing Choi' pac choi plant at harvest.

3.2. Plant Growth Parameters

SPAD and canopy size were measured at 33 and 41 DAG in the 2019 experiment (Table 1). There was a significant difference in SPAD among treatments at 33 DAG, but no difference was found at 41 DAG. At 33 DAG, non-grafted pac choi showed lower levels of leaf SPAD values than non- and self-grafted daikon radish. The lack of difference in SPAD between MQ/BK, MQ, and MQ/MQ indicated that BK as a rootstock did not impair accumulation of chlorophyll by MQ. The similar canopy size between MQ/BK, MQ, and MQ/MQ at both 33 and 41 DAG (Table 1) suggested that grafting with daikon radish did not reduce the leaf growth and expansion of pac choi.

Table 1. Relative chlorophyll content and canopy size of grafted, self-grafted, and non-grafted pac choi and daikon radish plants at 33 d after grafting (DAG) and 41 DAG in the 2019 experiment.

Treatment [z]	Relative Chlorophyll Content (SPAD)		Canopy (cm^2 $Plant^{-1}$)	
	33 DAG	41 DAG	33 DAG	41 DAG
BK	37.2 ± 0.8 a [y]	40.6 ± 0.7 a	-	-
BK/BK	36.7 ± 0.8 ab	40.3 ± 0.7 a	-	-
MQ	34.1 ± 0.8 c	40.7 ± 0.7 a	54.26 ± 2.71 a	78.32 ± 3.35 a
MQ/MQ	34.6 ± 0.8 bc	42.1 ± 0.7 a	58.90 ± 2.71 a	76.90 ± 3.35 a
MQ/BK	33.3 ± 0.8 c	40.0 ± 0.7 a	52.32 ± 2.71 a	75.42 ± 3.35 a
p value	0.006	0.194	0.150	0.813

[z] BK = Non-grafted 'Bora King' daikon radish; BK/BK = Self-grafted 'Bora King' daikon radish; MQ = Non-grafted 'Mei Qing Choi' pac choi; MQ/MQ = Self-grafted 'Mei Qing Choi' pac choi; MQ/BK = 'Mei Qing Choi' pac choi grafted onto 'Bora King' daikon radish. [y] Mean ± SE (standard error); means followed by the same letter are not significantly different at $p \leq 0.05$ according to Fisher's LSD test.

As shown in Figure 1B, above-ground pac choi and below-ground daikon radish taproot developed normally in MQ/BK plants. We observed cavities in the vascular bundle connections at the graft union area in grafted pac choi plants with the daikon radish rootstock but not in self-grafted daikon radish plants (Figure 1E,F), similar to what was reported in grafted Chinese cabbage/turnip plants [8]. In our experiments, grafting was carried out at 16 d after sowing (DAS) for daikon radish and 23 DAS for pac choi. According to Liu et al. [23], as early as 16 DAS, tuberization began in turnip (*B. rapa* subsp. *Rapa*) and the center part of the upper hypocotyl which accounted for about 50% of the cross section area was occupied by pith cells, with an actively dividing cambium circle and a thin xylem ring. When pac choi was grafted at 23 DAS, more than 50% of the center part of the upper hypocotyl of pac choi consisted of secondary xylem cells with highly lignified cell walls. This discrepancy in hypocotyl structure between scion and rootstock seedlings likely resulted in the formation of the cavity inside the graft union during healing as observed in our study (Figure 1E), especially considering that the cavity did not exist in self-grafted daikon radish (Figure 1F) or pac choi (Figure 1G).

Leaf number and taproot length and diameter were measured at harvest in both experiments (Table 2). In the 2016 pilot study, MQ/BK had more leaves than non-grafted BK, but did not differ significantly from MQ. In the 2019 experiment, MQ/BK had 35% and 26% more leaves than self- and non-grafted BK, respectively, but no difference was found between self- and non-grafted BK. Similar leaf numbers were observed for MQ/BK, MQ/MQ, and MQ. Total leaf area was also measured in the 2019 experiment and MQ/BK had smaller leaf area than all other treatments. In both 2016 and 2019, MQ/BK produced significantly shorter and smaller taproots than non-grafted BK and in 2019, MQ/BK was also smaller in taproot diameter than BK/BK (Table 2 and Figure 1D). Our results were consistent with Zheng et al. [24], who reported that the diameter of turnip was significantly smaller when grafted with rapeseed (*B. rapa* subsp. *oleifera*) than self-grafted turnip. BK/BK did not differ significantly from BK in taproot length but was 7% smaller in taproot diameter.

Table 2. Total leaf number and area and taproot length and diameter of grafted, self-grafted, and non-grafted pac choi and daikon radish plants at harvest in the 2016 pilot study and the 2019 experiment.

Treatment [z]	Leaf Number (no. $Plant^{-1}$)	Leaf Area (mm^2 $Plant^{-1}$)	Taproot Length (cm)	Taproot Diameter (mm)
		2016		
BK	18.9 ± 2.3 b [y]	-	10.9 ± 0.5 a	68.55 ± 3.98 a
MQ	26.6 ± 2.4 ab	-	-	-
MQ/BK	28.1 ± 2.4 a	-	8.2 ± 0.5 b	51.92 ± 3.56 b
p value	0.020		0.005	0.017
		2019		
BK	15.7 ± 0.6 b	2521.73 ± 60.62 a	9.2 ± 0.5 a	45.24 ± 0.71 a
BK/BK	14.7 ± 0.5 b	2401.06 ± 60.62 ab	8.4 ± 0.5 a	42.20 ± 0.71 b
MQ	20.4 ± 0.6 a	2423.26 ± 60.62 ab	-	-
MQ/MQ	19.5 ± 0.6 a	2311.67 ± 60.62 b	-	-
MQ/BK	19.8 ± 0.6 a	2080.12 ± 60.62 c	6.8 ± 0.5 b	34.94 ± 0.71 c
p value	<0.001	0.002	0.005	<0.001

[z] BK = Non-grafted 'Bora King' daikon radish; BK/BK = Self-grafted 'Bora King' daikon radish; MQ = Non-grafted 'Mei Qing Choi' pac choi; MQ/MQ = Self-grafted 'Mei Qing Choi' pac choi; MQ/BK = 'Mei Qing Choi' pac choi grafted onto 'Bora King' daikon radish. [y] Mean ± SE (standard error); means followed by the same letter are not significantly different at $p \leq 0.05$ according to Fisher's LSD test.

The primary root axis of radish consists of two anatomically distinct parts. The upper part originates from the hypocotyl whereas the lower part is true root tissue. Both lower and upper regions of the radish root thicken to form succulent tissue by increases in both cell number and cell size [25,26]. In this grafting experiment, the cut made on the daikon radish plant was in the thickening region of the hypocotyl as demonstrated by the longitudinal section of the graft union area of self-grafted daikon radish plant (Figure 1F). Very likely, grafting pac choi with radish shortens the hypocotyl part that could contribute to the formation of the taproot, leading to reduced taproot length compared with non-grafted radish, while self-grafting radish does not involve any loss of hypocotyl tissue. Furthermore, it has been found that in turnip the hypocotyl tissue is the main contributor to underground tuber development, and hypocotyl excision led to a lower expression level of genes controlling tuberization, leading to a substantial inhibition of tuber formation [24].

3.3. Gas-Exchange Parameters

Leaf transpiration rate, net CO_2 assimilation rate, intercellular CO_2 concentration, stomatal conductance, and instantaneous water use efficiency (iWUE) were compared for MQ/BK, MQ, MQ/MQ, BK, and BK/BK at 34 and 46 DAG in the 2019 experiment (Table 3). No difference in leaf transpiration rate was observed at 34 DAG, while at 46 DAG, MQ/BK had a similar transpiration rate as MQ and MQ/MQ, and all three treatments showed a 95% increase of transpiration rate on average than BK and BK/BK. Grafting significantly increased the net CO_2 assimilation rate of MQ/BK compared with MQ and MQ/MQ by 15% and 28%, respectively, at 34 DAG, while no difference was observed between MQ/BK and BK/BK. At 46 DAG, MQ/BK showed a net CO_2 assimilation rate that was 48% and 45% higher than MQ and MQ/MQ, respectively, but it did not differ significantly from BK/BK. MQ/BK also had a 21% higher net CO_2 assimilation rate than BK at 46 DAG. Very likely, MQ/BK had a stronger sink strength than MQ/MQ and MQ, which contributed to the higher photosynthetic rate observed [27]. Interestingly, at 34 DAG, MQ and MQ/MQ had similar intercellular CO_2 concentrations, which were significantly higher than that of other treatments. At 46 DAG, MQ/BK, MQ, and MQ/MQ had higher intercellular CO_2 concentration than BK and BK/BK and the same trend was observed for stomatal conductance at 46 DAG although no difference in stomatal conductance was detected at 34 DAG. At 34 DAG, MQ/BK had 36% and 53% higher iWUE than MQ and MQ/MQ, but did not differ from BK and BK/BK. However, at 46 DAG, MQ/BK, MQ, and MQ/MQ exhibited a similar level of iWUE which was significantly lower than that of BK and BK/BK.

Table 3. Leaf transpiration rate (E), net CO_2 assimilation rate (A), intercellular CO_2 concentration (Ci), stomatal conductance (gs), and instantaneous water use efficiency (iWUE) of grafted, self-grafted, and non-grafted pac choi and daikon radish plants at 34 d after grafting (DAG) and 46 DAG in the 2019 experiment.

Treatment [y]	E (mmol H_2O m^{-2} s^{-1})		A (μmol CO_2 m^{-2} s^{-1})		Ci (μmol CO_2 mol^{-1} air)		Gs (mol m^{-2} s^{-1})		iWUE (μmol CO_2 $mmol^{-1}$ H_2O) [z]	
	34 DAG	46 DAG	34 DAG	46 DAG	34 DAG	46 DAG	34 DAG	46 DAG	34 DAG	46 DAG
BK	4.81 ± 0.71 a [x]	2.80 ± 0.79 b	21.41 ± 0.75 a	17.22 ± 0.89 b	298.14 ± 11.10 b	251.33 ± 10.59 b	0.28 ± 0.05 a	0.15 ± 0.05 b	4.92 ± 0.56 a	7.26 ± 0.54 a
BK/BK	4.68 ± 0.71 a	4.18 ± 0.79 b	19.66 ± 0.75 ab	18.93 ± 0.89 ab	305.43 ± 11.10 b	273.09 ± 10.59 b	0.27 ± 0.05 a	0.25 ± 0.05 b	4.55 ± 0.56 ab	6.27 ± 0.54 a
MQ	4.95 ± 0.71 a	6.61 ± 0.79 a	16.10 ± 0.75 c	14.06 ± 0.89 c	330.36 ± 11.10 a	345.22 ± 10.59 a	0.32 ± 0.05 a	0.41 ± 0.05 a	3.59 ± 0.56 bc	2.49 ± 0.54 b
MQ/MQ	5.51 ± 0.71 a	6.33 ± 0.79 a	14.41 ± 0.75 c	14.31 ± 0.89 c	331.63 ± 11.10 a	346.80 ± 10.59 a	0.33 ± 0.05 a	0.40 ± 0.05 a	3.20 ± 0.56 c	2.51 ± 0.54 b
MQ/BK	4.35 ± 0.71 a	7.47 ± 0.79 a	18.44 ± 0.75 b	20.76 ± 0.89 a	299.39 ± 11.10 b	336.66 ± 10.59 a	0.25 ± 0.05 a	0.45 ± 0.05 a	4.88 ± 0.56 a	2.94 ± 0.54 b
p value	0.564	<0.001	<0.001	<0.001	0.011	<0.001	0.362	0.001	0.024	<0.001

[z] Instantaneous water use efficiency (iWUE) = net CO_2 assimilation rate (A)/transpiration rate (E). [y] BK = Non-grafted 'Bora King' daikon radish; BK/BK = Self-grafted 'Bora King' daikon radish; MQ = Non-grafted 'Mei Qing Choi' pac choi; MQ/MQ = Self-grafted 'Mei Qing Choi' pac choi; MQ/BK = 'Mei Qing Choi' pac choi grafted onto 'Bora King' daikon radish. [x] Mean ± SE (standard error); means followed by the same letter are not significantly different at $p \leq 0.05$ according to Fisher's LSD test.

Lower leaf transpiration rate and intercellular CO_2 concentration of radish compared with pac choi observed in the later growth stage could be owing to different leaf structures. Pac choi leaves are fleshy and glossy, while radish has trichomes on both upper and lower leaf surfaces [28]. It has been suggested that trichome density is negatively related to transpiration rate and CO_2 diffusion rate as trichomes can increase boundary layer resistance [29–31]. The trichomes on daikon radish leaves might have affected the leaf transpiration rates and intercellular CO_2 concentrations measured in this study. Further examination is needed to directly compare the intrinsic leaf structures of pac choi and daikon radish plants for their effects on gas exchange and gas exchange measurements.

Most water loss from leaves of intact plants is generally through open stomatal apertures [32], thus the higher stomatal conductance of MQ/BK was likely the driving factor for its lower iWUE compared with BK and BK/BK despite its higher net CO_2 assimilation rate. It has been found in tobacco (*Nicotiana tabacum* L.) that stomatal conductance did not always parallel photosynthetic capacity changes [33,34], which could partially explain the relatively high stomata conductance, but low net CO_2 assimilation rate observed in MQ and MQ/MQ.

3.4. Leaf and Taproot Harvest and Biomass Partition

In the 2016 pilot study, MQ and MQ/BK had 151% and 104% higher above-ground fresh weight (FW) than BK, and the former two did not differ significantly (Table 4). No difference was detected in above-ground dry weight (DW) between these three treatments. MQ/BK produced significantly lower below-ground FW and DW compared with BK, while similar levels of total FW and DW were observed between MQ/BK and BK.

Table 4. Above-ground, below-ground, and total fresh weight (FW) and dry weight (DW) of grafted, self-grafted, and non-grafted pac choi and daikon radish plants at harvest in the 2016 pilot study and the 2019 experiment.

Treatment [z]	Above-Ground FW (g Plant^{-1})	Below-Ground FW (g Plant^{-1})	Above-Ground DW (g Plant^{-1})	Below-Ground DW (g Plant^{-1})	Total FW (g Plant^{-1})	Total DW (g Plant^{-1})
			2016			
BK	240.07 ± 54.58 b [y]	332.15 ± 33.23 a	20.41 ± 2.52 a	19.84 ± 1.56 a	572.21 ± 67.48 a	41.08 ± 4.00 a
MQ	602.76 ± 48.82 a	-	25.50 ± 2.25 a	-	-	-
MQ/BK	490.63 ± 48.82 a	110.44 ± 29.72 b	24.42 ± 2.25 a	7.60 ± 1.20 b	601.07 ± 60.36 a	32.02 ± 3.10 a
p value	0.001	<0.001	0.331	<0.001	0.792	0.142
			2019			
BK	171.64 ± 6.62 d	103.47 ± 5.25 a	12.85 ± 0.25 a	6.58 ± 0.33 a	275.17 ± 4.44 b	19.45 ± 0.41 a
BK/BK	162.27 ± 6.62 d	77.55 ± 4.57 b	12.34 ± 0.25 a	5.33 ± 0.30 b	239.91 ± 4.44 c	17.67 ± 0.41 b
MQ	329.83 ± 6.62 a	-	10.09 ± 0.25 b	-	-	-
MQ/MQ	293.51 ± 6.62 b	-	8.98 ± 0.25 c	-	-	-
MQ/BK	269.08 ± 6.62 c	43.59 ± 3.45 c	9.43 ± 0.25 bc	2.24 ± 0.20 c	312.93 ± 4.44 a	11.68 ± 0.41 c
p value	<0.001	<0.001	<0.001	<0.001	<0.001	<0.001

[z] BK = Non-grafted 'Bora King' daikon radish; BK/BK = Self-grafted 'Bora King' daikon radish; MQ = Non-grafted 'Mei Qing Choi' pac choi; MQ/MQ = Self-grafted 'Mei Qing Choi' pac choi; MQ/BK = 'Mei Qing Choi' pac choi grafted onto 'Bora King' daikon radish. [y] Mean ± SE (standard error); means followed by the same letter are not significantly different at $p \leq 0.05$ according to Fisher's LSD test.

In the 2019 study, MQ/BK showed a significant reduction in above-ground FW compared with MQ and MQ/MQ by 18% and 8%, respectively (Table 4). MQ, MQ/MQ, and MQ/BK on average had 78% higher above-ground FW than the average of BK and BK/BK. However, BK and BK/BK had significantly higher above-ground DW than other treatments and MQ/BK did not differ significantly in above-ground DW from MQ and MQ/MQ. Grafting with pac choi significantly decreased the below-ground FW and DW of daikon radish compared with non- and self-grafted daikon radish. BK had higher below-ground FW and DW than BK/BK by 33% and 23%, respectively. MQ/BK exhibited a significantly higher total FW but a reduction in total DW compared with BK and BK/BK. The water content of both radish taproots and leaves was about 93%, whereas it reached 97% in the pac choi leaves, indicating that the pac choi leafy part had a disproportional contribution to the FW of MQ/BK. The differences in total FW and DW between 2016 and 2019 studies may be due to the different growing seasons and the time between grafting and harvest as the period from grafting to harvesting was 45% longer in 2016 compared with the 2019 study. Overall, the two experiments suggested that grafting between 'Mei Qing Choi' pac choi and 'Bora King' daikon radish had a greater impact on radish taproot development than the influence on pac choi leaf growth as the taproot DW of MQ/BK was significantly lower than BK, but the leaf DW was similar to that of MQ.

The grafting procedure per se negatively affected biomass accumulation in edible part in both self-grafted BK/BK and MQ/MQ compared with their non-grafted counterparts. Grafting can be viewed as mechanical wounding that triggers redistribution of the local resources or mobilization of resources from neighboring tissues to the injured part [35], which may divert the resources that could be used for plant growth. Moreover, wounding could induce jasmonic acid production, leading to suppression of mitosis [36], and thus reduce cell number and lead to reduced plant growth.

Gibberellins (GAs) have long been known to inhibit potato tuberization possibly by their involvement in photoperiodic control of tuber formation [37,38]. Grafting potato (*Solanum tuberosum* L.) with tomato (*Solanum lycopersicum* L.) decreased stolon (underground shoot) number and length as well as tuber number, but increased the gibberellic acid (GA_3) content of stolon and tuber compared with self-grafted potato [39]. The level of GAs has also been reported to play a vital role in carrot (*Daucus carota* L. var. *sativus*) elongation and expansion [40]. Leaf application of GA_3 inhibited tuberous root growth but improved shoot growth in radish, while leaf application of paclobutrazol, an inhibitor of gibberellin biosynthesis, improved taproot growth [41]. Auxin may also affect hypocotyl-tuber growth in turnip as show in in-vitro studies [23]. Peres et al. [38] grafted tomato mutants, which were incapable of certain hormone biosynthesis or photomorphogenesis, onto potato plants and suggested that failure to produce certain chemicals by the tomato mutant scion may have impeded formation of the potato tuber. Hence, modifications of signal molecules produced in and transported from pac choi may have led to reduced growth of the daikon radish taproot.

In most cases, plants partition photosynthates preferentially to vegetative organs in the early to middle growth stage and to reproductive or storage organs in late growth stage [42]. However, many root vegetable plants grow the vegetative biomass and develop the storage root at the same time, leading to a balance between them [42,43]. Grafting may disturb the source–sink balance between scion and rootstock [44]. Both the leaf apical meristem of pac choi and the taproot part of daikon radish are strong sinks [45] and possibly competed for photosynthates in the pac choi–daikon radish grafts. The source-sink relationship was likely altered to support the growth of pac choi leaves at a cost of reduction in radish taproot development in the grafted plants. The dry mass of the leafy part accounted for 81% of the total DW of MQ/BK, but for non-grafted daikon radish, the leafy part only accounted for 66% of the total DW.

For Chinese cabbage grafted onto turnip, the heading of the Chinese cabbage was restricted by the thickening of the turnip taproot, resulting in a small Chinese cabbage head when the turnip taproot was harvested [8]. This imbalance was attributed to the discrepancy in crop maturity and growth cycle requirement as the heading of the Chinese cabbage requires more time than the development of the turnip taproot. In our study, 'Mei Qing Choi' pac choi and 'Bora King' daikon radish were both fast-growing cultivars with 45 and 49 d to maturity, respectively (Johnny's Selected Seeds). However, we seeded the faster maturing pac choi 6 d earlier than the slower maturing daikon radish in order to better match their stem diameters for grafting. Further examinations are needed to elucidate the scion–rootstock interactions for grafting scenarios in which accumulated biomass of both scion (shoots) and rootstock (enlarged taproot) are harvested together for economic yield. In this special scenario, competition for water and nutrients, photosynthetic capacity, and photosynthate partitioning between above- and below-ground sinks are of particular importance. Interestingly, the reciprocal grafting experiment by Sugiura et al. [42] using *Raphanus sativus* genotypes with differential hypocotyl sink activities demonstrated the genotype-dependent autonomous regulation of the hypocotyl sink activity.

3.5. Mineral Nutrient Contents in Pac Choi Leaves and Daikon Radish Roots of Grafted Plants

Dried leaves of MQ, MQ/MQ, and MQ/BK, and dried taproots of BK, BK/BK, and MQ/BK from the 2019 experiment harvest were used to examine the macronutrient and micronutrient concentrations (Table 5) and accumulation (Table 6). Leaf N concentration did not differ significantly between MQ/BK and MQ. Interestingly, MQ/MQ had a significantly higher N concentration in the leaf tissue compared with MQ/BK. However, this seems contradictory to the finding that MQ/BK had a higher leaf photosynthetic rate (Table 3) than MQ/MQ. It needs to be pointed out that the entire above-ground leaf tissue was sampled for leaf nutrient analysis, whereas the most recently mature leaves were used for photosynthesis measurements. While MQ/BK had a lower level of leaf N concentration in the above-ground biomass, its higher leaf photosynthetic rate could be due to remobilization of N compounds from the older leaves to the most recently mature leaves that were used for photosynthesis measurement [46]. Compared with non-grafted pac choi, grafting with the daikon radish rootstock significantly decreased K and S concentrations in the leaf tissue of pac choi (by 21% and 45%, respectively), but it increased Zn concentration in the leaf tissue by 37%. MQ/BK had significantly higher concentrations of N (by 14%), K (by 30%), Mg (by 47%), Ca (by 38%), B (by 36%), and Zn (by 63%) in the taproot compared with the average of BK and BK/BK, while there were no differences between BK and BK/BK. By contrast, the concentration of S in the taproot of MQ/BK was reduced by 44% compared with BK and BK/BK. Overall, there was not a clear relationship between plant nutritional status and reduction of taproot in MQ/BK. Nutrient uptake of grafted pac choi–daikon radish in relation to scion-rootstock interactions is an intriguing area to explore.

Table 5. Mineral nutrient concentrations in leaves of grafted, self-grafted, and non-grafted pac choi and in taproots of grafted, self-grafted, and non-grafted daikon radish plants at harvest in the 2019 experiment.

Treatment [z]	N (mg g^{-1})	P (mg g^{-1})	K (mg g^{-1})	Mg (mg g^{-1})	Ca (mg g^{-1})	S (mg g^{-1})	B (μg g^{-1})	Zn (μg g^{-1})	Mn (μg g^{-1})	Fe (μg g^{-1})	Cu (μg g^{-1})
						Leaves					
MQ	51.5 ± 0.9 ab [y]	10.3 ± 0.3 a	92.1 ± 1.7 b	6.4 ± 0.2 a	30.2 ± 1.0 a	15.4 ± 0.4 a	44.6 ± 1.6 a	77.5 ± 3.5 b	180.4 ± 7.1 a	103.5 ± 6.8 a	0.7 ± 0.1 a
MQ/MQ	53.7 ± 0.9 a	10.5 ± 0.3 a	96.1 ± 1.7 a	6.7 ± 0.2 a	31.3 ± 1.0 a	15.4 ± 0.4 a	45.9 ± 1.6 a	76.7 ± 3.5 b	195.3 ± 7.1 a	101.7 ± 6.8 a	0.7 ± 0.1 a
MQ/BK	48.4 ± 0.9 b	9.7 ± 0.3 a	73.1 ± 1.7 c	6.4 ± 0.2 a	32.0 ± 1.0 a	8.5 ± 0.4 b	45.6 ± 1.6 a	106.5 ± 3.5 a	180.8 ± 7.1 a	120.1 ± 6.8 a	1.1 ± 0.1 a
p value	0.008	0.147	<0.001	0.352	0.437	<0.001	0.835	<0.001	0.175	0.070	0.057
						Taproot					
BK	28.2 ± 0.7 b	7.6 ± 0.3 a	51.1 ± 2.4 b	1.5 ± 0.1 b	3.3 ± 0.2 b	10.3 ± 0.3 a	20.7 ± 0.9 b	63.2 ± 3.8 b	27.0 ± 1.5 a	105.2 ± 17.9 a	0.9 ± 0.2 a
BK/BK	28.2 ± 0.7 b	7.4 ± 0.3 a	49.7 ± 2.4 b	1.4 ± 0.1 b	3.3 ± 0.2 b	10.6 ± 0.3 a	20.2 ± 0.9 b	65.0 ± 3.8 b	23.4 ± 1.5 a	92.2 ± 16.8 a	1.0 ± 0.2 a
MQ/BK	32.2 ± 0.7 a	8.1 ± 0.3 a	65.7 ± 2.4 a	2.2 ± 0.1 a	4.6 ± 0.2 a	5.9 ± 0.3 b	27.8 ± 0.9 a	104.3 ± 3.8 a	25.5 ± 1.5 a	90.9 ± 16.7 a	1.1 ± 0.2 a
p value	0.008	0.128	0.001	<0.001	<0.001	<0.001	<0.001	<0.001	0.201	0.664	0.677

[z] BK = Non-grafted 'Bora King' daikon radish; BK/BK = Self-grafted 'Bora King' daikon radish; MQ = Non-grafted 'Mei Qing Choi' pac choi; MQ/MQ = Self-grafted 'Mei Qing Choi' pac choi; MQ/BK = 'Mei Qing Choi' pac choi grafted onto 'Bora King' daikon radish. [y] Mean ± SE (standard error); means followed by the same letter are not significantly different at $p \leq 0.05$ according to Fisher's LSD test.

Table 6. Mineral nutrient contents accumulated in leaves of grafted, self-grafted, and non-grafted pac choi and in taproots of grafted, self-grafted, and non-grafted daikon radish plants at harvest in the 2019 experiment.

Treatment [z]	N (mg $Plant^{-1}$)	P (mg $Plant^{-1}$)	K (mg $Plant^{-1}$)	Mg (mg $Plant^{-1}$)	Ca (mg $Plant^{-1}$)	S (mg $Plant^{-1}$)	B (μg $Plant^{-1}$)	Zn (μg $Plant^{-1}$)	Mn (μg $Plant^{-1}$)	Fe (μg $Plant^{-1}$)	Cu (μg $Plant^{-1}$)
					Leaves						
MQ	519.6 ± 13.3 a [y]	104.2 ± 3.8 a	929.3 ± 25.1 a	64.6 ± 1.9 a	304.6 ± 10.8 a	155.0 ± 5.2 a	450.0 ± 21.1 a	781.9 ± 37.5 b	1819.5 ± 84.7 a	1043.8 ± 69.4 a	6.9 ± 1.2 a
MQ/MQ	481.6 ± 13.3 ab	93.8 ± 3.8 a	862.5 ± 25.1 a	59.9 ± 1.9 a	280.7 ± 10.8 a	138.1 ± 5.2 b	412.9 ± 21.1 a	688.1 ± 37.5 c	1753.8 ± 84.7 a	911.5 ± 69.4 a	6.4 ± 1.2 a
MQ/BK	456.2 ± 13.3 b	91.3 ± 3.8 a	690.4 ± 25.1 b	60.1 ± 1.9 a	300.7 ± 10.8 a	79.7 ± 5.2 c	429.8 ± 21.1 a	1003.4 ± 37.5 a	1707.4 ± 84.7 a	1133.1 ± 69.4 a	10.5 ± 1.2 a
p value	0.025	0.103	<0.001	0.123	0.172	<0.001	0.487	<0.001	0.636	0.081	0.081
					Taproot						
BK	185.3 ± 5.7 a	49.9 ± 1.5 a	333.8 ± 11.5 a	9.9 ± 0.3 a	21.8 ± 0.5 a	67.4 ± 2.4 a	135.3 ± 4.1 a	414.8 ± 15.9 a	178.4 ± 10.0 a	723.3 ± 106.9 a	6.0 ± 1.2 a
BK/BK	150.2 ± 5.7 b	39.3 ± 1.5 b	264.5 ± 11.5 b	7.7 ± 0.3 b	17.3 ± 0.5 b	56.4 ± 2.4 b	107.5 ± 4.1 b	346.6 ± 15.9 b	124.4 ± 10.0 b	503.8 ± 106.9 ab	5.5 ± 1.2 a
MQ/BK	72.1 ± 5.7 c	18.3 ± 1.5 c	148.2 ± 11.5 c	4.8 ± 0.3 c	10.2 ± 0.5 c	13.2 ± 2.4 c	62.9 ± 4.1 c	235.2 ± 15.9 c	57.8 ± 10.0 c	207.3 ± 106.9 b	2.5 ± 1.2 a
p value	<0.001	<0.001	<0.001	<0.001	<0.001	<0.001	<0.001	<0.001	<0.001	0.027	0.120

[z] BK = Non-grafted 'Bora King' daikon radish; BK/BK = Self-grafted 'Bora King' daikon radish; MQ = Non-grafted 'Mei Qing Choi' pac choi; MQ/MQ = Self-grafted 'Mei Qing Choi' pac choi; MQ/BK = 'Mei Qing Choi' pac choi grafted onto 'Bora King' daikon radish. [y] Mean ± SE (standard error); means followed by the same letter are not significantly different at $p \leq 0.05$ according to Fisher's LSD test.

With respect to leaf nutrient accumulation, MQ/BK had 37% greater Zn accumulation compared with the average of MQ/MQ and MQ (Table 6). However, MQ/BK decreased N, K, and S content by 12%, 26%, and 49%, respectively, compared with MQ. In terms of nutrient accumulation in the taproot, MQ/BK showed significantly lower accumulation of all measured minerals except Fe and Cu relative to BK and BK/BK, which was likely associated with the small size of the taproot in MQ/BK. The lower accumulation of nutrients in the taproot demonstrated that pac choi–daikon radish favored growth of the pac choi leaves at the expense of the radish taproot.

4. Conclusions

Successful grafts were produced between *B. rapa* var. *chinensis* and *R. sativus* var. *longipinnatus* in this study, resulting in a novel 'pac choi–daikon radish' product that may help save growing space and have added-value as perceived by farmers and consumers. More research is needed to optimize the seeding time and management of seedling production to help further improve the graft survival rate. Grafting pac choi with daikon radish did not severely impair the growth of the above-ground parts as grafted pac choi had similar SPAD value, canopy size, leaf number, and above-ground DW compared with non-grafted pac choi. Interestingly, grafting with radish increased the photosynthetic ability of the pac choi. However, grafting the daikon radish with pac choi decreased the taproot formation as reflected by the reduced length, diameter, FW and DW of the taproot. Future studies could explore different approaches such as cultivar selection and nutrient management to better balance the sizes of the above- and below-ground parts of this new pac choi–daikon radish product. Given the wide range of *B. rapa* var. *chinensis* and *R. sativus* var. *longipinnatus* cultivars, it would be interesting to explore different grafting combinations to characterize the range of graft performance. We only tested the graft performance under greenhouse conditions, and the grafted plants need to be further evaluated in field growing systems where biotic and abiotic stressors can be intensified. Generally, grafting between pac choi and daikon radish showed more negative impacts on mineral nutrient levels in radish taproots than in pac choi leaves. Sensory properties of the 'pac choi–daikon radish' product are unknown, and this aspect deserves further assessment. The inter-generic grafting between *B. rapa* var. *chinensis* and *R. sativus* var. *longipinnatus* could also provide a unique model system to further our understanding of scion-rootstock synergy and above- and below-ground sink competition in horticultural crops toward improving the use of grafting technology in sustainable vegetable production.

Author Contributions: Conceptualization, X.Z.; Data curation, T.G., Z.T.R. and K.E.B.; Formal analysis, T.G.; Funding acquisition, X.Z.; Investigation, T.G., Z.T.R. and K.E.B.; Methodology, T.G., Z.T.R., Z.E.B. and X.Z.; Project administration, X.Z.; Resources, X.Z. and J.K.B.; Supervision, X.Z. and J.K.B.; Writing—original draft, T.G. and X.Z.; Writing—review & editing, T.G., Z.T.R., K.E.B., Z.E.B., X.Z. and J.K.B. All authors have read and agreed to the published version of the manuscript.

References

1. Lee, J.M.; Kubota, C.; Tsao, S.J.; Bie, Z.; Echevarria, P.H.; Morra, L.; Oda, M. Current status of vegetable grafting: Diffusion, grafting techniques, automation. *Sci. Hortic.* **2010**, *127*, 93–105. [CrossRef]
2. Louws, F.J.; Rivard, C.L.; Kubota, C. Grafting fruiting vegetables to manage soilborne pathogens, foliar pathogens, arthropods and weeds. *Sci. Hortic.* **2010**, *127*, 127–146. [CrossRef]
3. Guan, W.; Zhao, X.; Hassell, R.; Thies, J. Defense mechanisms involved in disease resistance of grafted vegetables. *HortScience* **2012**, *47*, 164–170. [CrossRef]
4. Chen, A.; Komives, E.A.; Schroeder, J.I. An improved grafting technique for mature Arabidopsis plants demonstrates long-distance shoot-to-root transport of phytochelatins in Arabidopsis. *Plant Physiol.* **2006**, *141*, 108–120. [CrossRef]
5. Gurusamy, V.; Bett, K.E.; Vandenberg, A. Grafting as a tool in common bean breeding. *Can. J. Plant Sci.* **2010**, *90*, 299–304. [CrossRef]
6. Temperini, O.; Calabrese, N.; Temperini, A.; Rouphael, Y.; Tesi, R.; Lenzi, A.; Carito, A.; Colla, G. Grafting artichoke onto cardoon rootstocks: Graft compatibility, yield and Verticillium wilt incidence. *Sci. Hortic.* **2013**, *149*, 22–27. [CrossRef]
7. Oda, M.; Tsuji, K.; Nagaoka, M. Inter-generic, inter-specific and inter-varietal grafting in Cruciferae. *Acta Hortic.* **1990**, *319*, 425–430. [CrossRef]
8. Oda, M.; Nakajima, T. Adhesive grafting of Chinese cabbage on turnip. *HortScience* **1992**, *27*, 1136. [CrossRef]
9. Chen, Y.C.; Chang, W.C.; Wang, S.T.; Lin, S.I. Development of a grafting method and healing conditions to improve cabbage head quality. *HortTechnology* **2019**, *29*, 57–64. [CrossRef]
10. Yang, Y.W.; Tai, P.Y.; Chen, Y.; Li, W.H. A study of the phylogeny of *Brassica rapa*, *B. nigra*, *Raphanus sativus*, and their related genera using noncoding regions of chloroplast DNA. *Mol. Phylogenet. Evol.* **2002**, *23*, 268–275. [CrossRef]

11. Matsuzawa, Y.; Funayama, T.; Kamibayashi, M.; Konnai, M.; Bang, S.W.; Kaneko, Y. Synthetic *Brassica rapa-Raphanus sativus* amphidiploid lines developed by reciprocal hybridization. *Plant Breed.* **2000**, *119*, 357–359.
12. Lange, W.; Toxopeus, H.; Lubberts, J.H.; Dolstra, O.; Harrewijn, J.L. The development of Raparadish (x *Brassicoraphanus*, 2n = 38), a new crop in agriculture. *Euphytica* **1989**, *40*, 1–14.
13. Tonosaki, K.; Michiba, K.; Bang, S.W.; Kitashiba, H.; Kaneko, Y.; Nishio, T. Genetic analysis of hybrid seed formation ability of *Brassica rapa* in intergeneric crossings with *Raphanus sativus*. *Theor. Appl. Genet.* **2013**, *126*, 837–846.
14. Abràmoff, M.D.; Magalhães, P.J.; Ram, S.J. Image processing with ImageJ. *Biophotonics Intern.* **2004**, *11*, 36–42.
15. Chen, X.; Wang, J.; Shi, Y.; Zhao, M.Q.; Chi, G.Y. Effects of cadmium on growth and photosynthetic activities in pak choi and mustard. *Bot. Stud.* **2011**, *52*, 41–46.
16. He, Y.; Zhu, Z.; Yang, J.; Ni, X.; Zhu, B. Grafting increases the salt tolerance of tomato by improvement of photosynthesis and enhancement of antioxidant enzymes activity. *Environ. Exp. Bot.* **2009**, *66*, 270–278.
17. Gago, J.; de Menezes Daloso, D.; Figueroa, C.M.; Flexas, J.; Fernie, A.R.; Nikoloski, Z. Relationships of leaf net photosynthesis, stomatal conductance, and mesophyll conductance to primary metabolism: A multispecies meta-analysis approach. *Plant Physiol.* **2016**, *171*, 265–279.
18. Bumgarner, N.R.; Kleinhenz, M.D. *Grafting Guide: A Pictorial Guide to the Cleft and Splice Graft Methods as Applied to Tomato and Pepper*; The Ohio State University: Columbus, OH, USA, 2013; Available online: https://web.extension.illinois.edu/smallfarm/downloads/50570.pdf (accessed on 19 September 2020).
19. Garner, R.J.; Bradley, S. *The Grafter's Handbook*; Mitchell Beazley: London, UK, 2013.
20. Melnyk, C.W. Plant grafting: Insights into tissue regeneration. *Regeneration* **2017**, *4*, 3–14.
21. Hayashida, T.; Shibato, Y.; Hamachi, Y.; Yamato, Y.; Yamazaki, H.; Miura, H. Effect of light quality with different red/far-red photon flux density ratio on elongation of pak-choi (*Brassica chinensis*) seedlings grown under high temperatures. *J. Jpn. Soc. Hortic. Sci.* **2001**, *70*, 774–776.
22. Kwack, Y.; Kim, K.K.; Hwang, H.; Chun, C. Growth and quality of sprouts of six vegetables cultivated under different light intensity and quality. *Hortic. Environ. Biotechnol.* **2015**, *56*, 437–443.
23. Liu, M.; Bassetti, N.; Petrasch, S.; Zhang, N.; Bucher, J.; Shen, S.; Zhao, J.; Bonnema, G. What makes turnips: Anatomy, physiology and transcriptome during early stages of its hypocotyl-tuber development. *Hortic. Res.* **2019**, *6*, 38. [CrossRef]
24. Zheng, Y.; Luo, L.; Gao, Z.; Liu, Y.; Chen, Q.; Kong, X.; Yang, Y. Grafting induces flowering time and tuber formation changes in Brassica species involving FT signaling. *Plant. Biol.* **2019**, *21*, 1031–1038. [PubMed]
25. Tsuro, M.; Suwabe, K.; Kubo, N.; Matsumoto, S.; Hirai, M. Mapping of QTLs controlling root shape and red pigmentation in radish, *Raphanus sativus* L. *Breed. Sci.* **2008**, *58*, 55–61. [CrossRef]
26. Zaki, H.E.M.; Takahata, Y.; Yokoi, S. Analysis of the morphological and anatomical characteristics of roots in three radish (*Raphanus sativus*) cultivars that differ in root shape. *J. Hortic. Sci. Biotechnol.* **2012**, *87*, 172–178.
27. Bihmidine, S.; Hunter Iii, C.T.; Johns, C.E.; Koch, K.E.; Braun, D.M. Regulation of assimilate import into sink organs: Update on molecular drivers of sink strength. *Front. Plant Sci.* **2013**, *4*, 177. [CrossRef] [PubMed]
28. Abdel, C.G. Stomata and trichrome performances of four radish *Raphanus sativus* L. var. *sativus* cultivars grown in controlled cabinet under varying temperatures and irrigation levels. *Intl. J. Farm. Alli. Sci.* **2016**, *5*, 46–65.
29. Ehleringer, J.; Björkman, O.; Mooney, H.A. Leaf pubescence: Effects on absorptance and photosynthesis in a desert shrub. *Science* **1976**, *191*, 376–377.
30. Choinski, J.S., Jr.; Wise, R.R. Leaf growth development in relation to gas exchange in *Quercus marilandica* Muenchh. *J. Plant Physiol.* **1999**, *154*, 302–309. [CrossRef]
31. Hauser, M.T. Molecular basis of natural variation and environmental control of trichome patterning. *Front. Plant Sci.* **2014**, *5*, 1–7.
32. Thomson, G.E. Role of harvest technique and injuries in water loss from stored pak choi (*Brassica rapa* subsp. chinensis) heads. *N. Z. J. Crop Hortic.* **2005**, *33*, 111–115. [CrossRef]
33. Von Caemmerer, S.; Lawson, T.; Oxborough, K.; Baker, N.R.; Andrews, T.J.; Raines, C.A. Stomatal conductance does not correlate with photosynthetic capacity in transgenic tobacco with reduced amounts of Rubisco. *J. Exp. Bot.* **2004**, *55*, 1157–1166. [CrossRef]

34. Xu, Z.; Zhou, G. Responses of leaf stomatal density to water status and its relationship with photosynthesis in a grass. *J. Exp. Bot.* **2008**, *59*, 3317–3325. [CrossRef] [PubMed]
35. Rosenkranz, H.; Vogel, R.; Greiner, S.; Rausch, T. In wounded sugar beet (*Beta vulgaris* L.) taproot, hexose accumulation correlates with the induction of a vacuolar invertase isoform. *J. Exp. Bot.* **2001**, *52*, 2381–2385. [CrossRef] [PubMed]
36. Zhang, Y.I.; Turner, J.G. Wound-induced endogenous jasmonates stunt plant growth by inhibiting mitosis. *PLoS ONE* **2008**, *3*, e3699. [CrossRef] [PubMed]
37. Xu, X.; van Lammeren, A.A.M.; Vermeer, E.; Vreugdenhil, D. The role of gibberellin, abscisic acid, and sucrose in the regulation of potato tuber formation in vitro. *Plant Physiol.* **1998**, *117*, 575–584. [CrossRef] [PubMed]
38. Peres, L.E.P.; Carvalho, R.F.; Zsögön, A.; Bermúdez-Zambrano, O.D.; Robles, W.G.R.; Tavares, S. Grafting of tomato mutants onto potato rootstocks: An approach to study leaf-derived signaling on tuberization. *Plant Sci.* **2005**, *169*, 680–688. [CrossRef]
39. Zhang, G.; Mao, Z.; Wang, Q.; Song, J.; Nie, X.; Wang, T.; Zhang, H.; Guo, H. Comprehensive transcriptome profiling and phenotyping of rootstock and scion in a tomato/potato heterografting system. *Physiol. Plant* **2019**, *166*, 833–847. [CrossRef]
40. Wang, G.L.; Xiong, F.; Que, F.; Xu, Z.S.; Wang, F.; Xiong, A.S. Morphological characteristics, anatomical structure, and gene expression: Novel insights into gibberellin biosynthesis and perception during carrot growth and development. *Hortic. Res.* **2015**, *2*, 15028. [CrossRef]
41. Jabir, O.; Mohammed, B.; Benard Kinuthia, K.; Almahadi Faroug, M.; Nureldin Awad, F.; Muleke, E.M.; Ahmadzai, Z.; Liu, L. Effects of gibberellin and gibberellin biosynthesis inhibitor (paclobutrazol) applications on radish (*Raphanus sativus*) taproot expansion and the presence of authentic hormones. *Int. J. Agric. Biol.* **2017**, *19*, 779–786. [CrossRef]
42. Sugiura, D.; Betsuyaku, E.; Terashima, I. Manipulation of the hypocotyl sink activity by reciprocal grafting of two *Raphanus sativus* varieties: Its effects on morphological and physiological traits of source leaves and whole-plant growth. *Plant Cell Environ.* **2015**, *38*, 2629–2640. [CrossRef]
43. Chapin Iii, F.S.; Schulze, E.D.; Mooney, H.A. The ecology and economics of storage in plants. *Annu. Rev. Ecol. Syst.* **1990**, *21*, 423–447. [CrossRef]
44. Kyriacou, M.C.; Rouphael, Y.; Colla, G.; Zrenner, R.; Schwarz, D. Vegetable grafting: The implications of a growing agronomic imperative for vegetable fruit quality and nutritive value. *Front. Plant Sci.* **2017**, *8*, 741. [PubMed]
45. Yu, R.; Xu, L.; Zhang, W.; Wang, Y.; Luo, X.; Wang, R.; Zhu, X.; Xie, Y.; Karanja, B.; Liu, L. De novo taproot transcriptome sequencing and analysis of major genes involved in sucrose metabolism in radish (*Raphanus sativus* L.). *Front. Plant Sci.* **2016**, *7*, 585. [PubMed]
46. Sugiura, D.; Watanabe, C.K.A.; Betsuyaku, E.; Terashima, I. Sink-source balance and down-regulation of photosynthesis in *Raphanus sativus*: Effects of grafting, N and CO_2. *Plant Cell Physiol.* **2017**, *58*, 2043–2056. [PubMed]

8

Conventional Industrial Robotics Applied to the Process of Tomato Grafting using the Splicing Technique

José-Luis Pardo-Alonso [1], Ángel Carreño-Ortega [1,*], Carolina-Clara Martínez-Gaitán [2], Iacopo Golasi [3] and Marta Gómez Galán [1]

1 Department of Engineering, University of Almería, Research Center CIMEDES, Agrifood Campus of International Excellence (CeiA3), La Cañada de San Urbano, 04120 Almería, Spain; jolupa@ual.es (J.-L.P.-A.); mgg492@ual.es (M.G.G.)

2 Tecnova, Technological Center: Foundation for Auxiliary Technologies for Agriculture; Parque Tecnológico de Almería, Avda. de la Innovación, 23, 04131 Almería, Spain; cmartinez@fundaciontecnova.com

3 DIAEE-Area Fisica Tecnica, Università degli Studi di Roma "Sapienza", 00184 Rome, Italy; iacopo.golasi@uniroma1.it

* Correspondence: acarre@ual.es

Abstract: Horticultural grafting is routinely performed manually, demanding a high degree of concentration and requiring operators to withstand extreme humidity and temperature conditions. This article presents the results derived from adapting the splicing technique for tomato grafting, characterized by the coordinated work of two conventional anthropomorphic industrial robots with the support of low-cost passive auxiliary units for the transportation, handling, and conditioning of the seedlings. This work provides a new approach to improve the efficiency of tomato grafting. Six test rates were analyzed, which allowed the system to be evaluated across 900 grafted units, with gradual increases in the speed of robots work, operating from 80 grafts/hour to over 300 grafts/hour. The results obtained show that a higher number of grafts per hour than the number manually performed by skilled workers could be reached easily, with success rates of approximately 90% for working speeds around 210–240 grafts/hour.

Keywords: tomato grafting; splice grafting technique; agricultural robot; automated grafting; agricultural machinery

---

1. Introduction

The herbaceous graft is a growing technique that allows two pieces of living plant tissue to be joined together in such a way that they will unite and later grow and develop as a single composite plant [1]. This technique is widespread in Southeast Asia, the Mediterranean basin, and Europe for intensive cultivation in tomato greenhouses. With the use of grafting, plants with properties of agronomic interest are created, fundamentally seeking greater resistance to soil diseases and higher productivity in high-quality cultivars [2]. One of the technique's main disadvantages is its high cost of production. The seeds (of both the scions used and the added cost of the rootstock), the cost of labor, the supplies for each graft, the use of machinery and work tools, and post-graft care in the healing chambers are considered the most important factors in price determination [3–5].

It is estimated that the work of the grafting process itself can amount to approximately a quarter of the total costs associated per grafted plant; a third of these costs represent the total cost of the seeds, and the rest is essentially equally divided between the costs of materials and tools, the cost of the clip and the stay in the healing chamber, the energy costs, and the costs of the work of handling and transplantation personnel [6,7]. Proportionally, and with respect to the cost of the seeds, the scion represents 80% of the cost, compared to the 20% cost of the rootstock (Figure 1).

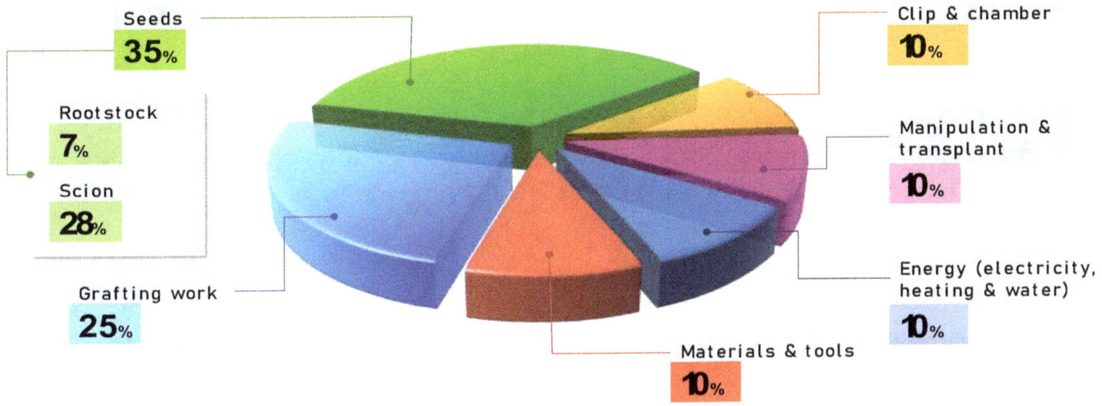

Figure 1. Estimated average cost distribution for a grafted plant. Average data assessed according to data collected in nurseries in Almería (Spain), [6,7].

On the other hand, the average cost of grafted tomato plants versus non-grafted plants varies considerably depending on several factors, mainly the productive scale of the nursery, the cost of labor, the production practices, and the cost of seeds employed, which can sometimes amount to more than 50% of the total costs [7]. In nurseries with a medium–high production volume, the costs of hand-grafted plants are estimated at approximately $0.67 for the USA, compared to $0.15 for non-grafted plants [8–10]. Similar prices are maintained for Asian countries, such as Japan and Korea [5], while for Spain and other European countries, the costs vary between €0.54 for hand-grafted plants compared to €0.18 for non-grafted plants [11]. These data corroborate that grafted plants can accumulate extra costs 3 to 4 times their cost without grafting.

Even so, the advantages of using grafted plants versus non-grafted plants, which include eliminating the common problems of soil pathogens that have traditionally been controlled by fumigation, have made the technique's use widespread and common in large regions of the world. Grafting has become the most effective and economically viable technique to address this problem [12], compared to other alternatives that have failed to provide a convincing ability to control these diseases, such as genetic improvement with resistance genes, greater crop rotation, soil solarization, the use of plastic mulch, biofumigation, the use of water vapor, crops without soil, the fallow technique, the use of trap plants, or the use of integrated biological control [13]. In Japan, Korea, and the rest of Southeast Asia, grafting is a common technique for the production of Solanaceae, especially in greenhouses, which constitute approximately 100% of the cultivated area [14]. Although its introduction in Europe and the Mediterranean basin occurred somewhat later, similar graft percentages are reached today. It is estimated that the cost/benefit ratio is 4.6 for grafted tomatoes, compared to 3.5 for non-grafted tomatoes [15].

Grafting is a task that requires considerable time, concentration, and dexterity, even for skilled workers. The delicate characteristics of the process and the biological requirements of the work seedlings, which need to be specially manipulated in a clean, warm, and humid environment, cause growing concern for plant producers due to the lack of available specialized personnel who are capable of facing intense workloads during short campaigns and with a high productive demand. Grafting requires up to three or four people and dedicated specific tasks within the process [3]. The shortage of skilled workers, along with an ageing agricultural population and an increasing demand for grafted plants, has made it necessary to automate grafting [16].

The need to use machinery in plant production to reduce the demand for human labor, expand production capacities, and improve product uniformity has been recognized for a long time [17]. In advanced agricultural countries, efforts are being made to develop and use automatic graft equipment due to the lack of labor in rural areas [18,19]. An improvement in grafting methods and techniques that reduce the cost of labor in grafting, its subsequent management, and transplants will contribute to the increased use of grafted plants worldwide [5,17].

There is an important tendency towards developing graft robots with a market potential, as opposed to manual grafting [18]. Splice grafting is a widely used method for Solanaceae, with the advantages of being easily mechanized and having well-defined and clear operations. The stem of the rootstock is cut, preferably below the cotyledons, at a specific angle. The scion, cut with the same angle, has a section that is more or less similar to the rootstock. Finally, by means of a special clamp or clip in the form of a tube, the two cuts are joined [1,20–22] (Figure 2).

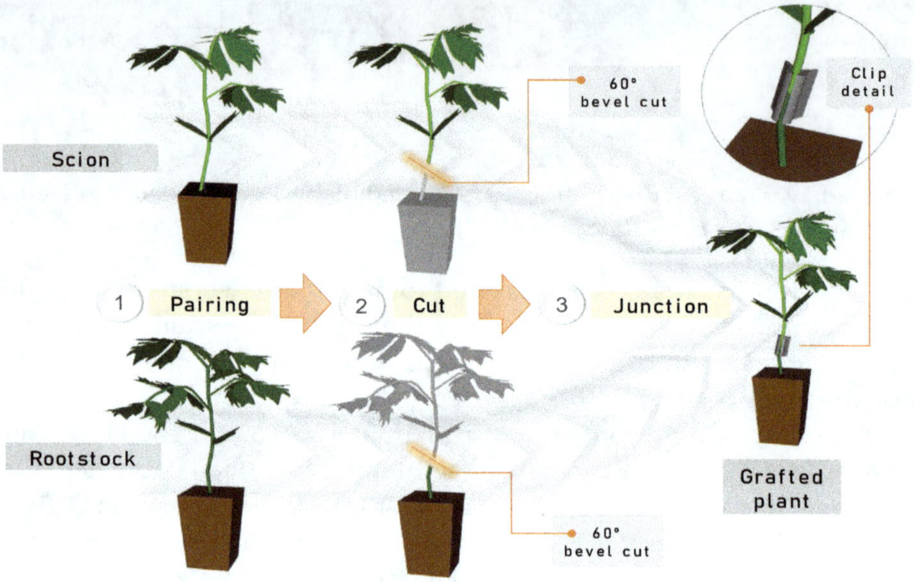

Figure 2. Sketch of the tomato grafting technique known as "tube grafting", "Japanese top-grafting", or "splice grafting".

Since the late 1980s and in the last three decades, there have been numerous attempts to invent equipment that reasonably succeeds in the automated grafting of horticultural plants. In the first two decades, the majority of developments came from Southeast Asia, while in the last decade, developments of mainly European origin have also been added [3,17,23]. This equipment has fundamentally been a semi-automated technology system, which facilitates the grafting task but requires up to two or three operators to function. Other developed systems are completely autonomous but enormously rigid in their performance, while at the same time being complex in their adaptability and operational requirements. Faced with these developments, and based on dedicated and specific automation, a study is presented herein of equipment based on conventional industrial robotic technology, supported by simple auxiliary equipment, which allows the productive requirements of the graft task to be met and can be easily adapted to other tasks and productive needs.

The price trend experienced in recent years in industrial robotics, which allows for the acquisition of small robotic units with similar initial investments at a cost no higher than the biannual cost associated with the minimum wage in developed countries [24,25], together with the use of passive auxiliary units with a low cost for transporting, cutting, and placing the binding clip on the seedlings, would allow for a rapid amortization of investment, which makes the study of this development alternative an area of interest.

The objective of this article was centred on the study and feasibility of automated grafting using a robotic cell based on the use of conventional industrial robotics, which allowed the grafting task to be faced with a greater system configurability and flexibility against the natural biological variability of the seedlings being used. This grafting system is supported by the use of simple and low-cost auxiliary equipment, which allows the task to be completed with tools external to the logistical tasks of the seedling trays, the cutting of the seedlings, and the dispensing and placement of the graft clip.

2. Materials and Methods

2.1. Device Description and Productive Process

The robotic equipment for grafting consists of transport devices, the manipulator itself, cutting mechanisms, and devices that facilitate bonding and grip [26]. True to this premise, the study equipment consisted of two anthropomorphic robots equipped with clips adapted for manipulating seedlings, with two seedling bevel cutting devices and a device for the forming, dispensing, and placement of the graft clips (Figure 3).

(a)

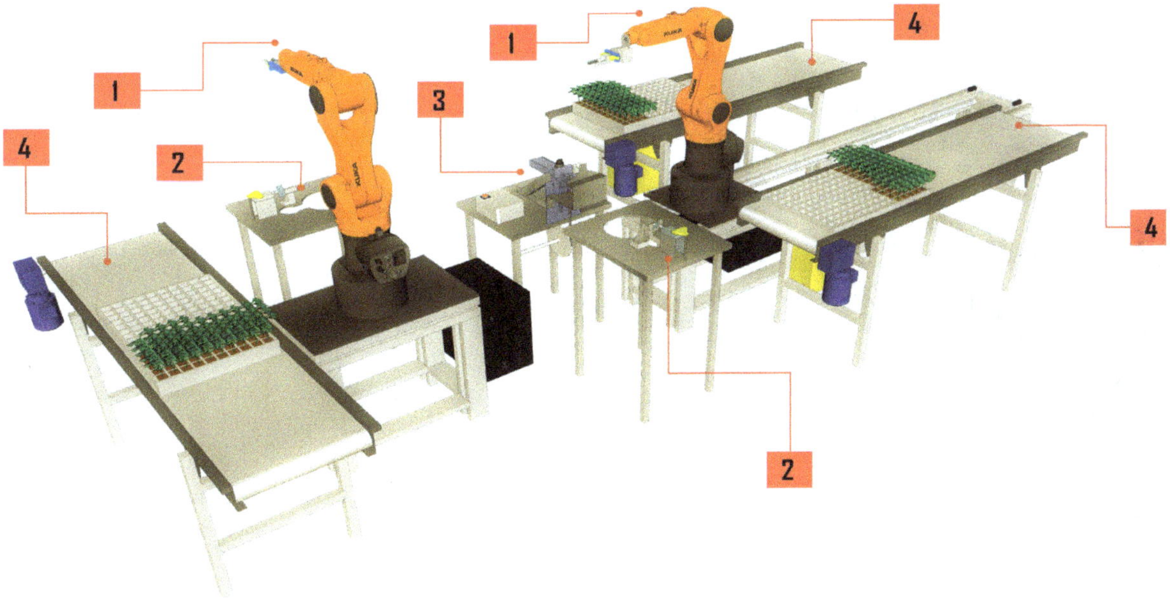

(b)

Figure 3. (**a**) General view of the robotic cell developed for grafting plants using the splicing technique. (**b**) Work cell sketch. (1) KUKA KR6 900 robots for manipulating the seedlings. (2) Cutting devices for rootstock and scion seedlings. (3) Forming, dispensing, and placement graft clip device. (4) Conveyor belts for seedling trays (rootstock, scion, and grafted plants). Demonstrative video of the system: https://youtu.be/9GvIDyrBBfo (accessed on 10 December 2019).

The seedling trays are loaded at the beginning of the process into conveyor belts suitable for this purpose, so that two seedling starter trays, loaded with rootstock and scion seedlings will be properly positioned in front of the working robots.

Each completed row of work is followed by an advance of the belt, which relocates the next row into the appropriate collection area. This process happens until the rows in the seedling starter trays (rootstock and scion) are finished. Likewise, there is a third conveyor belt with an empty output tray for the resulting seedlings to be placed on once grafted. This output conveyor belt also performs partial advances per row while working on that tray.

Once the trays reach the required locations, the two industrial robots, Kuka model Agilus R6 900 (developed by KUKA Roboter GmbH), work from their home position in a coordinated manner on the rootstock and scion to achieve graft completion. These robots have equivalent commercial equipment from other large manufacturers worldwide, such as the robot model IRB1200 (developed by ABB), the model MH5F (developed by Yaskawa), or the model LR Mate 200iD/7 L (developed by FANUC). All of them have similar load capacities, degrees of freedom, speeds, and working spaces, so their replacement would not lead to significant differences.

Each robot operates independently by handling each of the seedlings, which are obtained from the input trays. The rootstock and scion seedlings are approached in a simultaneous and coordinated manner: ① approach the input trays (AIT). This displacement is followed by a precision operation that separates the seedlings from their trays: ② grip and extraction (GE).

The final elements for the seedling manipulation consist of clamps composed of two fingers with an opening and closing parallel model MHZ2-32D from SMC, equipped with a padded extension zone of low-density (150–200 kg/m^3) polyurethane foam (neoprene) with high resilience for precise and firm seedling attachment. A photocell is located between the ends of these fingers, and a Sick LL3-TB02 optical fibre sensor detects the precise location of the stems of the seedlings, which can emerge at any position within the alveolus of the tray. The individual seedlings have a unique growth morphology, and the alveoli can even be empty (Figure 4).

(a) (b)

Figure 4. (**a**) Detailed view of the clamp or terminal element of the robots. (**b**) Robot and clamp sketch. (10) Industrial robots. (11) Clamping base. (12) Robot clamp. (13) Fingers with parallel opening and closing operation. (14) Padded area for holding seedlings. (15) Sensor for seedling detection.

After extracting the seedlings from their alveoli in the trays, the robotic arms carry the seedlings to the point where the cut will be made: ③ approach the cutting zone (AC). The system uses two equal pieces of equipment for cutting the seedlings to guarantee precision in the required cutting angle and the integrity of the dissected seedling. Both pieces of equipment are responsible for acting, one on the rootstock and another on the scion, allowing the cutting angle to be regulated and complementary between both plants. Prior to cutting, the robots insert the seedlings into a slot or channel located in front of the blade, where the stems are embedded to ensure the verticality of the stems during cutting. The cut is executed by activating a double-shank pneumatic cylinder, SMC model CXSM15-15, coupled to a terminal tool of a sharpened, disinfected, and interchangeable blade of stainless steel with a precise cut angle [27]. The cut is performed by a dry shock stroke of the blade against the seedling, which cleanly bisects the stem and ensures a clean cut: ④ cutting process (C). Meanwhile, an external blower separates the non-useful part of the treated seedling (Figure 5).

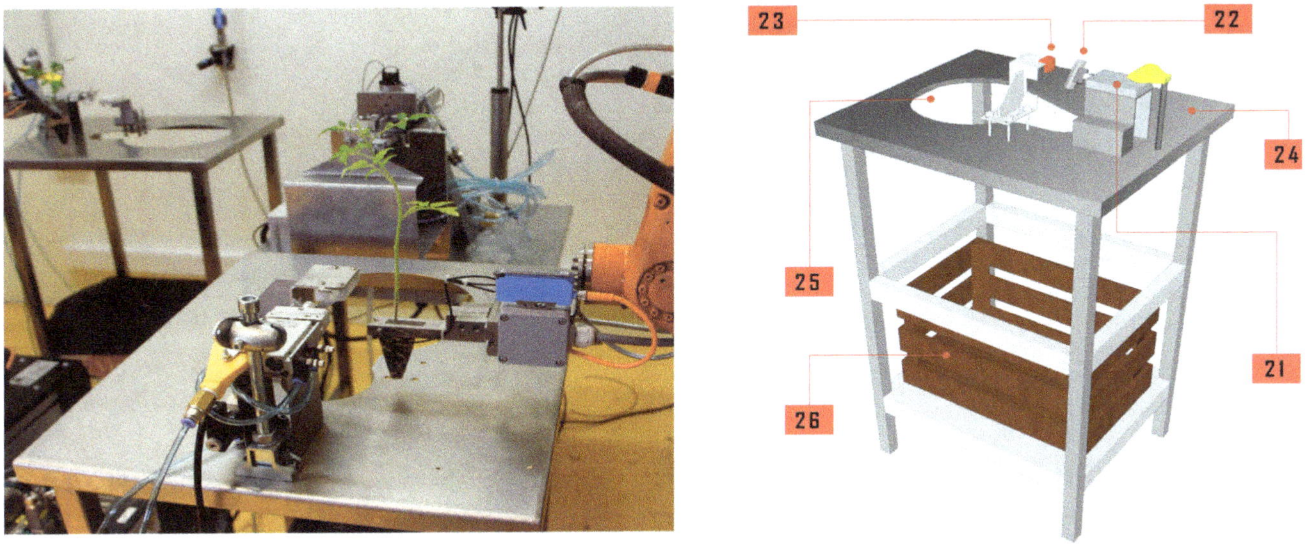

(a) (b)

Figure 5. (**a**) Detailed view of the cutting device. (**b**) Cutting device sketch. (21) Pneumatic cylinder to drive the blade. (22) Cutting blade. (23) Slot for seedling lace. (24) Worktable. (25) Hole for waste disposal. (26) Waste accumulation box. (26) Blower.

After the cut, the robotic arms carry the useful parts of the seedlings for the graft towards the bonding area: ⑤ approach the clip dispensing zone (ACD). The graft equipment that creates the bond is responsible for cutting plastic clips from a continuous roll and then dispensing and placing a clip on the seedlings to be grafted. Thus, this device contains two subsystems: one in charge of preparing the clip and another in charge of placing it.

The first clip preparation subsystem consists of a series of electropneumatic devices that act in a coordinated and sequential manner to perform the clip cutting process from a continuous roll of plastic tube, regulating the advance and thus controlling the length of the cut clips. To obtain a clip with the desired length, an SMC model CJPB10-5 microcylinder presses the plastic tube against the end of a second cylinder, an SMC model CXSM15-15, which advances in the vertical direction the exact length that the clip is desired to have. Finally, an SMC model CXSM15-15 cylinder, equipped with a sharp blade at its end, makes a clean cut with a sharp blow, creating the clip to be used.

The second subsystem contains a rotating cylinder, a Festo model DM-6-90-PA, which grips the cut clip, then tightens its wings and thus clamps and fully opens the clip. Finally, with an SMC model CXSM25-70 cylinder, the clip is brought closer by a precise horizontal movement to the junction point where the two seedlings to be grafted are located. At the point of clip placement, a passive fitting device is adapted to accommodate and locate the seedlings in front of the clip. To ensure a precise bond without unwanted seedling displacement, both parts remain in contact and can thus fasten the bond. Once the clip is placed on the seedlings, the device that kept the graft clip pinched loosens and moves away from the junction point, returning to its resting position (Figure 6).

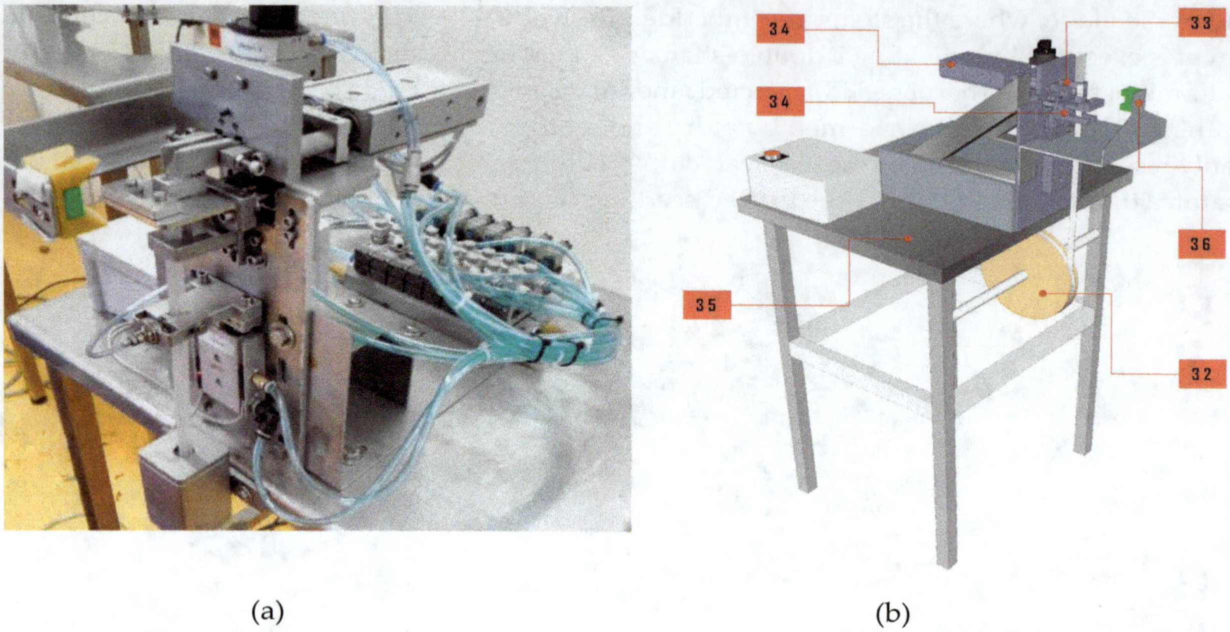

Figure 6. (**a**) Detailed view of the device that dispenses and places the graft clip. (**b**) Sketch of the device that dispenses graft clips. (31) Pneumatic cylinders and blade to make the grafting clip. (32) Continuous silicone roll for clip manufacture. (33) Tweezers for the grip and clip opening. (34) Pneumatic cylinder responsible for bringing the grafting clip closer to the junction point. (35) Worktable. (36) Plant placement device and grafting clip insertion point.

The graft is accomplished when both plants are placed in intimate contact with one another and the graft clip is pressed onto them: ⑥ clip preparation and placement (CPP). Once the clip is placed, the robot holding the scion releases the graft and withdraws, leaving the bonding area, while the other robot, which holds the completed graft by its lower part, moves the graft by the rootstock to the tray where the grafts will be left once finished: ⑦ approach the output trays (AOT). The graft is deposited in an alveolus of the output tray: ⑧ insertion on trays and release of plants (ITR).

Once the process is finished, both robots return to their resting positions, either as an end point or as a point of passage from where a new work cycle begins: ⑨ return home (RH). The process is repeated until the work trays are completed. The tray with the grafted seedlings is removed, and the grafted seedlings are subjected to a post-grafting process of healing, where their success is examined over 14 days: ⑩ post-graft losses (PGL). The entire system and grafting equipment described, as well as other secondary elements and auxiliary equipment, are managed and coordinated in a global manner through a central control unit, consisting of a PLC model CJ2 M, with an Omron CPU32.

During the grafting operation, 10 control points were established as singular intermediate points of reference in the process, which allowed us to record the partial times used and a distribution of failures during grafting (Figure 7).

The flowchart describing the operations and process described above allowed us to evaluate the validity and efficacy of conventional industrial robotics applied to tomato seedling grafting using external low-cost passive devices that facilitate grafting completion. The external devices act as a tool both in cutting the seedlings and in the dispensing and placing of the graft clip (Figure 8).

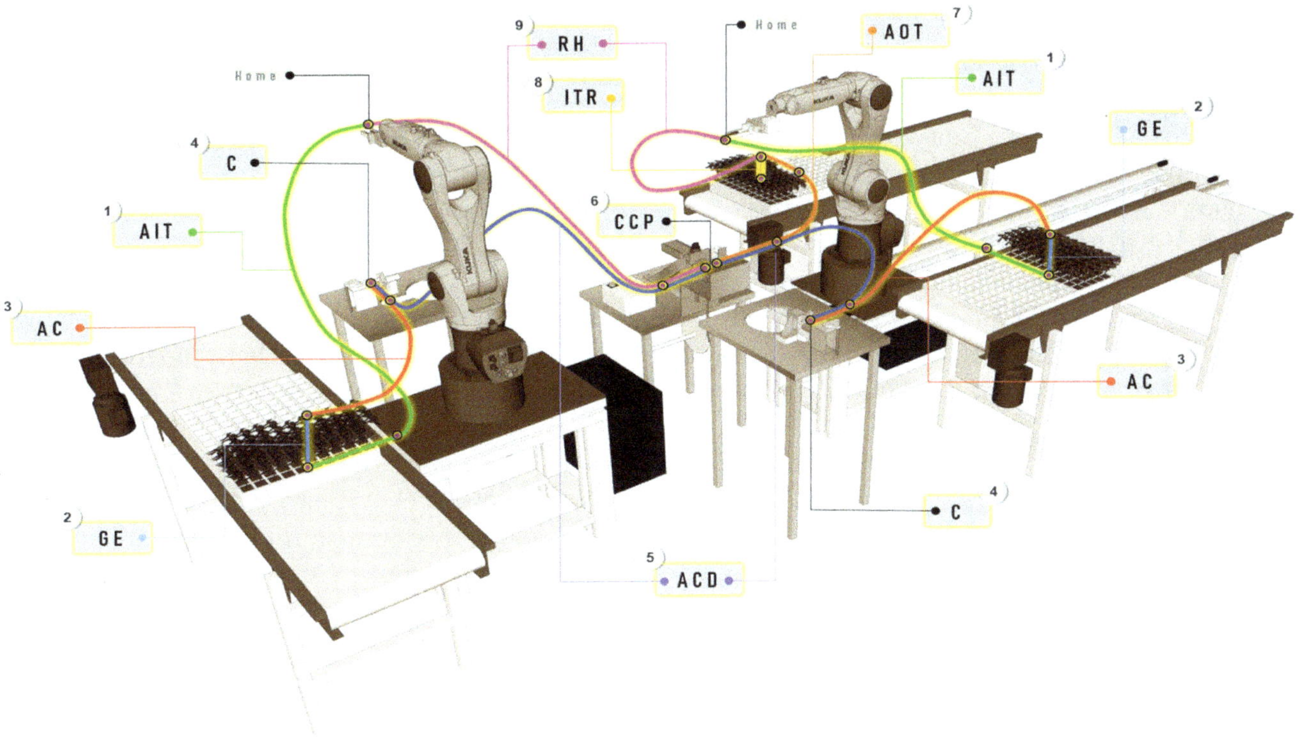

Figure 7. Travel details. Singular points: ① approach the input trays (AIT); ② grip and extraction (GE); ③ approach the cutting zone (AC); ④ cutting process (C); ⑤ approach the clip dispensing zone (ACD); ⑥ clip preparation and placement (CPP); ⑦ approach the output trays (AOT); ⑧ insertion on trays and release of plants (ITR); ⑨ return home (RH).

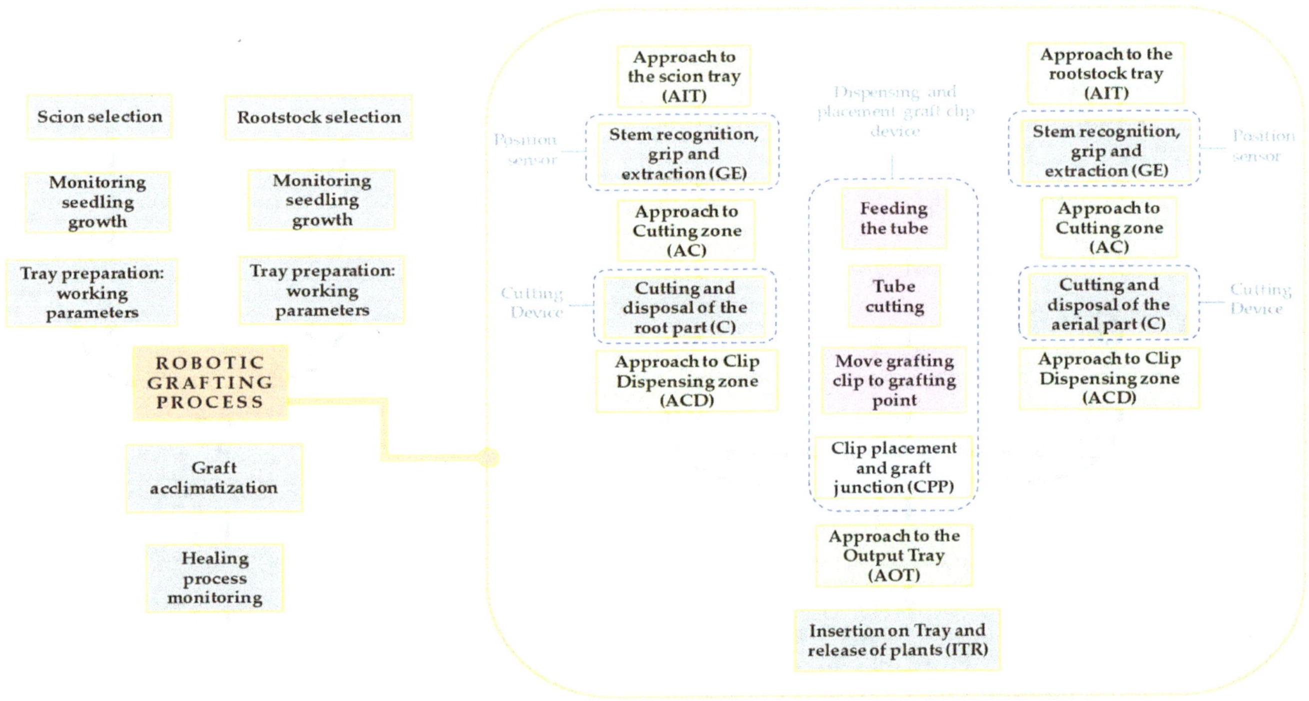

Figure 8. Flowchart of the developed automated grafting process.

2.2. Definition of Operating Conditions

The experiment was conducted at the Tecnova Technological Center (Centro Tecnológico Tecnova): Foundation for Agricultural Technologies of Agriculture, in Almería (36°52′38″N, 2°19′59″W) between the months of April and June 2017. The environment of Almería is a model of agricultural exploitation. Greenhouse growth of arable fruit crops has a high technical and economic performance, with tomato cultivation being of particular importance.

The rootstock used in the experiment was the interspecific hybrid "Maxifort" from De Ruiter Seed™, which is recommended for crops with better behavior at low temperatures and under high salinity conditions. The "Ventero" variety from De Ruiter Seed™ was used as an indeterminate hybrid tomato for branch harvest. Both types of seeds are routinely used in seedbeds to perform manual grafts using the "tomato on tomato" (T/T) technique, which demonstrates their prior compatibility with the robotic system.

The growth protocol developed in the nursery attempted to obtain plants that were grown and cared for until reaching a similar growth state between the rootstock and scion, with mature plants and those prepared for the graft having two–four well-defined true compound leaves [28] and stem diameters of at least 1.5 mm for the splicing method [29]. Therefore, stems with some natural variability characteristic of the development of each plant (between 1.5 and 2.5 mm in diameter in the area close to the cut for the scion, and between 2 and 3 mm in diameter for the area close to the cut for the rootstock) were worked with. Usually, in the automatic graft, the requirements in terms of growth and uniformity required for the rootstock and scion seedlings are as critical as in the manual graft [30], demanding an arduous previous task of pre-selection and pairing similar diameters between the linked seedlings. In our experiment, this work was eliminated because the seedlings were cut with a bevel at a 60° angle. From a certain cutting angle, between 50° and 70°, and provided that one works within the margins of natural variability between the previously established stems, the success rate of the graft was acceptable and higher than 95%. Therefore, the need to seek equal diameters of the workplaces was of lessened importance [27,31].

Prior to each experiment and for each tray, it was ensured that all the alveoli slots contained seedlings that met the previously established rootstock size criteria. The environmental conditions were regulated during the grafting process, with temperatures between 20 °C and 25 °C, conditions of relative humidity that were sometimes forced and were guaranteed to be above 75%, and stable, non-direct daylight luminosity conditions.

The data were collected for each test via filming and were then timed; the times until reaching each control point of the process until the graft was completed or the point of generation of each failure were evaluated.

Regarding the post-grafting conditions, the plants began to wilt immediately after cutting and grafting, so once each graft tray was finished, it was immediately introduced into a small healing chamber consisting of a tunnel slightly larger than the dimension of each tray and a low height, covered by a transparent film. This tunnel was placed inside a chamber in which the climatic conditions were controlled throughout the healing process. During the first 48 h, the plants were kept without illumination to reduce transpiration and evaporation. On the following days, the intensity of the light was increased, and a 14 h light photoperiod with a value ≈ 100 $\mu mol \cdot m^{-2} \cdot s^{-1}$ of PAR, (~ 3000 lux) of non-direct and diffuse light was established during the callus formation stage, from LED lights, corresponding to a value slightly above the compensation point because there is evidence that a high intensity of light prevents callus formation [1]. The level of illumination was gradually increased after several days. The temperature was established with a variable set point in the healing chamber between 23 °C and 30 °C, with an average of approximately 26 °C, slightly varying between the diurnal and nocturnal conditions. The relative humidity was initially established between 75% and 95% in an attempt to reduce the transpiration rate of the scions, avoiding high stress and thus preventing the drying of the graft [32]. The humidity level was gradually reduced in successive days to condition the grafts to the outdoors. The vapor pressure deficit (VPD), during the critical graft healing phase was around 0.8 kPa in the tunnel inside of the healing chamber, with the aim of decreasing transpiration. This value was gradually increased in the following days.

The success or failure of the final graft was evaluated by estimation and visual assessment performed daily across the 14 days after grafting, assessing the natural evolution of the graft, and analyzing other symptoms and external evidence that would determine its classification as either a success or a failure. In making this determination, key intermediate points were considered to mark an inflection of the task or singularity within the process, so that the successful or unsuccessful completion of this phase of the robotic process could be evaluated.

2.3. Experiments

In the process, different robot working speeds were tested, with the objective of determining the influence on the graft success. The robot speeds were constant within each test and ranged from 100 mm/s to 600 mm/s, with gradual increments of 100 mm/s. In total, six working speeds were tested.

A total of 900 grafts were prepared, divided into three experimental blocks consisting of a total of 300 grafted seedlings each, where a total of 50 grafts were performed for each of the six test rates. The sum of the experimental blocks, equal to one another, therefore consisted of a total of 150 resulting grafts for each of the six tested velocities.

Only the work times of the external processes on the seedlings (manipulations on the trays, cutting, and dispensing of the clip) were kept constant between the tests. It was understood that since the work times are based on pneumatic technology, they were optimized for the corresponding work speed at 3.5 bar (0.35 MPa).

Each experimental block, when was developed on a specific date, was treated under the same cultivation, manipulation, and post-graft healing conditions, with the aim of matching the development conditions between the three experimental blocks. In addition, for each of the experimental blocks, the positional order of each test rate was altered, thereby neutralizing the dependence of said factor.

The statistical analysis of the collected data was performed using the software Minitab v.18.1. The obtained results were subjected to an analysis of variance with a confidence level of 95%, and contrast tests were applied using Tukey's test (honestly significant difference, HSD test).

3. Results and Discussion

After performing the hypothesis test, we estimated that there were statistically significant differences between the grafting times used for the different tested work rates (Table 1).

For each velocity tested (between TS1 and TS6), the variability of time used in each test unit was mainly due to singularities that facilitated the manipulation and the development of the grafting work, to a greater or lesser extent. Singularities included the position of the alveolus in each row of work, the natural variability of the seedling emergence point within each alveolus, and the unique growth morphology of each seedling, among others. At low velocities, the difference for each work velocity was clear, given that the time taken to solve these singularities was less significant compared to the time spent in tasks not affected by these singularities. However, at high speeds, these factors became increasingly important and, to some extent, determined the time spent in each test unit.

Grouping the data by test speed, analyzing the failures associated with each control point that were recorded for the different velocities, and performing the hypothesis test, we estimated that, based on the Tukey's tests, there were statistically significant differences between the groups of different assay speeds (Table 2).

At low test rates, the success rate was higher, greater than 90% for speeds equal to or lower than TS3, and there was a significantly increasing graft failure for operating speeds equal to or greater than TS4. Low speeds, between TS1 and TS3, had similar behavior in terms of failures; therefore, we consider that their differences were derived from chance and not from the working speed itself. Therefore, it is clear that the TS3 production speeds are more attractive, given that they had a higher production ratio associated with a low failure rate.

The relationship linking the number of grafts/hour for each of the tested rates (TS1 to TS6) can be considered practically linear and was only altered by the random parameters derived, to a great extent, from the natural variability of the work seedling growth. Such nuances of correction barely affected the speed, which was largely marked by that established for both robots for each test, although the working speed of the auxiliary devices was kept constant. In addition, the percentage of successes/failures associated with each of the test rates (TS1 to TS6), considering their evolution, allowed us to observe a behavior similar to a quadratic function (Figure 9).

Table 1. Comparison between the mean times spent in grafting task for the six tested speeds. Grouping of comparisons applying Tukey's HSD test (honestly significant difference). Significance level $p < 0.05$.

TS (Test Speed)	N	Mean Time (s)	Variance	St. Dev	SS	St. Error	95% CI	Grouping (Tukey's HDS)
TS1	150	40.641	185.078	13.600	27576.61	0.617	(39.43; 41.85)	A
TS2	150	21.521	56.582	7.522	8430.76	0.617	(20.31; 22.73)	B
TS3	150	15.430	27.691	5.262	4125.97	0.617	(14.22; 16.64)	C
TS4	150	13.320	30.922	5.561	4607.40	0.617	(12.11; 14.53)	C
TS5	150	9.416	19.950	4.467	2972.62	0.617	(8.20; 10.63)	D
TS6	150	6.993	22.708	4.765	3383.49	0.617	(5.78; 8.20)	D

The different letters show significative differences.

Table 2. Comparison between the means for grafting failures for the six speeds tested. Grouping of comparisons applying Tukey's HSD test (honestly significant difference). Significance level $p < 0.05$.

TS Test Speed)	Speed (mm/s)	Check Points	Mean (Fails)	Variance	St. Dev	95% CI	Grouping (Tukey HDS)
TS1	100	10	1.50	3.16667	1.780	(0.022; 2.978)	A
TS2	200	10	1.60	3.37778	1.838	(0.122; 3.078)	A
TS3	300	10	1.50	3.16667	1.780	(0.022; 2.978)	A
TS4	400	10	2.20	6.17778	2.486	(0.722; 3.678)	AB
TS5	500	10	2.70	5.37778	2.319	(0.922; 3.878)	AB
TS6	600	10	4.70	11.34444	3.370	(3.220; 6.180)	B

The different letters show significative differences.

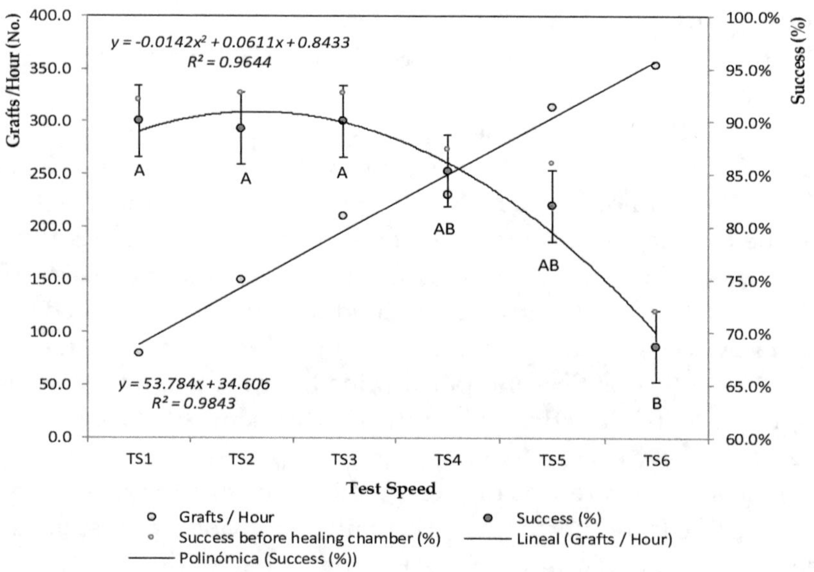

Figure 9. Grafts/hour versus grafting success for different test speeds (TS1 to TS6). Success rate includes the typical error. "Success before healing chamber" has been included to evaluate the influence on the global success of the grafting process.

At the working speeds of the TS3 and TS4 robots (210 and 240 grafts/hour, respectively), the number of grafts estimated for manual expert workers was already surpassed, i.e., a maximum of 150–240 grafts/hour [7,15], and an average of approximately no more than 1000 grafts per person per day [30,33,34]. In addition, the success rates for manual grafting are not usually very high, between 81% and 91% [35]. In part, this rate of failure may come from long hours in a hostile work environment, characterized by high humidity and temperature. The success rates achieved by the robotic graft were quite similar to those achieved by manual workers, reaching 90.0% for the TS3 speed of 300 mm/s and 85.3% for the TS4 speed of 400 mm/s.

For large cutting angles in splicing grafts, the difference between the diameters of the linked stems became less important as long as they stayed within certain a range [31]. For cuts at a 60° bevel, failure in the healing and grafting of the middle graft was between 3% and 6% of failures [27]. The difference between the number of successes considered after grafting and the number of successes recorded after the healing period in the chamber confirmed this parameter (less than 4%), and except for errors not visually detected in the grafting process, we associated the percentage of losses in the chamber to random problems in the healing process itself.

The established isolated intermediate control points allowed us to record the cause for each failed graft. Studying the origin of the graft failures associated with each velocity in detail, we determined that at low test rates, the failures detected in the grafting process in the GE phase (grip and extraction) were slightly more significant. This result may be because the seedling, when extracted from its alveolus at a lower velocity, experiences lower extraction acceleration, with a consequently greater resistance and grip to the alveolus walls. As a result, the roots tend to remain adhered and occasionally incur damage and tears; traction damage to the stem or alterations to the other plants in the tray may also occur, among other problems.

However, it was observed that, as the test speed increased, the number of failures in this phase decreased, as did the number of failures in the operations of adaptation and placement of the seedlings, such as the AC (approach the cutting zone) and ACD (approach the clip dispensing zone) phases, and in the tasks properly conducted by these tools: C (cutting process) and CPP (clip preparation and placement). In part, this was due to seedling management from certain working speeds, where there was a substantial acceleration in the displacements between points, and with it, the inertia experienced on the seedlings, which, together with their root ball (semi-compressed coconut fibre substrate) could suffer greater damage and tearing when experiencing such sudden changes of state. In addition, it was observed that high accelerations led to excessive seedling balance by the ends not held by the robot clamp (root ball and stem), causing the robot to lose its reference point at rest or to not return to it in time, thus spoiling the graft. The development of the grafting process is shown in Figure 10.

As the working speed increased, the times spent in the tasks performed by the external working tools became more important compared to the times spent in the displacement and pre-positioning of the seedlings in front of the tools. This factor is due to the fixed value of the velocities of the pneumatic devices in response to an incremental increase in the robot velocities (Figure 11).

At speeds equal to or lower than TS3, the recorded success rate was relatively good at approximately 90%, but it decreased substantially at higher speeds. The next tested speed, TS4, had a significantly different percentage of recorded failure, five points lower or 85.3%. It is important to assess the success associated with each work rate, because it is the factor that makes it feasible as an alternative to manual grafting, because the system evaluated is scalable in terms of systems and tools operating in parallel. That is, the clamp or end element could be adapted by cloning two or more gripping systems in parallel for the seedlings, and, to the same extent, the auxiliary devices or tools acting on the seedlings could be cloned, thus multiplying the number of plants grafted per hour, maintaining similar success rates. In addition, regarding these working speeds (TS3 and TS4), the number of grafts capable of being developed manually began to be exceeded. Therefore, when evaluating the grafts/success ratio, we estimated a better average behavior for velocities close to TS3 (210 grafts/hour).

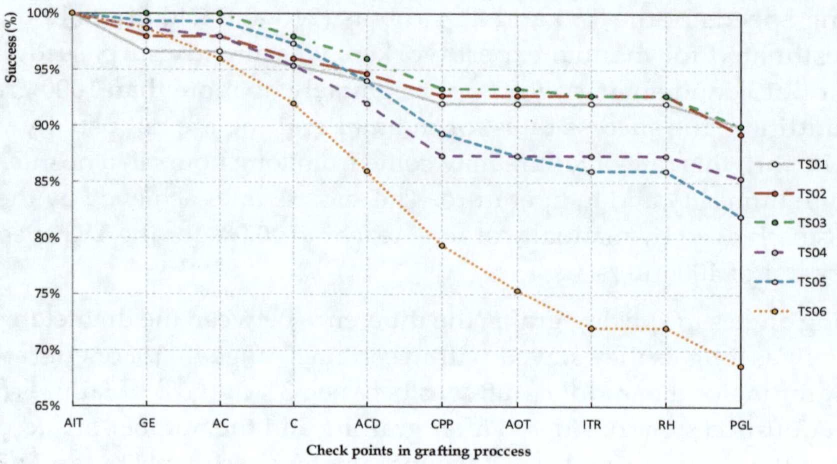

Figure 10. Success associated with each check point for the different test speeds, reflected in a percentage of successful plants in the grafting process. Check Points: approach the input trays (AIT), grip and extraction (GE), approach the cutting zone (AC), cutting process (C), approach the clip dispensing zone (ACD), clip preparation and placement (CPP), approach the output trays (AOT), insertion on trays and release of plants (ITR), return home (RH), and post-graft losses (PGL).

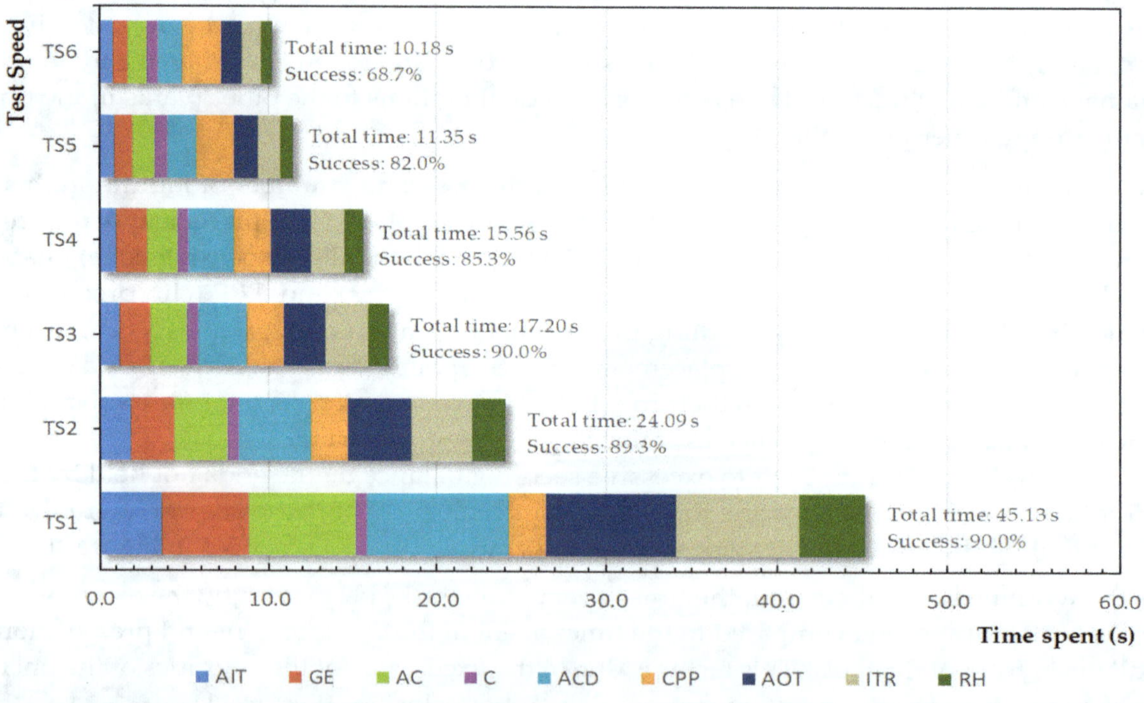

Figure 11. Average time spent in the development of each phase between control points. Check Points: approach the input trays (AIT), grip and extraction (GE), approach the cutting zone (AC), cutting process (C), approach the clip dispensing zone (ACD), clip preparation and placement (CPP), approach the output trays (AOT), insertion on trays and release of plants (ITR), and return home (RH).

Numerous studies have collected the test results of prototypes and commercial devices for the automated grafting of horticultural seedlings over the last three decades [3,5,7,23,36–44], and the results for dozens of pieces of equipment have been collected, mainly from Southeast Asia (Japan, Korea, and China mainly) and Europe.

Comparatively, we can refer to four factors that determine the convenience of and interest in the study equipment compared to other existing equipment: (a) the flexibility in terms of the horticultural family of work and the grafting method developed; (b) the degree of automation and the number of operators involved in the process; (c) system velocity (grafts/hour) and system efficiency; and (d) the price of the equipment.

(a) There are devices prepared exclusively for use with Solanaceae and others that allow working with Solanaceae and Cucurbitaceae. These pieces of equipment are mostly characterized by being inflexible. The study equipment, based on industrial robotics supported by low-cost external equipment, is specifically intended for splicing grafts, but enables easy and economic reconversion and readaptability, as it is able to work with other horticultural families and to apply other grafting methods. It even has the ability to perform other tasks in the seedbeds or productive environments.

(b) The existing prototypes and commercial devices present a different degree of automation, ranging from simple tools to help with grafting (cutting or dispensing the clip), to semi-automated equipment that requires the participation of two to three operators, to fully equipped devices. These are all automated and can be managed by a single operator. Nevertheless, they claim a high homogeneity in the work seedlings, requiring the important tasks of pre-sorting and pairing between seedlings, which to some extent tarnishes the autonomy of the system. The study equipment, by working with a high cutting angle, allowed us to avoid these previous tasks of searching and matching between the diameters of the workplaces and therefore enjoys a high degree of autonomy.

(c) Regarding the number of grafts per hour, the majority of studies present equipment ranging between 200 and 1200 grafts/hour. These values are much higher than those achieved using this equipment (210–240 grafts/hour), but these results are sufficient compared to manual grafting, and improvement is feasible by replicating terminal systems, allowing for working in parallel and over several plants simultaneously. In contrast, the system has a high success rate or efficiency of approximately 90%.

(d) The current price trend of industrial robotics, together with the use of simple, low-cost auxiliary equipment, allows us to estimate that the studied system is a rapidly amortizable system, with an initial investment in robots no greater than the biannual cost associated with the minimum interprofessional wage in countries [25], and an additional base cost of auxiliary and control equipment which is not higher than other interprofessional minimum wages (MW). This implies a total base investment of 5 MW (~€60,000). Faced with this, the estimated costs of high automation systems robotic equipment in high productivity environments (100 million plants per year) are estimated at investments above ~$7,500,000 [45]. Comparatively, and estimating our system working at approximately 230 grafts/hour, we would yield approximately 2 million grafts per year per piece of equipment, which would imply approximately 50 systems working in parallel to reach 100 million grafts. Such scaling would involve an investment of ~€3,000,000, well below the investment necessary for other robotic equipment.

4. Conclusions

The splicing technique, widely used for Solanaceae such as tomato, has the advantage of being simple and methodical, and therefore easily automatable. In the last three decades, there have been numerous attempts to develop equipment to deal with the automated grafting of horticultural plants, and the challenge of using conventional industrial robotics to perform the splicing graft process can provide a great opportunity. The robotic cell tested herein is based on two industrial robots and low-cost passive auxiliary units.

The use of low speeds, between 100 mm/s and 300 mm/s, allows ratios close to a 90% success rate to be maintained. At medium–high velocities, between 400 mm/s and 500 mm/s, success ratios were still acceptable above 80%. However, at a test speed of 600 mm/s, there was a considerable decrease in the success rate to less than 70%.

Consequently, we conclude that it is advisable to use a velocity close to 300 mm/s (90.0% success), which allows working at speeds higher than those estimated for manual expert workers, approximately 150–240 grafts/hour (with success rates between 81% and 91%). Decreasing the work rate below that point did not substantially improve the success rate.

Author Contributions: J.-L.P.-A. conceived and designed the experiments, performed the experiments, collected and analyzed the data, interpreted the results and developed the manuscript. Á.C.-O. conceived and designed the experiments, analyzed the data, interpreted the results and developed the manuscript. C.-C.M.-G. conceived, designed and performed the experiments. I.G. and M.G.G. provided constructive suggestions on experiment analysis.

Acknowledgments: This study was supported by Tecnova, Technological Center: Foundation for Auxiliary Technologies for Agriculture. The authors would like to acknowledge all the employees involved for their contributions to the experimental setting and data collection.

References

1. Hartmann, H.; Kester, D.E.; Davies, F.T.; Geneve, R.L. Principles of grafting and budding. In *Plant Propagation: Principles and Practices*, 8th ed.; Prentice Hall: Upper Saddle River, NJ, USA, 2010; Chapter 11.
2. Camacho-Ferre, F. El injerto en tomate como alternativa al bromuro de metilo. Experiencias con esta técnica en San Quintín. B.C.—México. II Agro Simp. *Int. Téc. Empres. Prod. y Tendencias* **2013**, 1–3. Available online: http://203.187.160.132:9011/simm.cicese.mx/c3pr90ntc0td/horta/docs/camachoFerre.pdf (accessed on 15 July 2019).
3. Lee, J.-M.; Oda, M. Grafting of Herbaceous Vegetable and Ornamental Crops, Horticultural Reviews. *Hortic. Rev.* **2003**, *28*, 61–124. [CrossRef]
4. Sakata, Y.; Ohara, T.; Sugiyama, M. The history and present state of the grafting of Cucurbitaceous vegetables in Japan. *III Int. Symp. Cucurbits* **2007**, 159–170. [CrossRef]
5. Lee, J.-M.; Kubota, C.; Tsao, S.J.; Bie, Z.; Echevarria, P.H.; Morra, L.; Oda, M. Current status of vegetable grafting: Diffusion, grafting techniques, automation. *Sci. Hortic.* **2010**, *127*, 93–105. [CrossRef]
6. Rivard, C.L.; Sydorovych, O.; O'Connell, S.; Peet, M.M.; Louws, F.J. An economic analysis of two grafted tomato transplant production systems in the United States. *Horttechnology* **2010**, *20*, 794–803. [CrossRef]
7. Lin, H.-S.; Chang, C.-Y.; Chien, C.-S.; Chen, S.-F.; Chen, W.-L.; Chu, Y.-C.; Chang, S.-C. Current Situation of Grafted Vegetable Seedling Industry and Its Mechanization Development in Taiwan. 2016, pp. 65–76. Available online: http://www.fftc.agnet.org/activities.php?func=view&id=20160113155600 (accessed on 15 July 2019).
8. Djidonou, D.; Gao, Z.; Zhao, X. Economic analysis of grafted tomato production in sandy soils in northern Florida. *Horttechnology* **2013**, *23*, 613–621. [CrossRef]
9. Rysin, O.; Rivard, C.; Louws, F.J. Is vegetable grafting economically viable in the United States: Evidence from four different tomato production systems. *Acta Hortic.* **2015**, *1086*, 79–86. [CrossRef]
10. Singh, H.; Kumar, P.; Chaudhari, S.; Edelstein, M. Tomato Grafting: A Global Perspective. *HortScience* **2017**, *52*, 1328–1336. [CrossRef]
11. De Miguel, A. *Use of Grafted Plants and I.P.M. Methods for the Production of Tomatoes in the Mediterranean Region*; Instituto Valenciano de Investigaciones Agrarias: Valencia, Spain, 2004; Available online: http://ec.europa.eu/clima/events/docs/0014/tomato_4_en.pdf (accessed on 15 July 2019).
12. De Miguel, A. Evolución del injerto de hortalizas en España. *Technol. Hortícola Hortic. Int.* **2009**, *10*, 72. Available online: http://www.horticom.com/revistasonline/horticultura/rhi72/10_17.pdf (accessed on 15 July 2019).
13. González, F.M.; Hernández, A.; Casanova, A.; Depestre, T.; Gómez, L.; Rodríguez, M.G. El injerto herbáceo: Alternativa para el manejo de plagas del suelo. *Rev. Protección Veg.* **2008**, *23*, 69–74, ISSN 2224-4697. Available online: http://www.shorturl.at/klAKU (accessed on 15 July 2019).
14. Hoyos Echevarria, P. Situación del injerto en horticultura en España: especies, zonas de producción de planta, portainjertos. *Ind. Hortícola* **2007**, *199*, 12–25. Available online: http://www.horticom.com/revistasonline/horticultura/rh199/12_25.pdf (accessed on 15 July 2019).
15. Ngo, Q.V. Grafted Tomato in Vietnam, From 0 to 7000 Ha/Year. 2016. Available online: http://www.fftc.agnet:htmlarea_file/activities/20160113155600/2016P032-7VN.pdf (accessed on 15 July 2019).
16. Kobayashi, K.; Suzuki, M.C. Grafting Robot. *J. Robot. Mechatron.* **1999**, *11*, 213–219. [CrossRef]
17. Tian, S.; Xu, D. Current status of grafting robot for vegetable. In Proceedings of the 2011 International Conference on Electronic & Mechanical Engineering and Information Technology, Harbin, China, 12–14 August 2011; Volume 4, pp. 1954–1957. [CrossRef]

18. Chiu, Y.C.; Chen, S.; Chang, Y.C. Development of a circular grafting robotic system for watermelon seedlings. *Appl. Eng. Agric.* **2011**, *10*, 95–102. [CrossRef]
19. Kim, H.M.; Hwang, S.J. Comparison of Pepper Grafting Efficiency by Grafting Robot. *Prot. Hortic. Plant Fact.* **2015**, *24*, 57–62. [CrossRef]
20. Oda, M. Grafting of Vegetables to Improve Greenhouse Production. Ph.D. Thesis, Food & Fertilizer Technology Center: College of Agriculture, Osaka Prefecture University, Sakai Osaka, Japan, 1998; pp. 1–11. Available online: http://www.agnet:library.php?func=view&id=20110803135029/ (accessed on 15 July 2019).
21. De Miguel, A.; Cebolla, V. Unión del injerto. *Terralia* **2005**, *53*, 50–60. Available online: https://www.terralia.com/terralias/view_report?magazine_report_id=365 (accessed on 15 July 2019).
22. Chiu, Y.; Chen, S.; Chang, Y.; Chou, L. Development of Robotic Grafting Systems for Fruit Vegetable Seedlings. In Proceedings of the FFTC & Tainan-DARES International Workshop on Grafting to Improve Fruit Vegetable Production, Tainan, Taiwan, 16–20 May 2016; pp. 77–85. Available online: http://www.fftc.agnet:library.php?func=view&style=type&id=20170331104933 (accessed on 15 July 2019).
23. Jinyuan, Z.; Yunsheng, T. Development Status of Internal and External Graft Machinery. *Council Agric. For.* **2015**, 99–106. Available online: https://book.tndais.gov.tw/Other/2015seedling/speech10.pdf (accessed on 15 July 2019).
24. Kuka Robots Ibérica. In Proceedings of the Conference Universidad de Vigo, Vigo, Spain, 27 October 2008; Available online: http://shorturl.at/yUVY9 (accessed on 15 July 2019).
25. Chiacchio, F.; Petropoulos, G.; Pichler, D. *The Impact of Industrial Robots on EU Employment and Wages: A Local Labour Market Approach*; Bruegel: Brussel, Belgium, 2018; pp. 1–18.
26. Feng-feng, W. Study on Grafting Machine of Camellia Seedling for Cleft Grafting. Ph.D. Thesis, Chinese Academy of Sciences, Beijing, China, 2011. Available online: https://www.dissertationtopic.net/doc/149228 (accessed on 15 July 2019).
27. Pardo-Alonso, J.-L.; Carreño-Ortega, Á.; Martínez-Gaitán, C.-C.; Callejón-Ferre, Á.-J. Combined Influence of Cutting Angle and Diameter Differences between Seedlings on the Grafting Success of Tomato Using the Splicing Technique. *Agronomy* **2018**, *9*, 5. [CrossRef]
28. Miles, C.; Flores, M.; Estrada, E. Injerto de Verduras Berenjenas y Tomates. In *Hoja de Datos de la Extensión, FS052ES*; Washington State University: Washington, DC, USA, 2013; pp. 1–4. Available online: https://s3.wp.wsu.edu/uploads/sites/2071/2014/04/Grafting-Eggplants-and-Tomatoes-SPAN-FS052ES.pdf (accessed on 15 July 2019).
29. Bumgarner, N.R.; Kleinhenz, M.D. *Grafting Guide: A Pictorial Guide to the Cleft and Splice Graft Methods as Applied to Tomato and Pepper*; Ohio State University, Research and Development Center: Columbus, OH, USA, 2014; Available online: http://www.walterreeves.com/wp-content/uploads/2010/11/tomato-grafting-guide.compressed.pdf (accessed on 15 July 2019).
30. Hassell, R.L.; Memmott, F.; Liere, D.G. Grafting methods for watermelon production. *HortScience* **2008**, *43*, 1677–1679. [CrossRef]
31. Bausher, M.G.; Road, S.R.; Pierce, F. Graft Angle and Its Relationship to Tomato Plant Survival. *HortScience* **2013**, *48*, 34–36. [CrossRef]
32. Johnson, S.J.; Miles, C.A. Effect of healing chamber design on the survival of grafted eggplant, tomato, and watermelon. *Horttechnology* **2011**, *21*, 752–758. [CrossRef]
33. Rivard, C.L. *Tomato Grafting for High Tunnel Production*; Kansas State University, Research and Extension: Manhattan, KS, USA, 2012; Available online: https://www.slideshare.net/UMNfruit/rivard-mn-ht2012a (accessed on 15 July 2019).
34. Tirupathamma, T.L.; Ramana, C.V.; Naidu, L.N.; Sasikala, K. Vegetable Grafting: A Multiple Crop Improvement Methodology. *Curr. J. Appl. Sci. Technol.* **2019**, *33*, 1–10. [CrossRef]
35. Sarah, A. *Masterson. Propagation and Utilization of Grafted Tomatoes in the Great Plains*; University of Alabama: Tuscaloosa, AL, USA, 2010; Available online: https://core.ac.uk/download/pdf/18529369.pdf (accessed on 15 July 2019).
36. Al-Razaq, A.H.A. Grafting techniques in vegetables crops: A review. *Plant Arch.* **2019**, *19*, 49–51. Available online: http://plantarchives:PDF%2019-1/49-51%20(4756).pdf (accessed on 15 July 2019).
37. Tian, S.; Ashraf, M.A.; Kondo, N.; Shiigi, T.; Momin, M.A. Optimization of Machine Vision for Tomato Grafting Robot. *Sens. Lett.* **2013**, *11*, 1190–1194. [CrossRef]

38. Yinghui, M.; Xiwen, L. *Root Pruning and Hole-Oblique Insertion Hypocotyl Automatic Grafting of Cucurbitaceous Vegetables*; Chinese Society of Agricultural Engineering: Beijing, China, 2011; pp. 2–7. Available online: http://www.tcsae.org/nygcxb/ch/reader/view_abstract.aspx?file_no=X201102006&flag=1 (accessed on 15 July 2019).
39. Song, G.; Linbin, J. Development of domestic and foreing vegetable grafting robot. *J. Northeast Agric. Univ.* **2007**, *38*, 847–851.
40. Oda, M. Use of Grafted Seedlings for Vegetable Production in Japan. *Acta Hortic.* **2008**, *770*, 15–20. [CrossRef]
41. Yamada, H. Research for Development of the Grafting Robot for Solanaceae. *Tech. Pap. Agric. Mach. Res. Assoc.* **2003**, *65*, 142–149. [CrossRef]
42. Lee, J.-M.; Bang, H.J.; Ham, H.S. Grafting of Vegetables. *Jpn. Soc. Agric. Mach. Food Eng.* **1998**, *67*, 1098–1104. [CrossRef]
43. Hwang, H.H. *Study on Development of Automatic Grafting System for Fruit Bearing Vegetable Seedling*; Escuela de Silvicultura de La Universidad Sungkyunkwan—Ministro de Agricultura y Silvicultura, Sungkyunkwan: Seoul, Korea, 1997.
44. Kurata, K. Cultivation of grafted Vegetables 2. Development of Grafting Robots in Japan. *HortScience* **1994**, *29*, 240–244. [CrossRef]
45. Lewis, M.D.; Kubota, C.; Tronstad, R. Scenario-Based Economic Analyses of Different Grafting Operation Sizes. Poster, (Table 2), 1000. 2012. Available online: http://www.vegetablegrafting:wp/wp-content/uploads/2012/11/Lewis_poster.pdf (accessed on 15 July 2019).

9

Interactive Effects of Grafting Techniques and Scion-Rootstocks Combinations on Vegetative Growth, Yield and Quality of Cucumber (*Cucumis sativus* L.)

Rana Shahzad Noor [1,2], Zhi Wang [1], Muhammad Umair [2], Muhammad Yaseen [3], Muhammad Ameen [4], Shoaib-Ur Rehman [5], Muzammil Usman Khan [5], Muhammad Imran [6], Waqar Ahmed [7] and Yong Sun [1,*]

1 Department of Agriculture, Biological, Environment and Energy Engineering, College of Engineering, Northeast Agricultural University, Harbin 150030, China; engr.rsnoor@uaar.edu.pk (R.S.N.); wangzhi@neau.edu.cn (Z.W.)

2 Faculty of Agricultural Engineering & Technology, PMAS-Arid Agriculture University, Rawalpindi 46000, Pakistan; umairkpr@uaar.edu.pk

3 Department of Agricultural Extension, College of Agriculture, University of Sargodha, Sargodha 40100, Pakistan; yaseen.baksh@uos.edu.pk

4 College of Engineering, Nanjing Agricultural University, Nanjing 210031, China; ameenkhan114@yahoo.com

5 Institute of Horticultural Sciences, University of Agriculture, Faisalabad 38040, Pakistan; shoaibhort@gmail.com (S.-U.R.); khancoca0@gmail.com (M.U.K.)

6 Faculty of Water Resources and Hydroelectric Engineering, Xi'an University of Technology, Xi'an 710048, China; 180414001@stu.xaut.edu.cn

7 Key Laboratory of Soybean Biology, Chinese Ministry of Education, Northeast Agricultural University, Harbin 150030, China; khuhrowaqar@gmail.com

* Correspondence: sunyong@neau.edu.cn

Abstract: The density of herbaceous crops creates a suitable environment to produce pathogens in the soil that intensify the attack of pathogens traditionally controlled by disinfectant, which are mostly prohibited and unlisted because of their toxicity. Grafting is an alternative technique to enhance abiotic stress tolerance and reduce root diseases due to soil-borne pathogens, thus enhancing crop production. This research study was conducted during the crop season of 2017 and 2018 in order to investigate the interactive effect of different grafting techniques of hybrid scion onto local rootstocks on plants survival, plant phenological growth, fruit yield and fruit quality under a controlled environment. The hybrid cucumber was also planted self-rooted. The cucumber (*Cucumis sativus* L.) cv. Kalaam F_1, Syngenta was grafted onto four local cucurbitaceous rootstocks; ridge gourd (*Luffa operculate* Cogn.), bitter gourd (*Momordica charantia* L.), pumpkin (*Cucurbita pepo* L.), bottle gourd (*Lagenaria siceraria* (Molina) Standl.) using splice grafting, tongue approach, single cotyledon and hole insertion grafting techniques and self-rooted hybrid cucumber under greenhouse conditions. The experimental results indicated that all local cucurbitaceous rootstocks showed a high compatibility with hybrid cucumber scion in the splice grafting method compared to other grafting and non-grafted methods. *Lagenaria siceraria* rootstocks were found highly compatible with cucumber cv Kalaam scion which gave significantly maximum plant survival rates (95%) due to high sap contents, high SPAD value, better vegetative growth and maximum fruit yield when compared with other rootstocks by employing the splice grafting method followed by tongue approach, single cotyledon and hole insertion grafting while the fruit quality of all rootstocks was observed to be similar. The non-grafted cucumber cv. Kalaam F_1 showed significant results of plant vegetative growth, fruit development and fruit quality and encountered grafting methods while the lowest result were associated with the hole insertion grafting method in all scion/rootstock combinations. The grafted plants have no significant effect on cucumber fruit dry matter and fruit quality while the fruit mineral compositions (N, P, K, Ca and Mg) were higher among grafted and non-grafted plant fruits. The results indicate that grafting hybrid cucumber onto four local cucurbitaceous rootstocks influenced growth, yield and fruit quality. Grafting can be alternative and control measure for soil-borne disease and to enhance cucumber production.

Keywords: cucumber; grafting techniques; rootstock-scion; soil-borne disease; resistant; tolerant crop growth; fruit yield; fruit quality

---

1. Introduction

Grafting entails a deliberate combination of parts of different plants of the same species by which vascular continuity is established [1]. The crown of the plant (Scion) adheres to the root part of the plant (rootstock) resulting in composite plant growth and development of a single plant (graft) [2]. The callus related to parenchyma cells develops from the plant tissue of the rootstock and scion around the joint portion to develop vascular connection [3]. Grafting is a horticultural technique applied in the sustainable agricultural practice for the protection of cucurbitaceous crops from soil-borne pathogens, nematodes, soil pH and salinity. This problem has been globally important since 2005 when the application of methyl bromide was prohibited [4]. The rapid multiplication of soil-borne pathogens and nematodes is due to the intensive use of soil and absence of crop rotation. This affects vegetable growth particularly under greenhouse conditions [5]. It is a big challenge to reduce the impact of soil pathogens for a sustainable agriculture production system [6] and monoculture is more susceptible than a diversified agricultural system [7]. The rootstocks resistant for nematodes are not available; nematodes cause 11 % annual average yield loss [8].

Grafting has become a technique with a high potential to improve the efficiency of modern and intelligent vegetable cultivation, and indicates adoptability and resistance under different stress situations [9]. The grafting of plants is carried out to develop tolerance against temperature variation, salt stress and heavy metal contents in the soil. The grafting scion on appropriate local rootstocks can medicate these issues [10–12]. The structural and biological development of graft between scion and rootstock has been studied and three basic phases have been observed, namely: combining scion and rootstock, callus creation around the joint and establishing continuity at joint through vascular re-differentiation [1,3,13,14]. These discussed events determine the success of the scion-rootstock combination, considering the role of plant growth hormones (auxin) in the grafting technique [15]. The graft may grow through the wound healing mechanism and formation of conductive vessels [16]. Therefore, the formation of vascular connection represents the last and most critical stage in wound healing because after healing, the transportation of water and solute start from the rootstock to the scion and the graft union develops a strength [13,17]. The time required for the completion of grafting stages is a little unpredictable [17] because an appropriate method for the assessment of development in grafting does not exist [18]. However, the development can be predicted by destructive and non-destructive methods, including the visual determination of graft [19], thermal camera image [19], cutting graft vertically at union to observe vascular system [20], electrical resistance measurement from scion to rootstock [19] asses tensile strength at scion-rootstock union [21,22], disruptive evaluation of hydraulic connection [23] and the nuclear magnetic resonance (NMR) method to show water flux inside vessels [23].

Cucurbitaceous crops are mostly grown during the warm season (greenhouse conditions, 21 °C to 32 °C). The temperature below 16 °C is not suitable for the germination of cucumber seeds [24]. Cucumber is grown during the warm season when vegetables produce good crops when grown under protection [24]. The above discussed problems severely influence the area and production under greenhouse conditions [25]. The grafting of scion on local rootstocks is the most effective solution to this problem. Cucumber (*Cucumis sativus* L.) grafted on different cucurbitaceous crop rootstocks are

Cucurbita maxima Duchesne, *C. moschata*, *C. ficifilia*, *Lagenaria* spp., *Luffa* spp. [26–28]. The selective method of grafting considerably depends upon crop type, grower decision and experience, as well as facilities availability [29]. Splice grafting [30,31], tongue grafting [32], single cotyledon [4,33] and hole insertion grafting [19,34]. Several studies indicated that the grafting technique showed productive findings such as increase growth, yield and tolerance against stress conditions [35–37]. These plant characteristics majorly linked with genotype of rootstock particularly plant growth and yield effected by rootstock [37,38]. The grafting of cucumber crop on different rootstock reduces precocity but the total yield significantly increased as compared with non-grafted plants [39].

The grafting had some undesirable effects such as deterioration in taste, changes in the fruit color, bigger fruit size. The grafting also resulted the increment in fructose content and sweetness of cucumber which indicates the significance of rootstock [40]. The collar size of grafted plants varies depending upon rootstock. Frequently, the diameter of the collar of the grafted plants is higher [39]. The concerning, uncertainties regarding scion-rootstock and environmental instructions. The best selection for rootstock based on production area, rootstock species and scion cultivars used under different conditions [41]. The combination of rootstock and scion must be carefully selected according to the specific calamitic and soil condition [42]. The adequate rootstock-scion combination could help to regulate the soil borne diseases, increase yield and improve fruit quality.

This motivates us to carry out a more detailed study of these significant parameters such as yield, earliness, taste and some fruit sensory characteristics of grafted cucumber plants on different cucurbit rootstocks. This study was carried out to investigate the effect of grafting a hybrid cucumber scion onto four local cucurbitaceous rootstocks by four different grafting technique under greenhouse conditions on plants survival, plant phenological growth, fruit yield and fruit quality in order to establish the most tolerant scion-rootstock combination and favorable method for cucumber grafting. The hybrid scion cucumber cv Kalaam F_1 and four rootstocks, namely ridge gourd (*Luffa operculata*), bitter gourd (*Momordica charantia*), pumpkin (*Cucurbita pepo*), bottle gourd (*Lagenaria siceraria*) were used as scion-rootstock combinations in order to check the compatibility of tongue grafting, splice grafting, single cotyledon grafting and hole insertion grafting.

2. Materials and Methods

This experimental study was conducted during the crop season of 2017 and 2018 at Floriculture Research Station, University of Agriculture, Faisalabad to evaluate the compatibility and suitability of grafting techniques for hybrid cucumber cv. Kalaam F_1 scion grafted onto four local cucurbitaceous rootstocks (Pumpkin, Ridge gourd, Bottle Gourd and Bitter gourd) under greenhouse conditions (Temperature 28 °C and 90% humidity). Screening and determining the resistant and tolerant rootstock against soil pathogenic characteristics were performed.

2.1. Grafting Plant Materials

The hybrid cucumber cv. *Kalaam* F_1 from Syngenta seed company was used as a scion. Four local rootstocks from cucurbitaceous species; ridge gourd (*Luffa operculata*), bitter gourd (*Momordica charantia*), pumpkin (*Cucurbita pepo*), bottle gourd (*Lagenaria siceraria*) were used as shown in Table 1.

2.2. Nursery of Scion and Rootstocks

Cucumber cv Kalaam F_1 scion were grafted on four local cucurbitaceous rootstocks by tongue grafting, splice grafting, single cotyledon grafting and hole insertion grafting. Scions were sown in a nursery greenhouse on 25th and 28th of July in 2017 and 2018 in a 200-hole tray and rootstocks were grown in seedling trays with 128 cells (280 × 540 × 40 mm). The seeds of four different rootstocks named as *Luffa operculata*, *Momordica charantia*, *Cucurbita pepo* and *Lagenaria siceraria* were sown after five days of scion seed sowing. Peat moss and vermiculite ratio 1:1 (v/v) was used as growing media because it sustained better equilibrium of moisture and air which was best for seedlings. 100 g ammonium sulphate, 150 g potassium sulphate, 400 g calcium superphosphate, 50 mL nutrient solution and 50 g of fungicide were mixed for each 50 kg of peat moss. The characteristics of the growing media before seed sowing were: pH-6.7, electrical conductivity EC-0.14 mS cm^{-1}, NO_3-15 mg dm^3, P-4 mg dm^3, K-31 mg dm^3, Ca-28 mg dm^3, Mg-17 mg dm^3 and total nitrogen 0.25% [43]. Nitrogen, phosphate and potassium fertilizer were also applied. The seeds were sown under 28 °C mean temperature and 80% relative humidity. The 400 scion plants and 400 rootstocks (100 for each rootstock) were prepared and 200 plants (100 scion and 100 rootstocks) were used in each grafting technique. The experimental treatments, their levels and replication were presented in Table 1. The scion-rootstock were shifted into plastic pots of 8 cm diameter with the same growing media. Before grafting, a laminar flow UV light was used to kill the microbes in the working place [43].

2.3. *Grafting Techniques and Procedure*

The four grafting techniques used were splice grafting, tongue grafting, single cotyledon and hole insertion grafting. The different aged seedlings were grafted under the different grafting methods. Grafting method required selection of seedlings that mainly based on its stem diameter. Each grafting technique was applied on 100 scion and 100 rootstocks (25 for each rootstock) with four treatments and five replications.

The scion and rootstock plants were taken out from the seedling tray. The grafting was performed and then the plants were shifted into disposable plastic pots with same growing media and characteristics. In tongue grafting technique (TAG), the plants were 17 and 15 days old (4 mm stem diameter) taken for root-stock and scion gave the best diameter match. The downward and upward slanting cut were made on scion stock and rootstock respectively [32,44,45]. In splice grafting, the 15 day old plants were preferably matched stem diameter that can fits with each other were best fits for that graft union. The rootstock and scion stock were cut slanted at 35°–45° downward and upward respectively [33,34]. The 14 days rootstock and 13 days scion stock fit best for this graft combination in single cotyledon grafting method. The scion was grafted on the stem of rootstock plant [36,37]. The hybrid cucumber scions of 13 days and pumpkin rootstocks of 10 days were best match for grafting. Rootstock was removed just above the false leaf, and the sharpened scion was inserted in it after making a hole between the false leaf interjections with the help of small narrow sterilized wooden sticks [20,38]. The union point was sealed with aluminum wrap and grafting clips. The plastic pots were enveloped with transparent plastic sheet in order to maintain the humidity level around the graft. The plastic sheet was cut from top side after three days in order to allow vertical growth. Grafting plants healing begun after five days and plastic sheet was taken off completely after six days. Plants were supplied water and fertilizer accordingly.

The plants were kept in screen house for eight days after twelve days of grafting. This whole procedure took 30–35 days after transplanting. Transplantation of grafted plants was performed on September 9th and 12th, 2017 and 2018. The grafted plants rows were of 10m in length and 1.2 m in width with 0.5 m distance between plants. Each single treatment consisted of twenty (20) plants in a row. The plants were transplanted keeping union of graft above the soil surface to prevent the development of roots from graft union that may cause infection and plant wilting. The conventional agricultural practices were adopted during the entire cucumber crop production. The cucumber fruit was harvest after 70 days from grafting (35 days after transplanting) and then cucumber fruit was harvested with two days interval. The following parameters were studied during this research study.

2.4. *Studied Characteristics*

The plant and fruit growth characteristics were studied to evaluate the compatibility of different grafting techniques applied on four local cucurbitaceous rootstocks and one hybrid cucumber scion.

2.4.1. Plant Vegetative Growth

The plant sustainability after grafting was measured till 30 days after grafting and the measurements were taken for plant vegetative growth 75 days after transplanting. Three plants were randomly selected from each treatment. The mean leaf area (cm^2) was measured through fresh weight method. The collected leaves were cleaned and weighted 0.001 g. The 20 disks with known weight were taken and the area of each disk was 1.0 cm^2.

$$\text{Leaf area } (cm^2) = \frac{\text{fresh weight of leaf No.10}}{\text{fresh weight of 20 diske}} \times 20 \times \text{area of disk} \quad (1)$$

The other plant vegetative parameter measured number of leaves, plant height (cm), shoot length (cm), stem diameter/thickness (cm), number of plant branches and chlorophyll content/SPAD values were using SPAD-501 (leaf chlorophyll meter) to estimate the leaves pigments or green color of leaves of 5th leaf (fully expanded) from the top of canopy [46]. At flowering stage flowering time (days), fruiting harvesting time (days) and survival rates (%) were observed. The plant survival rates and mortality were observed after 12 days from the grating by counting the number of plants wilted. The survival rates were recorded till harvesting stage.

2.4.2. Fruit Growth and Quality Analysis

The parameters studied for fruit growth and quality were total number of fruits/plants, fruit weight per cucumber (g), fruit shape index, total soluble solids (%), fruit dry matter (%), fruit chemical contents, mineral composition and fruit fresh weight (g), length of fruit (cm) & fruit diameter (cm).

The fruit shape index was a ratio between cucumber fruit length and diameter. Total soluble solids (T.S.S %) in fruit juice was measured using hand refractometer. Five samples were analyzed from each scion-rootstock combination [47]. Fruit dry matter (%) was measured through drying 100 g of fresh fruit weight in an oven at 70 °C till no further weight reduction condition reached. The contents of ascorbic acid in all grafted and non-grafted fruit were measured by 2,6-dichlorophenol-indophenol method [48]. Bradford G-250 reagent was used to calculate the soluble protein [49] while ninhydrin technique was used to measure total amino acids [50]. Soluble sugar was calculated by anthrone [51]. Fruit mineral composition by the atomic absorption spectrometer (Varian spectra 220) to calculate after dry fruit ashes (550 °C) and mixed in 1N HCl. The N and P were observed by Kjeldahl technique and vanado-molybdate phosphoric yellow color method. The fruit fresh weight (g), fruit length (cm) and fruit diameter (cm) were measured right after the first day of fruit development to harvesting (day 1–9).

2.5. Data Processing and Experimental Design

The experimental data collected were than analyzed statistically using one-way analysis of variance (ANOVA) to evaluate the level of significance. Tukey HSD at 5% level of significance was applied to check the significant difference between values of studied parameter. The complete statistical analysis was performed using Minitab 17 statistical software (Minitab Pty Ltd., Sydney, Australia).

Table 1. Treatments, their level and replication for different grafting techniques.

Grafting Technique	Cucurbitaceous Crop	Treatment Description (Scion-Rootstock)	
G_1 = Tongue grafting G_2 = Splice grafting G_3 = Single cotyledon grafting G_4 = Hole insertion Grafting G_n = Non-grafted	Scion; Cucumber cv. Kalaam F_1 Rootstock; Ridge gourd (*Luffa operculata*) Bitter gourd (*Momordica charantia*) Pumpkin (*Cucurbita pepo*) Bottle gourd (*Lagenaria siceraria*)	T_1 = Cucumber-Ridge gourd T_2 = Cucumber-Bitter gourd T_3 = Cucumber-Pumpkin T_4 = Cucumber-Bottle gourd T_c = Kalam F_1 (real rooted)	
		Replications	R_1 R_2 R_3 R_4 R_5

3. Results and Discussion

3.1. Plant Vegetative Growth

The results of vegetative growth and survival rate of cucumber (Cucumis sativus) plants grafted on different cucurbitaceous rootstocks were observed after 15 and 30 days of grafting in two crop season 2017 and 2018. The leaf area, number of leaves and SPAD values after the first 15 days of grafting, during both crop seasons of 2017 and 2018, were observed in all combination of cucumber (Kalaam F_1) scion and four local rootstocks; ridge gourd, bitter gourd, pumpkin and bottle gourd were presented in Figure 1. The mean leaf area (cm^2) of cucumber cv. Kalaam F_1 was found significantly maximum during both crop seasons 2017 and 2018 when grafted onto bottle gourd rootstock followed by ridge gourd, bitter gourd and pumpkin rootstock with the splice grafting (36.10, 37.65), single cotyledon grafting (32.40, 33.23), tongue grafting (28.23, 29.98) and hole insertion grafting (25.80, 26.13) respectively. The leaf area (21.43, 22.70) was found non-significant in non-grafting cucumber cv. Kalaam F_1. The results observed in both seasons didn't show any significant difference. The plants with bottle gourd rootstocks showed significantly maximum five plant leaves in tongue grafting, splice grafting and single cotyledon grafting during both crop seasons. The real rooted plants produced four plant leaves while the lowest figures were associated with pumpkin rootstock. Figure 1 illustrates that the SPAD values (chlorophyll content) were found significantly different in all grafting techniques and non-grafted plants during the vegetative growth stage. The splice grafting technique gave optimum results of chlorophyll content than other grafting methods. The significantly different SPAD values (49.31, 46.6) were found in cucumber cv. Kalaam F_1 grafted onto bottle gourd with splice grafting during 2017 and 2018 crop season respectively. The SPAD values in plant leaf of real rooted Kalaam F_1 and pumpkin rootstock in hole insertion grafting method were statistically non-significant.

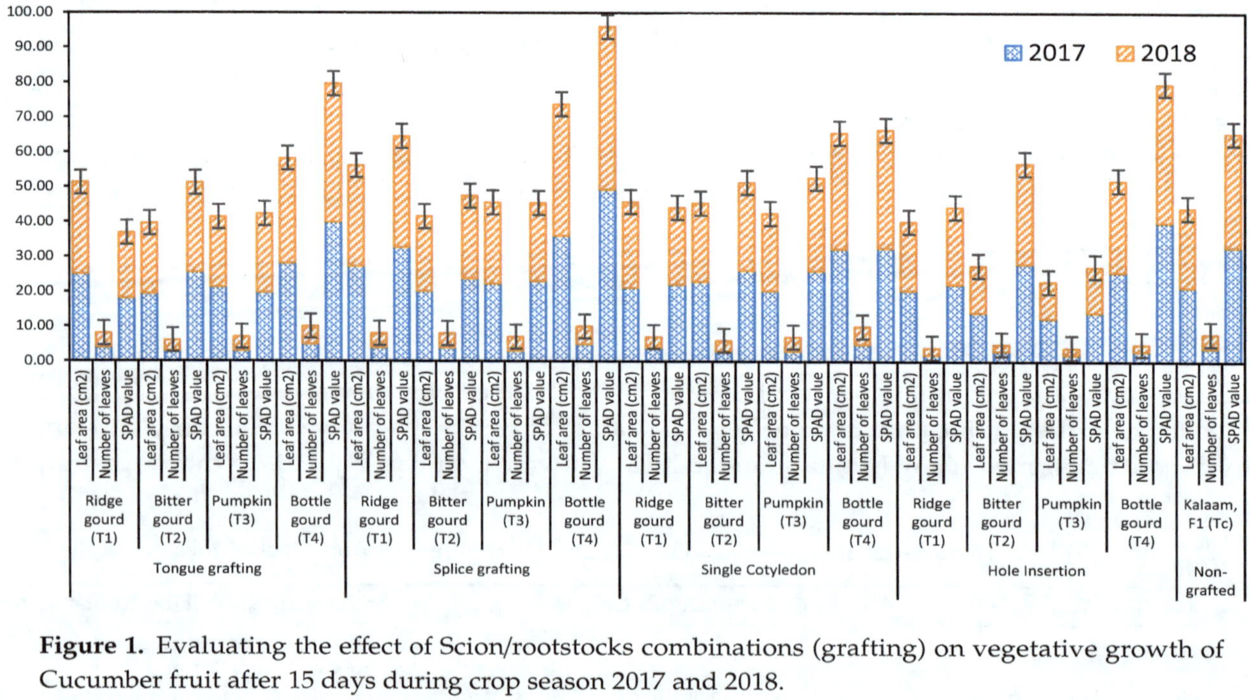

Figure 1. Evaluating the effect of Scion/rootstocks combinations (grafting) on vegetative growth of Cucumber fruit after 15 days during crop season 2017 and 2018.

After 30 days, the bottle gourd rootstock in splice grafting showed significantly higher mean leaf area (82.40, 85.28 cm^2), number of leaves (8, 8) and SPAD contents (49.31, 56.22) during both crop season 2017 and 2018 respectively as shown in Figure 2. The real rooted cucumber didn't show significant difference of leaf area (40.95, 44.65 cm^2), number of leaves (7, 7) and SPAD value (39.98, 42.8) during crop season of 2017 and 2018 while the rootstocks in hole insertion grafting showed non-significant results.

It could be found that the SPAD values and chlorophyll content have a strong correlation [42] and the combinations of scion/rootstocks had significantly affected the vegetative growth of cucumber [52].

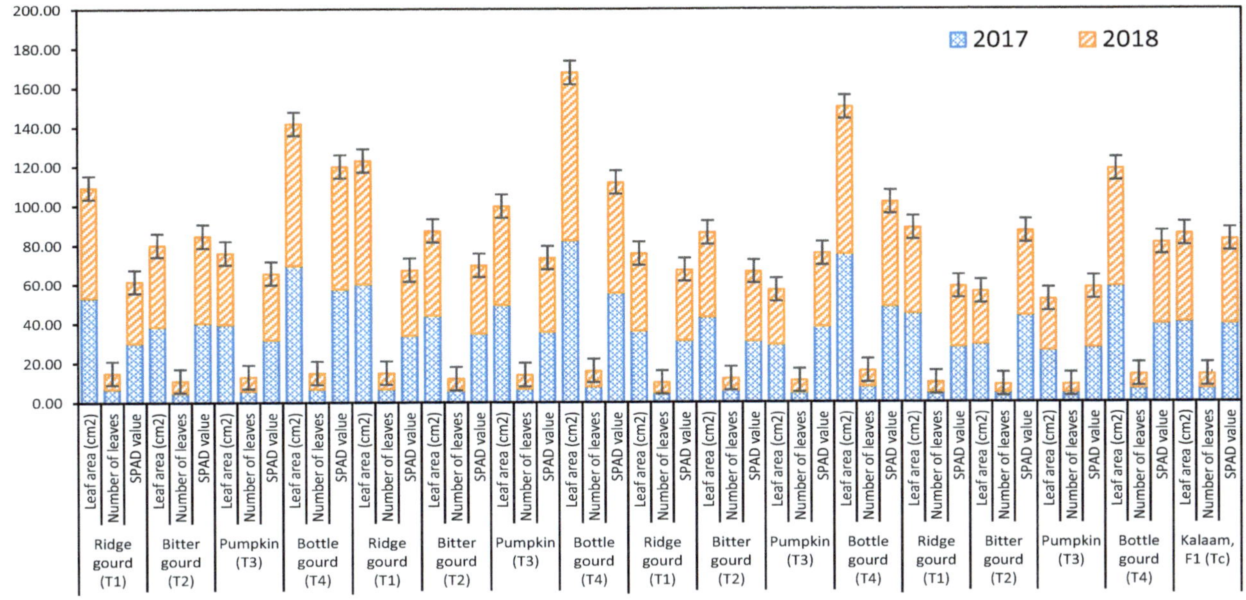

Figure 2. Evaluating the effect of Scion/rootstocks combinations (grafting) on vegetative growth of Cucumber fruit after 30 days during crop season 2017 and 2018.

Table 2 indicates plant mortality and the percentage of plant survival in the real rooted and grafted treatment of hybrid cucumber cv. Kalaam F_1 grafted on four local cucurbitaceous rootstocks (ridge gourd, bitter gourd, pumpkin and bottle gourd) with different grafting techniques during the 2017 and 2018 crop growing seasons. The statistical analysis presented in Table 2 shows that pumpkin rootstock in the hole insertion grafting method had a lower plant survival (42.00 ± 0.81 and 44.00 ± 1.29) in the 2017 and 2018 seasons respectively. These results showed unsuitability of hybrid Kalam F_1 with local rootstock. The significantly maximum plant survival rate of (96.00 ± 2.77) was obtained when grafted plants onto bottle gourd (T_4) in splice grafting during the first crop season, while in the second crop season bottle gourd (T_4) and ridge gourd (T_1) showed the same (95.00 ± 0.55) survival rates of grafted plants. Ridge gourd (T_1) and bitter gourd (T_2) in tongue grafting (G_1) and real rooted Kalaam, F_1 (T_c) didn't show any significant difference during first growing season while, in second crop season ridge gourd (T_1) and bottle gourd (T_4) in tongue grafting (G_1) and non-grafted Kalaam, F_1 (T_c) didn't show any significant difference in cucumber plants survival. The rootstocks under single cotyledon grafting method show significantly different results for plants survival. These results agreed with [53] who reported that the survival rate of cucumber plants increased under different grafting techniques.

The shoot length, plant height and stem diameter of cucumber (Cucumis sativus) plants grafted onto four local cucurbitaceous rootstocks with different grafting methods were presented in Figure 3. The bottle gourd rootstock had significant shoot length (cm) in tongue grafting (13.50, 13.30), splice grafting (12.20, 12.10) and single cotyledon grafting (11.34, 11.28) after first 15 days of grafting during 2017 and 2018 respectively while the hole insertion grafting showed non-significant results (4.15, 4.1) onto ridge gourd. The real rooted plant has (9.0, 9.20) shoot lengths. The plants with bottle gourd rootstock developed a significantly thick plant stem (0.19, 0.18 cm) in splice grafting and didn't show any significant difference of stem diameter during season 2017 and 2018 respectively. The real rooted hybrid cucumber had (0.15 and 0.16 cm) stem diameter after first 15 days of 2017 and 2018 seasons. In tongue grafting and single cotyledon grafting, the bottle gourd showed significant plant height (15.34, 15.50 cm), (16.10, 15.73 cm) while the pumpkin rootstock in hole insertion grafting had non-significant (6.55, 6.49 cm) plant height during 2017 and 2018 respectively.

Table 2. Effect of grafting hybrid cucumber cv. Kalaam F_1 onto local rootstocks on plant survival rate.

Grafting Technique	Rootstock Cultivar	Number of Grafted Plants	2017			2018		
			Mortality (%)		Plants Survival (%)	Mortality (%)		Plants Survival (%)
			Day, 15	Day, 30		Day, 15	Day, 30	
Tongue grafting (G_1)	Ridge gourd (T_1)	100	10.00 ± 0.12f	18.00 ± 0.53f	82.00 ± 1.08bc	8.00 ± 0.15f	17.00 ± 0.27hi	83.00 ± 2.31cd
	Bitter gourd (T_2)	100	13.00 ± 0.12e	20.00 ± 0.29ef	80.00 ± 1.60bc	14.00 ± 0.24d	19.00 ± 0.28h	81.00 ± 0.75cde
	Pumpkin (T_3)	100	20.00 ± 0.36c	35.00 ± 1.02b	65.00 ± 1.17f	19.00 ± 0.55c	34.00 ± 0.48c	66.00 ± 2.13hi
	Bottle gourd (T_4)	100	7.00 ± 0.13g	17.00 ± 0.42f	83.00 ± 1.95b	8.00 ± 0.12f	16.00 ± 0.49ij	84.00 ± 0.91cd
Splice grafting (G_2)	Ridge gourd (T_1)	100	4.00 ± 0.04h	7.00 ± 0.08gh	93.00 ± 0.61a	3.00 ± 0.10hi	5.00 ± 0.05k	95.00 ± 1.15a
	Bitter gourd (T_2)	100	3.00 ± 0.04h	8.00 ± 0.16g	92.00 ± 1.22a	2.00 ± 0.05i	7.00 ± 0.18k	93.00 ± 1.49ab
	Pumpkin (T_3)	100	7.00 ± 0.10g	17.00 ± 0.53f	83.00 ± 1.96b	5.00 ± 0.06g	14.00 ± 0.20j	86.00 ± 0.95bc
	Bottle gourd (T_4)	100	3.00 ± 0.07h	4.00 ± 0.09h	96.00 ± 2.77a	4.00 ± 0.10gh	5.00 ± 0.12k	95.00 ± 0.55a
Single Cotyledon grafting (G_3)	Ridge gourd (T_1)	100	11.00 ± 0.36f	22.00 ± 0.57e	78.00 ± 1.41bcd	9.00 ± 0.22f	23.00 ± 0.37g	77.00 ± 1.66def
	Bitter gourd (T_2)	100	10.00 ± 0.22f	26.00 ± 0.28d	74.00 ± 2.38cde	9.00 ± 0.13f	26.00 ± 0.38f	74.00 ± 1.18efg
	Pumpkin (T_3)	100	20.00 ± 0.58c	37.00 ± 0.90b	63.00 ± 1.07f	22.00 ± 0.65b	37.00 ± 0.75b	63.00 ± 1.57i
	Bottle gourd (T_4)	100	17.00 ± 0.51d	23.00 ± 0.62de	77.00 ± 2.22b-e	15.00 ± 0.42d	22.00 ± 0.57g	78.00 ± 0.56def
Hole Insertion grafting (G_4)	Ridge gourd (T_1)	100	14.00 ± 0.19e	30.00 ± 0.94c	70.00 ± 2.27def	11.00 ± 0.13e	28.00 ± 0.10ef	72.00 ± 1.20fgh
	Bitter gourd (T_2)	100	16.00 ± 0.26d	31.00 ± 0.33c	69.00 ± 1.49ef	14.00 ± 0.44d	31.00 ± 0.46d	69.00 ± 1.83ghi
	Pumpkin (T_3)	100	32.00 ± 0.34a	58.00 ± 1.62a	42.00 ± 0.81g	29.00 ± 0.78a	56.00 ± 0.79a	44.00 ± 1.29j
	Bottle gourd (T_4)	100	22.00 ± 0.52b	31.00 ± 0.60c	69.00 ± 0.44ef	20.00 ± 0.22c	29.00 ± 0.86de	71.00 ± 0.53fgh
Non-grafted (G_n)	Kalaam, F_1 (T_c)	100	7.00 ± 0.10g	18.00 ± 0.45f	82.00 ± 1.16bc	5.00 ± 0.12g	17.00 ± 0.40hi	83.00 ± 2.28cd

Means sharing same letters in a column are statistically non-significant ($p > 0.05$).

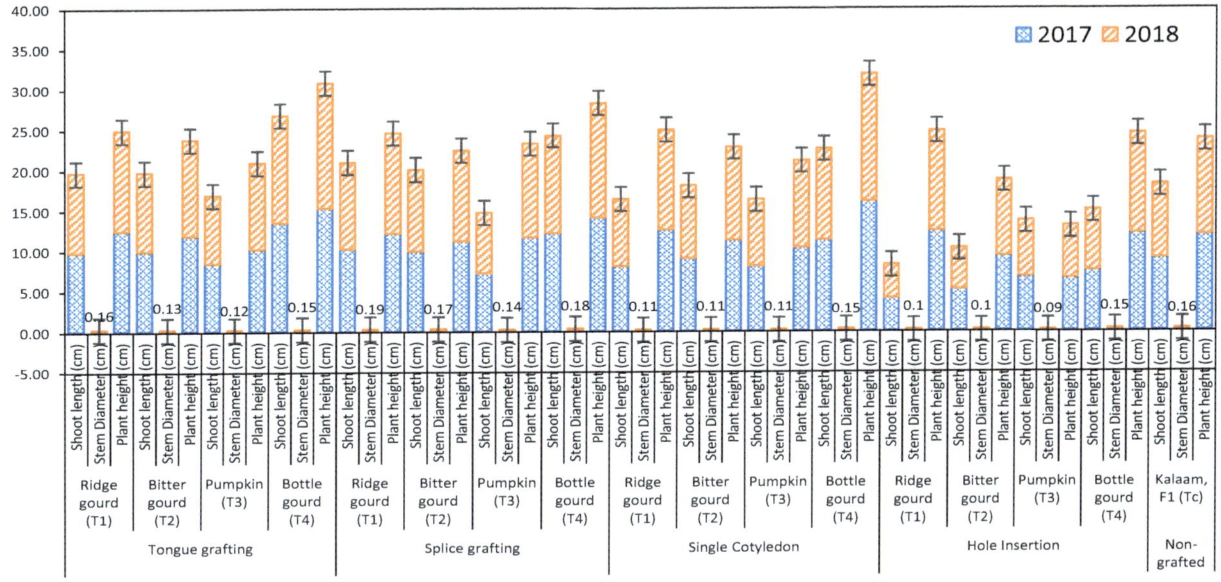

Figure 3. Effect of Scion/rootstocks combinations (grafting) on vegetative growth parameters of Cucumber fruit after 15 days during crop season 2017 and 2018.

Figure 4 presents shoot length, plant height and stem diameter of grafted cucumber (*Cucumis sativus*) plants onto cucurbitaceous rootstocks during the crop seasons of 2017 and 2018. After 30 days of grafting, The bottle gourd and ridge gourd rootstocks in splice grafting showed significant shoot length (21.90, 21.80 cm) and (20.98, 21.10 cm) while the hole insertion grafting showed non-significant results (9.10, 8.67 cm) onto pumpkin rootstock after 30 days of grafting during 2017 and 2018 respectively. The non-grafted plants didn't show any significant difference. The plants with bottle gourd rootstock developed significantly thick plant stems (0.37, 0.38 cm) in splice grafting. The bottle gourd in tongue grafting, ridge gourd and bitter gourd in splice grafting, bottle gourd in single cotyledon grafting and real rooted plants didn't show any significant difference of stem diameter during the seasons 2017 and 2018. The pumpkin rootstock in hole insertion grafting had (0.15 and 0.16 cm) non-significant stem diameter after 30 days of grafting during both 2017 and 2018 crop seasons. In splice grafting and single cotyledon grafting, the bottle gourd showed significant plant height (24.65, 24.50 cm), (24.0, 23.63 cm). The bottle gourd rootstock in tongue grafting, ridge gourd in splice grafting, bottle gourd in single cotyledon and real rooted plants didn't show any significant change of plant height while the rest of the rootstocks showed non-significant plant height during 2017 and 2018 respectively.

The data in Table 3 indicate that the scion/rootstock combinations (grafting) promoted cucumber plant growth and significantly increased the vigor of grafts plants compared to real rooted hybrid cucumber during both 2017 and 2018 crop periods. That was shown in plant height (cm), number of plant branches & leaves, stem thickness/diameter (cm), flowering time (days) and first fruit harvesting time (days). Grafting cucumber cv. Kalaam F_1 on cucurbitaceous rootstock (Bottle gourd) produced significantly maximum plant height, stem diameter and No. of leaves & nodes than rootstocks (Ridge gourd, Bitter gourd and Pumpkin). The sequence found between the grafting techniques was splice grafting, tongue grafting, single cotyledon grafting and hole insertion grafting. The splice grafting showed significantly maximum plant heights (622.0 ± 4.06 and 619.0 ± 4.58 cm), number of leaves (194.0 ± 6.32 and 183.0 ± 4.62), number of branches (13.00 ± 0.40 and 13.00 ± 0.58) and stem diameter (1.13 ± 0.03 and 1.10 ± 0.03 cm) for Lagenaria siceraria (bottle gourd) during crop season 2017 and 2018 respectively while, the lowest values were associated with pumpkin rootstock. Ridge gourd (T_1) and bitter gourd (T_2) in tongue grafting (G_1) & single Cotyledon grafting (G_3), ridge gourd (T_1) and bottle gourd (T_4) in hole insertion grafting (G_4) and real rooted Kalaam, F1 (T_c) didn't show any significant difference in plant height during first season while, in the second season most of the rootstocks didn't show any significant difference. Bitter gourd (T_2) pumpkin (T_3) bottle gourd (T_4) in hole Insertion grafting (G_4) showed maximum time for flowering during both 2017 and 2018 crop seasons respectively. Fruit harvesting time didn't show any significant difference for all grafting technique under both crop seasons. This could be interpreted that cultivation of grafted cucumber produced better plant vegetative growth compared with non-grafted plants in infested soil with nematode and soil salinity. The results are in line with [54] who described that vegetative growth (plant height, branches and leaves) was significantly higher in grafting than that of non-grafting. Hormone synthesis controlled by plant root could lead the plant growth and root to shoot ratios [55]. Zhang [56] stated that the vegetative growth (plant height, leaves, branches, leaf area and stem diameter), photosynthesis rate, yield, wilt resistance and root-knot nematode immunity were higher in scion/rootstock combination plants. The statistical analysis presented in Table 3 indicate the earliness of first female flower on grafted plants.

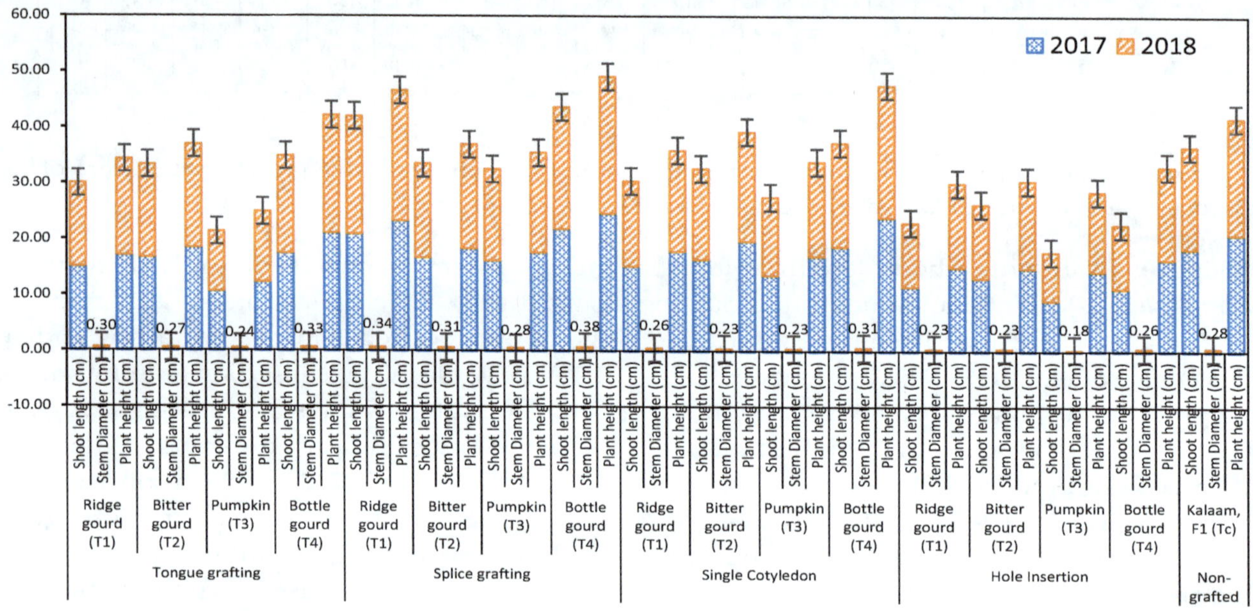

Figure 4. Effect of Scion/rootstocks combinations (grafting) on vegetative growth parameters of Cucumber fruit after 30 days during crop season 2017 and 2018.

The *Lagenaria siceraria*, *Luffa operculata* and *Momordica charantia* started flowering at 33rd day while pumpkin started on 34th day in splice grafting technique as shown in Table 3. The significantly different 62 days were taken by the plants for flowering in hole insertion grafting method. The non-grafted cucumber cv. Kalaam F_1 showed flowers after 60 days. Therefore, splice grafted plants showed early and high vigor compared with other grafts and non-grafted plants. The first fruit harvested after 12 days of flowering. The harvesting was carried out at the same time from all kind of grafted plants.

Ozarslandan [57] observed that all genotypes of *Lagenaria siceraria* (Bottle gourd) were liable to root-knot nematodes. The grafting of scion (Watermelon) onto rootstock (Bottle gourd) gave higher plant growth and fruit yield as compared to non-grafted fruit root-knot nematodes carry soil. It was concluded that bottle gourd rootstock tolerated nematodes with their rapid growth. In this study, splice grafted plants using all scion/rootstock combinations had significantly higher vegetative growth and fruit yield. According to our results, Lee [5] and Ioannou [58] observed that growth of scion/rootstocks were better than that of real rooted plants.

Cansev and Ozgur [42] described that, generally the growth and yield reduced significantly in real rooted (non-grafted) plants as compared with grafting techniques used for scion/rootstock combinations. The results suggested grafting of cucurbitaceous plants avoid yield reduction due to intensive cropping and soil borne disease. Cucumber grafting onto resistant rootstock was recommended technique in soil with high root-knot nematodes infestation and to produce more fruit yield from grafted plants [59].

Table 3. Effect of Scion/rootstocks combinations (grafting) on cucumber growth at harvesting stage.

Grafting Technique		2017						2018					
		Plant Height (cm)	Number/Plant		Stem Diameter (cm)	Flowering Time (Days)	First Fruit Harvesting Time (Days)	Plant Height (cm)	Number/Plant		Stem Diameter (cm)	Flowering Time (Days)	First Fruit Harvesting Time (Days)
			Branches/Plant	Leaves/Plant					Branches/Plant	Leaves/Plant			
Tongue grafting (G_1)	Ridge gourd (T_1)	570.0 ± 16.32a–d	10.00 ± 0.09c	160.0 ± 2.97cde	0.95 ± 0.03bc	34.00 ± 0.84ab	46.00 ± 1.07a	574.0 ± 15.11abc	11.00 ± 0.25bc	172.0 ± 4.77abc	0.93 ± 0.03bc	34.00 ± 0.58ab	46.00 ± 1.42a
	Bitter gourd (T_2)	566.0 ± 17.87a–d	9.00 ± 0.14d	144.0 ± 1.98efg	0.89 ± 0.02cd	35.00 ± 0.50ab	47.00 ± 0.76a	561.0 ± 9.79abc	10.00 ± 0.19cd	159.0 ± 3.08cde	0.86 ± 0.03cd	35.00 ± 0.92ab	47.00 ± 1.02a
	Pumpkin (T_3)	529.0 ± 10.35cd	9.00 ± 0.18d	157.0 ± 0.99cde	0.70 ± 0.01f–i	34.00 ± 0.37ab	46.00 ± 0.68a	555.0 ± 10.64abc	9.00 ± 0.21de	166.0 ± 2.57a–d	0.72 ± 0.02efg	34.00 ± 0.33ab	46.00 ± 0.70a
	Bottle gourd (T_4)	600.0 ± 13.87ab	10.00 ± 0.12c	177.0 ± 3.39b	1.00 ± 0.03b	34.00 ± 0.96ab	46.00 ± 0.56a	595.0 ± 13.81ab	10.00 ± 0.28cd	171.0 ± 2.14abc	1.01 ± 0.03ab	35.00 ± 1.14ab	47.00 ± 0.96a
Splice grafting (G_2)	Ridge gourd (T_1)	588.0 ± 12.96abc	11.00 ± 0.21b	163.0 ± 1.06bcd	1.00 ± 0.02b	33.00 ± 0.52b	45.00 ± 1.18a	597.0 ± 18.23ab	12.00 ± 0.33ab	177.0 ± 4.76ab	1.01 ± 0.02ab	33.00 ± 0.58b	45.00 ± 1.26a
	Bitter gourd (T_2)	567.0 ± 12.14a–d	11.00 ± 0.08b	171.0 ± 3.99bc	0.95 ± 0.01bc	34.00 ± 0.60ab	46.00 ± 1.42a	575.0 ± 16.23abc	11.00 ± 0.35bc	176.0 ± 2.97abc	0.93 ± 0.03bc	33.00 ± 0.54b	45.00 ± 0.70a
	Pumpkin (T_3)	546.0 ± 3.37bcd	10.00 ± 0.15c	159.0 ± 0.75cde	0.85 ± 0.02cde	34.00 ± 1.10ab	46.00 ± 0.69a	561.0 ± 10.87abc	11.00 ± 0.12bc	169.5 ± 2.71a–d	0.81 ± 0.02de	34.00 ± 0.65ab	46.00 ± 0.61a
	Bottle gourd (T_4)	622.0 ± 4.06a	13.00 ± 0.40a	194.0 ± 6.32a	1.13 ± 0.03a	33.00 ± 0.43b	45.00 ± 0.64a	619.0 ± 11.44a	12.60 ± 0.17a	182.5 ± 2.70a	1.10 ± 0.02a	33.00 ± 0.56b	45.00 ± 1.25a
Single Cotyledon grafting (G_3)	Ridge gourd (T_1)	540.0 ± 16.62bcd	8.00 ± 0.21e	131.0 ± 2.29ghi	0.80 ± 0.01def	34.00 ± 1.09ab	46.00 ± 1.40a	534.0 ± 7.80bc	10.00 ± 0.26cd	163.0 ± 3.15b–e	0.73 ± 0.01efg	34.00 ± 0.82ab	46.00 ± 1.49a
	Bitter gourd (T_2)	535.0 ± 11.66bcd	10.00 ± 0.10c	152.0 ± 0.21def	0.68 ± 0.02ghi	35.00 ± 0.82ab	47.00 ± 0.44a	543.0 ± 16.06bc	9.00 ± 0.16de	140.0 ± 2.99f	0.73 ± 0.02efg	35.00 ± 0.39ab	47.00 ± 1.03a
	Pumpkin (T_3)	530.0 ± 8.02cd	8.00 ± 0.10e	138.0 ± 3.09fgh	0.73 ± 0.02f–i	36.00 ± 0.21ab	48.00 ± 0.60a	541.0 ± 15.55bc	9.00 ± 0.11de	153.0 ± 4.35def	0.70 ± 0.02fg	36.00 ± 0.98ab	48.00 ± 0.50a
	Bottle gourd (T_4)	570.0 ± 16.36a–d	9.00 ± 0.17d	162.0 ± 4.53bcd	0.90 ± 0.02bcd	35.00 ± 0.67ab	47.00 ± 0.66a	575.0 ± 14.77abc	10.00 ± 0.22cd	171.6 ± 4.62abc	0.92 ± 0.02bc	35.00 ± 0.51ab	47.00 ± 0.83a
Hole Insertion grafting (G_4)	Ridge gourd (T_1)	544.0 ± 15.26bcd	8.00 ± 0.21e	127.0 ± 3.06hij	0.78 ± 0.02efg	36.00 ± 0.63ab	48.00 ± 0.94a	571.0 ± 11.11abc	10.00 ± 0.32cd	165.0 ± 3.43a-d	0.77 ± 0.01d–g	35.00 ± 0.60ab	47.00 ± 0.58a
	Bitter gourd (T_2)	509.0 ± 7.95d	7.00 ± 0.15f	116.0 ± 2.09ij	0.67 ± 0.01hi	37.00 ± 0.59a	49.00 ± 1.45a	521.0 ± 9.63c	8.00 ± 0.12e	138.0 ± 1.87f	0.70 ± 0.01fg	37.00 ± 0.85a	49.00 ± 1.00a
	Pumpkin (T_3)	412.0 ± 5.21e	5.00 ± 0.05g	111.0 ± 2.76j	0.63 ± 0.01i	37.00 ± 0.75a	49.00 ± 1.45a	401.0 ± 3.91d	4.00 ± 0.06f	103.0 ± 2.09g	0.67 ± 0.02g	37.00 ± 0.56a	49.00 ± 0.32a
	Bottle gourd (T_4)	565.0 ± 14.22a–d	9.00 ± 0.12d	163.0 ± 4.58bcd	0.75 ± 0.02e–h	35.00 ± 0.43ab	47.00 ± 1.12a	561.0 ± 6.74abc	8.00 ± 0.16e	162.0 ± 2.39b–e	0.78 ± 0.02def	37.00 ± 0.54a	49.00 ± 1.23a
Non-grafted (G_n)	Kalaam, F_1 (T_c)	552.0 ± 13.96bcd	8.00 ± 0.26e	133.0 ± 0.60gh	0.72 ± 0.02f-i	34.00 ± 0.65ab	46.00 ± 0.82a	571.0 ± 15.82abc	8.00 ± 0.09e	147.0 ± 4.36ef	0.70 ± 0.02fg	35.00 ± 0.90ab	47.00 ± 0.36a

Means sharing same letters in a column are statistically non-significant ($p > 0.05$).

The maximum early yield was produced by grafted plants rather than non-grafted plants [60]. Walters and Wehner [61] declared that root-knot nematodes were a major reason of cucumber yield reduction. Zhang [56] observed that grafted cucumber plants onto resistant rootstocks showed earlier flowering and first harvesting time than those of other rootstock and non-grafted plants.

Chen [62] reported that grafted cucumber onto local and resistant rootstocks had higher yields, better quality and a good control on soil pathogenic characteristics. The yield was significantly reduced cucumber fruit yield grown in non-treated soil compared to treated soil [46].

The damage of root-knot nematodes was highest in warm areas which resulted wilting plants under moisture stress [60]. These results agreed with Miguel and Maroto [63], they reported maximum yield in hybrid watermelon grafting and nematodes infested soil than self-rooted planed in fumigated (nematicide) soils. The local and resistant rootstock had enhanced the plant growth, flowering and yield of cucumber fruit [59]. The use of resistant cucurbitaceous rootstocks could increase the cucumber plant growth and fruit yield while the fruit quality was not different [64]. The grafted plants have no significant effect on fruit dry matter but improved nutritional concentration [65].

3.2. Fruit Growth and Quality

The effect of four different Scion/rootstocks combinations (grafting) on cucumber fruit, fruit fresh weight, fruit shape index, TSS and fruit dry matter was presented in Table 4. The splice grafting showed significantly maximum number of fruit per plant (14.60 ± 0.20 and 15.60 ± 0.38), Fruit weight/ cucumber (122.50 ± 2.42 and 121.29 ± 3.05 g), Fruit shape index (7.88 ± 0.22 and 7.84 ± 0.10), TSS (5.20 ± 0.11 and 5.11 ± 0.15%) and Fruit dry matter (4.62 ± 0.11 and 4.50 ± 0.12%) for *Lagenaria siceraria* (bottle gourd) during crop season 2017 and 2018 respectively while, the lowest values were associated with pumpkin rootstock. Ridge gourd (T_1) and bottle gourd (T_4) in tongue grafting (G_1), single Cotyledon grafting (G_3) and hole Insertion grafting (G_4) while ridge gourd (T_1) and bitter gourd (T_2) in splice grafting (G_2) didn't show any significant difference in number of fruits/plants. The real rooted cucumber showed significant different (10.30 ± 0.23) number of fruits per plant during both seasons.

This could be concluded that grafting hybrid cucumber onto local rootstocks gave significant increment in fruit length and diameter (fruit shape index) which results in more fruit yield. These results agreed with Al-Debei [43] who observed that when grafted cucumber cultivar on cucurbitaceous rootstock resulted in more vigorous cucumber plants. Ridge gourd (T_1), bitter gourd (T_2) and pumpkin (T_3) in tongue grafting (G_1), pumpkin (T_3) in splice grafting (G_2), ridge gourd (T_1), bottle gourd (T_4) i n single cotyledon grafting (G_3), bitter gourd (T_2) in hole insertion grafting (G_4) and Kalaam, F1 (T_c) in non-grafting (G_n) didn't show any significant difference in fruit weight during first season while in second season bottle gourd (T_4) in all grafting techniques showed significantly different results. The Kalaam, F_1 (T_c) also showed significant results for fruit weight.

Table 4. Effect of Scion/rootstocks combinations (grafting) on cucumber fruit growth and yield. TSS: Total soluble solids

Grafting Technique	Rootstock Cultivar	2017					2018				
		Number of Fruit/Plants	Fruit Weight/ Cucumber (g)	Fruit Shape Index	TSS (%)	Fruit Dry Matter (%)	Number of Fruit/Plants	Fruit Weight/ Cucumber (g)	Fruit Shape Index	TSS (%)	Fruit Dry Matter (%)
Tongue grafting (G_1)	Ridge gourd (T_1)	12.60 ± 0.17b	113.80 ± 1.49a–d	7.65 ± 0.24ab	4.70 ± 0.09b–e	4.35 ± 0.09a–c	13.00 ± 0.33b–d	115.90 ± 2.45abc	7.74 ± 0.12ab	4.74 ± 0.09ab	4.32 ± 0.07ab
	Bitter gourd (T_2)	12.00 ± 0.08bc	101.00 ± 3.24d–g	6.70 ± 0.20cde	4.47 ± 0.11c–g	4.30 ± 0.10abc	12.00 ± 0.22c–f	100.50 ± 0.72ef	6.77 ± 0.12def	4.43 ± 0.09bcd	4.2 5± 0.09abc
	Pumpkin (T_3)	10.00 ± 0.15ef	90.80 ± 2.12fgh	6.12 ± 0.06efg	4.00 ± 0.05ghi	3.80 ± 0.07def	11.50 ± 0.18ef	96.10 ± 1.85f	6.21 ± 0.10fg	4.04 ± 0.05de	3.76 ± 0.07cd
	Bottle gourd (T_4)	13.00 ± 0.28b	121.40 ± 3.36ab	6.88 ± 0.12b–e	5.00 ± 0.09ab	4.56 ± 0.11ab	13.10 ± 0.41bc	120.10 ± 2.42ab	6.97 ± 0.10cde	5.02 ± 0.06a	4.40 ± 0.14a
Splice grafting (G_2)	Ridge gourd (T_1)	13.00 ± 0.33b	116.90 ± 1.94abc	7.75 ± 0.19a	4.94 ± 0.07abc	4.40 ± 0.07abc	13.90 ± 0.25b	117.33 ± 1.03ab	7.71 ± 0.12ab	4.93 ± 0.13ab	4.43 ± 0.06a
	Bitter gourd (T_2)	13.00 ± 0.29b	107.00 ± 2.95cde	7.11 ± 0.11a–d	4.66±0.06b–e	4.45 ± 0.04abc	12.60 ± 0.27b–e	105.57 ± 1.80c–f	7.06 ± 0.23bcd	4.61 ± 0.09abc	4.40 ± 0.06a
	Pumpkin (T_3)	11.00 ± 0.13cde	89.60 ± 1.98gh	6.60 ± 0.19de	4.30 ± 0.06e–h	4.44 ± 0.12abc	11.70 ± 0.30def	97.45 ± 1.78f	6.67 ± 0.17def	4.22 ± 0.07cd	4.40 ± 0.10a
	Bottle gourd (T_4)	14.60 ± 0.20a	122.50 ± 2.42a	7.88 ± 0.22a	5.20 ± 0.11a	4.62 ± 0.11a	15.60 ± 0.38a	121.29 ± 3.05a	7.84 ± 0.10a	5.11 ± 0.15a	4.50 ± 0.12a
Single Cotyledon grafting (G_3)	Ridge gourd (T_1)	12.00 ± 0.32bc	111.00 ± 2.72a–d	7.50 ± 0.10abc	4.40 ± 0.12d–g	4.30 ± 0.06abc	13.00 ± 0.35b–d	116.00 ± 2.90abc	7.59 ± 0.20abc	4.45 ± 0.09bcd	4.26 ± 0.12abc
	Bitter gourd (T_2)	11.30 ± 0.24cd	95.40 ± 2.31efg	6.80 ± 0.12cde	4.18 ± 0.05fgh	4.15 ± 0.04b–d	12.00 ± 0.26c–f	98.39 ± 0.89f	6.79 ± 0.12def	4.21 ± 0.09cd	4.10 ± 0.10a–d
	Pumpkin (T_3)	10.00 ± 0.27ef	81.00 ± 1.38hi	5.72 ± 0.14fg	3.67 ± 0.09i	3.70 ± 0.05ef	11.00 ± 0.19fg	95.21 ± 1.37f	5.77 ± 0.06gh	3.65 ± 0.07e	3.60 ± 0.07de
	Bottle gourd (T_4)	12.00 ± 0.18bc	113.80 ± 2.70a–d	6.60 ± 0.13de	4.95 ± 0.06ab	4.32 ± 0.10abc	12.00 ± 0.14c–f	116.30 ± 3.43ab	6.67 ± 0.12def	4.91 ± 0.10ab	4.29 ± 0.11ab
Hole Insertion grafting (G_4)	Ridge gourd (T_1)	12.00 ± 0.35bc	102.80 ± 3.58def	6.93 ± 0.13bcd	4.12 ± 0.10ghi	4.15 ± 0.08b–d	12.80 ± 0.12b–e	110.20 ± 2.47b–e	7.04 ± 0.18b–d	4.20 ± 0.09cd	4.10 ± 0.10a–d
	Bitter gourd (T_2)	11.00 ± 0.11cde	88.50 ± 1.83gh	6.47 ± 0.11def	3.87 ± 0.06hi	3.86 ± 0.09def	11.50 ± 0.10ef	95.80 ± 1.24f	6.54 ± 0.17def	3.96 ± 0.05de	3.83 ± 0.07b–d
	Pumpkin (T_3)	9.30 ± 0.20f	74.90 ± 1.41i	5.40±0.09g	3.10 ± 0.08j	3.60 ± 0.05f	10.00 ± 0.19g	77.40 ± 1.36g	5.38 ± 0.08h	3.12 ± 0.05f	3.20 ± 0.09e
	Bottle gourd (T_4)	12.00 ± 0.13bc	109.20 ± 2.99bcd	6.31±0.15def	4.78 ± 0.14a–d	4.10 ± 0.04cde	12.60 ± 0.27b–e	112.90 ± 1.26a–d	6.30 ± 0.10efg	4.87 ± 0.15ab	4.02 ± 0.10a–d
Non-grafted (G_n)	Kalaam, F_1 (T_c)	10.30 ± 0.23def	100.80 ± 1.96d–g	6.50±0.15def	4.65 ± 0.11b–f	4.52 ± 0.08abc	10.00 ± 0.15g	103.60 ± 0.84def	6.45 ± 0.07d–g	4.62 ± 0.14abc	4.40 ± 0.12a

Means sharing same letters in a column are statistically non-significant ($p > 0.05$).

The results of fruit dry matter obtained from ridge gourd (T_1), bitter gourd (T_2) and pumpkin (T_3) in tongue grafting (G_1) and splice grafting (G_2), ridge gourd (T_1) and bottle gourd (T_4) in single cotyledon grafting (G_3), bitter gourd (T_2) in hole insertion grafting (G_4) and real rooted Kalaam, F1 (T_c) during first crop season 2017 were not statistically different. During second season, all rootstocks in splice grafting and real rooted Kalaam, F_1 didn't show any significant difference. These results are in line with Yetistir and Sari [66] who reported that grafted cucumber fruit with higher fruit fresh weight and dry matter had more nutritional concentration compared with low fresh weight cucumber fruit. Grafting of cucumber scion onto bottle gourd rootstock can enhance the reduction of cucumber fruit dry matter [67]. Ridge gourd (T_1) in tongue grafting (G_1), bitter gourd (T_2) in splice grafting (G_2) didn't show any significant difference. Bottle gourd (T_4) in tongue grafting (G_1) and single cotyledon grafting (G_3) similarly pumpkin (T_3) in tongue grafting (G_1) and ridge gourd (T_1) in hole insertion grafting (G_4) didn't show any significant difference for TSS in fruit juice during first crop season while in the second season, ridge gourd (T_1) in tongue grafting (G_1) and splice grafting (G_2) and bottle gourd (T_4) in single cotyledon grafting (G_3) and hole insertion grafting (G_4) didn't show any significant TSS contents but real rooted plants showed statistically significant TSS [54]. During first crop season bottle gourd (T_4) and ridge gourd (T_1) in splice grafting (G_2) didn't show any significant difference. The real rooted plants showed statistically different results in season 2018 but not in 2017.

Plant hormones regulate plant vegetative growth and reproductive development and are responsible for built root-shoot communication [68]. Plant with healthy rootstock produce more cytokinins into rising xylem sap which enhanced fruit yield [68]. The grafting changed hormones production and their impact on grafted plants. Flowering are controlled by plant hormones. It could be concluded that grafting increased the fruit yield since grafted plants have strong root system which improved disease immunity and photosynthesis. The plant growth of grafted cucumber and the resemblance of protective isozymes between grafted and non-grafted plants had a positive correlation [69].

The Table 5 indicated the effect of scion (cucumber cv. Kalaam F_1) grafted onto four local rootstocks (ridge gourd, bitter gourd, pumpkin and bottle gourd) with different grafting techniques on quality of cucumber fruit. The fruit quality was evaluated at 3, 6 and 9 days after pollination. It is clear from the Table that the contents of ascorbic acid in fruit decreased significantly with the fruit growth. The ascorbic acid, soluble protein and free amino acid decreased with periodically while soluble sugar content increased. The contents of ascorbic acid in cucumber cv. Kalaam F_1 fruit were higher initial days and then decrease gradually. Bottle gourd gave statistically significant results while the pumpkin showed lowest values. The ridge gourd, bitter gourd and real rooted plants didn't show any significant difference during both seasons. The cucumber cv. Kalaam F_1 fruit grafted onto bottle gourd had statistically significant soluble protein content during first and second seasons while pumpkin rootstock showed least values. The statistical difference was not found in other rootstocks. Free amino acid and soluble sugar didn't show any significant difference in all rootstocks during both seasons.

Table 5. Effect of Scion/rootstocks combinations (grafting) on Cucumber fruit quality.

Fruit Quality	DAP (Days)	2017					2018				
		Ridge Gourd (T_1)	Bitter Gourd (T_2)	Pumpkin (T_3)	Bottle Gourd (T_4)	Kalaam, F1 (T_c)	Ridge Gourd (T_1)	Bitter Gourd (T_2)	Pumpkin (T_3)	Bottle Gourd (T_4)	Kalaam, F1 (T_c)
Ascorbic Acid (mg/100 g FW)	3	52.70 ± 1.31ab	48.10 ± 0.96b	40.80 ± 1.27c	56.10 ± 0.90a	49.40 ± 0.31b	52.90 ± 1.21ab	48.50 ± 0.63b	41.50 ± 0.96c	55.70 ± 0.99a	49.10 ± 1.08b
	6	42.40 ± 0.78ab	39.10 ± 0.97b	31.30 ± 0.65c	44.00 ± 0.67a	38.80 ± 1.03b	42.00 ± 0.61ab	38.40 ± 0.65bc	32.60 ± 0.92d	43.50 ± 0.35a	38.30 ± 1.12c
	9	19.10 ± 0.60b	17.00 ± 0.47c	15.80 ± 0.20c	25.10 ± 0.48a	17.30 ± 0.29bc	19.20 ± 0.60b	17.20 ± 0.42bc	16.00 ± 0.24c	25.20 ± 0.60a	16.90 ± 0.51c
Soluble Protein (mg/g FW)	3	15.30 ± 0.21b	12.90 ± 0.37c	11.10 ± 0.29d	17.30 ± 0.47a	14.10 ± 0.33bc	15.60 ± 0.40b	13.20 ± 0.39c	11.50 ± 0.24d	17.20 ± 0.26a	13.50 ± 0.21c
	6	12.60 ± 0.14a	9.50 ± 0.23b	7.80 ± 0.13c	5.00 ± 0.06d	9.60 ± 0.17b	12.50 ± 0.13a	9.40 ± 0.26b	8.10 ± 0.10c	4.90 ± 0.03d	9.20 ± 0.16b
	9	10.10 ± 0.17b	7.50 ± 0.12d	6.70 ± 0.19e	12.40 ± 0.14a	8.50 ± 0.07c	10.40 ± 0.14b	7.70 ± 0.20c	6.80 ± 0.22d	12.30 ± 0.11a	8.00 ± 0.21c
Free amino acid (mg/100 g FW)	3	3.80 ± 0.06a	3.40 ± 0.09bc	3.10 ± 0.04c	3.70 ± 0.09ab	3.50 ± 0.06ab	3.70 ± 0.12a	3.30 ± 0.02bc	3.10 ± 0.08c	3.60 ± 0.05ab	3.10 ± 0.05c
	6	2.80 ± 0.06b	2.70 ± 0.02b	2.70 ± 0.02b	4.20 ± 0.09a	2.20 ± 0.02c	2.90 ± 0.03b	2.80 ± 0.09b	2.60 ± 0.08b	4.10 ± 0.09a	2.60 ± 0.08b
	9	2.20 ± 0.04b	2.20 ± 0.06b	2.10 ± 0.05bc	3.00 ± 0.02a	1.90 ± 0.05c	2.10 ± 0.03b	2.20 ± 0.06b	2.00 ± 0.06b	3.10 ± 0.07a	2.10 ± 0.04b
Soluble Sugar (mg/g FW)	3	5.00 ± 0.14a	4.50 ± 0.12ab	4.40 ± 0.14b	5.00 ± 0.12a	4.40 ± 0.07b	4.90 ± 0.16a	4.60 ± 0.08a	4.50 ± 0.07a	4.90 ± 0.16a	4.50 ± 0.08a
	6	5.90 ± 0.14a	5.60 ± 0.09ab	5.20 ± 0.06b	5.80 ± 0.18a	5.20 ± 0.11b	5.80 ± 0.06a	5.50 ± 0.12a	5.10 ± 0.04b	5.60 ± 0.04a	5.00 ± 0.04b
	9	6.80 ± 0.11a	6.20 ± 0.12b	6.10 ± 0.07b	7.00 ± 0.06a	6.30 ± 0.06b	6.60 ± 0.15ab	6.30 ± 0.14bc	6.00 ± 0.06c	6.90 ± 0.09a	6.10 ± 0.09bc

Means sharing same letters in a row are statistically non-significant ($p > 0.05$), DAP: days after pollination.

The Table 6 indicated the mineral composition of scion/rootstock combinations (grafts) and non-grafted cucumber cv. Kalaam F_1. Fruit N, P, K, Ca and Mg of hybrid cucumber grafted onto bottle gourd were significantly more among other rootstock and self-rooted cucumber fruits. The lowest values were associated with pumpkin rootstock. Scion/rootstock enhanced mineral circulation in plants. The contents of Potassium (K) and magnesium (Mg) in grafted fruits were statistically significant as compared to non-grafted fruits which was based on vigorous rootstock development responsible for absorbing water and nutrients effectively than other real roots [5]. The bottle gourd rootstock showed significantly maximum N, P, K, Ca and Mg contents during both crop seasons. Bitter gourd (T_2) and Kalaam, F_1 (T_c) didn't show any significantly difference during first season while, during second season all rootstocks had statistically different nitrogen contents. Ridge and bottle gourd didn't show any significant difference of phosphorus content during 2017 while results were statistically different for all rootstock in second season. Statistically no different results of K, Ca and Mg were found between both seasons. Although the scion/rootstock combinations determine fruit quality and nutritional contents, there was a significant influence of grafting found on fruit quality as compared to the original rooted plants. Therefore, the cucumber cv. Kalaam F1 grafted on four local root stocks (ridge gourd, bitter gourd, pumpkin and bottle gourd) have an improved mineral composition that can be used for commercial production.

Salehi-Muhammadi [70] described how the impact of rootstock on mineral composition depended upon the development of root genotypes which improved water and mineral transfer. The low performance of other rootstocks and non-grafted plants compared to the bottle gourd rootstock were justified by unsuitable grafting, which can cause improper scion growth or depression and mineral transfer from the union of grafting [42,71,72]. Several studies have been conducted to explain that effect in scion due to grafting techniques and the transport of genes from the local root system and hybrid scion through vascular bundles [73–75]. The rate of mineral transfer was higher in bottle gourd than other grafted and non-grafted plants, and tolerated a high crop load [76].

The fresh weight, length and diameter of cucumber fruit grafted on different cucurbitaceous rootstocks were presented in Figure 5. The observation carried out at the time of harvesting didn't show any significant difference in fruit weight, fruit length and fruit diameter during both crop seasons 2017 and 2018. The splice grafting of cucurbitaceous rootstock with hybrid cucumber fruit showed significantly maximum physical properties of cucumber fruit. The splice grafting showed significant fresh fruit weight (g) in bottle gourd (280.50 ± 3.39 and 277.00 ± 5.32), ridge gourd (252.00 ± 4.29 and 265.00 ± 7.31), bitter gourd (243.00 ± 3.34 and 252.00 ± 5.33) and pumpkin (227.00 ± 5.43 and 234.00 ± 1.68). All rootstocks showed significantly different fruit fresh weight during 2017 and 2018. The fruit length of cucumber showed statistically different results. Ridge gourd in tongue grafting and single cotyledon grafting while bitter gourd and pumpkin and ridge gourd and bottle gourd in splice grafting didn't show significant difference during first season but in second season the complete results obtained were statistically different. The fruit diameter was not statistically different during the first season but in the second season the results of fruit diameter were statistically different.

Table 6. Effect of Scion/rootstocks combinations on mineral contents in cucumber fruit.

Mineral Composition	2017					2018				
	N	P	K	Ca	Mg	N	P	K	Ca	Mg
Ridge gourd (T_1)	2.34 ± 0.03b	0.88 ± 0.01a	4.90 ± 0.11a	0.37 ± 0.00a	0.40 ± 0.00a	2.30 ± 0.05b	0.80 ± 0.02ab	4.85 ± 0.11a	0.33 ± 0.01a	0.38 ± 0.01a
Bitter gourd (T_2)	2.17 ± 0.04bc	0.80 ± 0.02b	4.40 ± 0.11b	0.30 ± 0.01b	0.34 ± 0.00b	2.10 ± 0.02cd	0.74 ± 0.02bc	4.34 ± 0.05b	0.27 ± 0.00b	0.31 ± 0.01b
Pumpkin (T_3)	2.10 ± 0.05c	0.68 ± 0.01c	4.27 ± 0.07b	0.27 ± 0.01b	0.27 ± 0.00c	2.00 ± 0.04d	0.70 ± 0.02c	4.21 ± 0.11b	0.22 ± 0.00c	0.25 ± 0.00c
Bottle gourd (T_4)	3.00 ± 0.05a	0.90 ± 0.01a	4.99 ± 0.09a	0.37 ± 0.01a	0.42 ± 0.01a	2.90 ± 0.04a	0.87 ± 0.01a	4.93 ± 0.13a	0.35 ± 0.00a	0.40 ± 0.00a
Kalaam, F_1 (Tc)	2.27 ± 0.04bc	0.73 ± 0.01c	4.27 ± 0.10b	0.30 ± 0.00b	0.33 ± 0.01b	2.20 ± 0.02bc	0.71 ± 0.02c	4.30 ± 0.08b	0.26 ± 0.01b	0.31 ± 0.01b

Means sharing same letters in a row are statistically non-significant ($p > 0.05$)

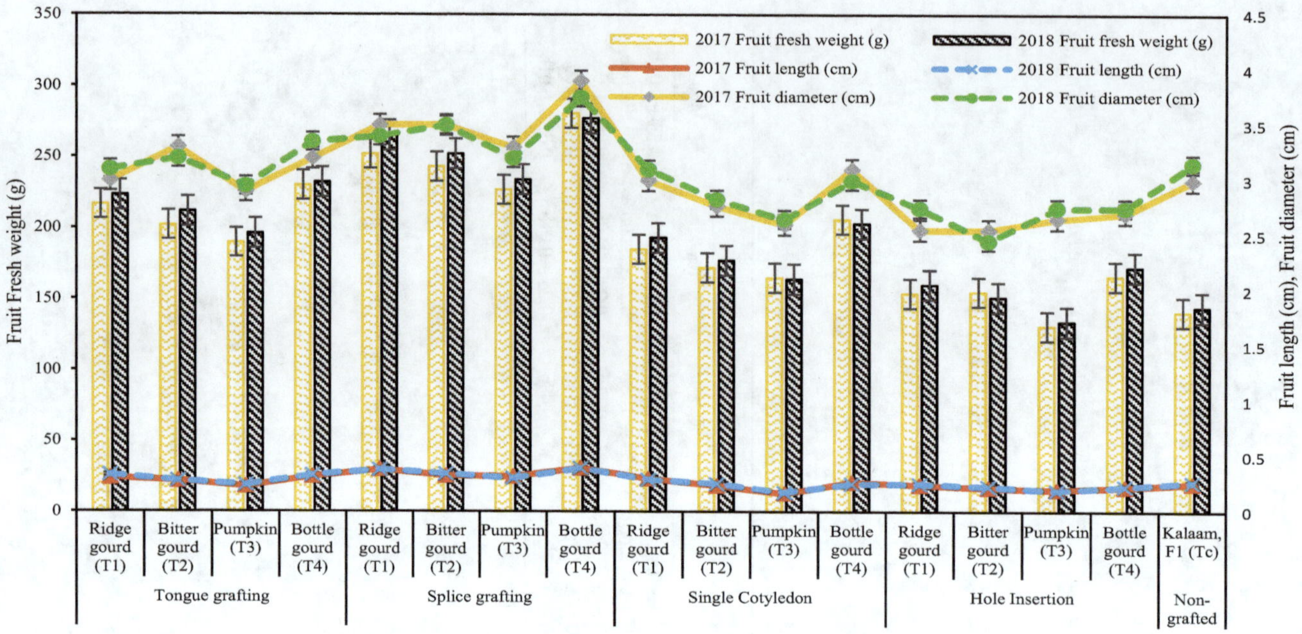

Figure 5. Effect of scion/rootstocks combinations (grafting) on physical properties of Cucumber fruit.

The effect of grafting cucumber cv. Kalaam F_1 onto local rootstocks on daily mean fresh weight of fruit, average length and average diameter of cucumber fruit during crop seasons 2017 and 2018 is presented in Figures 6–8. Figure 6 illustrates that after 4 days of pollination, the mean weight of fruit was increased significantly as shown in the Figure 6A–D. The fresh fruit weights (g) of all scion/rootstock combination in splice grafting were found significantly maximum than other grafted and real rooted cucumber plants during both 2017 and 2018 crop seasons. At day 9 after pollination, the fresh weight of splice grafting was (270 g) statistically significant for bottle gourd while the ridge gourd and bitter gourd (261 g and 24 7g) didn't show significant differences in fruit fresh weight. The pumpkin rootstock and real rooted cucumber cv. Kalaam F_1 had non-significant fruit fresh weight 161 and 138.25 grams. During both 2017 and 2018 crop seasons length of cucumber fruit (cm) improved significantly after pollination while from 3–8 days showed maximum fruit length in all scion/rootstock combinations as shown in Figure 7A–D. The fruit lengths in splice grafting were significantly longer as compared to that in other grafting and non-grafting methods. After 9 days of pollination for two years, bottle gourd had significantly higher length of cucumber while ridge gourd and bitter gourd didn't show significant differences of fruit length. The pumpkin and real rooted plants had non-significant mean fruit length. The diameter of cucumber fruit (cm) didn't increased significantly from day 1–3 and enhanced significantly from day 3–9 after pollination as shown in Figure 8A–D. Grafted plants had a significantly larger diameter of cucumber fruit while real rooted plants showed non-significant fruit diameter. After day 9 the diameter of cucumber plants in splice grafting were significantly higher 3.5, 3.3, 3.4 and 3.7 cm. The diameter of non-grafted cucumber fruit (3.1 cm) didn't show significant difference while the non-significant diameters of cucumber fruits were found 2.7, 2.6, 2.7 and 2.7 cm in hole insertion grafting of ridge gourd, bitter gourd, pumpkin and bottle gourd respectively. It could be concluded that the grafted plants had significantly higher weight, length and diameter of fresh fruit than those of fresh fruit in non-grafting. The results agreed with studies on citrus [77], grapes [78], lemon [79], mango [80] and melon [81] grafted on appropriate rootstocks.

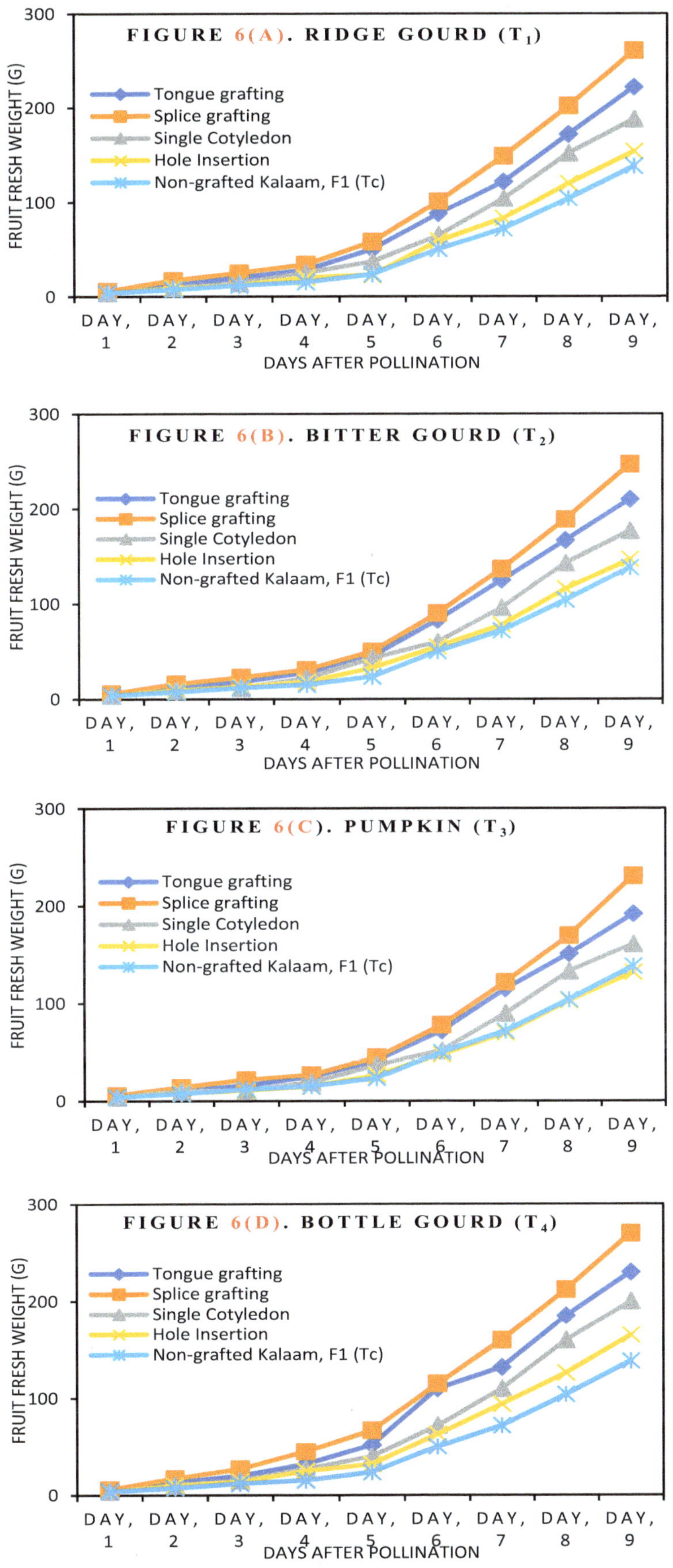

Figure 6. (**A–D**). Effect of scion/rootstocks combinations (grafting) on fresh weight of Cucumber fruit.

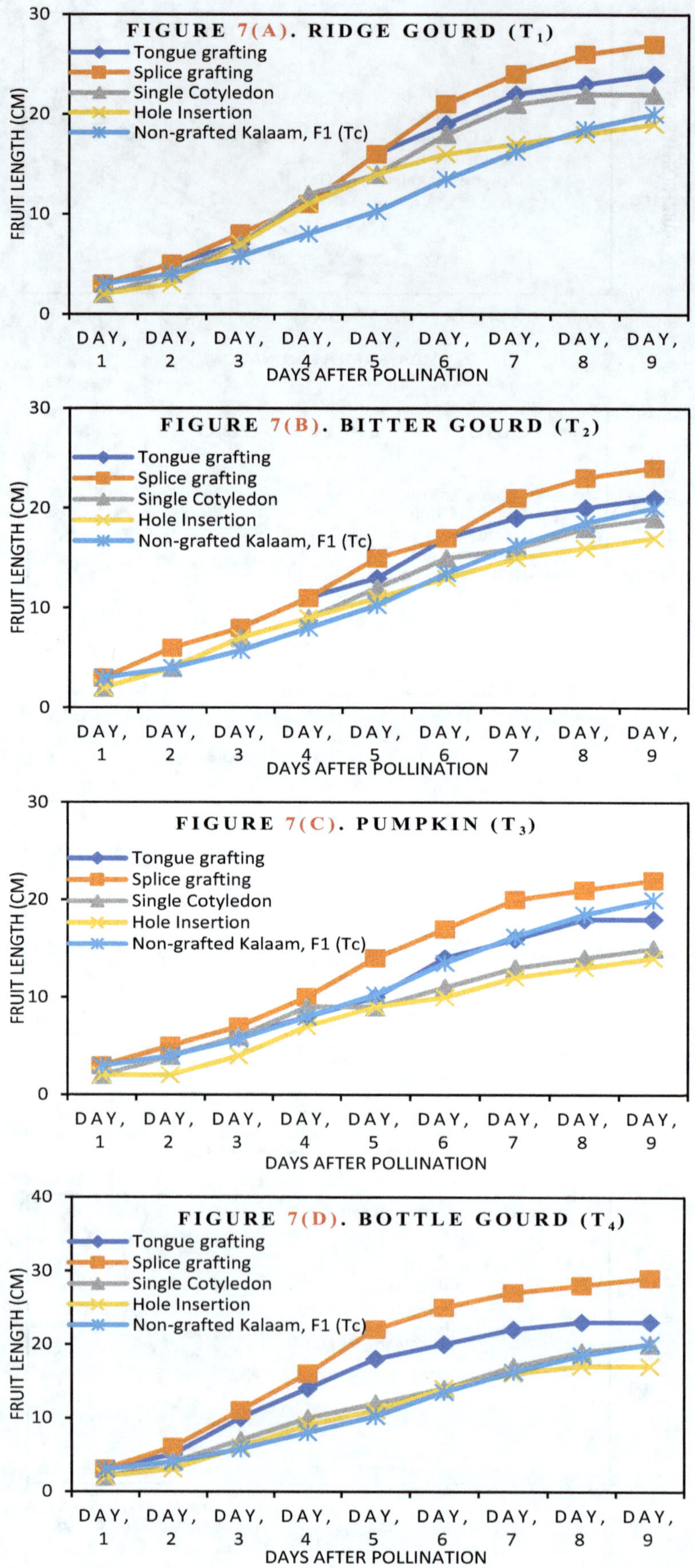

Figure 7. (**A–D**). Effect of grafting on Cucumber fruit development (fruit length) after pollination.

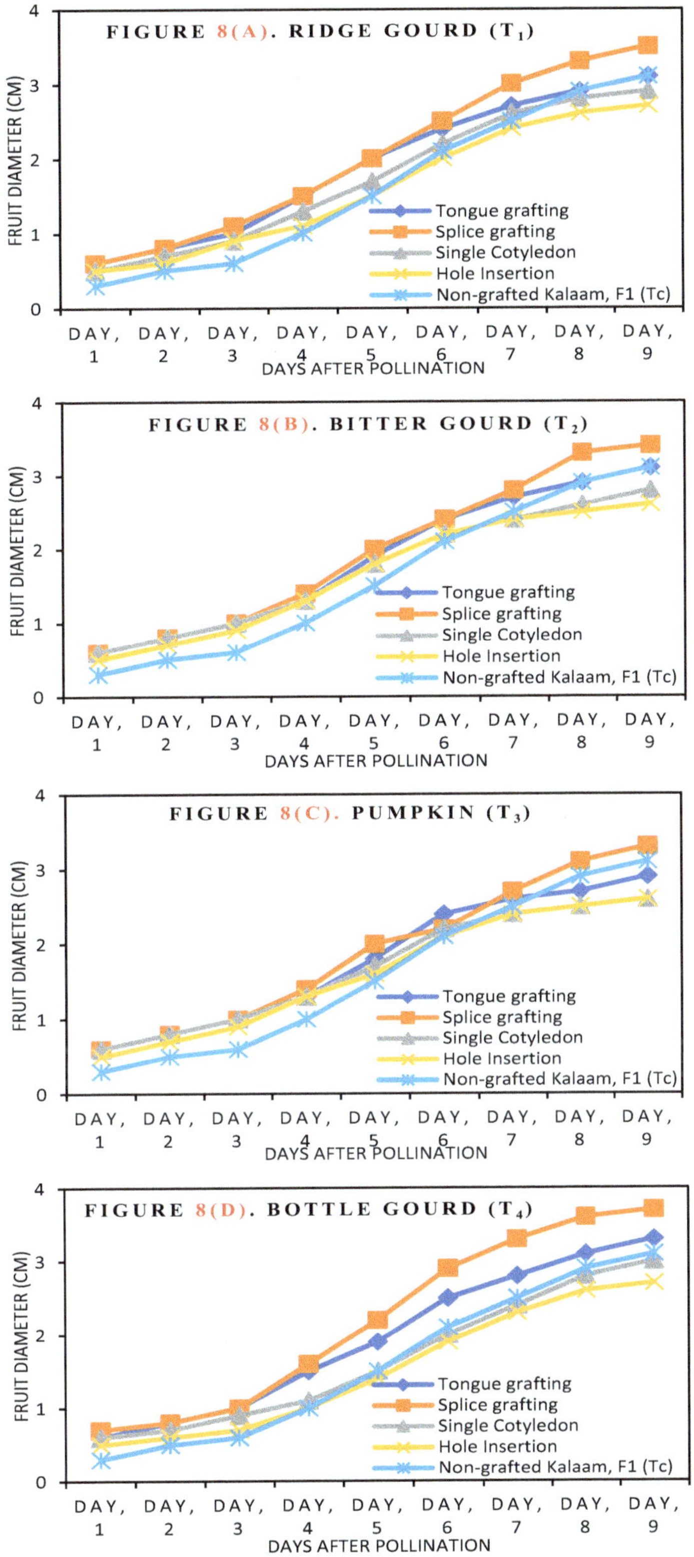

Figure 8. (A–D). Effect of grafting on Cucumber fruit development (fruit diameter) after pollination.

4. Conclusions

The pathological and soil-borne diseases particularly root-knot nematode represent major constraints for good seed germination, plant survival and growth, fruit development and fruit quality in cucumber production. There is no cucumber hybrid variety tolerant to nematode while fumigation of nematicides are expensive and can damage a sustainable agriculture system. Therefore, grafting is a real agricultural practice to control disease, and to improve cucumber production and quality in a suitable growing environment. The optimum plant vegetative growth, yield and fruit quality are based on both shoot and root genotypes. This study aimed to select and evaluate the resistant and tolerant scion/rootstock combination. The scion of hybrid cucumber cv. Kalaam F_1 and four local cucurbitaceous rootstocks were combined through tongue grafting (TAG), splice grafting, single cotyledon grafting and hole insertion grafting during 2017 and 2018. All scion/rootstocks combinations showed significantly maximum results in splice grafting than other grafting and non-grafted (self-rooted) plants. The hybrid cucumber cv. Kalaam F_1 grafted on bottle gourd rootstock gave significantly highest survival rate, plant vegetative growth, yield, fruit quality and resistance against soil-borne disease in splice grafting method followed by tongue approach, single cotyledon and hole insertion grafting. The non-significant values were associated with hole insertion grafting method in all scion/rootstock combinations. Grafting had no effect on cucumber fruit matter and fruit quality but fruit mineral composition (N, P, K, Ca and Mg) was improved significantly. Thus, grafting hybrid cucumber onto local rootstocks improved the growth, yield and quality of fruit, as well as the development of immunity to disease. The studies suggested that grafting of hybrid cucumber cv. Kalaam F_1 on local rootstocks grow well in infested soil to control soil-borne disease and to enhance cucumber production.

Author Contributions: Conceptualization, Design and Development, R.S.N.; Data collection, R.S.N., W.A. and M.I.; Formal Analysis, R.S.N., M.U.K. and M.A.; Investigation and Methodology, R.S.N., S.-U.R. and Z.W.; Supervision, M.U. and Y.S.; Visualization, R.S.N. and M.U.; Writing—original draft, R.S.N.; Writing review, Y.S., M.Y. and M.U.K.; Write-up editing, R.S.N.

Acknowledgments: The authors would like to acknowledge to all the services and technical support of Horticultural research station, University of Agriculture Faisalabad, Faculty of Agricultural Engineering and Technology PMAS-Arid Agriculture University Rawalpindi and Department of Agriculture, Biological, Environment and Energy Engineering, College of Engineering, Northeast Agricultural University, Harbin provided during research work setting, instrumentation, data collection and write-up compilation.

References

1. Pina, A.Y.; Errea, P. A review of new advances in mechanism of graft compatibility-incompatibility. *Sci. Hortic.* **2005**, *106*, 1–11. [CrossRef]
2. Mudge, K.; Janick, J.; Scoffield, S.; Goldschmidt, E.E. A history of grafting. *Hortic. Rev.* **2009**, *35*, 437–494.
3. Hartmann, H.T.; Kester, D.E.; Davies, F.T.; Geneve, R.L. Principles of grafting and budding. In *Plant Propagation: Principles and Practices*, 8th ed.; Pearson: London, UK, 2010; Chapter 11; pp. 415–463. ISBN 978-0-13-501449-3.
4. Lee, J.M.; Kubota, C.; Tsao, S.J.; Bie, Z.; Hoyos, E.P.; MorraL, O.M. Current status of vegetable grafting: diffusion, grafting techniques, automation. *Sci. Hortic.* **2010**, *127*, 93–105. [CrossRef]
5. Lee, J.M. Cultivation of grafted vegetables I. Current status, grafting methods, and benefits. *Hortic. Sci.* **1994**, *29*, 235–239.
6. Hartman, G.; Pawlowski, M.; Herman, T.; Eastburn, D. Organically Grown Soybean Production in the USA: Constraints and Management of Pathogens and Insect Pests. *Agronomy* **2016**, *6*, 16. [CrossRef]
7. Diacono, M.; Persiani, A.; Fiore, A.; Montemurro, F.; Canali, S. Agro-Ecology for Potential Adaptation of Horticultural Systems to Climate Change: Agronomic and Energetic Performance Evaluation. *Agronomy* **2017**, *7*, 35. [CrossRef]
8. St. Amand, P.C.; Wehner, T.C. Crop loss to 14 diseases in cucumber in North Carolina for 1983 to 1988. *Cuc. Genet. Coop. Rep.* **1999**, *14*, 15–17.
9. Pradeep, K.; Youssef, R.; Mariateresa, C.; Giuseppe, C. Vegetable Grafting as a Tool to Improve Drought Resistance and Water Use Efficiency. *Front. Plant Sci.* **2017**, *8*, 1130. [CrossRef]
10. Huang, Y.; Bie, Z.; He, S.; Hua, B.; Zhen, A.; Liu, Z. Improving cucumber tolerance to major nutrients induced salinity by grafting onto Cucurbita ficifolia. *Environ. Exp Bot.* **2010**, *69*, 32–38. [CrossRef]

11. Schwarz, D.; Rouphael, Y.; Colla, G.; Venema, J.H. Grafting as a tool to improve tolerance of vegetables to abiotic stresses: Thermal stress, water stress and organic pollutants. *Sci. Hortic.* **2010**, *127*, 162–171. [CrossRef]

12. Colla, G.; Rouphael, Y.; Rea, E.; Cardarelli, M. Grafting cucumber plants enhance tolerance to sodium chloride and sulfate salinization. *Sci. Hortic.* **2012**, *135*, 177–185. [CrossRef]

13. Moore, R. A model for graft compatibility-incompatibility in higher plants. *Am. J. Bot.* **1984**, *71*, 751–758. [CrossRef]

14. De Miguel, A.; Cebolla, V. *Terralia 53, Unión del Injerto*; Terralia: Madrid, Spain, October 2005; pp. 50–60. Available online: https://www.terralia.com/terralias/view_report?magazine_report_id=365 (accessed on 5 January 2018).

15. Kumar, R.M.S.; Gao, L.X.; Yuan, H.W.; Xu, D.B.; Liang, Z.; Tao, S.C.; Guo, W.B.; Yan, D.L.; Zheng, B.S.; Edqvist, J. Auxin enhances grafting success in Carya cathayensis (Chinese hickory). *Planta* **2018**, *247*, 761–772. [CrossRef]

16. Melnyk, C.W. Plant grafting: Insights into tissue regeneration. *Regeneration* **2017**, *4*, 3–14. [CrossRef]

17. De Velasco Alvarado, M.J. Anatomía y Manejo Agronómico de Plantas Injertadas en Jitomate. Master's Thesis, Universidad Autónoma de Chapingo, Chapingo, Mexico, May 2013. Available online: https://chapingo.mx/horticultura/pdf/tesis/TESISMCH2013050810128186.pdf (accessed on 1 June 2018).

18. Leonardi, C.; Romano, D. Recent issues on vegetable grafting. *Acta Hortic.* **2004**, *631*, 163–174. [CrossRef]

19. Yang, S.; Xiang, G.; Zhang, S.; Lou, C. Electrical resistance as a measure of graft union. *J. Plant Physiol.* **1993**, *141*, 98–104. [CrossRef]

20. Bletsos, F.A.; Olympios, C.M. Rootstocks and Grafting of Tomatoes, Peppers and Egg plants for Soil-borne Disease Resistance, Improved Yield and Quality. *Eur. J. Plant Sci. Biotechnol.* **2008**, *2*, 62–73. Available online: http://www.globalsciencebooks.info/Online/GSBOnline/images/0812/EJPSB_2(SI1)/EJPSB_2(SI1)62-73o.pdf (accessed on 1 June 2018).

21. Torii, T.; Kasiwazaki, M.; Okamoto, T.; Kitani, O. Evaluation of graft-take using a thermal camera. *Acta Hortic.* **1992**, *319*, 631–634. [CrossRef]

22. Oda, M.; Maruyama, M.; Mori, G. Water Transfer at Graft Union of Tomato Plants Grafted on to Solanum Rootstocks. *J. Jpn. Soc. Hortic. Sci.* **2005**, *74*, 458–463. [CrossRef]

23. De Miguel, A.; Cebolla, V. Terralia 53, Pages 50–60. Octubre 2005. Unión del Injerto. Available online: https://sci-hub.tw/http://hortsci.ashspublications.org/content/48/1/34.short (accessed on 5 May 2018).

24. Bausher, M.G. Road, South Rock, Pierce, Fort Graft Angle and Its Relationship to Tomato Plant Survival. *Hort. Sci.* **2013**, *48*, 34–36. Available online: http://hortsci.ashspublications.org/content/48/1/34.short (accessed on 6 June 2018).

25. Turquois, N.; Malone, M. Non-destructive assessment of developing hydraulic connections in the graft union of tomato. *J. Exp. Bot.* **1996**, *47*, 701–707. [CrossRef]

26. Xia, Y.; Sarafis, V.; Campbell, E.O.; Callaghan, P.T. Noninvasive imaging of water flow in plants by NMR microscopy. *Protoplasma* **1993**, *173*, 170–176. [CrossRef]

27. Hochmuth, R.; Davis, L.L.; Laughlin, W.L.; Simonne, E.H. *Evaluation of Organic Nutrient Sources in the Production of Greenhouse Hydroponic Basil*; Research Report 2003-08; University of Florida: Gainesville, FL, USA, June 2008; pp. 1–7.

28. Goh, K.M. Evaluation of potting media for commercial nursery production of container-grown plants. V. Patterns of release of nitrogen fertilizers in different media. *N. Z. J. Agr. Res.* **1979**, *22*, 163–171. [CrossRef]

29. Agricultural Statistics of Pakistan, 2017–2018. Available online: http://www.finance.gov.pk/survey/chapters_14/ (accessed on 1 December 2018).

30. Oda, M. *Grafting of Vegetable Crops*; Scientific Report of Agriculture and Biological Sciences; Osaka Prefecture University: Osaka, Japan, 2002; Volume 54, pp. 49–72.

31. Davis, A.R.; Perkins-Veazie, P.; Sakata, Y.; López-Galarza, S.; Maroto, J.V.; Lee, S.G.; Huh, Y.C.; Sun, Z.; Miguel, A.; King, S.R.; et al. Cucurbit grafting. *Criti. Rev. Plant Sci.* **2008**, *27*, 50–74. [CrossRef]

32. Sakata, Y.; Ohara, T.; Sugiyama, M. The history of melon and cucumber grafting in Japan. *Acta Hortic.* **2008**, *767*, 217–228. [CrossRef]

33. Ito, T. Present state of transplant production practices in Japanese horticultural industry. In Proceedings of the International Symposiumon Transplant Production Systems, Transplantant Production Systems, Yokohama, Japan, 21–26 July 1992; Kurata, K., Kozai, T., Eds.; Kluwer Academic Publishers: Amsterdam, The Netherlands, 1992.
34. Kurata, K.; Kozai, T. Transplant Production Systems. In Proceedings of the International Symposium on Transplant Production Systems, Yokohama, Japan, 21–26 July 1992; Kluwer Academic Publishers: Dordrecht, The Netherlands; Boston, MA, USA; London, UK; p. 335. Available online: http://www.springer.com/us/book/9780792317975 (accessed on 15 September 2018).
35. Lee, J.M.; Oda, M. Grafting of Herbaceous Vegetable and Ornamental Crops. In *Horticulture Reviews*; Janick, J., Ed.; John Wiley & Sons: Hoboken, NJ, USA, 2003; Volume 28, p. 453.
36. Nishiura, Y.; Murase, H.; Honami, N.; Taira, T. Development of Plug-in Grafting Robotic System Osaka Prefecture University. In Proceedings of the IEEE International Conference on Robotics and Automaton, Nagoya, Japan, 25–27 May 1995; pp. 2510–2517. Available online: http://ieeexplore.ieee.org/abstract/document/525636/ (accessed on 19 September 2018).
37. Oda, M. *Grafting of Vegetable Crops*; Osaka Prefecture University: Osaka, Japan, March 2002; Volume 54, pp. 49–72. Available online: http://repository.osakafu-u.ac.jp/dspace/bitstream/10466/1053/1/KJ00000052064.pdf (accessed on 22 June 2018).
38. Oda, M. Use of Grafted Seedlings for Vegetable Production in Japan. *Acta Hortic.* **2008**, *770*, 15–20. [CrossRef]
39. Lee, J.M.; Bang, H.J.; Ham, H.S. Grafting of Vegetables. *Jpn. Soc. Agric. Mach. Food Eng.* **1998**, *67*, 1098–1104. Available online: https://www.jstage.jst.go.jp/article/jjshs1925/67/6/67_6_1098/_pdf (accessed on 22 June 2018). [CrossRef]
40. Pavlou, G.; Vakalounakis, D.; Ligoxigakis, E. Control of root and stem rot of cucumber, caused by Fusarium oxysporum f. sp. radicis-cucumerinum, by grafting onto resistant rootstocks. *Plant Dis.* **2002**, *86*, 379–382. [CrossRef]
41. Huang, Y.; Tang, R.; Cao, Q.; Bie, Z. Improving the fruit yield and quality of cucumber by grafting onto the salt tolerant rootstock under NaCl stress. *Sci. Hortic.* **1999**, *122*, 26–31. [CrossRef]
42. Cansev, A.; Ozgur, M. Grafting cucumber seedlings on Cucurbita spp.: comparison of different grafting methods, scions and their performance. *J. Food Agric. Environ.* **2010**, *8*, 804–809.
43. Uysal, N.; Tuzel, Y.; Oztekin, G.B.; Tuzel, I.H. Effects of different rootstocks on greenhouse cucumber production. *Acta Hortic.* **2012**, *927*, 281–289. [CrossRef]
44. Moradipour, F.; Dashti, F.; Zahedi, B. Effect of grafting on yield and some vegetative characteristics of two greenhouse cucumber cultivar. *Iran. J. Hortic. Sci.* **2010**, *41*, 291–300.
45. Hoyos, P. Influence of different rootstocks on the yield and quality of greenhouses grown cucumbers. *Acta Hortic.* **2001**, *559*, 139–143. [CrossRef]
46. Lee, J.; Bang, H.; Ham, H. Quality of cucumber fruit as affected by rootstock. *Acta Hortic.* **1999**, *483*, 117–124. [CrossRef]
47. Sakata, Y.; Ohara, T.; Sugiyama, M. The history and present state of the grafting of Cucurbitaceous vegetables in Japan. *Acta Hortic.* **2007**, *731*, 159–170. [CrossRef]
48. Davis, A.R.; Perkins-Veazie, P.; Hassell, R.; Levi, A.; King, S.R.; Zhang, X. Grafting effects on vegetable quality. *HortScience* **2008**, *43*, 1670–1672. [CrossRef]
49. Sonneveld, C. Estimating quantities of water-soluble nutrients using a specific 1:2 by volume extract. *Soil Sci. Plant Anal.* **1990**, *21*, 1257–1265. [CrossRef]
50. Oda, M. *Grafting of Vegetables to Improve Greenhouse Production*; Extension Bulletin 480; Food and Fertilizer Technology Center: Taipei city, China, 1 December 1999; p. 11.
51. Ishibashi, K. Tongue-approach grafting in cucurbits. *Agr. Hort.* **1965**, *40*, 1899–1902. (In Japanese)
52. Lee, J.M.; Oda, M. Grafting of herbaceous vegetable and ornamental crops. In *Horticultural Reviews*; Janick, J., Ed.; John Wiley and Sons: New York, NY, USA, 2003; Volume 28, pp. 61–124.
53. Davis, R.F. Effect of Meloidogyne incognita on watermelon yield. *Nematropica* **2007**, *37*, 287–293.
54. Yadava, I.L. A rapid and nondestructive method to determine chlorophyll in intact leaves. *HortScience* **1986**, *21*, 1449.
55. Association of Official Analytical Chemists. *Official Methods of Analysis of the Association of Official Agriculture Chemists*, 12th ed.; Association of Official Agricultural Chemists: Washington, DC, USA, 1975; p. 870.

56. Ranganna, S. *Manual of Analysis of Fruit and Vegetable Products*; Tata McGraw-Hill Publishing: New Delhi, India, 1977; Volume 6, pp. 151–160.
57. Özarslandan, A.; Söğüt, M.A.; Yetişir, H.; Elekcıoğlu, I.H. Screening of bottle gourds (Lagenaria siceraria (Molina) Standley) genotypes with rootstock potential for watermelon production for resistance against Meloidogyne incognita (Kofoid & White, 1919) Chitwood and Meloidogyne javanica (Treub, 1885) Chitwood. *Turk. Entomol. Derg.* **2011**, *35*, 687–697.
58. Ioannou, N. Integrating soil solarization with grafting on resistant rootstocks for management of soil-borne pathogens of eggplant. *J. Hortic. Sci. Biotechnol.* **2001**, *7*, 396–401. [CrossRef]
59. Salata, A.C.; Bertolini, E.V.; Magro, F.O.; Cardoso, A.; Wilcken, S.R.S. Effect of grafting on cucumber production and reproduction of Meloidogyne javanica and M. incognita. *Hortic Bras.* **2012**, *30*, 590–594. [CrossRef]
60. Netscher, C.; Sikora, R.A. Nematode parasites of vegetables. In *Plant Parasitic Nematodes in Subtropical and Tropical Agriculture*; CAB International: Wallingford, UK, 1990; pp. 237–283.
61. Walters, S.A.; Wehner, T.C. 'Lucia', 'Manteo', and 'Shelby' root-knot nematode-resistant cucumber inbred lines. *HortScience* **1997**, *32*, 1301–1303. [CrossRef]
62. Chen, Z.D.; Wand, P.S.; Zhou, Y.; Ji, Y.L.; Wan, Z.J.; Peng, L. Effects of rootstock grafting on yield, quality and control of Meloidogyne incognita of cucumber (*Cucumis sativus* L.). *China Veget.* **2012**, *8*, 57–62.
63. Miguel, A.; Maroto, J.V. El injerto herbáceo en la sandía (Citrullus lanatus) como alternativa a la desinfecció n química del suelo. *Invest. Agrar. Prod. Prot. Veg.* **1996**, *11*, 239–253.
64. Zhu, J.; Bie, Z.L.; Huang, Y.; Han, X.X. Effect of grafting on the growth and ion concentrations of cucumber seedlings under NaCl stress. *Soil Sci. Plant Nutr.* **2008**, *54*, 895–902. [CrossRef]
65. Zhong, Y.Q.; Bie, Z.L. Effects of grafting on the growth and quality of cucumber fruits. *Acta Hortic.* **2007**, *761*, 341–347. [CrossRef]
66. Yetisir, H.; Sari, N. Effect of different rootstock on plant growth, yield and quality of watermelon. *Aust. J. Exp. Agric.* **2003**, *43*, 1269–1274. [CrossRef]
67. Huang, Y.; Bie, Z.L.; Liu, Z.X.; Zhen, A.; Wang, W.J. Protective role of proline against salt stress is partially related to the improvement of water status and peroxidase enzyme activity in cucumber. *Soil Sci. Plant Nutr.* **2009**, *55*, 698–704. [CrossRef]
68. Aloni, B.; Cohen, R.; Karni, L.; Aktas, H.; Edelstein, H. Hormonal signaling in rootstock-scion interactions. *Sci Hortic.* **2010**, *127*, 119–126. [CrossRef]
69. Zhang, J.; Shu, W.S. Mechanisms of heavy metal cadmium tolerance in plants. *J. Plant. Physiol. Mol. Biol.* **2006**, *32*, 1–8.
70. Salehi-Mohammadi, R.; Khasi, A.; Lee, S.G.; Huh, Y.C.; Lee, J.M.; Delshad, M. Assessing survival and growth performance of Iranian melon to grafting onto Cucurbita rootstocks. *Korean J. Hortic. Sci. Technol.* **2009**, *27*, 1–6.
71. Hartmann, H.T.; Kester, D.E.; Davies, F.T.; Geneve, R.L. *Plant Propagation: Principles and Practices*; Prentice Hall: Upper Saddle River, NJ, USA, 1997; p. 770.
72. Martínez-Ballesta, M.C.; Alcaraz-López, C.; Muries, B.; Mota-Cadenas, C.; Carvajal, M. Physiological aspects of rootstock-scion interactions. *Sci. Hortic.* **2010**, *127*, 112–118. [CrossRef]
73. Molnar, A.; Melnyk, C.W.; Bassett, A.; Hardcastle, T.J.; Dunn, R.; Baulcombe, D.C. Small silencing RNAs in plants are mobile and direct epigenetic modification in recipient cells. *Science* **2010**, *328*, 872–875. [CrossRef] [PubMed]
74. Dunoyer, P.; Brosnan, C.A.; Schott, G.; Wang, Y.; Jay, F.; Alioua, A.; Himber, C.; Voinnet, O. An endogenous, systemic RNAi pathway in plants. *Embryol. J.* **2010**, *29*, 1699–1712. [CrossRef]
75. Harada, T. Grafting and RNA transport via phloem tissue in horticultural plants. *Sci. Hortic.* **2010**, *125*, 545–550. [CrossRef]
76. Yamasaki, A.; Yamashita, M.; Furuya, S. Mineral concentrations and cytokinin activity in the xylem exudates of grafted watermelons as affected by rootstocks and crop load. *J. Jpn. Soc. Hortic. Sci.* **1994**, *62*, 817–826. [CrossRef]
77. Zekri, M. Citrus rootstocks affect scion nutrition, fruit quality, growth, yield and economical return. *Fruits* **2000**, *55*, 231–239. [CrossRef]
78. Economides, C.V.; Gregoriou, C. Growth, yield and fruit quality of nucellar frost 'Marsh' Grapefruit on fifteen rootstocks in Cyprus. *J. Amer. Soc. Hort.* **1993**, *118*, 326–329. [CrossRef]

79. Al-Jaleel, A.; Zekri, M.; Hammam, Y. Yield, fruit quality, and tree health of 'Allen Eureka' lemon on seven rootstocks in Saudi Arabia. *Sci. Hort.* **2005**, *105*, 457–465. [CrossRef]
80. Reddy, Y.T.N.; Kurian, R.M.; Ramachander, P.R.; Singh, G.; Kohli, R.R. Long-term effects of rootstocks on growth and fruit yielding patterns of 'Alphonso' mango (*Mangifera indica* L.). *Sci. Hort.* **2003**, *97*, 95–108. [CrossRef]
81. Traka-Mavronaa, E.; Metaxia, K.S.M.; Pritsaa, T. Response of squash (*Cucurbita* spp.) as rootstock for melon (*Cucumis melo* L.). *Sci. Hort.* **2000**, *83*, 353–362.

10

Partial Root-Zone Drying and Deficit Irrigation Effect on Growth, Yield, Water use and Quality of Greenhouse Grown Grafted Tomato

Branimir Urlić [1,*], Marko Runjić [1], Marija Mandušić [1], Katja Žanić [1], Gabriela Vuletin Selak [2], Ana Mateškovic [1] and Gvozden Dumičić [2]

1 Department of Applied Sciences, Institute for Adriatic Crops and Karst Reclamation, Put Duilova 11, 21000 Split, Croatia; marko.runjic@krs.hr (M.R.); marija.mandusic@krs.hr (M.M.); katja@krs.hr (K.Ž.); ana.mateskovic@krs.hr (A.M.)

2 Department of Plant Sciences, Institute for Adriatic Crops and Karst Reclamation, Put Duilova 11, 21000 Split, Croatia; gabriela.vuletin.selak@krs.hr (G.V.S.); gvozden.dumicic@krs.hr (G.D.)

* Correspondence: branimir@krs.hr

Abstract: The tomato is an important horticultural crop, the cultivation of which is often under influence of abiotic and biotic stressors. Grafting is a technique used to alleviate these problems. Shortage of water has stimulated the introduction of new irrigation methods: deficit irrigation (DI) and partial root-zone drying (PRD). This study was conducted in two spring–summer season experiments to evaluate the effects of three irrigation regimes: full irrigation (FI), PRD and DI on vegetative growth, leaf gas-exchange parameters, yield, water-use efficiency (WUE), nutrients profile and fruit quality of grafted tomatoes. In both years, the commercial rootstocks Emperador and Maxifort were used. In the first year, the scion cultivar Clarabella was grown on one stem and in the second year the cultivar Attiya was grown on two stems. Self-grafted cultivars were grown as a control. In both experiments, higher vegetative traits (leaf area and number, height, shoot biomass) were recorded in tthe plants grafted on commercial rootstocks. The stomatal conductance and transpiration rate were higher under FI. Under DI, transpiration was lowest and photosynthetic WUE was highest. Photosynthetic rate changed between irrigation treatments depending on plant type. In both years, the total yield was highest in grafted plants as result of more and bigger fruits per plant. In the 2nd year, grafted plants under FI had higher yield compared to PRD, but not to DI, while self-grafted plants did not differ between irrigation treatments. WUE was highest in DI and PRD treatments and in grafted plants. Leaf N, P, K and Ca was highest in tthe plants grafted on Emperador and Maxifort, while more Mg was measured in self-grafted plants. More Ca and Mg were recorded in tthe plants under DI and PRD. Fruit mineral concentrations were higher in tthe plants grafted on commercial rootstocks. Total soluble solids differed between irrigation regarding plant types, while fruit total acidity was higher in Emperador and Maxifort. In conclusion, our study showed that grafted plants could be grown under DI with minor yield reduction with 30–40% less water used for irrigation. Moderate DI could be used before PRD for cultivation of grafted tomato and double stemmed plants did not show negative effect on tomato yield so it can be used as standard under reduced irrigation.

Keywords: reduced irrigation; rootstocks; yield traits; leaf gas exchange

---

1. Introduction

The tomato (*Solanum lycopersicum* L.) is a leading vegetable and one of the most important horticultural crops. The world production of tomatoes is second to only potatoes, with an estimated production of 160 million tons [1]. Tomato production problems with abiotic and biotic stressors as results of intensive monoculture often create problems in tomato production. Tomato production losses caused by unfavorable growing conditions can be reduced by grafting onto specific rootstocks. Commercial vegetable grafting started at the beginning of the 20th century, with the primary intention to achieve tolerance to soil pathogens [2]. The substantial proportion of total tomato production in Europe and Asia currently include usage of grafted tomatoes [3]. In addition, the widespread use of grafting was expected to improve crop response to water, salt, nutrient deficiency and temperatures stresses and to improve fruit quality [4,5].

Water availability is decreasing worldwide—especially where agriculture uses between 50% and 90% of all water, such as semi-arid Mediterranean areas. In the context of climate change, improvement of irrigation practices is needed, by use of new sources (e.g., wastewater) or by application of new irrigation techniques [6]. Two main methods regarding the use of reduced irrigation are introduced, namely deficit irrigation (DI) and partial rootzone-drying (PRD) to improve water-use efficiency (WUE). These practices expose the plant to moderate drought stress, could increase abscisic acid (ABA) levels that leads to greater increase in WUE [7]. DI supplies less water to the entire rootzone than the amount lost by evapotranspiration, while PRD involves alternate wetting and drying of the root zones. PRD has been shown to improve DI and has resulted in substantial water savings, improved water-use efficiency and it is superior to DI in terms of yield maintenance in greenhouse or processing tomato [8,9].

Grafted plants showed better uptake of minerals and water than un-grafted plants due to vigorous root growth by the chosen rootstock [10]. Grafted tomato under deficit irrigation showed increased yield and WUE [11,12]. The tomato cultivar Boludo grafted on 144 tomato rootstocks showed higher shoot fresh weight under water deficit in 38% of combinations [13]. Tomato cultivars Belle and Clarabelle grafted on the rootstock He-man had similar vegetative growth and yield under PRD conditions in commercial greenhouse [14]. The breeding of commercial rootstocks like 'Maxifort' on the other hand was more directed to increase growth capacity and to alleviate soilborne diseases, instead of water economy [2], although it showed similar performance compared to drought tolerant cultivars [15]. Sink limitation is more pronounced in grafted tomato plants compared to non-grafted ones. One of the ways to avoid sink limitations is growing double-stemmed grafted plants. Grafting with two stems became the standard growing method for sustaining tomato production, as it decreased the costs per unit area by reducing the number of plants grown in a greenhouse by one-half [16,17].

The effects of grafting on tomato fruit quality are showing inconsistencies in of the results, mostly affected by the rootstock–scion combination [5]. Similarly, scarce reports showed that grafted plants under water stress differ in total soluble solids (TSS) and total acidity (TA) [11,14,18].

In the first study, we evaluated the effect of PRD on the growth, yield and quality of grafted tomato grown in a commercial greenhouse [14]. As proper evaluation of reduced irrigation methods apart PRD include DI, the two-year studies included the deficit irrigation treatments with similar amount of water applied as in PRD, but evenly to the whole root system. Since two stems are the standard practice for cultivation of grafted tomato, we also included stem number as a factor in our experiments. Finally, the purpose of our studies was to compare the responses of grafted tomato plants subjected to PRD or DI by evaluating vegetative growth, leaf gas-exchange parameters, yield traits, WUE and leaf and fruit mineral profile and fruit quality in a greenhouse located in a Mediterranean climate.

2. Material and Methods

Two experiments were conducted on tomatoes in spring–summer season in commercial greenhouses in the Split area (Mediterranean area of Croatia). The first experiment used plants grown on one stem, and in the second experiment two-stem plants were used.

2.1. Experiment I: Single Stem Plants (2016)

The first experiment was established in an unheated greenhouse in the Trogir (43°31′ N, 16°15′ E) used for intensive vegetable production for many years. The soil type was an alkaline clay with 8.09 pH

(H_2O), 7.51 pH (KCl), 4.4% soil organic matter and 120 mg of available P_2O_5, and 48 mg K_2O/100 g of soil. The greenhouse had side ventilation and the roof was 3 m high.

Tomato (*Solanum lycopersicum* L., cv. Clarabella F1, Rijk Zwaan, The Netherlands) plants were either self-grafted or grafted onto the rootstocks "Emperador" (*S. lycopersicum* × *S. habrochaites*, Rijk Zwaan) or "Maxifort" (*S. lycopersicum* × *S. habrochaites*, De Ruiter Seeds, Amsterdam North, The Netherlands). Both rootstocks, as noted by the seed companies, have high to medium vigor and are resistant to *Fusarium*, *Verticillium* and ToMV.

The scion seeds were sown on 17 January 2016 and rootstock seeds on 20 January 2016 in polystyrene plug trays with cell volume of 40 mL in an organic substrate (Brill Type 4, Brill Substrate, Georgsdorf, Germany). As scion and rootstocks have variable growth vigor and to ensure optimum stem diameter between scion and rootstock seedlings at grafting time the scion seeds were sown 3 days earlier than the rootstock. The trays with sown seeds were put in a heated greenhouse (day/night 25 °C).

The cv. 'Clarabella' seedlings were self-grafted and grafted onto both rootstocks at 25 days after sowing using the "splice grafting" method. Grafted seedlings were maintained under reduced light conditions (10% of the daily light intensity) at a relative humidity above 95% and temperature from 22 °C to 25 °C until callus formation. After callus formation, the seedlings were maintained as standard tomato transplant. Seedlings were grown in research greenhouse at the Institute for Adriatic Crops (Split, Croatia).

Tomato seedlings with four to five true leaves were transplanted 65 days after sowing (24 March 2016) in a two-row system (90 cm apart) with rows 60 cm apart with plants were spaced 50 cm in each row, for a total of 2.7 plants/m^2. The plants were drip irrigated with drippers (pressure-compensating emitters) set in opposite lines of row plants (15 cm from plants). After transplanting, plants were irrigated per the standard practice in the area. During the trial, plants were fertilized twice with Mg, as other nutrients had high available soil concentrations, as confirmed by N and K petiole sap analysis during growth phases.

Three irrigation treatments were started 30 days after transplanting: full irrigation (FI), deficit irrigation (DI) and partial-root zone drying (PRD). The soil moisture content was measured by tensiometers (Blumat digital, Leingarten, Germany) which were laboratory calibrated for conducted soil. The tensiometers were placed at 25–30 cm depth. In FI treatment, plants were watered to a soil moisture content of 65–75% of field capacity. DI treatments received 50% water used in FI by using drippers with half capacity than in FI. With PRD, half of the root system was irrigated at 65–75% of field water capacity (FWC), while other half of the roots were dried until soil moisture reached 35–40% of field capacity and then irrigation was shifted between two sides of the root system. PRD also received 50% water from FI. Different irrigation treatments were divided by placing PE folia to the depth of 45–50 cm to stop horizontal water movement. Taking into account water supplied before start of irrigation treatment, in total DI and PRD got 60% water supplied to FI that received 214 L per plant.

The plant height and the number of leaves (longer than 5 cm) were determined for 7 weeks after transplanting. Harvest started 70 days after transplanting (DAT) and lasted for 55 days—including 12 harvests of fruits as they matured (light red color) The average fruit weight and fruit number were recorded. The first four harvests were calculated as early yield. On the day of the last harvest, the aboveground parts of the subsample plants (3 plants per treatment) were removed, divided into leaves, stems and green fruits and weighed for fresh biomass (FM). After measuring the leaf area with leaf area meter LI-3000 (LI-COR, Bad Homburg, Germany), samples were put into an oven and dried at 70 °C to constant weight to obtain the DM. The yield divided by supplied water was used to find the yield WUE (WUEy).

2.2. Experiment II: Double Stem Plants (2017)

The greenhouse in the split was used for the second experiment (43°30′ N, 16°30′ E). The soil type was a clay loam with 8.71 pH (H_2O), 7.46 pH (KCl), 2.9% soil organic matter, 16.5% high active lime and 27 mg of available P_2O_5 and 31 mg K_2O/100 g of soil. Same rootstocks were used as in

Experiment I, and scion cultivar was Attiya (Rijk Zwaan, De Lier, the Netherlands) due to resistance to TSWV (Tomato spotted wilt virus) that influenced growth and yield in previous years on different tomato cultivars in used greenhouse. The scion and rootstocks seeds were sown on 19 January 2016 and 23 January 2016, respectively. Grafting was done on 17 February 2016; all other procedures for seedling production were done as in previous year. In this experiment, two types of seedlings were produced: self-grafted Attiya grown on one stem (ATT) and plants grown on two stems: self-grafted Attiya (AT), Attiya grafted on Emperador (EM) and Attiya grafted on Maxifort (MX), which in total gave four plant types. Two-stemmed seedlings were formed from side-shoots of cotyledons.

The tomato seedlings were transplanted 55 days after sowing (16 March 2016) in two systems: one stem seedlings as in previous years in a two-row system and two-stem seedlings in a one-row system with rows 120 cm apart. In each row, the plants were spaced 50 cm, which in both systems gave 2.7 stems/m^2.

Irrigation treatments started 50 DAT and included FI, DI and PRD. In this year, the soil moisture content was controlled by Maxi Rain soil moisture sensors (Elektronik Jeske, Windorf, Germany), which were set up to open electromagnetic valve when soil moisture was lower than 65% field water capacity and was irrigated until reached 80% FWC. DI treatment received 60% water supplied to FI using drippers with lower capacity. To better measure water needs in the PRD, this treatment had its own sensors for controlling irrigation and switching sides. The PRD used 50% of the water used in FI. In total—including water applied before irrigation treatment start—the DI used 70% and PRD 65% water of the FI. The FI in total received 233 L/plant, DI 170 L/plant and PRD 153 L/plant. House-made lysimeters (60 cm × 100 cm × 5 cm) were put below every irrigation treatment at depth of 60 cm to control possible leaching if plants were over-irrigated.

Plants were fertilized by irrigation system with N, K and Mg, depending on growth phases and plant needs. Plant height and leaf number were measured each week. Harvest started at 80 DAT (7 June) and lasted 45 days, consisting of 11 harvests. At the end, plants were divided to determine biomass partitioning, as in previous year.

Leaf nutrient concentrations were determined in the youngest fully developed leaves after the leaves were dried at 70 °C and then ground. The micro-Kjeldahl digestion method (Kjeltec System 1026, Tecator, Höganas, Sweden) was used to measure total leaf N concentration. Dry ash of grounded samples from a muffle furnace were dissolved in 2 mL HCl to extract the P, K, Ca and Mg. The K concentration was measured using a flame photometer (Model 410, Sherwood Scientific, Cambridge, UK). The vanadate-molybdate yellow color method using a UV-visible spectrophotometer was used to determine the P concentration (Cary 50 Scan, Varian, Palo Alto, CA, USA) at 420 nm. The Ca and Mg in solution were determined by atomic absorption spectrometry (SpectrAA 220, Varian, Palo Alto, CA, USA).

The quality parameters of fruits from each treatment were analyzed in the second experiment. For the tomato juice, the total soluble solids (TSS) content was determined by a DR 201–95 refractometer (Kruss optronic, Hamburg, Germany) and expressed in Brix at 20 °C. Acidity was determined by juice titration with 0.1-M NaOH was used for determination of acidity and results were expressed as citric acid. Gas-exchange parameters were measured using LI-6400 infrared gas analyzer (LI-COR, Inc., Lincoln, NE, USA) in youngest fully expanded leaves. Measurements were performed on six leaves per treatment 20 days after different irrigation techniques were applied in whole experiment. Measurements were conducted under constant light (PAR 750 μmol m^{-2} s^{-1}) and CO_2 concentration (400 μmol mol^{-1}). The environmental conditions in the greenhouse ranged from 22 °C to 33 °C for air temperature and from 33% to 42% for relative humidity (RH). The greenhouse light conditions (PAR) ranged from 300 to 1100 μmol m^{-2} s^{-1}. The transpiration rate (E) and photosynthetic rate (A) were determined from gas exchange measurements and were used to determine the photosynthetic/instantaneous water-use efficiency (PWUE) as the ratio between A and E [19].

The experiments were set up in a randomized block design, consisting of three replications. Each treatment (irrigation × plant type) was comprised of 12 plants. The data were evaluated by

ANOVA and when F-tests were significant, the means of the main factors (rootstock/plant type and irrigation technique) and their interactions were compared using the least significant difference test at $p \leq 0.05$. The data were statistically analyzed using StatView ver. 5.0 (SAS Institute, Cary, CA, USA).

3. Results

3.1. Plant Growth and Biomass Production

In the first year of the experiment, plants grown on one stem showed significant differences in vegetative parameters regarding rootstock type used (Table 1). Plants grafted on Emperador rootstock had highest leaf area, number of leaves, plant height and leaf and shoot dry biomass (DM) compared to self-grafted plants. The irrigation method and interaction between irrigation and graft did not show significant differences.

Table 1. Effect of rootstock type and irrigation treatment on tomato vegetative characteristics in the first year of the experiment with plants grown on one stem.

Treatments	Leaf			Shoot	Plant Height (70 DAT) (cm)
	Number (70 DAT)	Area (cm^2 plant^{-1})	DM (g plant^{-1})	DM (g plant^{-1}) *	
Rootstock (R)					
Self-grafted	24.4 b †	10,406 b	108.1 b	183.7 b	172.5 b
Emperador	26.7 a	19,117 a	188.6 a	296.9 a	184.1 a
Maxifort	26.2 ab	16,521 a	158.1 ab	248.7 ab	177.7 ab
Irrigation (I)					
Full	25.3	15,621	149.9	245.9	175.2
Deficit	26.7	14,990	144.4	234.1	180.8
PRD	26.2	15,433	160.5	249.2	178.3
I × R	ns	ns	ns	ns	ns

* DM—dry mass; † Significant differences between treatments (LSD test at $p \leq 0.05$) are indicated with different letters within columns. ns—non-significant.

In the second year, statistical analyses for vegetative growth are presented in Table 2. Results are presented per stem for two-stemmed grafted plants with plants of one stem. Similar to the first experiment year, most measured traits were significantly affected by rootstock type. Additionally, the leaf number showed differences in irrigation technique applied. Leaf and shoot DM per stem was significantly highest in Emperador and lowest in Attiya. No difference in these traits were found between one-stem Attiya and Maxifort. It can be concluded that Maxifort produces two times more DM per whole plant.

Table 2. Effect of rootstock type and irrigation treatment on tomato vegetative characteristics in the second year of the experiment with plants grown on one and two stem.

Treatments	Leaf			Shoot	Plant Height (60 DAT) (cm)
	Number (60 DAT)	Area (cm^2 plant^{-1})	DM (g stem^{-1})	DM (g stem^{-1})	
Rootstock (R)					
One-stem SG *	25.8 b †	14,039 a	118.5 b	177.8 b	166.2 b
Two-stem SG	22.2 c	7758 b	77.3 c	124.4 c	154.8 b
Emperador	27.3 a	11,669 a	147.8 a	252.4 a	199.8 a
Maxifort	26.7 ab	13,916 a	117.5 b	195.5 b	192.8 a
Irrigation (I)					
Full	25.5 ab	12,308	117.3	189.8	178.1
Deficit	26.1 a	12,064	111.2	178.3	181.6
PRD	24.8 b	11,164	117.3	194.6	177.9
I × R	ns	ns	ns	ns	ns

* SG—self-grafted plants; DM—dry mass; † significant differences between treatments (LSD test at $p \leq 0.05$) are indicated with different letters within columns. ns—non-significant.

3.2. Leaf Gas Exchange

The leaf gas-exchange parameters measured 20 days after initiation of reduced irrigation treatments (DI + PRD) are in Table 3. Stomatal conductance (gs), intercellular CO_2 (Ci), transpiration rate (E) and photosynthetic WUE (PWUE) were significantly affected by irrigation rate. The values of gs, Ci and E were highest under FI and lowest under DI treatment, while PWUE had highest value under DI that differed from FI and PRD. In addition, only stomatal conductance differed ($p \leq 0.05$) between rootstocks, with plants grafted on Emperador having the highest values. The effect of the interaction of the rootstock/plant type × irrigation was recorded for photosynthetic rate (A), photosynthetic WUE (Figure 1A,B) and intercellular CO_2 (data not shown). The highest A was measured for one stem self-grafted Attiya under DI, while lowest was found for same plants under FI. On average, PWUE was highest at DI and as shown by interaction did not differed between plant types under DI, while differences in two other irrigation treatments was influenced by rootstock.

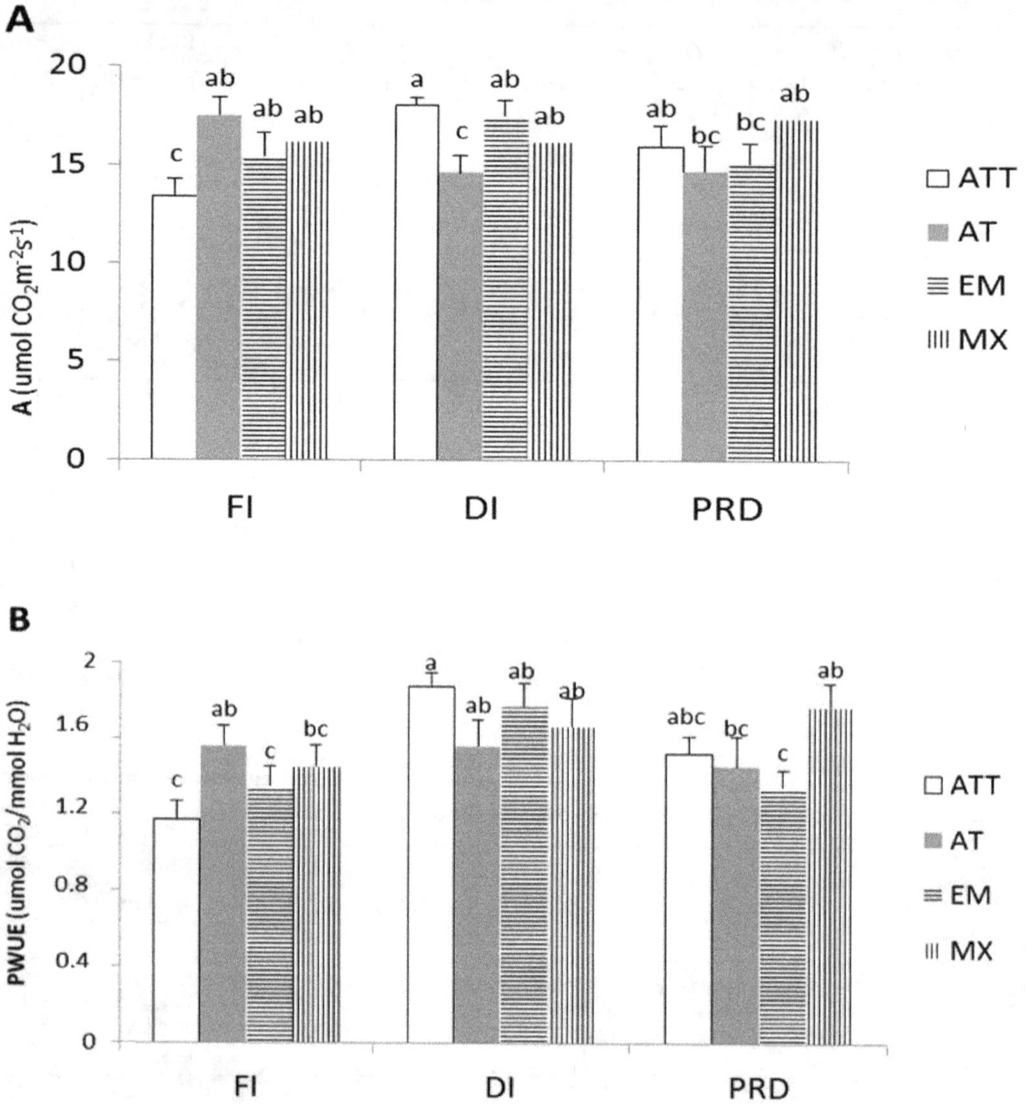

Figure 1. Photosynthetic rate (**A**,**B**) photosynthetic water-use efficiency—PWUE of grafted tomato plants grown with one or two stems under three irrigation techniques. Vertical bars represent SE values ($n = 6$). a—significant difference by the LSD test at $p \leq 0.05$ are indicated by different letters above column; ATT—one-stem Attiya; AT—two-stem Attiya; EM—Emperador, MX—Maxifort.

Table 3. Leaf gas-exchange parameters and photosynthetic water-use efficiency of grafted tomatoes grown with one and two stems under three irrigation techniques in the second year of the experiment.

Treatments	Photosynthetic Rate (A) (μmol CO_2 m^{-2} s^{-1})	Stomatal Conductance (gs) (mol H_2O m^{-2} s^{-1})	Intercellular CO_2 (Ci) (μmol CO_2 mol^{-1})	Transpiration Rate (E) (mmol H_2O m^{-2} s^{-1})	Photosynthetic WUE (PWUE) (μmol CO_2/mmol H_2O)
Rootstock (R)					
One-stem SG	15.8*	0.81 b	336.8	10.9	1.45
Two-stem SG	15.6	0.85 b	339.5	10.8	1.44
Emperador	16.0	0.92 a	341.5	11.4	1.40
Maxifort	16.6	0.84 b	335.9	10.7	1.55
Irrigation (I)					
Full	15.8	0.93 a	342.0 a	11.7 a	1.35 b
Deficit	16.5	0.79 b	334.7 b	10.1 c	1.63 a
PRD	15.8	0.85 b	338.7 ab	10.9 b	1.45 b
I × R	*	ns	*	ns	*

* Significant differences between treatments (LSD test at $p \leq 0.05$) are indicated with different letters within columns. ns—non-significant.

3.3. *Yield and Water-Use Efficiency*

The effects of rootstock type and irrigation technique on tomato yield traits and WUE in the first experiment are presented in Table 4, Average fruit weight, fruit number per plant and total yield were affected by rootstock and were higher in grafted plants than in self-grafted ones. Early yield was significantly affected by irrigation technique and highest values were noted for DI plants differing only with plants cultivated under PRD regime. WUE was affected by rootstock and irrigation. As expected, plants under DI and PRD had almost double values of WUE. Emperador and Maxifort plants significantly differed from self-grafted ones and on average had 40% higher WUE values.

Table 4. Yield parameters and water-use efficiency of grafted tomato plants grown with one stem under three irrigation techniques in the first year of the experiment.

Treatments	Fruit Mean Weight (g)	Number Plant^{-1}	Early Yield Plant^{-1} (g)	Yield Plant^{-1} (g)	WUEy (g L^{-1})
Rootstock (R)					
Self-grafted	243 b *	19.3 b	761	4749 b	34.1 b
Emperador	293 a	24.1 a	821	7089 a	50.0 a
Maxifort	304 a	23.1 a	847	7064 a	40.9 a
Irrigation (I)					
Full	278	21.0	805 ab	5953	27.8 b
Deficit	279	43853.0	938 a	6583	53.5 a
PRD	283	43912.0	687 b	6366	51.8 a
I × R	ns	ns	ns	ns	ns

* Significant differences between treatments (LSD test at $p \leq 0.05$) are indicated with different letters within columns. ns—non-significant.

Yield parameters and WUE results for second year are shown in Table 5. As in the first year, plants grafted on commercial rootstocks had significantly ($p \leq 0.001$) higher fruit mean weight and also number of fruits per stem. Early yield was affected ($p \leq 0.001$) by plant type and highest was noted for self-grafted Attiya grow with one stem. Yield WUE was significantly ($p \leq 0.001$) higher for Emperador and Maxifort grafted plants and also for plants grown under both types of reduced irrigation (DI + PRD).

Total yield results for second year are shown in Figure 2. Although, the interaction between factors was not significant, it was important to compare yield of same plant type under different irrigation. This analysis showed that grafted plants with commercial rootstocks grown under FI had highest total yield that differed from same plants type grown under PRD, but not significantly different from DI. In addition, one stem self-grafted plants from all irrigation regimes did not differed to each other and two stem ones had lowest yield had under PRD.

Table 5. Yield parameters and water-use efficiency of grafted tomato plants grown with one and two stems under three irrigation techniques in the second year of the experiment.

Treatments	Fruit Mean Weight (g)	Number stem^{-1}	Early Yield Stem^{-1} (g)	Yield Stem^{-1} (g)	WUEy (g L^{-1})
Rootstock (R)					
One-stem SG	210 b *	17.6 b	1415 a	3696 b	41.1 b
Two-stem SG	226 b	14.8 c	786 b	3347 b	36.8 b
Emperador	261 a	20.4 a	604 b	5279 a	58.1 a
Maxifort	271 a	19.7 a	699 b	5342 a	59.5 a
Irrigation (I)					
Full	256	18.4	772	4769 a	40.8 b
Deficit	239	18.7	799	4436 ab	52.2 a
PRD	239	17.5	949	4187 b	54.9 a
I × R	ns	ns	ns	ns	ns

* Significant differences between treatments (LSD test at $p \leq 0.05$) are indicated with different letters within columns. ns—non-significant.

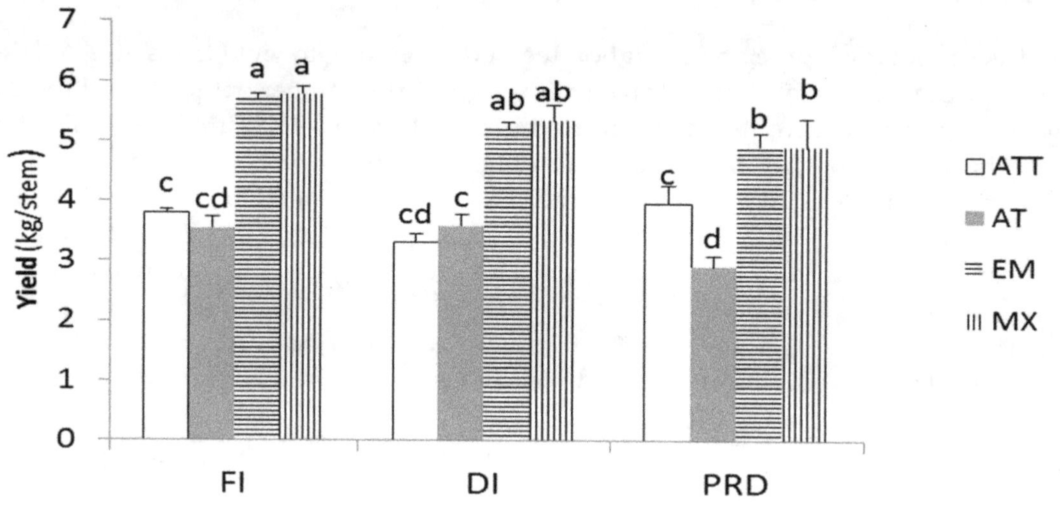

Figure 2. Total yield (kg stem^{-1}) of grafted tomato plants grown with one or two stems under three irrigation techniques. Vertical bars represent SE values ($n = 4$). Different letters above column indicate a significant difference by the LSD test at $p < 0.05$. ATT—one-stem Attiya; AT—two-stem Attiya; EM—Emperador; MX—Maxifort.

3.4. Mineral Content and Fruit Quality

In the second experiment, all leaf mineral concentrations significantly differed by rootstock type, while Ca and Mg were affected also with irrigation method (Table 6). Plants grafted on commercial rootstocks had highest values for N, P and K ($p \leq 0.001$). Ca was lowest in self-grafted plants grown with one-stem, while Mg was highest in same plants. Plants grown under both reduced irrigation techniques had significantly ($p \leq 0.01$) highest Ca and Mg leaf concentrations. Fruit mineral concentration in same year was significantly affected by rootstock. Plants grafted on commercial rootstocks had higher values than both types of self-grafted plants (Table 7).

The effects of rootstock and irrigation on fruit quality traits are presented in Table 6. Both TSS and TA differed by rootstock type ($p < 0.05$). There were no differences among irrigation techniques in these traits, but the effect of the interaction of the rootstock × irrigation was found for TSS ($p < 0.05$) (Figure 3). Highest TSS was recorded in one-stem Attiya grown under DI and differ from values same plants under FI and PRD. Plants grafted on Emperador had higher TSS under PRD than other irrigation treatments. TA was higher in fruits grown on commercial rootstocks.

Table 6. Leaf mineral concentrations of self-grafted and grafted tomatoes grown under three irrigation techniques in the second year.

Treatments	N	P	K	Ca	Mg
	g kg^{-1} DW				
Rootstock (R)					
One-stem SG	28.9 b	2.01 b	15.0 b	23.1 b	5.34 a
Two-stem SG	31.5 b	2.32 ab	14.2 b	29.3 a	4.98 ab
Emperador	36.5 a	3.23 a	21.0 a	28.3 a	4.5 bc
Maxifort	37.7 ab	3.31 a	21.8 a	28.5 a	4.45 c
Irrigation (I)					
Full	33.9	2.82	17.5	24.7 b	4.41 b
Deficit	32.8	2.73	17.6	28.3 a	5.17 a
PRD	34.2	2.7	18.9	29.0 a	4.87 a
I × R	ns	ns	ns	ns	ns

Significant differences between treatments (LSD test at $p \leq 0.05$) are indicated with different letters within columns. ns—non-significant.

Table 7. Effect of rootstock type and irrigation treatment on tomato fruits mineral concentrations and quality parameters in the second year of the experiment with plants grown on one and two stems.

Treatments	N	P	K	Ca	Mg	TSS (Brix^o)	TA (g L^{-1})
	g kg^{-1} DW						
Rootstock (R)							
One-stem SG	18.2 b	2.08 b	29.9 b	1.22 b	1.37 c	4.7 a	4.5 b
Two-stem SG	18.1 b	2.51 b	30.5 b	1.39 ab	1.41 bc	4.4 b	4.6 b
Emperador	22.1 a	3.26 a	37.7 a	1.59 a	1.61 a	4.3 b	5.6 ab
Maxifort	19.9 a	3.19 a	35.4 a	1.51 a	1.57 ab	4.4 b	5.9 a
Irrigation (I)							
Full	20.1	2.81	33.9	1.46	1.55	4.3	5
Deficit	19.5	3.04	33.1	1.42	1.43	4.5	5.8
PRD	19.4	2.56	33.5	1.47	1.50	4.5	4.8
I × R	ns	ns	ns	ns	ns	*	ns

* Significant differences between treatments (LSD test at $p \leq 0.05$) are indicated with different letters within columns. ns—non-significant.

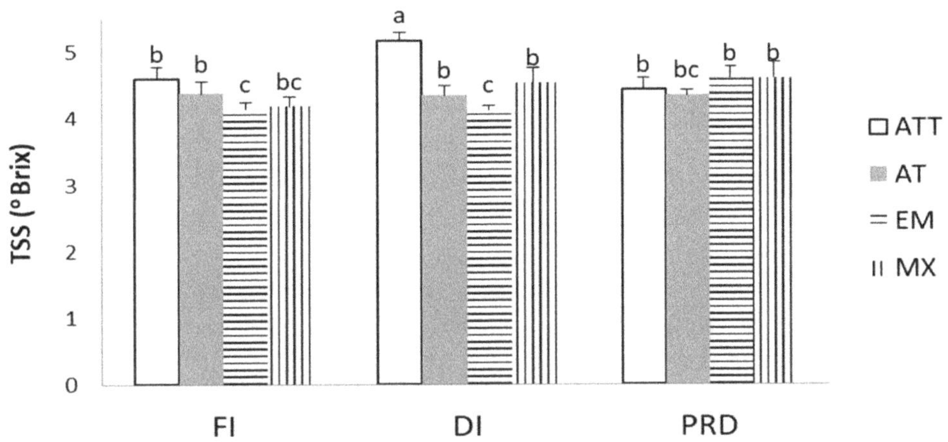

Figure 3. Total soluble solids (TSS) of tomato fruits sampled from different grafted plants grown with three irrigation techniques. Vertical bars represent SE values ($n = 4$). Different letters above column indicate a significant difference by the LSD test at $p \leq 0.05$. ATT—one-stem Attiya; AT—two-stem Attiya; EM—Emperador; MX—Maxifort.

4. Discussion

In general, the use of commercial rootstocks resulted in highly improved plant vigor in terms of excessive vegetative growth. In our studies, the plant height and leaf number measured at 60 or 70 DAT were highest with plants grafted on Emperador and Maxifort rootstocks; the difference was even more pronounced in the second trial. Comparing height with leaf number, it can be assumed that interleaf nodes interval was not influenced by grafting or irrigation. These findings partially differed from other studies [14,20]. Similarly, vegetative biomass production (as leaf area, leaf and shoot DM) was bigger in grafted plants. For example, shoot DM in year one was 40% and 60% higher in Maxifort and Emperador grafted plants compared to self-grafted ones. Similar values were found in the second year on stem basis, but when we include all plant production (both stems) vegetative biomass is at least two or three times higher. Vigorous rootstocks had enough capacity to provide satisfactory vegetative growth by roots that can supply needed water and nutrients for assimilates production. On average, leaf area was not influenced by irrigation, as others found that tomato under DI and PRD had smaller leaf area than control and explained as soil drying affected roots reaction and production of chemical signals, i.e., changed ABA concentration or xylem sap pH that leads to stomatal closure and decreases leaf expansion growth [21].

Interactive effect between rootstock type and irrigation treatment showed that plants grafted on commercial rootstocks did not differ in photosynthetic rate (A), while both types of self-grafted ones differed depending on the applied irrigation technique. It can be concluded that grafted plant had better assimilative processes. The optimization of A under water stress could be modified by the rootstock through action on biochemical and biophysical processes [22]. Stomatal conductance and intercellular CO_2 measured 20 days after starting irrigation treatments was lower under DI and PRD. These effects of reduced irrigation in some cases were noted later after initiation of irrigation [23], while in another study differences between treatments disappeared with time [24]. Valerio et al. [23] showed that lower stomatal conductance was related with leaf ABA accumulation, more ABA reduced stomatal conductance. Reducing stomatal conductance is a typical response to soil drying as stomatal closure is primary response to water deficit so plants could better contol water loss due to transpiration [7]. Stomatal closure reduced transpiration rate which was more pronounced in DI. Although, stomatal conductance was similar in both DI and PRD, it was expected that transpiration will be similar in both of them suggesting the response is mostly to the overall amount of water supplied to roots [21]. In contrast, in our study DI received 5% more water than PRD, so it seems that hydraulic signal is an important factor because plants under PRD on the wet side of the root can absorb enough water to keep higher level of transpiration [23]. Photosynthetic WUE had highest values under DI as result of lowest transpiration rate and similar photosynthetic rate to other irrigation. This can lead to lower biomass production as was noted in our study as reduced shoot biomass (although not significant) under DI and others found similar [21].

Rootstocks may affect tomato productivity positively or negatively, although in most cases yield increased both under non and stress conditions and depended on rootstock/scion combination [24]. In our experiment, plants grafted on commercial rootstocks had highest yield as a result of more and bigger fruits per plant, as was found in other studies [11,17,25]. Enhanced fruit production could be clearly related with higher plant biomass [15]. Early yield was different between years, in the first year highest early yield was noted under DI which can be related to a more pronounced water stress that hastened fruit ripening in this treatment. Topcu et al. [26] found higher tomato early yield in PRD than DI plants in experiment with more water reduction (50%) comparing our 40%. In experiment with two stems (second year), early yield was highest in one-stem-grafted Attiya what is result of longer period of growth for two stems plants because they were trained as side-shoots from cotyledons. In both experiments, cultivars grafted on commercial rootstocks had highest total yield under all irrigation treatments. In the second year comparing yield of these plants, it was found that under FI yield did not differ from DI but differ from PRD (Figure 1). It seems that rootstocks due to its vigor have enough capacity for water uptake to sustain yield under DI. It is important to notice that growing on two stems (2nd experiment) did not reduce yield markedly when comparing with one stem plants (1st experiment), although different cultivars were used what should be taken into account. Rahmatian et al. [16] found dry matter allocation was not influenced by grafting or stem numbers and that good balance between vegetative and generative growth can depend on rootstocks. Other studies done with ungrafted greenhouse and processing tomato mostly obtained higher yield using PRD than DI [8] or similar to DI and FI [24].

WUE is the main indicator of plant water relations and is regulated by physiological mechanisms. In both years WUEy calculated as ratio between yield and water applied per treatment was higher in rootstock-grafted plants and as expected under PRD and DI. In these treatments higher fruit yield and lower water use resulted in improved WUE. It was not shown that PRD improved WUE better than DI, which means that irrigation volume is more important than used irrigation technique in determining yield or all crop growth as was suggested before [21,23,24], although other found irrigation technique can be more important [9]. Comparing two experiments it can be seen that in double stemmed plants (2nd experiment) WUE was higher leading to conclusion water use was optimized. In addition, in 1st Experiment WUEy was much lower in FI than in the second year what can be related to use of different soil moisture meters: tensiometers and soil sensors. The tensiometers was used for hand-operated irrigation, which could have led to overirrigation in the first year. It was shown that automatic operated tensiometers was more effective, which can be compared with sensors with automatic valves in our study [27].

The leaf mineral concentrations of P and K were under range of sufficiency while others were in range (N and Mg) or above (Ca) as proposed for greenhouse tomato. Grafting is considered as an effective tool for improving nutrient uptake and use efficiency in vegetables, although those were observed under optimal nutrient status in the root zone. N, P and K had higher concentrations in the plants grafted on commercial rootstocks what was expected and already confirmed in other studies that showed that nutrient uptake depends on rootstock–scion combinations [28]. Higher leaf P in grafted plants were reported for grafted eggplant and watermelon [29,30]. Grafted plants had more vigorous root system, which could be reason for increase in active uptake of P that has low mobility in soils. Self-grafted and grafted plants had low leaf K (under sufficiency range) because fertilization was not intensive as in commercial production. Potassium is nutrient normally required in the largest amount in tomato production. Grafting promote better growth and K uptake even under low K supply as was shown by Schwarz et al. [31]. These nutrients (N, P, K) concentrations were not affected by water supply rate, although opposite was shown for N in other studies for PRD or DI in non-grafted tomatoes [32]. Increase in K concentrations under water stress was found in some non-grafted and grafted tomatoes explaining that K accumulation improves stomatal resistance which improve drought tolerance [33]. In other case, decrease in grafted tomato leaf K was noted with increase in water stress level [12].

Regarding Ca^{2+} and Mg^{2+}, a significant increase in tomato Ca leaf concentration was found due to grafting what is in line with other reports [29,31]. In addition, both DI and PRD resulted in more leaf Ca than plants under FI. It was found that tomatoes under PRD had increased Ca uptake due to higher plant water status and lower stomatal conductance [34]. Higher Ca uptake induced by grafting are important for the tomato fruits due to the possibility of blossom-end rot incidence. Different than Ca, in grafted tomato was found lower leaf Mg what is in line with previous studies and could be also influenced by rootstock and cultivars used [5,14]. It seems that grafting somehow decrease Mg uptake in grafted vegetables, but reason it is not yet clear. Possible higher Ca uptake reacts antagonistically to Mg uptake, which could be related to specific transport systems [35]. Under reduced irrigation treatments higher leaf Mg was measured and same was found for mini watermelons [30]. Mg^{2+} ion has largest hydrated radius among cations and this property makes Mg^{2+} bind weakly to negatively charged soil colloids and root cell walls [36], which could lead to decreased Mg uptake under FI conditions due to leching in sub-root zones. The fruit mineral concentrations was influenced by rootstock type showing that highest values in the plants grafted on Emperador and Maxifort. Other found effect of rootstock, but also influence of water stress on fruit minerals [18].

Higher TSS was affected by plant type with highest values in one stem plants. Interactive analysis showed it is mostly result of highest values of same plant type under DI. The enhanced TSS in that treatment could be result of water stress, although it is not clear why similar was not found in self-grafted two stem plants. Self-grafted double stemmed plants possible use more assimilates for additional vegetative growth [16]. Grafted plants had lower TSS what is often found even when used different cultivars and rootstocks [25]. For grafted plants vigorous roots can be additional sinks for assimilates and also better water uptake can result in dilution effect of fruits sugars [10]. Under PRD all plant types had similar TSS so it can be concluded that self and grafted plants with this irrigation type changed mechanisms responsible for results recorded under DI. Grafting on commercial rootstocks decreased Mg leaf content, which can possible lead to latent Mg deficiency influencing carbohydrate partitioning requiring for obtaining maximum yield and ensuring sugar accumulation in fruits [37]. In our study, both rootstocks increased the TA. Their increase by grafting was also found in many other experiments under different conditions [5]. Grafting under regular and low K resulted in higher TA, independent of K in fruits [31], while in our study K in fruits grafted on both rootstocks was higher compared to self-grafted plants. It is known that K concentration in fruits can be positively related with acid content, although further investigations are needed.

5. Conclusions

In the present experiment, we evaluated growth, gas-exchange parameters, yield, WUE and leaf mineral concentrations and fruit quality of self-grafted and tomato grafted on two commercial rootstocks cultivated in greenhouses in Mediterranean climate under three irrigation techniques: FI, DI and PRD. First year plants were grown with one stem and in the second with two stems.

In conclusion, these studies for the first time demonstrates the effects of parallel usage of different reduced irrigation techniques on grafted tomato vegetative and generative traits. Grafting onto commercial rootstocks improved plants growth and yield both in cultivation with one or two stems. Grafted plants under DI had minimal yield reduction compared to FI in double stemmed plants. WUE was highly improved with grafting and application of PRD and DI. That was more pronounced in experiment with two stem plants and could be result of different biomass partitioning and irrigation scheduling based on soil moisture sensors. Leaf mineral concentrations were higher in grafted plants as possible better uptake of vigorous rootstocks while Mg was reduced what imply contrasting rootstock–scion interactions. These findings indicate that grafted plants can be grown under moderate DI before PRD and that two stem plants could be used under that irrigation regime.

Author Contributions: Conceptualization, B.U., G.D., K.Ž.; methodology, B.U., M.R.; validation, G.D., G.V.S.; formal analysis, B.U., G.D.; investigation, B.U., M.R., G.V.S., M.M., A.M.; resources, B.U., G.D.; data curation, M.R., M.M., A.M.; writing—original draft preparation, B.U.; writing—review and editing, K.Ž., G.D.; visualization, B.U.; supervision, B.U., G.D., M.R.; project administration, B.U.; funding acquisition, B.U. All authors have read and agreed to the published version of the manuscript.

Acknowledgments: The authors are thankful to Silvia Milišić, Jelena Dumanić and Željko Bilić for their help maintaining experiments, collecting data and providing chemical analyses. In addition, special thanks to families Hrabar and Janković for providing usage of greenhouses and help in conducting experiments.

References

1. FAOSTAT. 2018. Available online: http://www.fao.org/faostat/en/#data/QC (accessed on 14 July 2018).
2. Rivard, C.L.; Louws, F.J. Grafting to manage soilborne diseases in heirloom tomato production. *HortScience* **2008**, *43*, 2104–2111. [CrossRef]
3. Grieneisen, M.L.; Aegerter, B.J.; Stoddard, C.S.; Zhang, M. Yield and fruit quality of grafted tomatoes, and their potential for soil fumigant use reduction. A meta-analysis. *Agron. Sustain. Dev.* **2018**, *38*, 29. [CrossRef]

4. Schwarz, D.; Rouphael, Y.; Colla, G.; Venema, J.H. Grafting as a tool to improve tolerance of vegetables to abiotic stresses: Thermal stress, water stress and organic pollutants. *Sci. Hortic.* **2010**, *127*, 162–171. [CrossRef]
5. Kyriacou, M.C.; Rouphael, Y.; Colla, G.; Zrenner, R.; Schwarz, D. Vegetable grafting: The implications of a growing agronomic imperative for vegetable fruit quality and nutritive value. *Front. Plant Sci.* **2017**, *8*, 741. [CrossRef] [PubMed]
6. Lovelli, S.; Perniola, M.; Scalcione, E.; Troccoli, A.; Ziska, L.H. Future climate change in the Mediterranean area: Implications for water use and weed management. *Ital. J. Agron.* **2012**, *7*, 44–49. [CrossRef]
7. Liu, F.; Shahnazari, A.; Andersen, M.N.; Jacobsen, S.E.; Jensen, C.R. Effects of deficit irrigation (DI) and partial root drying (PRD) on gas exchange, biomass partitioning, and water use efficiency in potato. *Sci. Hortic.* **2006**, *109*, 113–117. [CrossRef]
8. Kirda, C.; Topcu, S.; Cetin, M.; Dasgan, H.Y.; Kaman, H.; Topaloglu, F.; Derici, M.R.; Ekici, B. Prospects of partial root zone irrigation for increasing irrigation water use efficiency of major crops in the Mediterranean region. *Ann. Appl. Biol.* **2007**, *150*, 281–291. [CrossRef]
9. Giuliani, M.M.; Nardella, E.; Gagliardi, A.; Gatta, G. Deficit irrigation and partial root-zone drying techniques in processing tomato cultivated under Mediterranean climate conditions. *Sustainability* **2017**, *9*, 2197. [CrossRef]
10. Martínez-Ballesta, M.C.; Alcaraz-López, C.; Muries, B.; Mota-Cadenas, C.; Carvajal, M. Physiological aspects of rootstock–scion interactions. *Sci. Hortic.* **2010**, *127*, 112–118. [CrossRef]
11. Ibrahim, A.; Wahb-Allah, M.; Abdel-Razzak, H.; Alsadon, A. Growth, yield, quality and water use efficiency of grafted tomato plants grown in greenhouse under different irrigation levels. *Life Sci. J.* **2014**, *11*, 118–126.
12. Al-Harbi, A.; Hejazi, A.; Al-Omran, A. Responses of grafted tomato (*Solanum lycopersiocon L.*) to abiotic stresses in Saudi Arabia. *Saudi J. Biol. Sci.* **2016**, *24*, 1274–1280. [CrossRef] [PubMed]
13. Albacete, A.; Andújar, C.; Dodd, I.; Giuffrida, F.; Hichri, I.; Lutts, S.; Thompson, A.; Asins, M. Rootstock-mediated variation in tomato vegetative growth under drought, salinity and soil impedance stresses. *Acta Hortic.* **2015**, *24*, 141–146. [CrossRef]
14. Urlić, B.; Runjić, M.; Žanić, K.; Mandušić, M.; Selak, G.V.; Pasković, I.; Dumičić, G. Effect of partial root-zone drying on grafted tomato in commercial greenhouse. *Hortic. Sci.* **2020**, *47*, 36–44. [CrossRef]
15. Fullana-Pericàs, M.; Conesa, M.À.; Ribas-Carbó, M.; Galmés, J. The Use of a Tomato Landrace as Rootstock Improves the Response of Commercial Tomato under Water Deficit Conditions. *Agronomy* **2020**, *10*, 748. [CrossRef]
16. Rahmatian, A.; Delshad, M.; Salehi, R. Effect of grafting on growth, yield and fruit quality of single and double stemmed tomato plants grown hydroponically. *Hortic. Environ. Biotechnol.* **2014**, *55*, 115–119. [CrossRef]
17. Soare, R.; Dinu, M.; Babeanu, C. The effect of using grafted seedlings on the yield and quality of tomatoes grown in greenhouses. *Hortic. Sci.* **2018**, *45*, 76–82. [CrossRef]
18. Sánchez-Rodríguez, E.; Leyva, R.; Constán-Aguilar, C.; Romero, L.; Ruiz, J.M. Grafting under water stress in tomato cherry: Improving the fruit yield and quality. *Ann. Appl. Biol.* **2012**, *161*, 302–312. [CrossRef]
19. Hatfield, J.L.; Dold, C. Water-use efficiency: Advances and challenges in a changing climate. *Front. Plant Sci.* **2019**, *10*, 103. [CrossRef]
20. Khah, E.M.; Kakava, E.; Mavromatis, A.; Chachalis, D.; Goulas, C. Effect of grafting on growth and yield of tomato (*Lycopersicon esculentum* Mill.) in greenhouse and open-field. *J. Appl. Hortic.* **2006**, *8*, 3–7. [CrossRef]
21. Tahi, H.; Wahbi, S.; Wakrim, R.; Aganchich, B.; Serraj, R.; Centritto, M. Water relations, photosynthesis, growth and water-use efficiency in tomato plants subjected to partial rootzone drying and regulated deficit irrigation. *Plant Biosyst.* **2007**, *141*, 265–274. [CrossRef]
22. Fullana-Pericàs, M.; Conesa, M.À.; Pérez-Alfocea, F.; Galmés, J. The influence of grafting on crops' photosynthetic performance. *Plant Sci.* **2019**, *295*, 110250. [CrossRef]
23. Valerio, M.; Lovelli, S.; Sofo, A.; Perniola, M.; Scopa, A.; Amato, M. Root and leaf abscisic acid concentration impact on gas exchange in tomato (*Lycopersicon esculentum* Mill) plants subjected to partial root-zone drying. *Ital. J. Agron.* **2017**, *11*, 12. [CrossRef]
24. Nardella, E.; Giuliani, M.M.; Gatta, G.; De Caro, A. Yield response to deficit irrigation and partial root-zone drying in processing tomato (*Lycopersicon esculentum* Mill.). *J. Agric. Sci. Technol.* **2012**, *2*, 209.

25. Turhan, A.; Ozmen, N.; Serbeci, M.S.; Seniz, V. Effects of grafting on different rootstocks on tomato fruit yield and quality. *Hortic. Sci.* **2011**, *38*, 142–149. [CrossRef]
26. Topçu, S.; Kirda, C.; Dasgan, Y.; Kaman, H.; Çetin, M.; Yazici, A.; Bacon, M.A. Yield response and N-fertiliser recovery of tomato grown under deficit irrigation. *Eur. J. Agron.* **2007**, *26*, 64–70. [CrossRef]
27. Muñoz-Carpena, R.; Li, Y.C.; Klassen, W.; Dukes, M.D. Field comparison of tensiometer and granular matrix sensor automatic drip irrigation on tomato. *HortTechnology* **2005**, *15*, 584–590. [CrossRef]
28. Martínez-Andújar, C.; Albacete, A.; Martínez-Pérez, A.; Pérez-Pérez, J.M.; Asins, M.J.; Pérez-Alfocea, F. Root-to-shoot hormonal communication in contrasting rootstocks suggests an important role for the ethylene precursor aminocyclopropane-1-carboxylic acid in mediating plant growth under low-potassium nutrition in tomato. *Front. Plant Sci.* **2016**, *7*, 1782. [CrossRef] [PubMed]
29. Leonardi, C.; Giuffrida, F. Variation of plant growth and macronutrient uptake in grafted tomatoes and eggplants on three different rootstocks. *Eur. J. Hortic. Sci.* **2006**, *71*, 97–101.
30. Rouphael, Y.; Cardarelli, M.; Colla, G.; Rea, E. Yield, mineral composition, water relations, and water use efficiency of grafted mini-watermelon plants under deficit irrigation. *HortScience* **2008**, *43*, 730–736. [CrossRef]
31. Schwarz, D.; Öztekin, G.B.; Tüzel, Y.; Brückner, B.; Krumbein, A. Rootstocks can enhance tomato growth and quality characteristics at low potassium supply. *Sci. Hortic.* **2013**, *149*, 70–79. [CrossRef]
32. Wang, Y.S.; Liu, F.L.; Andersen, M.N.; Jensen, C.R. Improved plant nitrogen nutrition contributes to higher water use efficiency in tomatoes under alternate partial root-zone irrigation. *Funct. Plant Biol.* **2010**, *37*, 175–182. [CrossRef]
33. Sánchez-Rodríguez, E.; Leyva, R.; Constán-Aguilar, C.; Romero, L.; Ruiz, J.M. How does grafting affect the ionome of cherry tomato plants under water stress? *Soil Sci. Plant Nutr.* **2014**, *60*, 145–155. [CrossRef]
34. Sun, Y.; Feng, H.; Liu, F. Comparative effect of partial root-zone drying and deficit irrigation on incidence of blossom-end rot in tomato under varied calcium rates. *J. Exp. Bot.* **2013**, *64*, 2107–2116. [CrossRef] [PubMed]
35. Gransee, A.; Führs, H. Magnesium mobility in soils as a challenge for soil and plant analysis, magnesium fertilization and root uptake under adverse growth conditions. *Plant Soil* **2013**, *368*, 5–21. [CrossRef]
36. Maguire, M.E.; Cowan, J.A. Magnesium chemistry and biochemistry. *Biometals* **2002**, *15*, 203–210. [CrossRef]
37. Verbruggen, N.; Hermans, C. Physiological and molecular responses to magnesium nutritional imbalance in plants. *Plant Soil* **2013**, *368*, 87–99. [CrossRef]

11

Arsenic Accumulation in Grafted Melon Plants: Role of Rootstock in Modulating Root-To-Shoot Translocation and Physiological Response

Enrica Allevato [1,†], Rosario Paolo Mauro [2,†], Silvia Rita Stazi [3,*], Rosita Marabottini [1], Cherubino Leonardi [2], Anita Ierna [4] and Francesco Giuffrida [2]

1 DIBAF, University of Tuscia, Via S.C. De Lellis snc, 01100 Viterbo, Italy; eallevato@unitus.it (E.A.); marabottini@unitus.it (R.M.)
2 Di3A, University of Catania, Via Valdisavoia 5, 95123 Catania, Italy; rosario.mauro@unict.it (R.P.M.); cherubino.leonardi@unict.it (C.L.); francesco.giuffrida@unict.it (F.G.)
3 DSCF, University of Ferrara, Via Borsari 46, 44121 Ferrara, Italy
4 CNR-IBE, Via P. Gaifami 18, 95126 Catania, Italy; anita.ierna@cnr.it
* Correspondence: silviarita.stazi@unife.it
† The authors share the first Authorship.

Abstract: The bio-agronomical response, along with the arsenic (As) translocation and partitioning were investigated in self-grafted melon "Proteo", or grafted onto three interspecific ("RS841", "Shintoza", and "Strong Tosa") and two intraspecific hybrids ("Dinero" and "Magnus"). Plants were grown in a soilless system and exposed to two As concentrations in the nutrient solution (0.002 and 3.80 mg L^{-1}, referred to as As− and As+) for 30 days. The As+ treatment lowered the aboveground dry biomass (−8%, on average), but the grafting combinations differed in terms of photosynthetic response. As regards the metalloid absorption, the rootstocks revealed a different tendency to uptake As into the root, where its concentration varied from 1633.57 to 369.10 mg kg^{-1} DW in "Magnus" and "RS841", respectively. The high bioaccumulation factors in root (ranging from 97.13 to 429.89) and the low translocation factors in shoot (from 0.015 to 0.071) and pulp (from 0.002 to 0.008) under As+, showed a high As mobility in the substrate–plant system, and a lower mobility inside the plants. This tendency was higher in the intraspecific rootstocks. Nonetheless, the interspecific "RS841" proved to be the best rootstock in maximizing yield and minimizing, at the same time, the As concentration into the fruit.

Keywords: *Cucumis melo* L.; arsenic; grafting; translocation; bioaccumulation

---

1. Introduction

Melon (*Cucumis melo* L.) is one of the major Cucurbit species, playing an important role in irrigated farmlands of the Mediterranean area [1]. In these regions, the water intended for irrigation could be contaminated with heavy metals as result of weathering of soil minerals and human activities [2]. Irrigation with contaminated water can affect food quality and safety for the presence of metalloids and/or heavy metals harmful for human health [3,4]. Among these, arsenic (As) is a persistent, non-degradable metalloid widely present in the environment both for geogenic or anthropogenic reasons [5]. Most As compounds are odorless, tasteless, and water-soluble, creating a serious health risk because of their carcinogenic potential [6]. This metalloid exists in plant organs both as inorganic and organometallic species, whose concentration and oxidation states are dependent on the type and amounts of sorbents in the substrate, pH, redox potential (Eh), and soil microbial activity [7,8].

Arsenic is non-essential and generally toxic to plants. At high concentrations (depending on the species) it interferes with multiple metabolic processes, leading to growth and yield disturbance, and even plant death [9,10]. The As(III) root uptake (i.e., the dominating form in anaerobic root environments) occurs by passive transmembrane transport involving members of the nodulin 26-like intrinsic protein family of plant aquaporins [11]. As(V), which dominates in aerobic root environments, enters plant roots via phosphate (Pi) transporters, as the oxyanion chemical structure of As(V) is analogous to Pi [12]. As a consequence, As(V) compete with Pi absorption and P-dependent metabolic processes during ATP synthesis, with subsequent disturbance of major biological functions [13,14].

The efficiency of As translocation from roots to shoots via xylem loading affects the plants' As tolerance and their proneness to accumulate this metalloid into the edible fraction, so posing potential risks to human health [14].

In horticultural systems, vegetable grafting is a multipurpose technique improving crops performances and product quality under both optimal and suboptimal growth conditions [15–17]. It has been proposed as a means to reduce the heavy metals uptake and translocation to the shoots and the edible parts [18,19], although the mechanisms underlying such impediment are still unclear [20]. Thus, investigating the use of melon rootstocks could be useful for understanding the behavior of different grafting combinations, with the view to improve crop performances and product safety in As-polluted areas.

Considering the above, the objective of this research was to study the effects of different rootstock genotypes on As uptake, accumulation, and partitioning, as well as on agronomical and physiological response of melon plants subjected to a high concentration of the metalloid in the root environment. To this end, the present experiment was performed to investigate: (i) If As concentration in the nutrient solution influences its uptake and translocation in melon plants; (ii) if and how the different rootstocks can mitigate the As accumulation within plant organs; (iii) the possible role of different rootstock genotypes in modulating the bio-agronomical response of melon plants to the elevated As concentration in the root zone.

2. Materials and Methods

2.1. Experimental Site, Plant Material, and Management Practices

The experiment was conducted in 2014, in a greenhouse situated in the coastal area of Eastern Sicily (37°24′26″ N, 15°03′37″ E, 6 m a.s.l.). The local climate is semi-arid/Mediterranean, with mild and wet winters, and hot, dry summers. A 1000 m^2, east–west-oriented, multi-aisle greenhouse was used, having a steel tubular structure and covered with polycarbonate slabs. Mean air temperature, relative humidity and global radiation inside the greenhouse (two sets of sensors in the center of each experimental plot) were recorded on a data logger (CR10 X; Campbell Scientific Ltd., Loughborough, UK). Melon cv. "Proteo" F_1 (Syngenta Seed, Basel, Switzerland) belonging to the Reticulatus group was used as scion. Five F_1 rootstock genotypes were included in the study, of which 2 were intraspecific, namely "Dinero" (Syngenta Seed) and "Magnus" (Agris), and 3 were interspecific (*Cucurbita maxima* Duchesne × *C. moschata* Duchesne ex Poir.), namely "RS841" (DeRuiter Seeds, Oxnard, CA, USA) "Shintoza" (Fenix, Belpasso, Italy) and 'Strong Tosa' (Syngenta Seed). Self-grafted "Proteo" was used as control. Splice-grafted plantlets were used, whereas plastic clips were applied to secure the creation of the graft union. Plantlets were obtained from a specialized nursery and transplanted at the stage of 3 true leaves on April 22, in 5 L pots filled with perlite (3–5 mm). Pots were placed in troughs (5 per main plot) of an open soilless system, placed at 5 cm from the soil surface and with a distance of 50 cm between pots and 100 cm between troughs, obtaining a plant density of 2 plants m^{-2}. During the trial, the crop was fertigated with a nutrient solution having the following composition, including the starting well water (mmol L^{-1}): 11.2 NO_3^-, 0.3 NH_4^+, 1.3 $H_2PO_4^-$, 6.6 K^+, 0.9 SO_4^{2-}, 3.4 Ca^{2+}, 2.5 Mg^{2+}. The concentration of microelements (μmol L^{-1}) was: 15 Fe^{3+}, 10 Mn^{2+}, 0.75 Cu^{2+}, 5 Zn^{2+}, 30 B^{3+}, 0.5 Mo^{6+}. The pH was maintained at 5.9 by adding H_2SO_4 (95% concentration, 1.83 kg L^{-1}). Bumblebees were introduced into the greenhouse to maximize pollination. Stems were left to grow horizontally, whereas plants were pruned by removing the lateral shoots. Only 1 lateral ramification per plant was left, bearing 2 fruits. In order to evaluate the effects of studied factors at the same date, the trial was stopped when the first fruit was fully ripe (i.e., 45 days after transplant). Fifteen days after transplanting, the nutrient solution was differentiated to obtain 2 As concentrations, a control solution (0.002 mg L^{-1} As, which was the concentration in the starting fertigation solution, hereafter As−) and an As-enriched one (3.8 mg L^{-1} As, hereafter As+) obtained by adding sodium arsenate (Na_2HAsO_4 $7H_2O$, 24% As content). The As+ treatment was chosen to simulate the working condition of a soil having 55 ppm of As, ~7.0% of which was bioavailable (a common situation in As contaminated soils). The nutrient solutions were supplied using a drip irrigation system with one emitter per plant (4 L h^{-1}). The amount of nutrient solution supplied at each irrigation was quantified on a weekly basis, according to the volume of the substrate exploited by the roots and the corresponding water contained in the substrate at the intervals from −10 to −50 hPa of matrix potential (12 mL 100 mL^{-1}).

The experiment was arranged in a randomized split-plot design with three replicates, assigning the As concentration of the nutrient solution to the main plots, and the rootstock combination to the subplots. The overall experimental area inside the greenhouse was 450 m^2 (15.0 × 30.0 m), including 900 plants (324, excluding border plants), divided into 36 net experimental units (2 As levels × 6 grafting combinations × 3 replicates) each containing 9 plants.

2.2. *Leaf Relative Chlorophyll (Chl) Content and Gas Exchange Measurements*

Forty-five days after transplanting plants were checked for leaf relative Chl content, through a portable Chl meter (SPAD 502; Minolta Camera, Osaka, Japan). Before measurements, the instrument was calibrated according to manufacturer's instructions. All readings were taken from the adaxial side of the tallest fully expanded leaf. To minimize possible interactions with either plant water status and natural irradiance level [21,22], measurements were made in the morning, starting at 08:00 h (local solar time). Instantaneous leaf photosynthetic rate (A_N), stomatal conductance (gs), and leaf transpiration rate (E) were also measured from 11:00 to 13:00 inside a 6.25 cm^2 leaf chamber of a portable photosynthesis system (LCi; ADC BioScientific Ltd., Hoddesdon, UK). Photosynthetic water use efficiency (WUE) was calculated as the ratio A_N/E [23]. During measurements, leaf temperature was 27.4 ± 2 °C, while mean photosynthetic photon flux density was in the range of 1500 ± 100 µmol photons m^{-2} s^{-1}. Duplicate measurements were taken from four plants per sub-plot.

2.3. *Plant Growth and Development Measurements*

On the same date of physiological measurements, all the plants within each replicate were harvested and their main fractions (leaves, stem and fruits, regardless of the ripening stage) weighed separately. The number of leaves per plant (L_N) was determined, while plant leaf area (L_A) was measured using an Image Analysis System (DeltaT Devices Ltd., Cambridge, UK), then subsamples of raw materials were kept in a thermo-ventilated oven at 70 °C (Binder, Milan, Italy) until constant weight, in order to determine their dry weight (DW). From the original data frame, the leaf area ratio (LAR, the ratio between the area and total plant biomass) and leaf weight ratio (LWR, the dry weight of leaves to whole plant dry weight) were calculated.

2.4. *Arsenic and Phosphorous Determination in Plant Tissues*

Roots, shoots, and fully ripe fruits (1 per plant) were analyzed to determine the quantity of the total As and phosphorus (P). To this end, about 200 mg of samples were subjected to acid digestions and to As and P determinations, which were performed according to Stazi et al. [19]. The reagents were super pure for trace analysis. The accuracy of the measurements was assessed using SRM 1573a as standard reference materials trace metals. Total As quantification were performed using an inductively-coupled plasma optical emission spectrometer (ICP-OES) with an axially viewed configuration (8000 DV, PerkinElmer, Shelton, CT, USA) equipped with an ultrasonic nebulizer. The As detection limit for employed technique was 0.1 µg L^{-1}. With the aim of understanding the metabolic pathway followed by this element once absorbed by the plant, we measured the amount of trivalent and pentavalent

As. Inorganic As species were extracted without the addition of hydrogen peroxide according to Rintala et al. [24] with some modifications. In brief, 200 mg of roots were homogenized and digested with 10 mL of a mixture of HNO_3 (1%, v/v) and left to react overnight. The samples were subjected to microwave-assisted extraction according to the following program—3 min from 25 to 55 °C (step 1), 10 min at 55 °C (step 2), 2 min from 55 to 75 °C (step 3), 10 min at 75 °C (step 4), 2 min from 75 to 95 °C (step 5) and 30 min at 95 °C (step 6). Samples were then quantitatively transferred into tubes and centrifuged for 15 min at 10,000 rpm at 4 °C. The supernatant was filtered with a 0.22 μm PVDF filter. The concentration of As(III) was determined directly with an ICP-OES equipped with a hydride generation system. The total concentration of inorganic arsenic species [As(III)+As(V)] were obtained after reducing As(V) to As(III) through 5.0% (w/v) ascorbic acid and potassium iodide in hydrochloric acid, and the content of As(V) was calculated from the difference between total As concentration and that of As(III) [25].

2.5. *Bioaccumulation and Translocation Factors*

Arsenic bioaccumulation factor (BAF) (i.e., the ability of a plant to accumulate this element from water) was calculated on a DW basis, as the ratio among its concentration in root (BAF_{root}), shoot (BAF_{shoot}), and pulp (BAF_{pulp}) and the corresponding concentration in the nutrient solution. The root-to-plant translocation factor (TF) defines the movement and distribution of As from roots to the aerial part of the plant. The TF was calculated on a DW basis, as the ratio between the As concentration in shoot (TF_{shoot}) and fruit pulp (TF_{pulp}), and the corresponding concentration in roots at the end of the experiment [26–28].

2.6. *Statistical Procedures*

All data were subjected to Shapiro–Wilk and Levene's test, in order to check for normal distribution and homoscedasticity, respectively, then to a 'As concentration × rootstock' analysis of variance (ANOVA), according to the split-plot experimental layout adopted in the greenhouse. Percentage data were Bliss' transformed before the ANOVA (untransformed data are reported and discussed), while multiple means comparison was performed through Fisher's protected least significant difference (LSD) test ($p = 0.05$). The As concentrations in root, shoot, and pulp were subjected at principal component analysis (PCA) to verify the interaction between the different factors able to synthetize the considered variables. All calculations were performed using the Excel (Microsoft Corporation, Redmond, WA, USA) and JMP 11.0 statistical software package (SAS Institute, Cary, NC, USA).

2.7. *Microclimate Conditions Inside the Greenhouse*

During the experiment, the average mean temperature progressively increased from 18.6 to 26.4 °C (on day 12 and 36 after transplanting, respectively), while relative humidity showed an opposite trend, as it passed from 63.7% to 40.5% (on day 11 and 33 after transplanting, respectively). Solar radiation amply paralleled the trend of mean temperature, as it progressively increased passing from 10.45 to 15.02 MJ m^{-2} (on day 11 and 36 after transplanting, respectively). As regards the spatial variability among main plots, the differences in terms of mean temperature, relative humidity, and solar radiation never exceeded 0.2 °C, 2.1%, and 0.8 MJ m^{-2}, respectively.

3. Results

3.1. *Aboveground Plant Biomass and Partitioning*

Whole plant biomass was affected by the main factors and their interaction (Table 1). In As− treatment this variable was higher in "Proteo" grafted onto "Strong Tosa" and "RS841" (309 g $plant^{-1}$, on average), whereas in response to As+ treatment it significantly decreased in "Proteo" grafted onto "Dinero" (from 251.7 to 198.7 g $plant^{-1}$, −21%) and "Strong Tosa" (from 317.0 to 239.3 g $plant^{-1}$, −25%) (Table 1). Among the plant components, stem biomass resulted 30 and 24 g $plant^{-1}$ in As− and As+, respectively (−20%) (Table 1). Passing from As− to As+, there was a general decrease of the leaf biomass too. This response was more marked in "Proteo" grafted onto "Dinero" (from 66.4 to 42.0 g plant^{-1}, −37%), 'Magnus' (from 56.7 to 44.5 g plant^{-1}, −22%), 'RS841' (from 66.3 to 52.1 g plant^{-1}, −21%), and "Strong Tosa" (from 58.0 to 44.4 g plant^{-1}, −23%), whereas no significant difference was recorded when "Proteo" was self-grafted or grafted onto "Shintoza" (Table 1). When grown in the As+, fruit biomass showed a significant variation in "Proteo" grafted onto 'Magnus' and "Strong Tosa", where it varied by +33% (from 119.9 to 159.1 g plant^{-1}) and −26% (from 231.1 to 170.9 g plant^{-1}), respectively (Table 1).

Table 1. Aboveground biomass production and partitioning (g dry weight plant^{-1}) in melon plant as affected by As concentration in the nutrient solution and rootstock. Different letters within main factors indicate significance at Fisher's protected least significant difference (LSD) test (p = 0.05). NS—not significant.

Variable		Rootstock						Mean	$LSD_{interaction}$ (p = 0.05)
		"Proteo" (Control)	"Dinero"	"Magnus"	"RS841"	"Shintoza"	"Strong Tosa"		
Plant	As−	235.0	251.7	207.0	300.3	247.7	317.0	**259.8 a**	45.1
	As+	220.0	198.7	226.3	278.7	268.3	239.3	**238.6 b**	
	Mean	**227 bc**	**225 c**	**217 c**	**289 a**	**258 ab**	**278 a**		
Stem	As−	29.7	31.2	31.4	29.6	30.8	28.0	**30.1 a**	NS
	As+	22.2	21.9	22.7	25.0	28.4	23.39	**24.0 b**	
	Mean	**25.9 a**	**26.5 a**	**27.1 a**	**27.3 a**	**29.6 a**	**25.9 a**		
Leaf	As−	56.8	66.4	56.7	66.3	55.8	58.0	**60.0 a**	10.3
	As+	47.6	42.0	44.5	52.1	58.0	44.4	**48.1 b**	
	Mean	**52.2 a**	**54.2 a**	**50.6 a**	**59.2 a**	**56.9 a**	**51.2 a**		
Fruit	As−	148.8	154.0	119.9	204.6	161.1	231.1	**169.9 a**	37.3
	As+	150.3	134.6	159.1	201.5	182.0	170.9	**166.4 a**	
	Mean	**149.6 c**	**144.3 c**	**139.5 c**	**203.0 a**	**171.5 b**	**201.0 a**		

3.2. Leaf Growth Variables

All leaf growth variables were significantly affected by the 'As concentration × rootstock' interaction (Table 2). With the As increase in the nutrient solution, both L_N and L_A showed the strongest reduction in the grafting combinations "Proteo"/"Dinero" (−43 leaves plant^{-1} and −34.4 dm^2 plant^{-1}, respectively), "Proteo"/"Magnus" (−42 leaves plant^{-1} and −27.2 dm^2 plant^{-1}), and "Proteo"/'Strong Tosa' (−27 leaves plant^{-1} and −19.3 dm^2 plant^{-1}) (Table 2). Similarly, LAR and LWR showed a significant reduction passing from As− to As+ solution, with the strongest drops recorded in the grafting combinations "Proteo"/"Dinero" (−7.6 cm^2 g^{-1} DW and −0.054 g g^{-1} DW, respectively) and "Proteo"/"Magnus" (−16 cm^2 g^{-1} DW and −0.073 g g^{-1} DW) (Table 2).

Table 2. Leaf growth variables in melon plant as affected by As concentration in the nutrient solution and rootstock. Different letters within main factors indicate significance at Fisher's LSD test (p = 0.05).

Variable		Rootstock						Mean	$LSD_{interaction}$ (p = 0.05)
		"Proteo" (Control)	"Dinero"	"Magnus"	"RS841"	"Shintoza"	"Strong Tosa"		
L_N (n. plant^{-1})	As−	163	178	187	158	157	176	**170 a**	22
	As+	159	135	145	140	162	149	**148 b**	
	Mean	**161 a**	**156 a**	**166 a**	**149 a**	**160 a**	**162 a**		
L_A (dm^2 plant^{-1})	As−	74.1	91.5	86.9	75.9	83.6	86.3	**83.1 a**	14.5
	As+	66.9	57.1	59.7	63.3	83.3	67.0	**66.2 b**	
	Mean	**70.5 a**	**74.3 a**	**73.3 a**	**69.6 a**	**83.4 a**	**76.7 a**		
LAR (cm^2 g^{-1} DW)	As−	31.7	36.4	42.4	25.3	34.2	27.2	**32.9 a**	7.1
	As+	30.7	28.8	26.4	22.7	31.5	28.3	**28.1 b**	
	Mean	**31.2 ab**	**32.6 ab**	**34.4 a**	**24.0 c**	**32.8 a**	**27.7 bc**		
LWR (g g^{-1} DW)	As−	0.243	0.267	0.270	0.223	0.230	0.183	**0.236 a**	0.034
	As+	0.217	0.213	0.197	0.190	0.217	0.187	**0.203 b**	
	Mean	**0.230 ab**	**0.240 a**	**0.233 a**	**0.207 ab**	**0.223 bc**	**0.185 c**		

LN—number of leaves; LA—leaf area; LAR—leaf area ratio; LWR—leaf weight ratio.

3.3. *Leaf Relative Chl Content and Gas Exchanges*

Leaf relative Chl content did not differ in relation to rootstock, while in all grafting combinations it increased by 7% (from 55.3 to 59.3 SPAD units) in the As+ solution (Table 3). The highest A_N values in As− solution were recorded in "Proteo" grafted onto "Magnus", "RS841", and 'Strong Tosa' (27.4 μmol CO_2 m^{-2} s^{-1}, on average). Nevertheless under As+ treatment, the grafting combination "Proteo"/"RS841" showed a significant A_N reduction (from 27.7 to 22.7 μmol CO_2 m^{-2} s^{-1}, −18%), whereas an opposite trend was recorded in self-grafted "Proteo" (from 25.3 to 27.7 μmol CO_2 m^{-2} s^{-1}, +10%) (Table 3). Passing from As− to As+, gs significantly varied in the grafting combinations "Proteo"/"Dinero" (from 273 to 385 μmol CO_2 m^{-2} s^{-1}, +41%), "Proteo"/"Shintoza" (from 190 to 245 μmol CO_2 m^{-2} s^{-1}, +29%), and "Proteo"/"Magnus" (from 275 to 207 μmol CO_2 m^{-2} s^{-1}, −25%), with the control showing stable gs values across nutrient solutions (Table 3). "RS841" rootstock maximized the WUE values in As−, but showed the highest decrease in response to As+ (being reduced from 3.5 to 3.0 μmol CO_2 μmol^{-1} H_2O m^{-2} s^{-1}, −12%), while an opposite trend was recorded in self-grafted "Proteo" (from 3.0 to 3.4 μmol CO_2 μmol^{-1} H_2O m^{-2} s^{-1}, +14%). A similar response was recorded when "Proteo" was grafted onto 'Strong Tosa' (from 2.8 to 3.3 μmol CO_2 μmol^{-1} H_2O m^{-2} s^{-1}, +17%) and "Magnus" (from 2.8 to 3.6 μmol CO_2 μmol^{-1} H_2O m^{-2} s^{-1}, +26%) (Table 3).

Table 3. Leaf relative Chl content and gas exchange variables in melon plant as affected by As concentration in the nutrient solution and rootstock. Different letters within main factors indicate significance at Fisher's LSD test (p = 0.05). NS—not significant.

Variable		Rootstock						Mean	$LSD_{interaction}$ (p = 0.05)
		"Proteo" (Control)	"Dinero"	"Magnus"	"RS841"	"Shintoza"	"Strong Tosa"		
SPAD	As−	55.7	56.4	54.7	56.1	52.9	55.9	**55.3 b**	NS
	As+	57.5	59.3	58.1	61.7	58.6	60.8	**59.3 a**	
	Mean	**56.6 a**	**57.8 a**	**56.4 a**	**58.9 a**	**55.7 a**	**58.4 a**		
A_N (μmol CO_2 m^{-2} s^{-1})	As−	25.3	25.0	26.7	27.7	25.1	27.7	**26.2 a**	2.2
	As+	27.7	23.2	25.6	22.7	25.4	29.1	**25.6 a**	
	Mean	**26.5 b**	**24.1 c**	**26.1 b**	**25.2 bc**	**25.2 bc**	**28.4 a**		
Gs (μmol CO_2 m^{-2} s^{-1})	As−	240	273	275	255	190	330	**261 a**	43
	As+	247	385	207	230	245	300	**269 a**	
	Mean	**243 b**	**329 a**	**241 b**	**243 b**	**218 b**	**315 a**		
WUE (μmol CO_2 μmol^{-1} H_2O m^{-2} s^{-1})	As−	3.0	3.0	2.8	3.5	3.0	2.8	**3.0 b**	0.3
	As+	3.4	2.8	3.6	3.0	3.2	3.3	**3.2 a**	
	Mean	**3.2 a**	**2.9 b**	**3.2 a**	**3.3 a**	**3.1 ab**	**3.1 ab**		

A_N—net assimilation rate; gs—stomatal conductance; WUE—water use efficiency.

3.4. *Arsenic Accumulation in Plant Fractions*

Arsenic accumulation in plant fractions was significantly affected by the main factors and their interaction (Table 4). In As− treatment, the highest As_{root} was recorded in the rootstocks "Magnus" and "Proteo" (11.75 mg kg DW^{-1}, on average), while the lowest one in "RS841" (6.26 mg kg DW^{-1}). In the As+ treatment, As_{root} increased in all the tested rootstocks, with the highest concentration recorded in "Magnus" (1633.57 mg kg^{-1} DW, +131 fold than control), while the lowest one in "RS841" (369.10 mg kg^{-1} DW, +58 fold) (Table 4). The As+ treatment promoted As_{shoot} in a rootstock-dependent way too, since passing from As− to As+, "RS841" showed the highest As_{shoot} increase (by 25.61 mg kg^{-1} DW, +43 fold than control), whereas 'Strong Tosa' showed the least one (by 21.75 mg kg^{-1} DW, +36 fold) (Table 4). By increasing the As concentration in the nutrient solution, As_{pulp} increased in all the grafting combinations, with the highest rise recorded in "Proteo" grafted onto "Shintoza" and "Dinero" (by 3.09 mg kg^{-1} DW, on average, equal to +14 fold than control) and the least one recorded in self-grafted "Proteo" (by 1.86 mg kg^{-1} DW, +2 fold) (Table 4).

Table 4. Total arsenic accumulation (mg kg^{-1} DW) in different organs of melon plant as affected by As concentration in the nutrient solution and rootstock. Different letters within main factors indicate significance at Fisher's LSD test (p = 0.05).

Variable		Rootstock						Mean	$LSD_{interaction}$ (p = 0.05)
		"Proteo" (Control)	"Dinero"	"Magnus"	"RS841"	"Shintoza"	"Strong Tosa"		
As_{root}	As−	11.84	7.10	11.65	6.26	8.09	8.35	**8.88 b**	77.33
	As+	1031.53	1090.73	1633.57	369.10	732.75	805.88	**943.93 a**	
	Mean	**521.68 b**	**548.91 b**	**822.61 a**	**187.68 d**	**370.42 c**	**407.12 c**		
As_{shoot}	As−	0.71	0.86	0.77	0.59	0.47	0.60	**0.67 b**	1.73
	As+	23.71	25.50	23.81	26.20	23.90	22.35	**24.24 a**	
	Mean	**12.21 ab**	**13.18 a**	**12.29 ab**	**13.39 a**	**12.18 ab**	**11.48 b**		
As_{pulp}	As−	0.86	0.14	0.46	0.29	0.30	0.60	**0.44 b**	0.25
	As+	2.72	3.16	2.99	2.98	3.46	3.19	**3.08 a**	
	Mean	**1.79 ab**	**1.65 b**	**1.72 ab**	**1.64 b**	**1.88 a**	**1.89 a**		

3.5. Arsenic Speciation in Root

Arsenic speciation in root was significantly affected by 'As concentration × rootstock' interaction (Table 5). In As− treatment, $As_{inorganic}$ in root accounted from 40.1% ("Shintoza") to 60.1% ("Proteo"); however, in the As+ solution, this variable significantly decreased in the intraspecific rootstocks, namely "Proteo", "Dinero", and "Magnus" (−14.57%, on average), while it increased in the interspecific hybrids "RS841", "Shintoza", and 'Strong Tosa' (+34.93%, on average) (Table 5). In As− treatment, $AsIII_{root}$ widely differed among rootstocks, ranging from 0.01 to 1.94 mg kg^{-1} DW in "RS841" and "Proteo", respectively. In the As+ treatment it showed the highest rise in 'Strong Tosa' (+577.24 mg kg^{-1} DW; i.e., +3207 fold than control) and "Shintoza" (+287.39 mg kg^{-1} DW; i.e., +3193 fold), and the least one in "RS841" (+82.25 mg kg^{-1} DW; i.e., +8225 fold) (Table 5). In As− treatment, AsV_{root} showed the highest value in "Proteo" (5.17 mg kg^{-1} DW) and the least one in "Dinero" and "RS841" (2.88 mg kg^{-1} DW, on average) (Table 5). When plants were grown in the As+ solution, such variable showed the highest increase in "Magnus" (by 345.94 mg kg^{-1} DW; i.e., +80 fold than control), followed by "Proteo" and "Dinero" (by 255.88 mg kg^{-1} DW, on average; i.e., +64 fold), and the least one in 'Strong Tosa' (by 41.85 mg kg^{-1} DW; i.e., +14 fold) (Table 5).

Table 5. Inorganic As (percentage of total), As(III), and As(V) (mg kg^{-1} DW) in root of melon plant as affected by As concentration in the nutrient solution and rootstock. Different letters within main factors indicate significance at Fisher's LSD test (p = 0.05).

Variable		Rootstock						Mean	$LSD_{interaction}$ (p = 0.05)
		"Proteo" (Control)	"Dinero"	"Magnus"	"RS841"	"Shintoza"	"Strong Tosa"		
$As_{inorganic}$	As−	60.1	54.1	48.0	46.0	40.1	38.2	**47.8 a**	12.7
	As+	48.6	38.9	31.0	80.7	71.2	77.2	**57.9 b**	
	Mean	**54.3 bc**	**46.5 cd**	**39.5 d**	**63.4 a**	**55.6 ab**	**57.7 ab**		
$AsIII_{root}$	As−	1.94	0.96	1.25	0.01	0.09	0.18	**0.74 b**	71.89
	As+	235.98	168.31	150.07	82.26	287.48	577.42	**250.25 a**	
	Mean	**118.96 bc**	**84.64 cd**	**75.66 cd**	**41.14 d**	**143.79 b**	**288.80 a**		
AsV_{root}	As−	5.17	2.88	4.35	2.87	3.15	3.01	**3.57 b**	102.15
	As+	264.86	254.95	350.29	215.72	233.87	44.86	**227.42 a**	
	Mean	**135.02 a**	**128.91 a**	**177.32 a**	**109.30 a**	**118.51 a**	**23.94 b**		

3.6. Arsenic Bioaccumulation and Translocation Factors

All the bioaccumulation and translocation factors proved to be higher under the control growing conditions, displaying a sharp decrease in response to the As+ treatment (Table 6). Shifting from As− to As+, the strongest BAF_{root} reduction was found in "Proteo" and "Magnus" (−5646.22 and −5394.68, respectively, corresponding to −95% and −93%), while the least one in "Dinero" and "RS841" (−3260.82 and −3033.38, respectively, corresponding to −92% and −97%) (Table 6). Differently, "Dinero" caused the sharper BAF_{shoot} decrease in response to As+ treatment (−422.82, i.e., −98%), followed by "Magnus" (−377.37, i.e., −98%), while "Shintoza" generated the smallest variation (−229.6, i.e.,

−97%) (Table 6). With reference to BAF_{pulp}, "Proteo" and "Strong Tosa" generated the strongest drop in response to As+ (by −426.92 and −298.47, respectively, corresponding to −99.8% and −99.7%), whereas a less marked decrease was recorded when "Proteo" was grafted onto "Shintoza" (−150.45, i.e., −99.4%), "RS841" (−145.92; i.e., −99.5%) and "Dinero" (−68.05, i.e.; −98.8%) (Table 6). The increase in As concentration in the root environment caused the highest TF_{shoot} reductions in "Proteo" grafted onto "Dinero" (−0.097; i.e., −80.2%) and "Magnus" (−0.051, i.e., −77.3%), while the least ones in the grafting combinations "Proteo"/"Shintoza" (−0.025; i.e., −43.1%) and "Proteo"/"RS841" (−0.024; i.e., −25.3%) (Table 6). "Strong Tosa", "Magnus" and "Shintoza" proved the highest TF_{pulp} reduction in response to the As stress (−0.069, −0.038, and −0.032, respectively, corresponding to −94.5%, −95.0%, and −86.5%), whereas "Dinero" showed the least one (−0.007; i.e., −84.2%) (Table 6).

Table 6. Arsenic bioaccumulation and translocation factors (adimensional) in shoot and pulp of melon plants as affected by As concentration in the nutrient solution and rootstock. Different letters within main factors indicate significance at Fisher's LSD test (p = 0.05).

Variable		Rootstock						Mean	$LSD_{interaction}$ (p = 0.05)
		"Proteo (Control)	**"Dinero"**	**"Magnus"**	**"RS841"**	**"Shintoza"**	**"Strong Tosa"**		
BAF_{root}	As−	5917.67	3547.86	5824.57	3130.51	4045.17	4176.02	**4440.30 a**	370.18
	As+	271.45	287.04	429.89	97.13	192.83	212.07	**248.40 b**	
	Mean	**3094.56 a**	**1917.45 c**	**3127.23 a**	**1613.82 d**	**2119.00 bc**	**2194.05 b**		
BAF_{shoot}	As−	355.46	429.53	383.63	295.42	235.89	301.22	**333.52 a**	50.54
	As+	6.24	6.71	6.26	6.89	6.29	5.88	**6.38 b**	
	Mean	**180.85 bc**	**218.12 a**	**194.95 ab**	**151.16 cd**	**121.09 d**	**153.55**		
BAF_{pulp}	As−	427.64	68.88	229.97	146.70	151.36	299.31	**220.64 a**	15.50
	As+	0.72	0.83	0.79	0.78	0.91	0.84	**0.81 b**	
	Mean	**214.18 a**	**34.86 e**	**115.38 c**	**73.74 d**	**76.13 d**	**150.07 b**		
TF_{shoot}	As−	0.060	0.121	0.066	0.095	0.058	0.074	**0.079 a**	0.014
	As+	0.023	0.024	0.015	0.071	0.033	0.028	**0.032 b**	
	Mean	**0.042 cd**	**0.072 b**	**0.040 d**	**0.083 a**	**0.046 cd**	**0.051 c**		
TF_{pulp}	As−	0.073	0.019	0.040	0.048	0.037	0.073	**0.048 a**	0.007
	As+	0.003	0.003	0.002	0.008	0.005	0.004	**0.004 b**	
	Mean	**0.038 a**	**0.011 d**	**0.021 c**	**0.028 b**	**0.021 c**	**0.039 a**		

3.7. Phosphorus Accumulation in Plant Fractions

Phosphorus accumulation in plant fractions was significantly affected by the main factors and their interaction (Table 7). Passing from As− to As+ treatment, P_{root} increased mostly in self-grafted "Proteo" (from 3527 to 31,390 mg kg^{-1} DW, +790%), followed by "Dinero" (from 3566 to 17,059 mg kg^{-1} DW, +378%), "Magnus" (from 14,362 to 22,938 mg kg^{-1} DW, +60%), and "Shintoza" (from 12,535 to 19,210 mg kg^{-1} DW, +53%) (Table 7). Conversely, the As+ treatment significantly reduced P_{shoot} as compared to the standard nutrient solution, with the highest drop recorded in "Proteo" grafted onto "Dinero" and "Magnus" (from 7001 to 4923 mg kg^{-1} DW, on average, −30%) (Table 7). Passing from As− to As+, P_{pulp} significantly increased in "Proteo" grafted onto "Dinero", "Magnus", and 'Strong Tosa' (by 4472, 1334, and 1203 mg kg^{-1} DW, respectively, corresponding to +758%, +35%, and +30%), whereas did not show any significant variation in the other graft combinations (Table 7).

Table 7. Phosphorus accumulation (mg kg^{-1} DW) in different organs of melon plants as affected by As concentration in the nutrient solution and rootstock. Different letters within main factors indicate significance at Fisher's LSD test ($p = 0.05$).

Variable		Rootstock						Mean	$LSD_{interaction}$ ($p = 0.05$)
		"Proteo" (Control)	"Dinero"	"Magnus"	"RS841"	"Shintoza"	"Strong Tosa"		
P_{root}	As−	3527	3566	14362	12959	12535	20875	**11304 b**	3111
	As+	31390	17059	22938	10379	19210	22343	**20553 a**	
	Mean	**17458 bc**	**10312 d**	**18650 b**	**11669 d**	**15872 c**	**21609 a**		
P_{shoot}	As−	7038	7131	6871	6039	6465	6538	**6681 a**	564
	As+	6434	5064	4782	5563	5861	5786	**5582 b**	
	Mean	**6736 a**	**6098 b**	**5827 b**	**5801 b**	**6163 b**	**6162 b**		
P_{pulp}	As−	4745	590	3827	4119	4110	3996	**3565 b**	521
	As+	5151	5062	5161	3797	4561	5199	**4822 a**	
	Mean	**4948 a**	**2826 d**	**4494 b**	**3958 c**	**4335 b**	**4597 ab**		

3.8. Principal Component Analysis

The first two principal components gave eigenvalues equal to 2.75 (PC1) and 0.21 (PC2), and together accounted for 98.9% of the total variance (Table 8). The As accumulation into the different plant fractions positively contributed to PC1, with correlation coefficient ranging from 0.556 to 0.589, while PC2 was strongly and positively correlated to As_{root} (0.830) and negatively correlated to As_{shoot} and As_{pulp} (−0.367 and −0.419, respectively). The resulting PCA scatterplot showed a clear separation among plants grown in As− (on the negative side of PC1) and those grown in As+ (on the right side of PC1). The first group clustered together, with no separation between different plant organs, while the second group showed a further separation in two sub-clusters, in the second and fourth quadrant of the centroid, respectively (Figure 1). The first sub-cluster grouped on the positive side of PC2 the grafting combinations (self-grafted "Proteo", "Proteo"/"Magnus", and "Proteo"/"Dinero") accumulating As mainly in root, while the second sub-cluster grouped those combinations ("Proteo"/'Strong Tosa', "Proteo"/"Shintoza", and "Proteo"/"RS841"), accumulating As in the aerial parts of the plant, namely shoot and pulp (Figure 1).

Table 8. Correlation coefficients for each trait with respect to the first two principal components, eigenvalues, and relative and cumulative proportions of the explained variance.

Trait	Principal Component Coefficients	
	First	**Second**
As_{root}	0.556	0.830
As_{shoot}	0.589	−0.367
As_{pulp}	0.586	−0.419
Eigenvalue	2.75	0.21
Explained Variability (%)	91.91	7.07

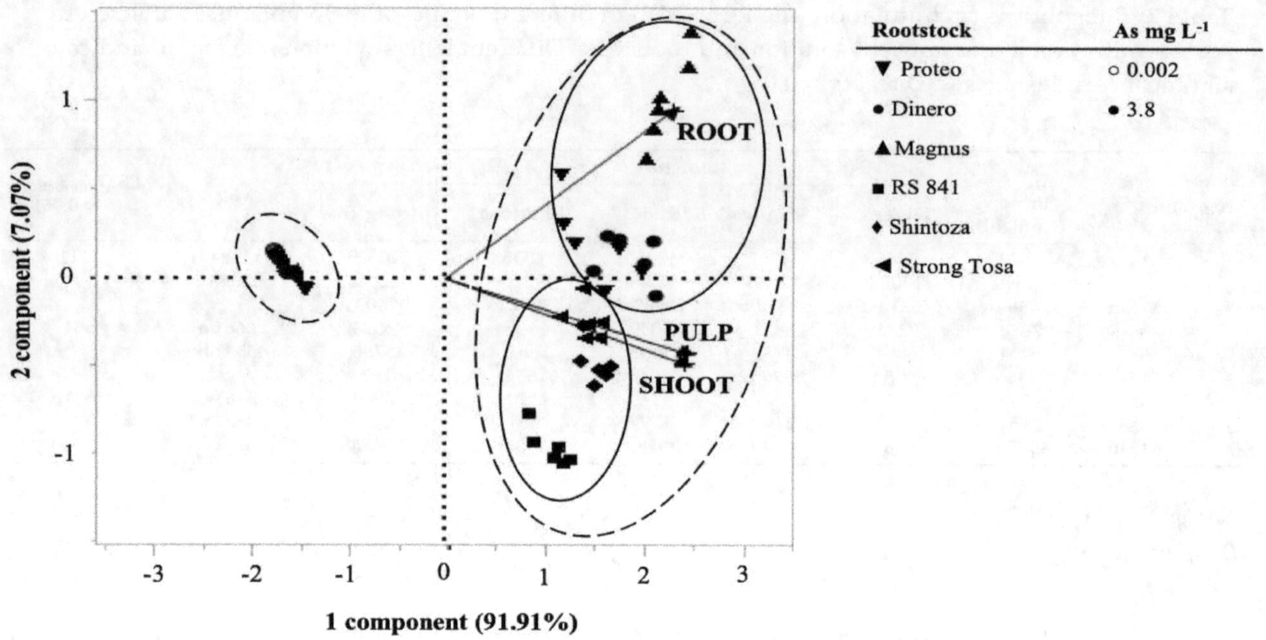

Figure 1. Principal Component Analysis scatter-plot based on total As concentration in root, shoot and pulp of melon plant "Proteo" as a function of the grafting combination and As concentration of the nutrient solution. Dotted lines group together treatments at different concentrations of As in the nutrient solution, while solid lines group together the grafting combinations.

4. Discussion

The environmental As contamination is a worldwide health threat due to the toxic and carcinogenic nature of this metalloid [29]. From an agricultural viewpoint, plants exposure to As− stress can cause morphological, physiological, and biochemical changes, leading to altered photosynthesis, stunted growth, reduced crop productivity, and worsened toxicological profile of the edible fractions [30]. In our experiment, the bioaccumulation factors values revealed that, even at a low As concentration in the nutrient solution, melon behaves as an hyperaccumulator plant, following the gradient root > shoot > fruit generally reported in literature [31]. However, when exposed to an As+ solution, all the grafting combinations acted to limit the As entrance into the plant, as can be inferred from the average reduction of BAF_{root} (−94.4%), BAF_{shoot} (−98.1%), and BAF_{pulp} (−99.6%). Nonetheless plants exposure to As stress highlighted evidences of systemic stress, consisting in a decrease of the whole aboveground dry biomass per plant, which mainly mirrored the reduction of both, stem and leaf dry biomass per plant. The reduced plant growth in response to As toxicity originates from complex alterations involving enzymes activity, induction of oxidative stresses, or altered nutrient uptake and balance into the plant [32,33]. In the present experiment melon plants subjected to As stress showed a decreased ability to maintain the normal growth equilibrium among plant fractions, by partially losing (~20%) the ability of photosynthates investment into leaf biomass, so indicating the main photosynthetic organs as an elective target of As-induced alterations. Leaf area, leaf area ratio, and leaf weight ratio were predominantly affected by the As toxicity, more than leaf number, indicating leaf cells proliferation and elongation as primarily affected by the As stress, more than leaf cells differentiation. According to Koyama and Kikuzawa [34] and Ropokis et al. [35], leaf area is a trait having a central role in determining the level of nutrient uptake, via the rate of whole plant photosynthesis and transpiration. In this sense, the decreased leaf area we recorded suggests the triggering of morpho-physiological modifications reducing the As entrance inside the plant, probably by modulating the plant's transpiration and nutrient demand.

Regarding photosynthesis, As accumulation in leaf tissues is responsible for key physiological events such as chlorophyll degradation and leaf necrosis, decreased activity of the enzymes involved in photosynthetic metabolism, disturbance of photosynthates transport, and stomatal behavior [36,37]. This, apparently, was not reflected in our experiment where, besides the absence of any leaf necrosis (data not shown), there was a general increase in leaf relative chlorophyll content, as can be inferred from SPAD readings. This result is consistent with those of Zu et al. [37], which have reported that As is able to promote chlorophyll content in leaf tissues, becoming harmful only above a threshold concentration.

The analysis of net photosynthetic rate revealed different degrees of sensibility to the As toxicity among grafting combinations, with self-grafted "Proteo" experiencing a slight stimulation under As stress, a feature opposite to that displayed by "Proteo"–"RS841" combination. Both trends were associated to correspondingly similar patterns of water use efficiency, but not to any correlated change in stomatal conductance, indicating that mechanisms beyond stomatal behavior are involved in determining the photosynthetic response of As-stressed melon plants. Interestingly, under severe As stress, the A_N variations induced by the different rootstocks were not correlated to any corresponding variation of the aboveground biomass per plant, so that A_N and plant growth were apparently not related to each other. Indeed, under As+ treatment, both whole plant and fruit biomass resulted generally higher in "Proteo" grafted onto the *C. maxima* × *C. moschata* hybrids (namely "RS841", "Shintoza" and "Strong Tosa"; i.e., those rootstocks conferring no A_N enhancement under As-enriched nutrient solution). This seems to suggest a differential alteration of the energy balance of the plants, likely attributable to different energy dissipation pathways involved in the As stress response [38]. The data of As speciation and partitioning we found seem to confirm such hypothesis. Indeed, when exposed to the As+ treatment, the intraspecific rootstocks showed the highest total As concentration in roots, but the lowest one under its inorganic form (from 31% to 49%), meaning that up to 69% of their As_{root} was present in organic and complexed form. This is the As form whose translocation from root is hampered, because of bonds with thiol groups of root-synthetized phytochelatins and subsequent sequestration into root cells vacuole [39]. On the contrary, the interspecific rootstocks subjected to the As+ treatment, showed the lowest total As concentration in their root, demonstrating their superior ability to buffer the As entrance into the plant from the nutrient solution. However, they showed also the highest inorganic As_{root} incidence (up to 81%; i.e., the most mobile As form from the root organ) [39,40]. Inorganic As(III) is the main species suitable for transport trough xylem vessels and/or root efflux, while As(V) is rapidly reduced in roots to As(III) and then transported from the root cortical cells to the xylem vessels [40,41]. Accordingly, when growing under conditions of As stress, all the interspecific rootstocks had the highest TF_{shoot} and TF_{pulp} values, with "Shintoza" and "Strong Tosa" having also the highest concentration of As(III) in their roots. On the contrary, "Magnus" and "Dinero" confirmed the lowest TF_{shoot} and TF_{pulp} values, respectively, since the highest incidence of organic As into their roots.

Several authors have reported that the As uptake and translocation kinetics are mainly dependent on its concentration in the root environment and plant species [42,43], with the diversity of the root systems having a central role in determining differences among genotypes [44]. The PCA scatter plot highlighted that the As concentration in the root environment had a pivotal role in triggering different behaviors among rootstocks. Indeed, when exposed to the standard nutrient solution, all the grafting combinations clustered together on the negative side of PC1 (indicating lower values of As accumulation in plant tissues), without differences in terms of As partitioning inside the plant fractions. The opposite was noticed under conditions of imposed As stress, so that the grafting combinations clustered on the positive side of PC1, with a further distinction along the PC2 between rootstocks promoting the As accumulation mainly in roots (i.e., the intraspecific rootstocks) and those promoting its accumulation in shoot and pulp, namely "Shintoza" and "Strong Tosa". "RS841" slightly diverged from these latter owing to its ability to contain the As accumulation into the pulp.

The present study took into account the uptake and translocation of P too, given its peculiar interaction with As, as well as for its role in improving the conversion of solar energy into new plant biomass [45]. The complex interrelation between arsenate and phosphate in the substrate–plant systems has brought, up to now, no univocal results in literature, since their interaction can be either synergistic or antagonistic, depending on the growing conditions [46]. In our experiment, both P concentration and partitioning inside the plant were significantly modified by the As level in the nutrient solution, so that in As-stressed plants prevailed an increased P concentration in root and pulp, while a lowered one in shoot. This tendency was more marked when "Proteo" was grafted onto the intraspecific combinations. To this end, the protective role of P against the As-induced lipid peroxidation of cellular membranes has been suggested [47]. In this view, the selective increase in P concentration inside the plant fractions could be the result of a melon response aimed at protecting both root (the plant organ firstly exposed to the As pollution) and reproductive structures from the As injury, this latter feature having been formerly described in rise [48]. On the other hand, the decreased P concentration in shoot tissues seems well related to the decreased L_A, LAR, and LWR per plant recorded under conditions of As stress, so suggesting a possible role of P metabolism in buffering the transpiration-driven As entrance inside the plant.

5. Conclusions

The outcome of this experiment shows that As uptake and translocation into melon plants were both influenced by its concentration in the nutrient solution and by the genetic background of the rootstock. All tested plants, when grown in an As-enriched solution, showed a high As mobility in the substrate–plant continuum, resulting As accumulators most of all in root and shoot. Intra- and interspecific rootstocks differently influenced the As_{root} accumulation inside the plant, with the interspecific hybrids proving a superior ability to limit the metalloid entrance into the roots. When subjected to As stress, melon plants acted to limit the metalloid translocation from root to shoot and pulp, so indicating the triggering of physiological mechanisms aimed at limiting the As diffusion inside the plant. To this end, the intraspecific rootstocks displayed a better ability to retain the toxic metalloid into the root system, most of all in an organic form. This feature suggests a higher efficiency of the root chelation mechanisms of As, occurring, in turn, at the expenses of the photosynthetic balance. The interspecific *C. maxima* × *C. moschata* rootstocks gave the best results in terms of fruit biomass under conditions of As stress, so suggesting the possibility to exploit their superior bio-agronomical potential with the aim to improve melon performances in heavily As-polluted areas. "RS841" proved to be the best rootstock in the view of maximizing yield and minimizing, at the same time, the As concentration into the fruit. Thus, the significant differences we found suggest the possibility to deepen this area of research, with the aim of identifying rootstocks genotypes for their superior agronomic and toxicological response to high As contamination in the root zone.

Author Contributions: C.L. and S.R.S. conceived the experiments and proposed the experimental design. F.G. and R.P.M. carried out the experiments. R.M. and E.A. carried out chemical experiments. R.P.M., A.E., S.R.S., R.M. interpreted data and wrote the manuscript. F.G. and A.I. critically revised the manuscript.

References

1. Calatayud, Á.; San Bautista, A.; Pascual, B.; Maroto, J.V.; López-Galarza, S. Use of Chlorophyll Fluorescence Imaging as Diagnostic Technique to Predict Compatibility in Melon Graft. *Sci. Hortic. (Amsterdam)* **2013**, *149*, 13–18. [CrossRef]
2. Gupta, S.; Satpati, S.; Nayek, S.; Garai, D. Effect of Wastewater Irrigation on Vegetables in Relation to Bioaccumulation of Heavy Metals and Biochemical Changes. *Environ. Monit. Assess.* **2010**, *165*, 169–177. [CrossRef] [PubMed]
3. Muchuweti, M.; Birkett, J.W.; Chinyanga, E.; Zvauya, R.; Scrimshaw, M.D.; Lester, J.N. Heavy Metal Content of Vegetables Irrigated with Mixtures of Wastewater and Sewage Sludge in Zimbabwe: Implications for Human Health. *Agric. Ecosyst. Environ.* **2006**, *112*, 41–48. [CrossRef]
4. Verkleij, J.A.C.; Golan-Goldhirsh, A.; Antosiewisz, D.M.; Schwitzguébel, J.P.; Schröder, P. Dualities in Plant Tolerance to Pollutants and Their Uptake and Translocation to the Upper Plant Parts. *Environ. Exp. Bot.* **2009**, *67*, 10–22. [CrossRef]

5. Meharg, A.A.; Macnair, M.R. Relationship between Plant Phosphorus Status and the Kinetics of Arsenate Influx in Clones of *Deschampsia cespitosa* (L.) Beauv. That Differ in Their Tolerance to Arsenate. *Plant Soil* **1994**, *162*, 99–106. [CrossRef]
6. Hughes, M.F.; Beck, B.D.; Chen, Y.; Lewis, A.S.; Thomas, D.J. Arsenic Exposure and Toxicology: A Historical Perspective. *Toxicol. Sci.* **2011**, *123*, 305–332. [CrossRef]
7. Carbonell-Barrachina, A.A.; Aarabi, M.A.; DeLaune, R.D.; Gambrell, R.P.; Patrick, W.H. The Influence of Arsenic Chemical Form and Concentration on Spartina Patens and Spartina Alterniflora Growth and Tissue Arsenic Concentration. *Plant Soil* **1998**, *198*, 33–43. [CrossRef]
8. Stazi, S.R.; Marabottini, R.; Papp, R.; Moscatelli, M.C. Arsenic in Soil: Availability and Interactions with Soil Microorganisms. In *Heavy Metal Contamination of Soils*; Sherameti, I., Varma, A., Eds.; Springer International Publishing: Cham, Switzerland, 2015; pp. 113–126. [CrossRef]
9. Garg, N.; Singla, P. Arsenic Toxicity in Crop Plants: Physiological Effects and Tolerance Mechanisms. *Environ. Chem. Lett.* **2011**, *9*, 303–321. [CrossRef]
10. Finnegan, P.M.; Chen, W. Arsenic Toxicity: The Effects on Plant Metabolism. *Front. Physiol.* **2012**, *3*, 1–18. [CrossRef]
11. Maurel, C.; Verdoucq, L.; Luu, D.T.; Santoni, V. Plant Aquaporins: Membrane Channels with Multiple Integrated Functions. *Annu. Rev. Plant Biol.* **2008**, *59*, 595–624. [CrossRef]
12. Allevato, E.; Stazi, S.R.; Marabottini, R.; D'Annibale, A. Mechanisms of arsenic assimilation by plants and countermeasures to attenuate its accumulation in crops other than rice. *Ecotoxicol. Environ. Saf.* **2019**, *185*, 109701. [CrossRef]
13. Abbas, G.; Murtaza, B.; Bibi, I.; Shahid, M.; Niazi, N.K.; Khan, M.I.; Amjad, M.; Hussain, M.; Natasha. Arsenic Uptake, Toxicity, Detoxification, and Speciation in Plants: Physiological, Biochemical, and Molecular Aspects. *Int. J. Environ. Res. Public Health* **2018**, *15*, 59. [CrossRef] [PubMed]
14. Verbruggen, N.; Hermans, C.; Schat, H. Mechanisms to Cope with Arsenic or Cadmium Excess in Plants. *Curr. Opin. Plant Biol.* **2009**, *12*, 364–372. [CrossRef] [PubMed]
15. Kyriacou, M.C.; Rouphael, Y.; Colla, G.; Zrenner, R.; Schwarz, D. Vegetable Grafting: The Implications of a Growing Agronomic Imperative for Vegetable Fruit Quality and Nutritive Value. *Front. Plant Sci.* **2017**, *8*, 1–23. [CrossRef]
16. Rouphael, Y.; Kyriacou, M.C.; Colla, G. Vegetable Grafting: A Toolbox for Securing Yield Stability under Multiple Stress Conditions. *Front. Plant Sci.* **2018**, *8*, 10–13. [CrossRef] [PubMed]
17. Sabatino, L.; Iapichino, G.; D'Anna, F.; Palazzolo, E.; Mennella, G.; Rotino, G.L. Hybrids and Allied Species as Potential Rootstocks for Eggplant: Effect of Grafting on Vigour, Yield and Overall Fruit Quality Traits. *Sci. Hortic. Amsterdam* **2018**, *228*, 81–90. [CrossRef]
18. Kumar, P.; Edelstein, M.; Cardarelli, M.; Ferri, E.; Colla, G. Grafting Affects Growth, Yield, Nutrient Uptake, and Partitioning under Cadmium Stress in Tomato. *HortScience* **2015**, *50*, 1654–1661. [CrossRef]
19. Stazi, S.R.; Cassaniti, C.; Marabottini, R.; Giuffrida, F.; Leonardi, C. Arsenic Uptake and Partitioning in Grafted Tomato Plants. *Hortic. Environ. Biotechnol.* **2016**, *57*, 241–247. [CrossRef]
20. Savvas, D.; Colla, G.; Rouphael, Y.; Schwarz, D. Amelioration of Heavy Metal and Nutrient Stress in Fruit Vegetables by Grafting. *Sci. Hortic.* **2010**, *127*, 156–161. [CrossRef]
21. Hoel, B.O.; Solhaug, K.A. Effect of irradiance on chlorophyll estimation with the Minolta SPAD-502 leaf chlorophyll meter. *Ann. Bot.* **1998**, *82*, 389–392. [CrossRef]
22. Martínez, D.E.; Guiamet, J.J. Distortion of the SPAD 502 chlorophyll meter readings by changes in irradiance and leaf water status. *Agronomie* **2004**, *24*, 41–46. [CrossRef]
23. Al-hamdani, S.H.; Murphy, J.M.; Todd, G.W. Stomatal conductance and CO_2 assimilation as screening tools for drought resistance in sorghum. *Can. J. Plant Sci.* **1991**, *71*, 689–694. [CrossRef]
24. Rintala, E.M.; Ekholm, P.; Koivisto, P.; Peltonen, K.; Venäläinen, E.R. The intake of inorganic arsenic from long grain rice and rice-based baby food in Finland—Low safety margin warrants follow up. *Food Chem.* **2014**, *150*, 199–205. [CrossRef] [PubMed]
25. Welna, M.; Pohl, P.; Szymczycha-Madeja, A. Non-chromatographic Speciation of Inorganic Arsenic in Rice by Hydride Generation Inductively Coupled Plasma Optical Emission Spectrometry. *Food Anal. Methods* **2019**, *12*, 581–594. [CrossRef]
26. Liu, J.; Zhang, X.H.; Li, T.Y.; Wu, Q.X.; Jin, Z.J. Soil characteristics and heavy metal accumulation by native plants in a Mn mining area of Guangxi, South China. *Environ. Monit. Assess.* **2014**, *186*, 2269–2279. [CrossRef]

27. de Campos, F.V.; de Oliveira, J.A.; da Silva, A.A.; Ribeiro, C.; dos Santos Farnese, F. Phytoremediation of arsenite-contaminated environments: Is *Pistia stratiotes* L. a useful tool? *Ecol. Indic.* **2019**, *104*, 794–801. [CrossRef]
28. Pandey, J.; Verma, R.K.; Singh, S. Screening of most potential candidate among different lemongrass varieties for phytoremediation of tannery sludge contaminated sites. *Int. J. Phytoremediat.* **2019**, *21*, 600–609. [CrossRef]
29. Saldaña-Robles, A.; Abraham-Juárez, M.R.; Saldaña-Robles, A.L.; Saldaña-Robles, N.; Ozuna, C.; Gutiérrez-Chávez, A.J. The Negative Effect of Arsenic in Agriculture: Irrigation Water, Soil and Crops, State of the Art. *Appl. Ecol. Environ. Res.* **2018**, *16*, 1533–1551. [CrossRef]
30. Stazi, S.R.; Allevato, E.; Marabottini, R. Contaminazione Dei Prodotti Agricoli Da Arsenico: Assimilazione e Strategie Di Mitigazione. *Italus Hortus* **2016**, *23*, 19–32.
31. Ma, L.Q.; Komar, K.M.; Tu, C.; Zhang, W.; Cai, Y.; Kennelley, E.D. A Fern That Hyperaccumulates Arsenic. *Nature* **2001**, *409*, 579. [CrossRef]
32. Gomes, M.P.; Carvalho, M.; Carvalho, G.S.; Marques, T.C.L.L.S.M.; Garcia, Q.S.; Guilherme, L.R.G.; Soares, A.M. Phosphorus Improves Arsenic Phytoremediation by *Anadenanthera peregrina* by Alleviating Induced Oxidative Stress. *Int. J. Phytoremediat.* **2013**, *15*, 633–646. [CrossRef] [PubMed]
33. Mehmood, T.; Bibi, I.; Shahid, M.; Niazi, N.K.; Murtaza, B.; Wang, H.; Ok, Y.S.; Sarkar, B.; Javed, M.T.; Murtaza, G. Effect of Compost Addition on Arsenic Uptake, Morphological and Physiological Attributes of Maize Plants Grown in Contrasting Soils. *J. Geochem. Explor.* **2017**, *178*, 83–91. [CrossRef]
34. Koyama, K.; Kikuzawa, K. Is Whole-Plant Photosynthetic Rate Proportional to Leaf Area? A Test of Scalings and a Logistic Equation by Leaf Demography Census. *Am. Nat.* **2009**, *173*, 640–649. [CrossRef] [PubMed]
35. Ropokis, A.; Ntatsi, G.; Kittas, C.; Katsoulas, N.; Savvas, D. Impact of Cultivar and Grafting on Nutrient and Water Uptake by Sweet Pepper (*Capsicum annuum* L.) Grown Hydroponically Under Mediterranean Climatic Conditions. *Front. Plant Sci.* **2018**, *9*, 1–12. [CrossRef] [PubMed]
36. Rahman, A.M.; Hasegawa, H.; Mahfuzur Rahman, M.; Nazrul Islam, M.; Majid Miah, M.A.; Tasmen, A. Effect of Arsenic on Photosynthesis, Growth and Yield of Five Widely Cultivated Rice (*Oryza sativa* L.) Varieties in Bangladesh. *Chemosphere* **2007**, *67*, 1072–1079. [CrossRef]
37. Zu, Y.Q.; Sun, J.J.; He, Y.M.; Wu, J.; Feng, G.Q.; Li, Y. Effects of Arsenic on Growth, Photosynthesis and Some Antioxidant Parameters of Panax Notoginseng Growing in Shaded Conditions. *Int. J. Adv. Agric. Res.* **2016**, *4*, 78–88.
38. Srivastava, S.; Akkarakaran, J.J.; Sounderajan, S.; Shrivastava, M.; Suprasanna, P. Arsenic Toxicity in Rice (*Oryza sativa* L.) Is Influenced by Sulfur Supply: Impact on the Expression of Transporters and Thiol Metabolism. *Geoderma* **2016**, *270*, 33–42. [CrossRef]
39. Zhao, F.J.; Ma, J.F.; Meharg, A.A.; McGrath, S.P. Arsenic Uptake and Metabolism in Plants. *New Phytol.* **2009**, *181*, 777–794. [CrossRef]
40. Raab, A.; Schat, H.; Meharg, A.A.; Feldmann, J. Uptake, Translocation and Transformation of Arsenate and Arsenite in Sunflower (*Helianthus annuus*): Formation of Arsenic-Phytochelatin Complexes during Exposure to High Arsenic Concentrations. *New Phytol.* **2005**, *168*, 551–558. [CrossRef]
41. Pickering, I.J.; Prince, R.C.; George, M.J.; Smith, R.D.; George, G.N.; Salt, D.E. Reduction and Coordination of Arsenic in Indian Mustard 1. *Plant Physiol.* **2000**, *122*, 1171–1177. [CrossRef]
42. Ouyang, Y. Phytoextraction: Simulating Uptake and Translocation of Arsenic in a Soil-Plant System. *Int. J. Phytoremediat.* **2005**, *7*, 3–17. [CrossRef]
43. Wang, J.; Zhao, F.J.; Meharg, A.A.; Raab, A.; Feldmann, J.; McGrath, S.P. Mechanisms of Arsenic Hyperaccumulation in Pteris Vittata. Uptake Kinetics, Interactions with Phosphate, and Arsenic Speciation. *Plant Physiol.* **2002**, *130*, 1552–1561. [CrossRef] [PubMed]
44. Chaturvedi, I. Effects of Arsenic Concentration and Forms on Growth and Arsenic Uptake and Accumulation by Indian Mustard (*Brassica juncea* L.) Genotyopes. *J. Cent. Eur. Agric.* **2006**, *7*, 31–40.
45. Ierna, A.; Mauro, R.P.; Mauromicale, G. Improved Yield and Nutrient Efficiency in Two Globe Artichoke Genotypes by Balancing Nitrogen and Phosphorus Supply. *Agron. Sustain. Dev.* **2012**, *32*, 773–780. [CrossRef]
46. Tu, C.; Ma, L.Q. Effects of Arsenate and Phosphate on Their Accumulation by an Arsenic-Hyperaccumulator *Pteris vittata* L. *Plant Soil* **2003**, *249*, 373–382. [CrossRef]
47. Gunes, A.; Pilbeam, D.J.; Inal, A. Effect of Arsenic-Phosphorus Interaction on Arsenic-Induced Oxidative Stress in Chickpea Plants. *Plant Soil* **2009**, *314*, 211–220. [CrossRef]

12

Appraisal of Salt Tolerance under Greenhouse Conditions of a *Cucurbitaceae* Genetic Repository of Potential Rootstocks and Scions

Giuseppe Carlo Modarelli [1], Youssef Rouphael [1], Stefania De Pascale [1], Gölgen Bahar Öztekin [2], Yüksel Tüzel [2], Francesco Orsini [3,*] and Giorgio Gianquinto [3]

1 Department of Agricultural Sciences, University of Naples Federico II, 80055 Portici, Italy; giuseppecarlo.modarelli@unina.it (G.C.M.); youssef.rouphael@unina.it (Y.R.); depascal@unina.it (S.D.P.)
2 Department of Horticulture, Faculty of Agriculture, Ege University, 35100 Bornova-Izmir, Turkey; golgen.oztekin@ege.edu.tr (G.B.Ö.); yuksel.tuzel@ege.edu.tr (Y.T.)
3 DISTAL—Department of Agricultural and Food Sciences, Alma Mater Studiorum—University of Bologna, 40127 Bologna, Italy; giorgio.gianquinto@unibo.it
* Correspondence: f.orsini@unibo.it

Abstract: Soil salinization due to climate change and intensive use of water and soil is increasing exponentially. *Cucurbitaceae* species are cultivated worldwide and the identification of salinity tolerant genotypes to be used as rootstock or scion for securing yield stability in salt affected agricultural areas is a research priority. In the present greenhouse study, we assessed the response to salinity (0 mM a non-salt control and 150 mM NaCl dissolved in the nutrient solution) in the seedlings of 30 genotypes of cucurbits grown in a floating hydroponic system. The species tested included 16 genotypes of *Cucumis melo* L. (CM1-16), 6 *Citrullus vulgaris* Schrad. (CV1-6), 2 interspecific hybrids of *Cucurbita maxima* Duch. × *Cucurbita moschata* Duch. (CMM-R1 and 2), 4 bottle gourd (*Lagenaria siceraria* (Molina) Standl. (LS1-4)), 1 *Cucurbita moschata* Duch. (CMO51-17), and 1 luffa (*Luffa cylindrica* Mill. (LC1)) species. Results highlighted different morphological and physiological traits between the species and genotypes and a different response to salt stress. We identified *C. maxima* × *C.moscata* interspecific hybrid CMM-R2, melon genotypes CM6, CM7, CM10, and CM16 together with watermelon genotypes CV2 and CV6 and bottle gourd LS4 as salt tolerant genotypes and possible candidates as salt resistant rootstock to be introduced in grafting programs.

Keywords: NaCl; *Citrullus vulgaris* Schrad; *Luffa cylindrica* Mill; *C. maxima* Duch. × *C. moschata* Duch.; seedlings; morpho-physiological traits; grafting

---

1. Introduction

Soil and water resource salinization is one of the main abiotic stress factors that reduce plant growth and crop productivity worldwide [1]. It has been estimated that the total land affected by salinization covers approximately 412 million ha and mainly occur in arid and semiarid regions of more than 100 countries in all continents [2]. Generally, a saline environment influences every aspect of crop physiology and growth by causing water deficits due to the low water potential in the root medium, plant toxic ions uptake (e.g., Na^+, Cl^-, SO_4^{2-}), reduction in the uptake/or transport to the shoot of K^+, Ca^{2+}, and Mg^{2+} [1]. Moreover, plants grown in saline environments experience reduction of photosynthetic capacity [3] and growth in terms of number of leaves, shoot length, decrease in photosynthetic pigment content due to the negative effect of Na solutes within plant cells, and a decrease in fresh and dry matter content [4].

Breeding programs to combine salt resistance traits from different germplasm are increasing among seed companies. However, they currently require complicated evaluation and result in delays in the release of new varieties in the market due to the complexity of the salinity traits [5].

Furthermore, plant responses to salinity derive from complex and multifaceted mechanisms [6] and salt tolerance traits involve several physiological and genetic features that overall limit the success of resistance and/or tolerance trait transfer into commercial varieties [1]. To reduce or avoid production losses due to salt stress, a sustainable and fast solution is grafting highly productive genotypes onto potentially salt tolerant rootstocks [1]. Vegetable grafting is a widely used technique in Japan, Korea, the Mediterranean basin, as well as in several European countries to avoid biotic (i.e., soilborne and foliar pathogens, weeds, and arthropods) or abiotic stressors like drought, flooding, heavy metals contamination, sub optimal temperatures, nutritional deficiencies, and salinity. The salt tolerance of grafted plants is influenced by both scion and rootstock [6]. Generally, the use of salt tolerant rootstocks allows the mitigation of the detrimental effects of salinity and guarantees stable yields during the growing cycle through specific morphological, biochemical, metabolic, and physiological mechanisms. Accordingly, under saline conditions, grafted plants tend to accumulate more biomass in the root system, thus allowing for mitigated salinity effects by increasing the root/shoot ratio [1,7–9]. Additionally, to cope with saline stress, grafted plants adopt strategies like salt exclusion in the shoot and retention of salt ions in the root system [1,7]. Besides these strategies, grafting promotes, at the cellular level, a better maintenance of potassium homeostasis, together with accumulation of compatible solutes and osmolytes in the cytosol, along with compartmentation of salt ions in the vacuole through the activation of the antioxidant defense system, and induction of hormones mediated changes in plant growth [1,7,10,11].

Cucurbitaceae species like melon (*Cucumis melo* L.), cucumber (*Cucumis sativus* L.), and watermelon (*Citrullus vulgaris* Schrad.) are generally considered salt sensitive or moderately sensitive crops, while being often cultivated in areas undergoing soil and water salinization [12,13]. In cucurbits, grafting commercial varieties into salt tolerant rootstocks was shown to reduce production losses by improving their photosynthetic capacity [1,3]. Watermelon is commonly grafted on bottle gourd (*Lagenaria siceraria* (Molina) Standl.), interspecific hybrids between *C. maxima* and *C. moschata*, and wild watermelon (*Citrullus* spp.) [14]. Cucumber is generally grafted on bottle gourd, luffa (*Luffa cylindrica* Mill.), *Cucurbita* interspecific hybrids, and *Cucumis* spp [15]. Finally, melon is generally grafted on interspecific hybrids between *C. maxima* Duch. and *C. moschata* Duch. and *Cucumis melo* L. rootstocks and in some cases on luffa [5]. However, their response to salinity, as rootstock vary between the genotypes, and completed screening to identify salt tolerant rootstock varieties to be adopted in commercial grafting program have not yet been carried out.

In the light of these observations, our aim was to evaluate the response to salinity stress in different melons (*Cucumis melo* L.), watermelon (*Citrullus vulgaris* Schrad.), interspecific hybrids of *C. maxima* Duch. × *C. moschata* Duch., bottle gourd (*Lagenaria siceraria* (Molina) Standl.), *Cucurbita moschata* Duch. *cv* Plovdivski 51–17, and luffa (*Luffa cylindrica* Mill.) *Cucurbitaceae* rootstock and scion genotypes, in terms of plant growth and photosynthetic pigments to identify salt tolerant genotypes to be used in commercial grafting programs.

2. Materials and Methods

2.1. Growth Conditions, Plant Material, and Salinity Treatments

The experiment was carried out in a polyethylene covered double span nursery greenhouse at the Department of Horticulture, Faculty of Agriculture, Ege University (38°27′16.2″ N, 27°13′17.8″ E) in Bornova, Izmir Turkey during the spring 2014 growing season.

Plant material included 16 melon varieties (*Cucumis melo* L.), 6 watermelons (*Citrullus vulgaris* Schrad.), 2 interspecific hybrids of *Cucurbita maxima* Duch. × *Cucurbita moschata* Duch., 4 bottle gourd (*Lagenaria siceraria* (Molina) Standl.) varieties, 1 luffa (*Luffa cylindrica* Mill.) that were obtained from the Aegean Agricultural Research Institute Department of Biodiversity and Genetic Resources (Menemen, Izmir-Turkey) and 1 *Cucurbita moschata* Duch. *cv* Plovdivski 51–17 obtained from the "Maritsa" Vegetables Crops Research Institute (Plovdiv, Bulgaria) (Table 1). Two salt treatments were considered, by applying 0 (non-saline control) and 150 mM of sodium chloride (NaCl) dissolved in the nutrient solution. Eighty seeds per genotype were sown in polystyrene trays in a mixture of peat and perlite (75:25 *v:v*) on 5 May 2014 and placed in a germination room for three days (temperature of 24 °C, 60% RH) and then placed in an unheated nursery greenhouse (mean daily temperature of 28 °C) for two weeks. Seedlings were fertigated with commercial fertilizer twice per day until the start of the experiment.

On 22 May 2014, forty seedlings of each species/genotypes were divided per each salinity treatment and placed in separate floating hydroponic systems. The nutrient solution was composed as follows: nitrate 13.14 mM, phosphorus 0.94 mM, potassium 5.83 mM, calcium 3.79 mM, iron 35.8 μM, boron 37 μM, copper 1.6 μM, molybdenum 0.5 μM. The electrical conductivity (EC) of the nutrient solution was 2.0 dS m^{-1}. The nutrient solution was replaced every two days and aerated with air pumps. In the saline treatment (150 mM NaCl), salinity stress was induced by progressively dissolving 50 mM of NaCl every two days in the nutrient solution until the final concentration of 150 mM was achieved within one week.

Table 1. List of the tested species and the relative genotypes used in this study.

Species	Genotype		Species	Genotype	
	Original Code	Working Code		Original Code	Working Code
Cucumis melo L.	TR31586	CM1	*Citrullus vulgaris* Schrad.	TR40374	CV1
	TR40563	CM2		TR64141	CV2
	TR43722	CM3		TR43211	CV3
	TR45883	CM4		TR66066	CV4
	TR47822	CM5		TR43342	CV5
	TR48527	CM6		TR80748	CV6
	TR48611	CM7	*Lagenaria sicerania* (Molina) Standl.	Macis	LS1
	TR49583	CM8		TR62066	LS2
	TR51531	CM9		TR79616	LS3
	TR51763	CM10		TR82049	LS4
	TR61583	CM11	*Cucurbita maxima* Duch. × *Cucurbita moschata* Duch.	Nun 9075	CMM-R1
	TR61626	CM12		RS841	CMM-R2
	TR61851	CM13	*Cucurbita moschata* Duch.	Plovdivski 51-17	CMO 51-17
	Kirkagac	CM14	*Luffa cylindrica* Mill.		LC1
	Arava	CM15			
	Cesme	CM16			

2.2. Plant Growth Measurements

Plant growth was measured when salinity treatment reached its final concentration of 150 mM (12 days after the start of the NaCl treatment, at 29 days after sowing, DAS) on 3 plants per *Genotype × NaCl concentration* combination measuring the number of leaves, the shoot diameter, recorded with an electronic caliper, shoot and root length with a ruler. Shoot, root, and leaf fresh weight (FW) were recorded with an electronic balance and dry weight (DW) was recorded after drying the samples for 48 h at 70 °C. Dry matter (DM) was calculated as percentage of DW/FW.

2.3. Photosynthetic Pigments Determination

Chlorophyll *a*, *b* and carotenoid contents were determined at 29 DAS on 3 plants per each *Genotype × NaCl concentration* combination by grinding 250 mg of leaf tissue from fully expanded leaves with quartz crystals in 30 mL of acetone (80% in vol). Leaf extracts were read at 450, 645, and 663 nm with a Varian Carry 100UV-Vis spectrophotometer (Varian Inc, Palo Alto, CA, USA), pigments content was calculated according to Arnon [16] and values were expressed as mg g^{-1} FW.

2.4. Electrolyte Leakage Analysis

Electrolyte leakage (i.e., membrane permeability) was measured at 29 DAS collecting three leaf disks (10 mm diameter each) from fully expanded leaves on 3 plants per *Genotype × NaCl concentration* combination and immersed in 50 mL of distilled water. EC was measured immediately (1 min -) (EC_1) and after 60 min of shaking (EC_{60}). The samples were then autoclaved (121 °C) for 25 min, and total conductivity (EC_T) of bathing solution was measured after cooling. Electrolyte leakage was calculated and expressed as a percentage (%) according to Blum and Ebercon [17] using the following Equation:

$$\text{Electrolyte leakage (\%)} = (EC_{60} - EC_1)/EC_T.$$

2.5. Statistical Analysis

A completely randomized design was adopted for the experiment. The experiment was conducted on a total of 40 plants per *Genotype × NaCl concentration* combination and each analysis was carried out on at least 3 plants per *Genotype × NaCl concentration* randomly selected. Data were analyzed by ANOVA using SPSS 25 software package (www.ibm.com/software/analytics/spss) and means were compared using Duncan post hoc test ($p \leq 0.05$). The cluster heatmap was generated using the ClustVis online software [18] using Euclidean distance as the similarity measure and hierarchical clustering with complete linkage on the genotype percent variation, $\log_{(x+1)}$ transformed, between 150 and 0 mM of NaCl on all the analyzed parameters. The principal component analysis (PCA) was conducted using the Primer 6 software package (PRIMER-e, Albany, New Zealand) on the percentage variation of all the morphological and physiological analyzed parameters to highlight differences between genotypes to the increase of NaCl concentration in the nutrient solution. Five components were extracted by the PCA analysis. The PCA output includes treatment component scores as well as variable loadings.

3. Results

3.1. Plant Growth and Biomass Production

The statistical analysis revealed intrinsic genotypic differences, and a genotypic specific response to the NaCl concentration in the nutrient solution in terms of number of leaves, shoot diameter, shoot length, and root length (Table 2). Compared to non-saline seedlings, under 150 mM NaCl, the number of leaves decreased significantly only in the melon genotype CM5 (−31%) while it did not vary significantly in the other genotypes (Figure 1a). Shoot diameter between the genotypes grown under 0 mM NaCl was higher in the interspecific hybrid CMM-R1 (on average 4.6 mm) and bottle gourd genotypes LS1, LS3, and LS4 (on average 4.66, 4.73, and 4.73 mm respectively), whereas bottle gourd LS2, watermelon CV3, and melon (CM9, CM10, CM14, and CM15) seedlings turned out to be thinner (Figure 1b). The genotype and the NaCl concentration significantly affected shoot diameter (Table 2); as a result, 150 mM of NaCl increased shoot diameter by 47% and 36% in CM10 and CV3 genotypes, respectively, while it decreased by −26% in CMM-R1 *C. maxima × C. moscata* hybrid and it did not vary in the other genotypes (Figure 1b). Shoot length was significantly affected by the genotypes (Table 2). Particularly, among seedlings grown in the absence of salt stress, melon genotypes (CM3, CM7, CM8, CM9, CM13, and CM16) produced the longest shoots, while the shortest were observed in bottle gourd LS2, watermelon (CV3, CV4, CV5, CV6) and the interspecific hybrids CMM-R1 and CMM-R2. (Figure 1c). A genotypic response to the NaCl concentration was also observed in shoot length (Table 2). Accordingly, compared to seedlings grown under 0 mM NaCl, shoot length in saline grown seedlings decreased significantly in all the tested genotypes except for the watermelon (CV3, CV4, and CV6), bottle gourd (LS1, LS2) genotypes, and the luffa (LC1) (Figure 1c). Seedling root length was significantly different among the genotypes. In fact, under 0 mM NaCl, melon CM5 developed the longest root while melon CM16 and watermelon CV3 the shortest (Figure 1d). On the other hand, when 150 mM NaCl was applied, no significant differences in root length compared to control conditions were observed in the studied genotypes.

Table 2. Leaf number (n plant^{-1}), shoot and root length (cm), shoot diameter (mm), shoot and root dry weight (DW, g plant^{-1}), and dry matter content (DM, %) in seedlings of sixteen *Cucumis melo* L., six *Citrullus vulgaris* Schrad., two *C. maxima* × *C. moschata*, four *Lagenaria siceraria* (Molina) Standl. different genotypes, and *Cucurbita moschata* Duch. *cv* 51–17 and *Luffa cylindrica* Mill. grown at 0 or 150 mM of NaCl. (1 week after the beginning of the treatment). Non significance or significance differences at $p \leq 0.05$, 0.01, or 0.001 are indicated as: ns, ** and *** respectively.

	Leaf Number	Shoot Length	Shoot Diameter	Root Length	Shoot DW	Root DW	Shoot DM	Root DM
Genotype (G)	**	***	**	**	**	**	**	**
Salt (NaCl)	**	**	ns	ns	**	ns	**	ns
G × NaCl	**	**	**	***	**	**	**	**

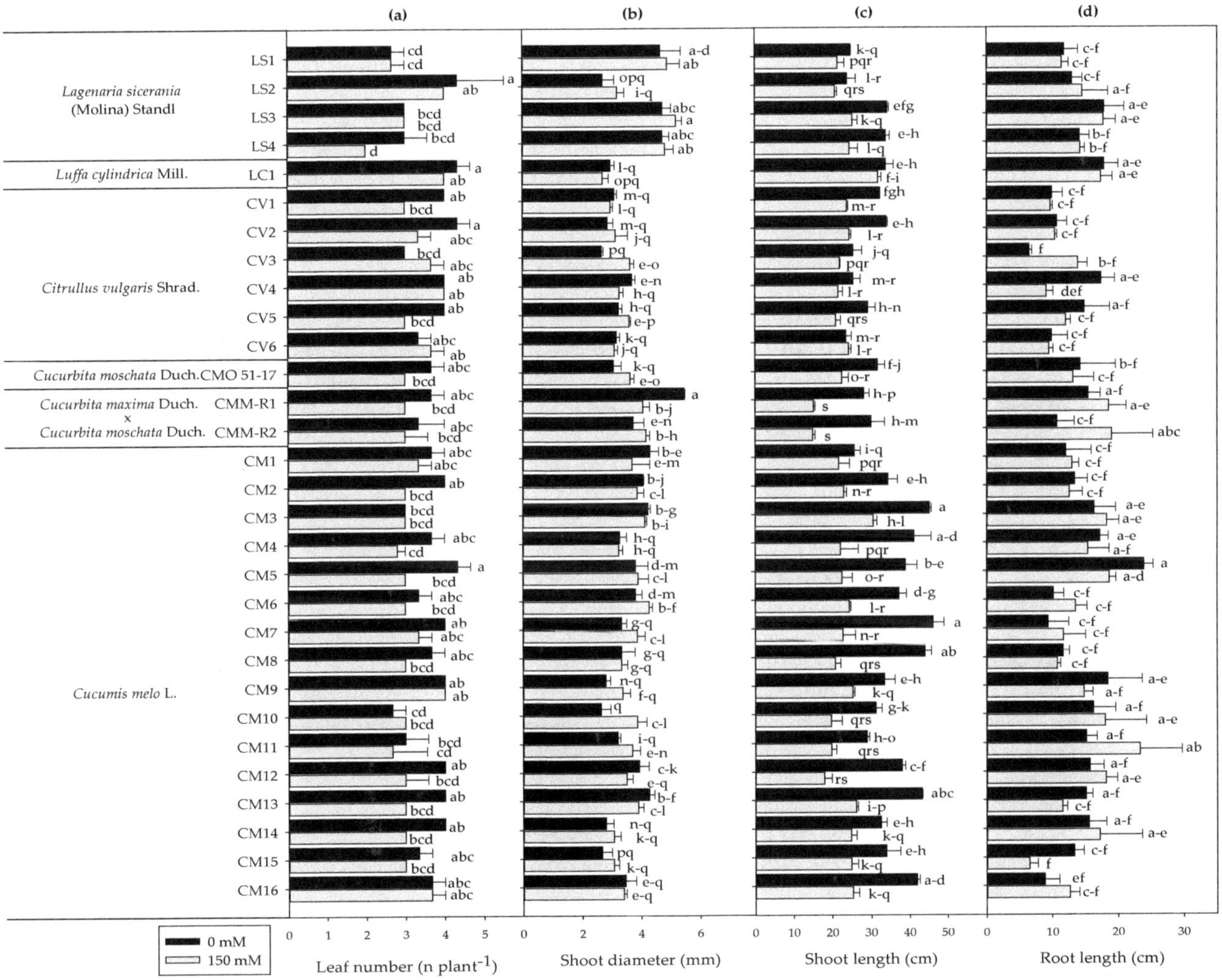

Figure 1. Plant growth, after one week of NaCl treatment, in term of number of leaves (**a**), shoot diameter (**b**), shoot length (**c**), and root length (**d**) in seedlings of sixteen *Cucumis melo* L, six *Citrullus vulgaris* Schrad., two *C. maxima* × *C. moschata*, four *Lagenaria siceraria* (Molina) Standl. different genotypes, and *Cucurbita moschata* Duch. *cv* 51–17 and *Luffa cylindrica* Mill. Mean values ± standard errors; $n = 3$ followed by different letters within each parameter are significantly different based on Duncan post hoc ($p < 0.05$).

3.2. Plant Biomass Production

The statistical analysis revealed intrinsic genotypic differences, and a genotypic specific response to the NaCl concentration in terms of shoot and root dry weight and dry matter content (Table 2). Under non-saline conditions, melon genotypes CM12 and CM13, together with CMM-R1, produced heavier shoots while watermelon CV3 the lightest (Figure 2a). As compared to 0 mM NaCl, shoot dry weight between the genotypes was differently affected by 150 mM NaCl. A significant decrease in shoot dry weight was observed in melon (CM2, CM3, CM5, CM7, CM8, CM9, CM12, and CM13) as well as in CMM-R1, CMO 51–17, watermelon (CV1, CV2), and bottle gourd LS2, while it increased in CMM-R2, watermelon CV3, and bottle gourd LS1 and did not vary in the other genotypes (Figure 2a). In the absence of salinity, root dry weight was higher in watermelon CV4, followed by bottle gourd (LS3 and LS4), while the lowest root dry weights were recorded in melon genotypes (CM6, CM15, and CM16) (Figure 2b). When 150 mM NaCl was applied, root dry weight only decreased in CV4 (−79%), while it was not affected in the other genotypes (Figure 2b). Shoot dry matter content differed between the genotypes at both NaCl concentrations. Under 0 mM NaCl, the highest biomass accumulation was observed in watermelon CV6 while the lowest was associated with melon CM1 (Figure 2c). Furthermore, 150 mM (NaCl) resulted in significant increases in shoot dry matter in the bottle gourd genotypes (LS1, LS3, and LS4), together with watermelon (CV2, CV3, and CV5), melon CM1, and both the CMM-R1 and CMM.R1 interspecific hybrids (with the strongest accumulation registered in CMM-R2) (Figure 2c). Differences in root dry matter percentage were associated with genotypes and their interaction with salinity (Table 2). Moreover, under 0 mM NaCl, the highest root biomass content was observed in melon (CM2, CM3, and CM5), while the lowest in bottle gourd (LS1, LS3, and LS4) (Figure 2d). Compared to 0 mM NaCl, root dry matter content decreased in seedlings grown under 150 mM of NaCl of CMO 51–17 (−26%), melon CM3 (−8%), CM10 (−23%), CM11 (−20%), and the interspecific hybrids (CMM-R1: −18%) and (CMM-R2: −48%), while it significantly increased in melon (CM1, CM4, CM5, CM6, CM7, CM8, CM15, and CM16), together with watermelon (CV1, CV2, CV3, CV4, and CV5) (Figure 2d).

3.3. Photosynthetic Pigments

Intrinsic difference in photosynthetic pigments profile and content were observed between the genotypes (Table 3). Among non-salinized seedlings, chlorophyll (*a*, *b*, and total) and carotenoids content was higher in melon (CM11 and CM12), while the lowest concentration was observed in CM5. Compared to non-saline seedlings, chlorophyll *a* content significantly increased under salt stress in CMM-R1 and CMM-R2 (+47% and +61%), while it decreased in CMO 51–17 (−40%) and LC1 (−49%) and it did not vary in the other genotypes (Figure 3a). On the other hand, chlorophyll *b* content increased in CMM-R1 (+38%) and CMM-R2 (+64%) seedlings grown under saline conditions, while it decreased in LC1 (−51%), and was not affected by salinity in any of the other genotypes (Figure 3b). As compared to non-saline conditions, the total chlorophyll and carotenoid content increased under 150 mM NaCl in CMM-R1 (+44% and +52%, respectively), CMM-R2 (+62% and +54%), while they decreased in LC1 (−50% and −28%) and it did not vary for the other genotypes (Figure 3c,d).

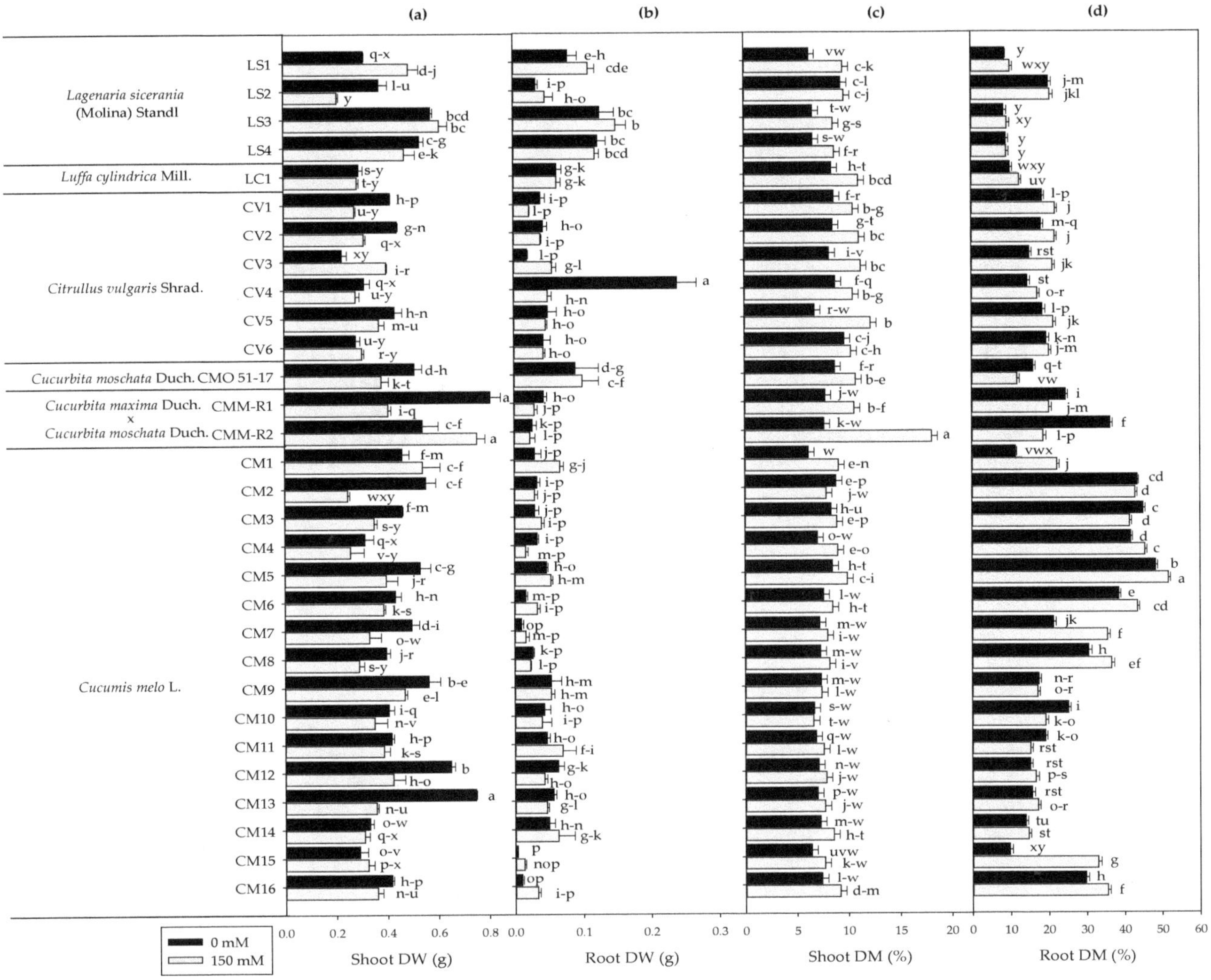

Figure 2. Plant biomass accumulation, after one week of NaCl treatment, in term of shoot dry weight (**a**), root dry weight (**b**), shoot dry matter content (**c**), and root dry matter content (**d**) in seedlings of sixteen *Cucumis melo* L., six *Citrullus vulgaris* Schrad., two *C. maxima* × *C. moschata*, four *Lagenaria siceraria* (Molina) Standl. different genotypes, and *Cucurbita moschata* Duch. *cv* 51–17 and *Luffa cylindrica* Mill. Mean values ± standard errors; $n = 3$ followed by different letters within each parameter are significantly different based on Duncan post hoc ($p < 0.05$).

Table 3. Chlorophyll *a*, *b*, total carotenoids content, and electrolyte leakage in leaves of seedlings of sixteen *Cucumis melo* L., six *Citrullus vulgaris* Schrad, two *C. maxima* × *C. moschata*, four *Lagenaria siceraria* (Molina) Standl. different genotypes, and *Cucurbita moschata* Duch. *cv* 51–17 and *Luffa cylindrica* Mill. grown at 0 or 150 mM of NaCl. (1 week after the beginning of the treatment). Non significance or significance differences at $p \leq 0.05$ or 0.01 are indicated as: ns and ** respectively.

	Chlorophyll *a* (mg g^{-1})	Chlorophyll *b* (mg g^{-1})	Total Chlorophyll (mg g^{-1})	Carotenoids (mg g^{-1})	Electrolyte Leakage (%)
Genotype (G)	**	**	**	**	**
Salt (NaCl)	ns	ns	ns	ns	**
G × NaCl	**	**	**	**	**

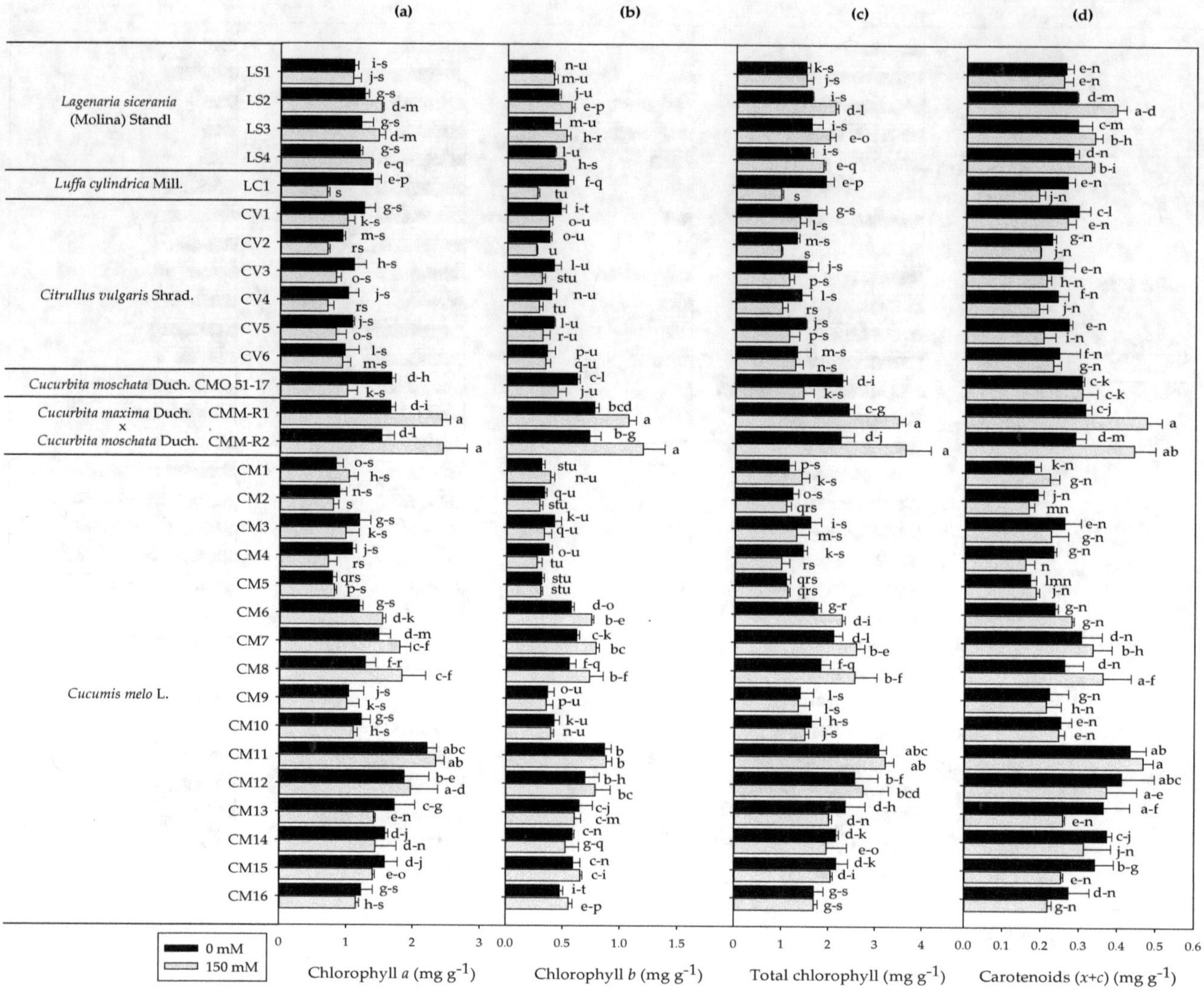

Figure 3. Photosynthetic pigment content, after one week of NaCl treatment, in term of chlorophyll *a* (**a**), chlorophyll *b* (**b**), total chlorophyll (**c**), and carotenoids ($x + n$) (**d**) in seedlings of sixteen *Cucumis melo* L., six *Citrullus vulgaris* Schrad., two *C. maxima* × *C. moschata*, four *Lagenaria siceraria* (Molina) Standl. different genotypes, and *Cucurbita moschata* Duch. *cv* 51–17 and *Luffa cylindrica* Mill. Mean values ± standard errors; $n = 3$ followed by different letters within each parameter are significantly different based on Duncan post hoc ($p < 0.05$).

3.4. Electrolyte Leakage

Electrolyte leakage was significantly different between the genotypes (Table 3). Under 0 mM NaCl, CMM-R2, LS1, and CM2 showed the highest value of electrolyte leakage, while the lowest electrolyte leakage values were observed in CV4, LC1, LS4, and CM10 (Figure 4). Electrolyte leakage values were generally increased by salinity, although significant differences from control conditions were only observed in LS2 (+184%), CV13 (+243%), CM10 (+4662%), and CM13 (+509%) (Figure 4).

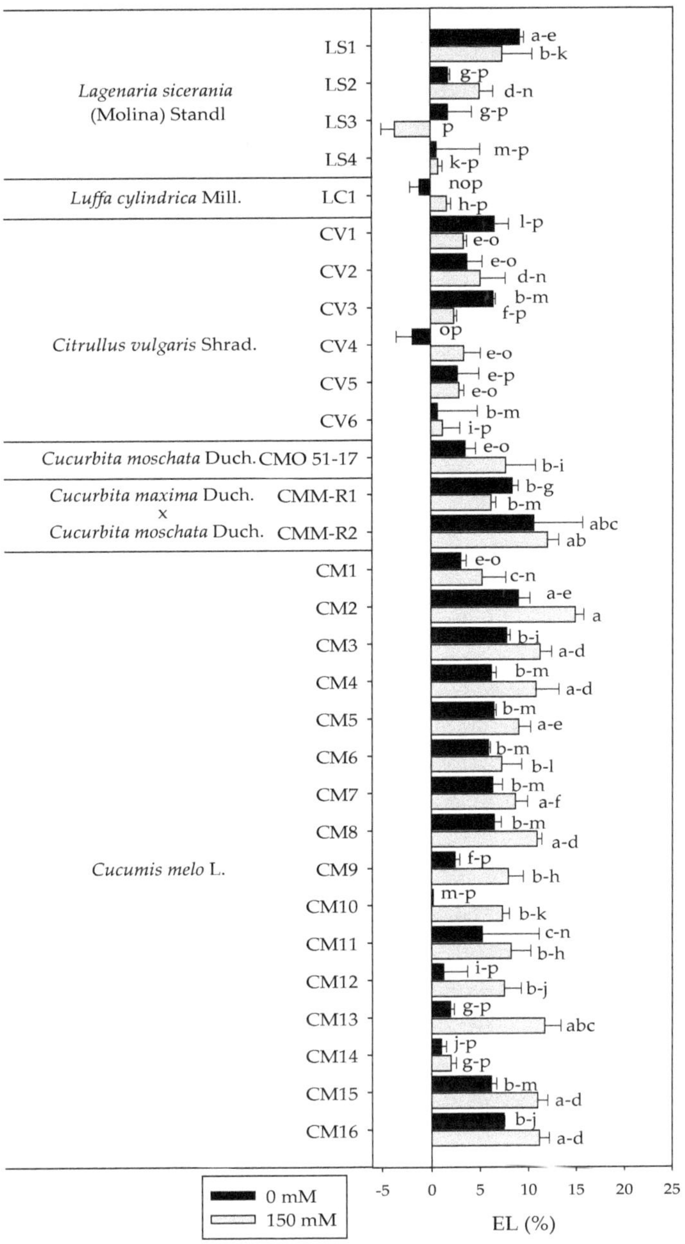

Figure 4. Electrolyte leakage (EL) in leaves, after one week of NaCl treatment, in seedlings of sixteen *Cucumis melo* L., six *Citrullus vulgaris* Schrad., two *C. maxima* × *C. moschata*, four *Lagenaria siceraria* (Molina) Standl. different genotypes, and *Cucurbita moschata* Duch. *cv* 51–17 and *Luffa cylindrica* Mill. Mean values ± standard errors; $n = 3$ followed by different letters within each parameter are significantly different based on Duncan post hoc ($p < 0.05$).

3.5. Cluster Heat Map and Principal Component Analysis

To obtain a detailed overview and to better distinguish the morpho-physiological changes induced by salinity on the tested genotypes, a cluster heat map and a principal component analysis (PCA) were conducted considering the percentage variation between 150 and 0 mM NaCl for all the aforementioned measured parameters.

The cluster heatmap shows that the main clustering factor was the genotypes regardless of the species, highlighting common response traits to the increase of NaCl between the different genotypes (Figure 5). The genotypes formed two main clusters. The first was occupied by the watermelon genotype CV2 that clustered alone, mainly as a consequence of an increase in leaf number, root length, shoot diameter, and shoot and root dry weight, together with a decrease in electrolyte leakage and photosynthetic pigment content (Figure 5). The second main cluster is split into two sub-groups. In order of distance, the first sub-group arranges bottle gourd LS2, LS3, and LS4 with melon CM8 and the interspecific hybrids CMM-R1 and CMM-R2, together into several further sub-sets characterized by similar positive variations in photosynthetic pigment content and negative variation in electrolyte leakage and root dry matter content (Figure 5). The second sub-group contains most of the genotypes and is organized into two further subsets, one of which has two divisions. The first subset is determined by similar variation in shoot and root dry matter content and electrolyte leakage grouping melon (CM1, CM6, and CM7, CM15 and CM16) and watermelon CV3 and bottle gourd LS1 genotypes together (Figure 5). Furthermore, the CM1 genotype showed the highest positive variation in shoot diameter, while CM6 and CM7 showed the highest increase in dry root weight (Figure 5). The decrease observed under saline conditions in photosynthetic pigments, root length, and dry weight, together with shoot length and diameter and leaf number accounted for the formation of the second subset. This is further subdivided into two divisions, the first of which contains watermelon (CV1, CV4, and CV5), melon (CM4, CM5, CM11, CM12, and CM13), and luffa (LC1) genotypes. Among these genotypes, LC1, CM4, and CV5 showed the highest decrease in photosynthetic pigment contents, shoot diameter and root length, and dry weight. A decrease in root dry weight was also observed in CM11, CM13, and watermelon CV1 (Figure 5). The second and last subdivision is explained mainly by a decrease in shoot and root dry matter and places, together melon (CM2, CM3, CM9, and CM10), watermelon CV6, and *C. moscata* CMO 51–17. Among the genotypes, melon (CM2, CM3, and CM10) showed the highest decrease in shoot dry matter, while melon CM3, the highest increase in electrolyte leakage and CV6 also showed the highest increase in shoot length (Figure 5).

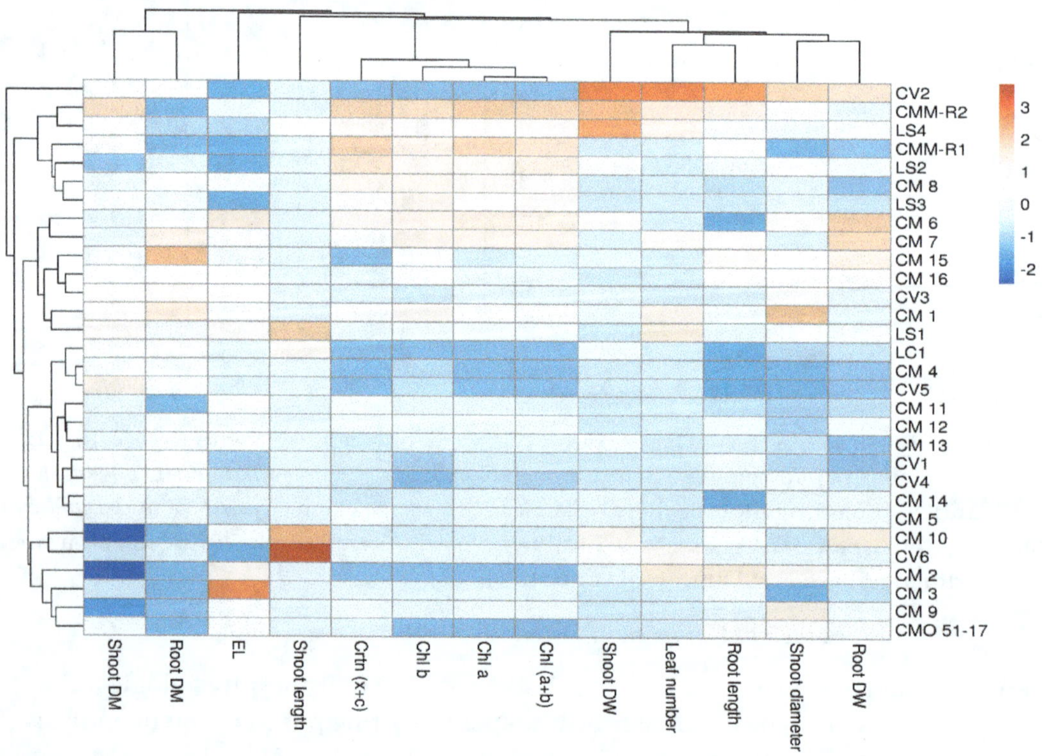

Figure 5. Cluster heat map analysis of the percentage variation to the increase of NaCl level of all the analyzed parameters in seedlings of sixteen *Cucumis melo* L, six *Citrullus vulgaris* Schrad., two *C. maxima* × *C. moschata*, four *Lagenaria siceraria* (Molina) Standl. different genotypes, and *Cucurbita moschata* Duch. *cv* 51–17 and *Luffa cylindrica* Mill. grown in a floating system in a nursery greenhouse. It was generated using the ClustVis online software [18] with Euclidean distance as the similarity measure and hierarchical clustering with complete linkage.

The PCA of the relative percentage variation of all the analyzed parameters in seedlings grown at 150 mM NaCl as compared with the control (0 mM NaCl) highlighted that the first three principal components (PCs) were related with eigenvalues higher than 1 and explained 64.5% of the total variance, with PC1, PC2, and PC3 accounting for 32.6%, 20.1%, and 11.9% respectively (data not shown). Both the species and the genotype contributed to the separation of PC1 and PC2, as highlighted in the PCA output, revealing common trait variations among genotypes regardless of the species (Figure 6). The melon genotypes CM3, CM4, CM5, CM11, CM13, and CM14, together with the watermelon genotypes CV1, CV3, and CV4, were concentrated in the upper right quadrant of the PCA output and positively correlated to an increase in electrolyte leakage (Figure 6). The watermelon genotypes CV2 and CV5, together with CMO 51–17, luffa LC1, bottle gourd genotype LS1, and melon genotypes CM2, CM10, and CM15 were concentrated in the negative lower right quadrant of PCA output and were correlated to an increase under saline conditions of shoot length and root dry matter content (Figure 6). The interspecific hybrid CMM-R1, together with melon CM8, CM9, CM12, and bottle gourd LS2 and LS3 were concentrated in the upper left quadrant of the PCA output (Figure 6) and together with the interspecific hybrid CMM-R2 (located in the lower left quadrant), were characterized with an increase in photosynthetic pigment content under saline conditions. In the left lower quadrant depicted genotypes CM1, CM6, CM7, and CM16, together with bottle gourd LS4 and watermelon CV6 and were characterized with increase in leaf number, shoot diameter, shoot dry weight, and dry biomass accumulation and root length under saline conditions (Figure 6).

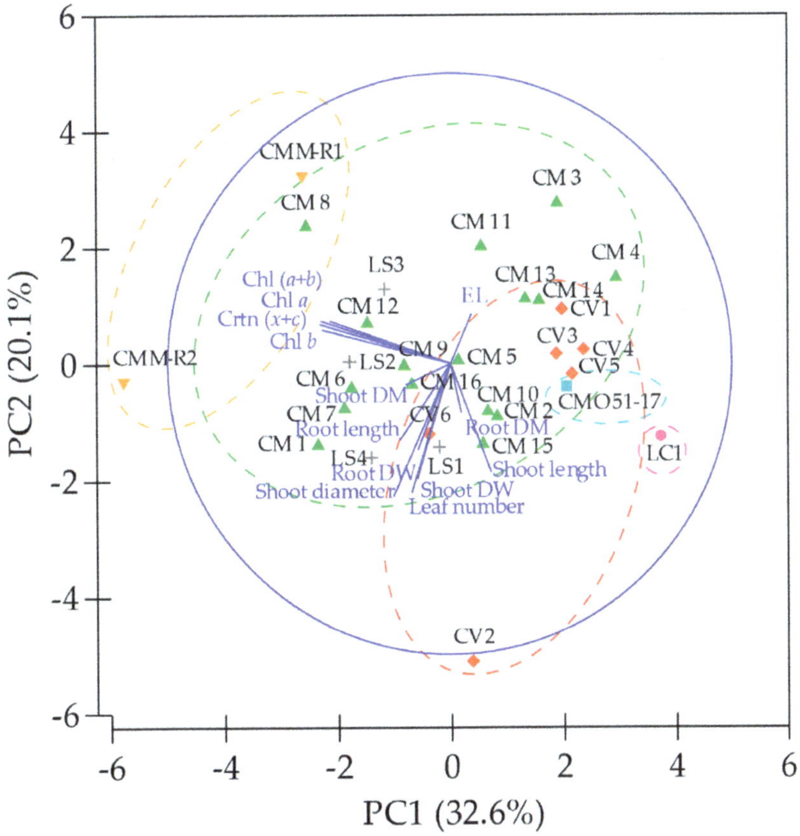

Figure 6. Principal component (PC) loading plot and scores of principal component analysis of the percentage of variation between 150 mM and 0 mM of NaCl of all the parameter analyzed in seedlings of sixteen *Cucumis melo* L. (CM), six *Citrullus vulgaris* Schrad (CV), two *C. maxima* × *C. moschata* (CMM-R), four *Lagenaria siceraria* (Molina) Standl. (LS) different genotypes, and *Cucurbita moschata* Duch. *cv* 51–17 and *Luffa cylindrica* Mill. grown in a floating system in a nursery greenhouse.

4. Discussion

It is well known that plants' response to different NaCl concentrations differs between species and even within genotypes of the same species. In this study, high variability in various morphological and physiological traits between the tested genotypes even from the same species were observed. Furthermore, different responses to NaCl were identified, confirming the high variability within the assessed germplasm collection.

According to Munns [19], a saline environment in the root zone triggers the activation of various mechanisms in plant physiology, morphology, and metabolism. This takes place in two main phases. Firstly, the presence of NaCl in the soil creates osmotic stress in the roots due to different solute concentration, causing the activation of multiple metabolic pathways that in the short term lead to stomata closure to reduce evapotranspiration and preserve water, and net photosynthesis reduction [1,20,21]. Morphological change primarily includes the growth inhibition of aerial organs like shoots and a reduction in the number of leaves. In the current experiment, 150 mM of NaCl significantly reduced the number of leaves only in melon (CM5), though a decreasing trend was observed in most genotypes. Shoot growth was significantly inhibited by salinity in most genotypes with the exception of watermelon (CV3, CV5, CV6) and luffa (LC1), which turned out to be more salt tolerant for this trait. The root system serves as an interphase between plant and soil, and its anatomy determines root performance [21]. Generally, root growth in a saline environment is inhibited less than the aerial organs [22]. In our study, root growth was not significantly affected by salt stress as the main factor, however, our data highlight a genotypic-specific response to salt concentration since root growth was promoted in salt-treated seedlings of CMM-R1, CMM-R2, and watermelon (CV3) and decreased in watermelon (CV6) and melon (CM14).

Salinity stress leads to a decrease in plant biomass, mainly in the epigean organs and this is ascribed to a decrease in CO_2 fixation by photosynthesis due a reduction in stomatal conductance, together with nutrient disorders caused by the toxic action of Na^+ and Cl^- ions inhibiting Ca^{2+} and K^+ uptake [1,21–23].Under conditions of 150 mM NaCl, shoot dry weight was negatively affected only in melon (CM2, CM5, CM7, CM8, CM12, and CM13), *C. moscata* (CMO 51–17), watermelon (CV1 and CV2), and in the interspecific hybrid CMM-R1, the latter was the most salt sensitive compared with the other genotypes. In particular, watermelon (CV3), the interspecific hybrid CMM-R2, and bottle gourd (LS1) significantly increased shoot dry weight in response to salinity. On the other hand, shoot dry matter accumulation increased in saline conditions in bottle gourd (LS1, LS3, LS4) and in watermelon (CV2, CV3, and CV5) and melon (CM1) and both the interspecific hybrids CMM-R1 and CMM-R2. Many salinity tolerance traits in grafted plants are associated with the root system. A salt tolerant rootstock genotype alleviates the deleterious shoot growth inhibition, allocating more biomass to the root system, increasing the root surface able to uptake water and nutrients, resulting in a higher growth rate and biomass accumulation [24,25]. Accordingly, with 150 mM NaCl, root dry weight did not decrease in any of the genotypes except for watermelon (CV5), and root dry matter accumulation was significantly enhanced in melon (CM1, CM4, CM5, CM6, CM7, CM8, CM15, and CM16) as well as watermelon (CV1, CV2, CV3, CV4, and CV5), while it decreased in melon (CM3, CM10, and CM11) and CMM-R1 and CMM-R2. Several studies have reported the growth depressing effect of salt on melon [26–28], cucumber [3,29], mini-watermelon [30], tomato [31], and lettuce [32] grown hydroponically under greenhouse conditions. The second phase of the salt stress starts with the accumulation of sodium and chloride ions in the leaves and their degrading action on cell membranes and chloroplast membranes. This degrades the tonoplast together with chlorophyll molecules because of chlorophyllase enzyme [33] as well as the interference of salt ions with pigment-protein complexes which accelerate leaf senescence and abscission [34]. Contradictory information is available in literature on salt stress related to chlorophyll contents among the same species and to others. In fact, studies on melon *cv.* Parnon and *cv.* "Tempo F1", under saline conditions, reported a decrease in chlorophyll content [35,36] in line with studies on pumpkin [37], cucumber [38], tomato [31], canola and wheat [39,40]. Conversely, other authors observed an increase of photosynthetic pigments in different melon genotypes [41–43] in line with studies on other species like hot pepper [44], sunflower [45], and sesame [46].

Sanoubar and collaborators in 2016 [47] indicated that chlorophyll concentration in stressed tissues can be construed as an index of tissue tolerance to NaCl. Accordingly, our results indicate a good tolerance in most of the genotypes, in particular in the interspecific hybrids CMM-R1 and CMM-R2, where photosynthetic pigments significantly increased under saline condition. Contrarily, in luffa, the decreasing chlorophyll content in response to salinity could be associated with greater sensitivity to salt stress [21]. Adaptation mechanisms induced by salt stress generally influence the light harvesting complex due to a faster decrease of chlorophyll *a* content as compared to chlorophyll *b*, which leads to an increase in the chlorophyll *a*/*b* ratio [34]. In our experiment, the chlorophyll *a*/*b* ratio was modified under saline condition only in CMO 51–17 and watermelon (CV1) where chlorophyll *a* decreased and chlorophyll *b* did not vary. Different hypotheses have been formulated to explain the increase in photosynthetic pigments in response to salinity. The first considers the morphological modifications associated with the decrease of leaf area due the smaller cell size, resulting in an enhancement of photosynthetic pigments concentration [45,46]. A second hypothesis, proposed by Garcia-Valenzuela and collaborators [48], associated the increase of photosynthetic pigments to the faster response that is generally observed in the biosynthesis of pigments as compared to the cell growth rate [48]. To date, however, most the credited hypothesis relates the increase observed in photosynthetic pigments to a short term acclimatization response to salinity and that a prolonged exposure would in any case be detrimental [49]. The ability to isolate or exclude sodium and chloride is a key factor to consider as a salt tolerance trait [27,28]. The longer the salt stress is prolonged, the higher the accumulation of sodium and chloride inside the plant is. To cope with this osmotic disorder, plants exclude uptake of these compounds primarily through root exclusion, and subsequently through compartmentation, by isolating salt in vacuoles. However, the higher the ability to isolate (salt) in the vacuole is, the less the cell membrane is damaged. Moreover, if the concentration increases inside the cytosol, plants start to synthesize different organic solutes to maintain the osmotic turgor and to reduce the deleterious effect of salt due to cell membrane degradation and the increase of reactive oxygen species (ROS) [19,21,50]. In our experiment, electrolyte leakage increased dramatically only in LS2, CV4, CM10, and CM13, revealing a high index of internal damage due to the presence of sodium and chloride inside the leaf cells of these genotypes which must be considered a sensitive response to salt stress.

5. Conclusions

Salinity stress leads to a decrease in plant growth with more pronounced deleterious effect in the shoot rather than the roots. Similar physiological and morphological salt adaptation traits were observed in different genotypes from different species. In light of the above consideration, we identified *C. maxima* × *C. moscata* interspecific hybrid CMM-R2, melon genotypes (CM6, CM7, CM10, and CM16), together with watermelon (CV2 and CV6) and bottle gourd (LS4) as salt tolerant genotypes and possible candidates as salt resistant rootstock to be introduced in grafting programs. Contrarily, luffa and melon (CM1, CM2, CM3, CM4, CM5 CM8, CM9, CM11, CM12, CM13, CM14, and CM15), watermelon (CV1, CV3, CV4, and CV5), together with CMO 51–17 and bottle gourd LS1 proved salt sensitive. In conclusion, this study provides information to growers, scientists, extension specialists, and breeders on the behavior of the tested genotypes. Further research on these genotypes is needed to clarify their performance in saline environments in the long term and their compatibility with commercial grafted varieties.

Author Contributions: Conceptualization, G.C.M., F.O., Y.T., G.B.Ö., G.G.; methodology, G.C.M., G.B.Ö., Y.T., F.O.; validation, G.C.M., Y.R., F.O.; formal analysis, G.C.M.; investigation, G.C.M., Y.R., F.O.; resources, F.O., G.B.Ö., Y.T.; data curation, G.C.M., Y.R., F.O.; writing—original draft preparation, G.C.M.; writing—review and editing, F.O., Y.R., S.D.P.; visualization, G.C.M., Y.R., F.O.; supervision, S.D.P., Y.R., Y.T., G.B.Ö., G.G.; project administration, G.B.Ö.; funding acquisition, Y.R., G.G., S.D.P. All authors have read and agreed to the published version of the manuscript.

Acknowledgments: The authors are thankful to Sofia Siepi, Elyar Saeedi and Gamze Çılgın for their help collecting data and technical assistance during the experiment.

References

1. Colla, G.; Rouphael, Y.; Leonardi, C.; Bie, Z. Role of grafting in vegetable crops grown under saline conditions. *Sci. Hortic.* **2010**, *127*, 147–155. [CrossRef]
2. Montanarella, L.; Badraoui, M.; Chude, V.; Baptista Costa, I.D.S.; Mamo, T.; Yemefack, M.; Singh Aulakh, M.; Yagi, K.; Young Hong, S.; Vijarnsorn, P.; et al. *Status of the World's Soil Resources: Main Report*; FAO: Rome, Italy, 2015.
3. Rouphael, Y.; Cardarelli, M.; Rea, E.; Colla, G. Improving melon and cucumber photosynthetic activity, mineral composition, and growth performance under salinity stress by grafting onto Cucurbita hybrid rootstocks. *Photosynthetica* **2012**, *50*, 180–188. [CrossRef]
4. Balkaya, A.; Yildiz, S.; Horuz, A.; Doğru, S.M. Effects of Salt Stress on Vegetative Growth Parameters and Ion Accumulations in Cucurbit Rootstock Genotypes. *J. Crop Breed. Genet.* **2016**, *2*, 11–24.
5. King, S.R.; Davis, A.R.; Zhang, X.; Crosby, K. Genetics, breeding and selection of rootstocks for Solanaceae and Cucurbitaceae. *Sci. Hortic.* **2010**, *127*, 106–111. [CrossRef]
6. Edelstein, M.; Ben-Hur, M.; Cohen, R.; Burger, Y.; Ravina, I. Boron and salinity effects on grafted and non-grafted melon plants. *Plant Soil* **2005**, *269*, 273–284. [CrossRef]
7. Singh, H.; Kumar, P.; Kumar, A.; Kyriacou, M.C.; Colla, G.; Rouphael, Y. Kumar Grafting Tomato as a Tool to Improve Salt Tolerance. *Agronomy* **2020**, *10*, 263. [CrossRef]
8. Schwarz, D.; Öztekin, G.B.; Tüzel, Y.; Brückner, B.; Krumbein, A. Rootstocks can enhance tomato growth and quality characteristics at low potassium supply. *Sci. Hortic.* **2013**, *149*, 70–79. [CrossRef]
9. Barbieri, G.; Vallone, S.; Orsini, F.; Paradiso, R.; De Pascale, S.; Zakharov, F.; Maggio, A. Stomatal density and metabolic determinants mediate salt stress adaptation and water use efficiency in basil (*Ocimum basilicum* L.). *J. Plant Physiol.* **2012**, *169*, 1737–1746. [CrossRef]
10. Zhu, J.; Bie, Z.; Huang, Y.; Han, X. Effect of grafting on the growth and ion concentrations of cucumber seedlings under NaCl stress. *Soil Sci. Plant Nutr.* **2008**, *54*, 895–902. [CrossRef]
11. Huang, Y.; Bie, Z.; Liu, P.; Niu, M.; Zhen, A.; Liu, Z.; Lei, B.; Gu, D.; Lu, C.; Wang, B. Reciprocal grafting between cucumber and pumpkin demonstrates the roles of the rootstock in the determination of cucumber salt tolerance and sodium accumulation. *Sci. Hortic.* **2013**, *149*, 47–54. [CrossRef]
12. Xiong, M.; Zhang, X.; Shabala, S.; Shabala, L.; Chen, Y.; Xiang, C.; Nawaz, M.A.; Bie, Z.; Wu, H.; Yi, H.; et al. Evaluation of salt tolerance and contributing ionic mechanism in nine Hami melon landraces in Xinjiang, China. *Sci. Hortic.* **2018**, *237*, 277–286. [CrossRef]
13. Sarabi, B.; Bolandnazar, S.; Ghaderi, N.; Ghashghaie, J. Genotypic differences in physiological and biochemical responses to salinity stress in melon (*Cucumis melo* L.) plants: Prospects for selection of salt tolerant landraces. *Plant Physiol. Biochem.* **2017**, *119*, 294–311. [CrossRef] [PubMed]
14. Davis, A.R.; Perkins-Veazie, P.; Sakata, Y.; López-Galarza, S.; Maroto, J.V.; Lee, S.-G.; Huh, Y.-C.; Sun, Z.; Miguel, A.; King, S.R.; et al. Cucurbit Grafting. *Crit. Rev. Plant Sci.* **2008**, *27*, 50–74. [CrossRef]
15. Sakata, Y.; Ohara, T.; Sugiyama, M. The hystory of melon and cucumber grafting in Japan. *Acta Hortic.* **2008**, *767*, 217–228. [CrossRef]
16. Blum, A.; Ebercon, A. Cell Membrane Stability as a Measure of Drought and Heat Tolerance in Wheat. *Crop Sci.* **1981**, *21*, 43–47. [CrossRef]
17. Arnon, D.I. Copper enzymes in isolated chloroplasts. Polyphenoloxidase in beta vulgaris. *Plant Physiol.* **1949**, *24*, 1–15. [CrossRef]
18. Metsalu, T.; Vilo, J. ClustVis: A web tool for visualizing clustering of multivariate data using Principal Component Analysis and heatmap. *Nucleic Acids Res.* **2015**, *43*, W566–W570. [CrossRef]
19. Munns, R.; Passioura, J.; Colmer, T.D.; Byrt, C. Osmotic adjustment and energy limitations to plant growth in saline soil. *New Phytol.* **2019**, *225*, 1091–1096. [CrossRef]
20. Moradi, F.; Ismail, A.M. Responses of Photosynthesis, Chlorophyll Fluorescence and ROS-Scavenging Systems to Salt Stress During Seedling and Reproductive Stages in Rice. *Ann. Bot.* **2007**, *99*, 1161–1173. [CrossRef]

21. Acosta-Motos, J.R.; Ortuño, M.F.; Bernal-Vicente, A.; Diaz-Vivancos, P.; Sánchez-Blanco, M.J.; Hernández, J.A. Plant Responses to Salt Stress: Adaptive Mechanisms. *Agronomy* **2017**, *7*, 18. [CrossRef]
22. Yetişir, H.; Uygur, V. Plant growth and mineral element content of different gourd species and watermelon under salinity stress. *Turk. J. Agric. For.* **2009**, *33*, 65–77.
23. Carillo, P.; Raimondi, G.; Kyriacou, M.C.; Pannico, A.; El-Nakhel, C.; Cirillo, V.; Colla, G.; De Pascale, S.; Rouphael, Y. Morpho-physiological and homeostatic adaptive responses triggered by omeprazole enhance lettuce tolerance to salt stress. *Sci. Hortic.* **2019**, *249*, 22–30. [CrossRef]
24. Geilfus, C.-M.; Zörb, C.; Mühling, K.H. Salt stress differentially affects growth-mediating β-expansins in resistant and sensitive maize (*Zea mays* L.). *Plant Physiol. Biochem.* **2010**, *48*, 993–998. [CrossRef]
25. Ahmed, I.M.; Dai, H.; Zheng, W.; Cao, F.; Zhang, G.; Sun, D.; Wu, F. Genotypic differences in physiological characteristics in the tolerance to drought and salinity combined stress between Tibetan wild and cultivated barley. *Plant Physiol. Biochem.* **2013**, *63*, 49–60. [CrossRef] [PubMed]
26. Colla, G.; Rouphael, Y.; Cardarelli, M.; Massa, D.; Salerno, A.; Rea, E. Yield, fruit quality and mineral composition of grafted melon plants grown under saline conditions. *J. Hortic. Sci. Biotechnol.* **2006**, *81*, 146–152. [CrossRef]
27. Orsini, F.; Sanoubar, R.; Öztekin, G.B.; Kappel, N.; Tepecik, M.; Quacquarelli, C.; Tüzel, Y.; Bona, S.; Gianquinto, G. Improved stomatal regulation and ion partitioning boosts salt tolerance in grafted melon. *Funct. Plant Boil.* **2013**, *40*, 628–636. [CrossRef] [PubMed]
28. Sanoubar, R.; Orsini, F.; Gianquinto, G. Ionic partitioning and stomatal regulation. *Plant Signal. Behav.* **2013**, *8*, e27334. [CrossRef]
29. Colla, G.; Rouphael, Y.; Rea, E.; Cardarelli, M. Grafting cucumber plants enhance tolerance to sodium chloride and sulfate salinization. *Sci. Hortic.* **2012**, *135*, 177–185. [CrossRef]
30. Colla, G.; Rouphael, Y.; Cardarelli, M.; Rea, E. Effect of Salinity on Yield, Fruit Quality, Leaf Gas Exchange, and Mineral Composition of Grafted Watermelon Plants. *HortScience* **2006**, *41*, 622–627. [CrossRef]
31. Gong, B.; Wen, D.; Vandenlangenberg, K.; Wei, M.; Yang, F.; Shi, Q.; Wang, X. Comparative effects of NaCl and NaHCO3 stress on photosynthetic parameters, nutrient metabolism, and the antioxidant system in tomato leaves. *Sci. Hortic.* **2013**, *157*, 1–12. [CrossRef]
32. Lucini, L.; Rouphael, Y.; Cardarelli, M.; Canaguier, R.; Kumar, P.; Colla, G.; Lucini, L. The effect of a plant-derived biostimulant on metabolic profiling and crop performance of lettuce grown under saline conditions. *Sci. Hortic.* **2015**, *182*, 124–133. [CrossRef]
33. Parida, A.K.; Das, A.B.; Das, P. NaCl stress causes changes in photosynthetic pigments, proteins, and other metabolic components in the leaves of a true mangrove,Bruguiera parviflora, in hydroponic cultures. *J. Plant Boil.* **2002**, *45*, 28–36. [CrossRef]
34. Munns, R. Plant Adaptations to Salt and Water Stress. *Adv. Bot. Res.* **2011**, *57*, 1–32.
35. Mavrogianopoulos, G.; Spanakis, J.; Tsikalas, P. Effect of carbon dioxide enrichment and salinity on photosynthesis and yield in melon. *Sci. Hortic.* **1999**, *79*, 51–63. [CrossRef]
36. Kaya, C.; Tuna, A.L.; Ashraf, M.; Altunlu, H. Improved salt tolerance of melon (*Cucumis melo L.*) by the addition of proline and potassium nitrate. *Environ. Exp. Bot.* **2007**, *60*, 397–403. [CrossRef]
37. Sevengor, S.; Yasar, F.; Kusvuran, S.; Ellialtioglu, S. The effect of salt stress on growth, chlorophyll content, lipid peroxidation and antioxidative enzymes of pumpkin seedling. *Afr. J. Agric. Res.* **2011**, *6*, 4920–4924.
38. Tiwari, J.K.; Munshi, A.D.; Kumar, R.; Pandey, R.N.; Arora, A.; Bhat, J.S.; Sureja, A.K. Effect of salt stress on cucumber: Na+–K+ ratio, osmolyte concentration, phenols and chlorophyll content. *Acta Physiol. Plant.* **2009**, *32*, 103–114. [CrossRef]
39. Shahbaz, M.; Noreen, N.; Perveen, S. Triacontanol modulates photosynthesis and osmoprotectants in canola (*Brassica napus* L.) under saline stress. *J. Plant Interact.* **2013**, *8*, 350–359. [CrossRef]
40. Shahbaz, M.; Ashraf, M.; Athar, H.-R. Does exogenous application of 24-epibrassinolide ameliorate salt induced growth inhibition in wheat (*Triticum aestivum* L.)? *Plant Growth Regul.* **2008**, *55*, 51–64. [CrossRef]
41. Romero, L.; Belakbir, A.; Ragala, L.; Ruiz, J.M. Response of plant yield and leaf pigments to saline conditions: Effectiveness of different rootstocks in melon plants (*Cucumis melo* L.). *Soil Sci. Plant Nutr.* **1997**, *43*, 855–862. [CrossRef]

42. Akrami, M.; Arzani, A. Physiological alterations due to field salinity stress in melon (*Cucumis melo* L.). *Acta Physiol. Plant.* **2018**, *40*, 1–14. [CrossRef]
43. Sivritepe, H. Özkan; Sivritepe, N.; Eris, A.; Turhan, E. The effects of NaCl pre-treatments on salt tolerance of melons grown under long-term salinity. *Sci. Hortic.* **2005**, *106*, 568–581. [CrossRef]
44. Ziaf, K.; Amjad, M.; Pervez, M.A.; Iqbal, Q.; Rajwana, I.A.; Ayyub, M. Evaluation of different growth and physiological traits as indices of salt tolerance in hot pepper (*Capsicum annuum* L.). *Pak. J. Bot.* **2009**, *41*, 1797–1809.
45. Rivelli, A.R.; De Maria, S.; Pizza, S.; Gherbin, P. Growth and physiological response of hydroponically-grown sunflower as affected by salinity and magnesium levels. *J. Plant Nutr.* **2010**, *33*, 1307–1323. [CrossRef]
46. Bazrafshan, A.H.; Ehsanzadeh, P. Growth, photosynthesis and ion balance of sesame (*Sesamum indicum* L.) genotypes in response to NaCl concentration in hydroponic solutions. *Photosynthetica* **2014**, *52*, 134–147. [CrossRef]
47. Sanoubar, R.; Cellini, A.; Veroni, A.M.; Spinelli, F.; Masia, A.; Antisari, L.V.; Orsini, F.; Gianquinto, G. Salinity thresholds and genotypic variability of cabbage (*Brassica oleracea* L.) grown under saline stress. *J. Sci. Food Agric.* **2015**, *96*, 319–330. [CrossRef]
48. García-Valenzuela, X.; Garcìa-Moya, E.; Rascón-Cruz, Q.; Herrera-Estrella, L.; Aguado-Santacruz, G. Chlorophyll accumulation is enhanced by osmotic stress in graminaceous chlorophyllic cells. *J. Plant Physiol.* **2005**, *162*, 650–661. [CrossRef]
49. Wang, X.; Chang, L.; Wang, B.; Wang, D.; Li, P.; Wang, L.; Yi, X.; Huang, Q.; Peng, M.; Guo, A. Comparative proteomics of Thellungiella halophila leaves from plants subjected to salinity reveals the importance of chloroplastic starch and soluble sugars in halophyte salt tolerance. *Mol. Cell. Proteomics* **2013**, *12*, 2174–2195. [CrossRef]
50. Shalata, A.; Neumann, P.M. Exogenous ascorbic acid (vitamin C) increases resistance to salt stress and reduces lipid peroxidation. *J. Exp. Bot.* **2001**, *52*, 2207–2211. [CrossRef]

13

Solanum aethiopicum gr. *gilo* and its Interspecific Hybrid with *S. melongena* as Alternative Rootstocks for Eggplant: Effects on Vigor, Yield and Fruit Physicochemical Properties of Cultivar 'Scarlatti'

Leo Sabatino [1,*], Giovanni Iapichino [1], Giuseppe Leonardo Rotino [2], Eristanna Palazzolo [1], Giuseppe Mennella [3] and Fabio D'Anna [1]

1 Dipartimento Scienze Agrarie, Alimentari e Forestali, Università di Palermo, Viale delle Scienze, 90128 Palermo, Italy; giovanni.iapichino@unipa.it (G.I.); eristanna.palazzolo@unipa.it (E.P.); fabio.danna@unipa.it (F.D.)

2 Consiglio per la Ricerca in Agricoltura e l'Analisi dell'Economia Agraria CREA-GB, Centro di Ricerca Genomica e Bioinformatica, Via Paullese, 26836 Montanaso Lombardo (LO), Italy; giuseppeleonardo.rotino@crea.gov.it

3 Consiglio per la Ricerca in Agricoltura e l'Analisi dell'Economia Agraria CREA-OF, Centro di Ricerca Orticoltura e Florovivaismo, Via Cavalleggeri, 84098 Pontecagnano-Faiano, Italy; giuseppe.mennella@crea.gov.it

* Correspondence: leo.sabatino@unipa.it

Abstract: Grafting is generally considered effective in ameliorating vegetable crop tolerance to biotic and abiotic stresses. The use of interspecific hybrid as rootstock for eggplant may represent a valid alternative approach to enhance eggplant performance. However, studies on the effects of different rootstocks on eggplant plant vigor, yield, and fruit quality traits often show conflicting results. Thus, an experiment was performed in two spring–summer growing seasons (2014 and 2015) by grafting eggplant 'Scarlatti' F_1 hybrid on two accessions of *S. aethiopicum* gr. *gilo* and on the interspecific hybrid *S. melongena* × *S. aehtiopicum* gr. *gilo* in comparison to the most common eggplant rootstock *S. torvum*. Results indicate that *S. melongena* × *S. aethiopicum* gr. *gilo* interspecific hybrid and *S. torvum* improved grafting success, plant vigor, early flowering and yield in 'Scarlatti' F_1 scion. All rootstocks tested did not negatively influence fruit apparent quality traits and fruit quality composition. Moreover, fruit glycoalkaloids content remained below the recommended threshold value. These findings suggest that the use of *S. melongena* × *S. aethiopicum* gr. *gilo* interspecific hybrid as rootstock may be a good alternative to the most commonly used *S. torvum*.

Keywords: wild eggplant relative; interspecific hybrid; scion/rootstock combination; plant vigour; yield; fruit quality attributes

---

1. Introduction

Eggplant (*Solanum melongena* L.) is the world's sixth most important vegetable after tomato, watermelon, onion, cabbage, and cucumber [1] and the most important *Solanum* crop native to the Old World [2]. Sicily is one of the most important production areas within the Mediterranean Basin that is considered a secondary center of diversification [3]. Soilborne diseases arising from continuous cropping are major problems that limit productivity in eggplant [4]. Therefore, improving soilborne resistance is one of the major scientific and economic challenges in eggplant. Grafting has been attempted as an effective mean to control soilborne diseases and abiotic stresses while simultaneously not adversely affect plant growth or even to improve vigor and yield [4]. Among the abiotic stresses that affect vegetable crops, salinity continues to be a main factor in reducing vegetable crop yield and profits in many arid and semi-arid regions. In this respect, Colla et al. [5] demonstrated the effectiveness of grafting to improve salinity tolerance in cucumber. However, the use of specific rootstocks may offer many other advantages such as drought resistance [6] or heavy metal tolerance [7].

Solanum torvum Sw. is the rootstock commonly used for grafting, especially in the most intensive protected cultivation. However, its use has been limited due to lack of rapid and homogeneous synchronized seed germination [8]. Studies have shown that grafting can affect yield and fruit quality in eggplant [9,10]. Kyriacou et al. [11], in their review, reassessed that grafting itself and the prevalence of particular type of commercial rootstocks influence vegetable fruit quality and, partially, storability. They also reported that grafting significantly affects morphometric traits, textural characteristics, sweetness and acidity, as well as functional compounds in eggplant. However, according to Kyriacou et al. [11], current reports on the changes conferred by grafting on eggplant fruit quality provide conflicting information probably due to the environment in which experiment were run. In particular, they stressed both possible rootstock–scion interactions, underscoring graft combinations and different stemming from failure to standardize fruit harvest maturity. In this respect, Gisbert et al. [9] and Sabatino et al. [10] found that the use of interspecific hybrid rootstocks derived from compatible crosses of eggplant with related species can be a valuable approach to improve/preserve eggplant production.

Solanum aethiopicum gr. *gilo* (scarlet eggplant), a close relative species of *S. melongena*, has been considered a noticeable resource for eggplant genetic improvement and as potential rootstock. It presents traits of interest, including resistance to *F. oxysporum* f. sp. *melongenae*, *R. solanacearum* [12,13], and root-knot nematodes [14]. Since scarlet eggplant is a cultivated species, it does not display those undesirable traits that are commonly present in the wild relatives of eggplant such as small fruit production, presence of prickles, and high saponins and glycoalcaloids concentrations [15]. A great genetic and morphological variability has been evidenced in *S. aethiopicum* [16]; therefore, it is particularly relevant to characterize the accessions for the presence and expressivity of the traits of interest before starting a breeding program. Interspecific sexual and somatic hybrids between *S. melongena* and *S. aethiopicum* gr. *gilo* have been produced and although they present a high degree of sterility, recombination between the two genomes has been demonstrated [17] and backcross generations to *S. melongena* with useful introgressions of *S. aethiopicum* gr. *gilo* obtained [13,18,19]. The aim of this work was to explore the possibility to use two accessions of *S. aethiopicum* gr. *gilo* and an initial highly fertile introgression line (BC_1) from the somatic hybrid between *S. melongena* and *S. aethiopicum* gr. *gilo* as rootstock for eggplant. The introgression line was resistant to *Fusarium oxysporum* f.sp. *melongenae* and is the founder of the series of improved eggplant lines resistant to *Fusarium* wilt carrying the resistant locus *RfoSa1* [18]. The effects of these three genetic materials when used as rootstock, on vigor, yield, and fruit quality traits of eggplant 'Scarlatti' F_1 hybrid were evaluated. The results are compared with those obtained from ungrafted, self-grafted, and *S. torvum* rootstock grafted plants. Therefore, our aim was also to validate the assumption that using interspecific hybrid rootstocks may be a good approach to enhance eggplant performance because the presence of the eggplant genome together that one of the wild/allied species may improve the overall affinity between rootstock and scion exploiting the advantage of grafting.

2. Materials and Methods

2.1. Plant Material

The eggplant F_1 hybrid 'Scarlatti' (black cylindrical shape) (EnzaZaden, The Netherlands) was used as the scion cultivar as well as the rootstock and ungrafted control. Five rootstocks were evaluated, which included materials corresponding to the three species *S. melongena* (F_1 hybrid 'Scarlatti'), *Solanum torvum*, and *Solanum aethiopicum* gr. *gilo* (two accessions, named accession 1 and accession 2), and one interspecific hybrid of *S. melongena* × *S. aethiopicum* gr. *gilo*, which is a double haploid line obtained from another culture of the tetraploid backcrosses from the somatic hybrid eggplant cv

Dourga(+) *S. aethiopicum* gr. *gilo* accession 2 with a tetraploid plant of the eggplant line DR2 [13,18]. The materials used as rootstocks originated from the germplasm collection of the Consiglio per la Ricerca in Agricoltura e l'Analisi dell'Economia Agraria CREA-GB, Centro di Ricerca Genomica e Bioinformatica (Montanaso Lombardo, Italy).

2.2. Seedlings Production and Grafting

For the production of the grafted plant material, on 10 February 2014 and 9 February 2015, *S. aethiopicum* gr. *gilo* and *S. torvum* rootstock seeds were planted in 40-cell seedling trays (cell volume of 83 cm^3), under a temperature regime of 25 °C/18 °C (day/night) in a propagation greenhouse. After 20 days, seeds of the F_1 'Scarlatti' eggplant were planted in 104-cell trays (cell volume of 33 cm^3) under the same temperature regime and planting method as the rootstocks. Due to the faster germination and growth, the *S. melongena* × *S. aethiopicum* gr. *gilo* hybrid rootstock was sown simultaneously to the F_1 'Scarlatti' hybrid. Trays were watered manually every day to maintain the substrate at water holding capacity. Seventy-five days after the sowing of *S. torvum* and *S. aethiopicum* gr. *gilo* (accessions 1 and 2), all seedlings had reached an adequate diameter for grafting. The eggplant cultivar Scarlatti was grafted onto 'Scarlatti' rootstocks (self-grafted), *S. torvum*, *S. aethiopicum* gr. *gilo* (accession 1 and accession 2), and *S. melongena* × *S. aethiopicum* gr. *gilo* rootstocks using the tube grafting method described by Lee [20] and modified by Miceli et al. [21]. Plants at the 3–4 leaf stage were used as rootstocks. The grafted plantlets were misted, incubated within a plastic tunnel in a greenhouse, and maintained at a temperature of 20 °C and a humidity rate of 95% for 7 days. After 7 days, the grafted plantlets were acclimatized to the natural conditions of the greenhouse by slowly dropping the humidity (RH 70–80%) for 3 days, until they were ready for transplant.

2.3. Cultivation Conditions

'Scarlatti' plants ungrafted, self-grafted, and grafted onto *S. melongena*, *S. torvum*, and accession 1 and 2 of *S. aethiopicum* gr. *gilo* rootstocks were transplanted on 5 May 2014 and 4 May 2015 on a Typic Rhodoxeralf soil in the experimental farm of the Department of Agricultural, Alimentary and Forest Sciences (SAAF) (longitude 13°19′ E, latitude 38°09′ N), of the University of Palermo, Italy. The field trials were conducted in a sandy clay loam soil (46.5% sand, 22.3% silt, 31.2 clay) at pH 7.2 in a completely randomized design with 4 replications of 10 plants (40 plants per treatment). In both years, the preceding crop was cauliflower. Plants were spaced 1.0 m between rows and 0.5 m apart within the row and drip irrigated. Fertilization was applied with drip irrigation throughout the growing cycle and consisted of 250 kg nitrogen ha^{-1}, 150 kg phosphorous pentoxide ha^{-1}, and 250 kg potassium oxide ha^{-1} (Yara, Oslo, Norway). Standard horticultural practices for eggplant production in the Mediterranean environment were adopted [22].

Average daily temperature during the experimental period from May to August of 2014 and 2015 was obtained from the meteorological station of the experimental farm of the Department SAAF (Figure S1). In terms of temperatures, the weather during the experimental period in 2014 and 2015 was similar to the long term average. Nevertheless, the average monthly temperatures showed the highest negative deviation in June (1.8 and 2.2 °C in 2014 and 2015, respectively).

2.4. Grafting Success, Biometric Parameters, Yield, and Apparent Fruit Quality Evaluation

Grafting success was recorded after two weeks from grafting and was calculated on 100 grafted seedlings for each rootstock used. Plant height and root collar diameter (via a digital caliper) at 50, 80, and 110 days after transplanting (DAT), number of leaves at 50 DAT, and aboveground biomass produced at the end of fruit harvest (including total yield and vegetative part produced (weight of the plant at the end of harvests plus vegetative part removed by pruning after plant branching)) were determined. First flower formation (expressed as DAT) were also recorded.

Immediately after harvesting fruits were weighed. Total yield (kg plant^{-1}), marketable yield (kg plant^{-1}), and number of marketable fruits per plant were collected. Average marketable fruit weight (g) was calculated.

Color (a* and b* parameters -CIELab) was measured on four replications of five fruits per rootstock–scion combination. The records were taken on two opposite point of eggplant fruit skin (equatorial zone) by a colorimeter (Chroma-meter CR-400, Minolta Corporation, Ltd., Osaka, Japan). Hue angle (H°) was calculated as follows: H° = arctan(b*/a*).

Fruit firmness was determined by measuring its resistance to the plunger of a digital penetrometer (Trsnc, Italy). Each fruit was subjected to a compression in two opposite point in the equatorial part using a 6 mm diameter stainless steel cylinder probe. The mean peak force was calculated in Newton (N).

Apparent fruit quality traits of 'Scarlatti' eggplant fruit were measured in four replications of ten representative commercially mature fruits from non-grafted and self-grafted plants, and from plants derived from 'Scarlatti' scions grafted onto *S. torvum* and accession 1 and 2 of *S. aethiopicum* gr. *gilo* hybrid rootstocks. Fruit length/width ratios were calculated. Several traits were measured in an arbitrary scale according to the European Eggplant Genetic Resources Network (EGGNET) descriptors [23]. These traits included fruit curvature (1 = none; 9 = U-shaped), fruit cross section (1 = circular; 9 = very irregular), fruit calyx length (1 = very short (>10%); 9 = very long (>75%)), and fruit calyx prickles (0 = none; 9 = very many (>30)). In addition to these EGGNET descriptors, seed index (0 = none; 5 = very many (>80) seeds visible in a longitudinal fruit section) was measured.

2.5. Pulp Browning, Soluble Solid Content, and Chemicals

The colorimeter was also used to determine the lightness of fruit pulp by measuring L* value (0 = black and 100 = white). Fruits were sectioned in the equatorial part and the color of the pulp was measured immediately after cutting (L_0) and after 30 min (L_{30}) in two areas (central and lateral) of the section. The oxidation potential was estimated using Larrigaudiere et al. [24] method with little modifications as in part suggested by Concellòn et al. [25]. The oxidation potential was expressed as $\Delta L_{30} = (L_{30} - L_0)$.

Sampling for the quality analysis of the fruits was carried out as described by Sabatino et al. [10] and Sabatino et al. [26]. Thus, 3–5 commercially mature fruits for each replication from the second and third harvest were used; only healthy fruits were chosen. Care was taken to ensure that each sample contained the same percentage weight of apical, middle, and distal parts of the fruits. Qualitative fruit characteristic analyses were conducted on fruits harvested from labeled fruits (the flowers were labeled at the fruit set stage) and all fruits were harvested after 35 days from labeling (fruit commercial maturity stage).

Sample of the fruit pulp were squeezed by hand with a garlic squeezer. The juice was filtered and soluble solids content (SSC) was measured using a digital refractometer (MTD-045nD, Three-In-One Enterprises Co. Ltd. Taiwan).

Fruit dry matter percentage was determined in samples dried at 105 °C until constant weight as 100% × (dry weight/fresh weight).

Proteins, metals, total anthocyanins, chlorogenic acid, and glycoalkaloids were determined only in the second trial (2015). Protein concentration was determined from N content obtained from the Kjeldahl method. In particular, a sample rate was subjected to acid-catalyzed mineralization to turn the organic nitrogen into ammoniacal nitrogen. The ammoniacal nitrogen was then distilled in an alkaline pH. The ammonia formed during this distillation was collected in a boric acid solution and determined through titrimetric dosage. The protein concentration was reported as N × 6.25. Phosphorus content were assessed using colorimetry [27]. Ca, Mg, and K were determined using atomic absorption spectroscopy following wet mineralization [28].

Polyphenols were extracted and analyzed according to Stommel and Whitaker [29] with minor modifications. The analyses were performed through a Waters E-Alliance HPLC system constituted by a 2695 separations module with quaternary pump, auto sampler, and a 2996 photodiode array detector; data were acquired and analyzed with Waters Empower software on a personal computer. A binary mobile phase gradient of methanol in 0.01% aqueous phosphoric acid was used according to this procedure: 0–15 min, linear increase from 5 to 25% methanol; 15–28 min, linear increase from 25 to 50% methanol; 28–30 min, linear increase from 50 to 100% methanol; 30–32 min, 100% methanol; 32–36 min, linear decrease from 100 to 5% methanol; 36–43 min, 5% methanol. The flow rate was 0.8 mL/min. Quantification of chlorogenic acid (CA), carried out after a RP-HPLC separation, was based on absorbance at 325 nm relative to the sesamol internal standard and an external standard of authentic CA (Sigma-Aldrich, St. Louis, MO). The results were expressed as mg·100 g^{-1} of fw.

The extraction and the analysis of anthocyanins were carried out on 200 mg of lyophilized and powdered peel as reported in Mennella et al. [30]. Briefly, the chromatographic separations were performed at a flow rate of 0.8 mL/min and at 0.1 absorbance units full scale (AUFS). Purified delphinidin-3-rutinoside (D3R, Polyphenols Laboratories AS, Sandnes, Norway) was used as external standard in RP-HPLC analyses, with a different retention time (23.9 min) compared to delphinidin-3-(p-coumaroylrutinoside)-5-glucoside (nasunin), that was eluted at a longer retention time (25.8 min for cis-nasunin and 26.1 min for trans-nasunin, respectively). As for nasunin quantification, a partially purified standard was used according to Lo Scalzo et al. [31]. The results were expressed as mg·100 g^{-1} of peel fw; the limit of detection was 2.00 mg·100 g^{-1} of peel dw.

Glycoalkaloids were extracted as described by Birner [32] with some modifications. Glycoalkaloid extraction was detected from 0.5 g lyophilized samples and powdered flesh tissue by 95% ethanol. The analyses were carried out by means of RP-HPLC according to Kuronen et al. [33] using partially purified solasonine and solamargine as the external standard. The data were expressed as mg·100 g^{-1} fw; the limit of detection was 0.03 mg·100 g^{-1} of dw.

2.6. Statistical Analysis

Data were statistically analyzed by a two-way analysis of variance (ANOVA) (year (Y) × rootstock (R)) using SPSS software package version 14.0 (StatSoft, Inc., Chicago, IL, USA). Proteins, metals, total anthocyanins, chlorogenic acid and glycoalkaloids, which were determined only in the second trial (2015), were analyzed via one-factor analysis of variance (ANOVA) using a fixed-effects model for the effect of rootstock treatment. For data expressed in percentage, the arcsin transformation before ANOVA analysis [$\emptyset = \arcsin(p/100)^{1/2}$] was applied. To separate treatment means within each measured parameter, Tukey HSD test was performed at $p = 0.05$.

3. Results

The interaction between rootstock and year was always not significant; therefore, for the sake of simplicity it was omitted from the tables and graphs. Similarly, the factor year was statistically significant only for first flower formation and is reported only in the correspondent table.

3.1. Grafting Success and Plant Biometric Parameters

Rootstock significantly affected grafting success, number of leaves at 50 DAT and aboveground biomass (Table 1).

Table 1. Year and rootstock effects on grafting success and plant biometric parameters of 'Scarlatti' F_1 scion.

Treatments	Grafting Success (%)	No. Leaves 50 DAT (No.)	Aboveground Biomass (kg)	First Flower Formation (DAT)
Rootstock				
'Scarlatti' ungrafted	-	17.5 b	4.5 ab	53.1 a
S. torvum	99.3 a	24.3 ab	5.1 a	49.0 d
'Scarlatti' self-grafted	99.6 a	25.4 a	4.9 a	52.9 ab
Hybrid *S. melongena* × *S. aethiopicum*	99.0 a	21.1 ab	4.6 ab	49.3 d
S. aethiopicum (accession 1)	93.6 b	27.3 a	3.7 c	51.9 bc
S. aethiopicum (accession 2)	89.1 b	17.6 b	4.0 bc	51.5 c
Year				
2014	96.5 a	22.3 a	4.4 a	52.0 a
2015	95.8 a	22.1 a	4.5 a	50.6 b
Significance				
Rootstock	***	**	***	***
Year	NS	NS	NS	***

DAT, days after transplanting. Values within a column and a year followed by the same letter are not significantly different at $p \leq 0.05$ (Tukey HSD Test). The significance is designated by asterisks as follows: *, statistically significant differences at *p*-value below 0.05; **, statistically significant differences at p-value below 0.01; ***, statistically significant differences at *p*-value below 0.001; NS = not significant.

The tube grafting method proved highly efficient with success percentages that ranged from 89.1 to 99.6% among the materials used (Table 1). No significant differences were found in terms of success rate among 'Scarlatti', *S. torvum*, and *S. melongena* × *S. aethiopicum* gr. *gilo* hybrid rootstocks, which showed percentages of graft success ranging from 99% (*S. melongena* × *S. aethiopicum* gr. *gilo* hybrid) to 99.6% ('Scarlatti'). In contrast, *S. aethiopicum* gr. *gilo* (accessions 1 and 2) had a significantly lower percentage of success (93.6 and 89.1%, respectively) with respect to the other rootstocks (Table 1). No visible disaffinity signals at the grafting zone were observed for any rootstock–scion combination.

The mean number of leaves at 50 DAT varied between 17.5 and 27.3 leaves per plant for the 'Scarlatti' ungrafted and *S. aethiopicum* gr. *gilo* (accession 1) rootstock, respectively (Table 1). 'Scarlatti' scions grafted onto *S. melongena* ('Scarlatti') and *S. aethiopicum* gr. *gilo* (accession 1) rootstocks had a higher number of leaves at 50 DAT than those ungrafted or grafted onto *S. aethiopicum* gr. *gilo* (accession 2). However, eggplant plants grafted onto *S. torvum* and *S. melongena* × *S. aethiopicum* gr. *gilo* hybrid did not significantly differ from those grafted onto the other rootstocks tested.

Aboveground biomass produced ranged from 5.1 kg for 'Scarlatti' grafted on *S. torvum* rootstock to 3.7 kg for those grafted on *S. aethiopicum* gr. *gilo* (accession 1) rootstock (Table 1). Plants grafted onto *S. torvum* rootstock did not significantly differ compared to those grafted onto *S. melongena* (self-grafted plants), whereas the lowest aboveground biomass values were collected from eggplant plants grafted onto *S. aethiopicum* gr. *gilo* (accessions 1 and 2) (3.7 and 4.0 kg).

Regardless of the rootstock, the second trial (2015) gave the shortest time of first flower formation (50.6 DAT), whereas, irrespective of the year, plants grafted onto *S. melongena* × *S. aethiopicum* gr. *gilo* hybrid and *S. torvum* gave the shortest time of first flower formation (49.3 and 49.0 DAT, respectively). 'Scarlatti' ungrafted had the longest first flower formation (53.1 DAT) (Table 1).

Rootstock significantly influenced plant height at 50, 80, and 110 DAT (Figure 1).

Plants grafted onto *S. aethiopicum* gr. *gilo* (accession 1) had the highest plant height, whereas plants grafted onto *S. aethiopicum* gr. *gilo* (accession 2) had the lowest values (Figure S2).

At 80 DAT, no significant differences were found in plant height among 'Scarlatti', *S. torvum*, *S. aethiopicum* gr. *gilo* (accession 1), and *S. melongena* × *S. aethiopicum* gr. *gilo* hybrid rootstocks. *S. aethiopicum* gr. *gilo* (accession 2) had a significantly lower plant height value with respect to the other rootstocks (Figure 1). However, plant height at 50 and 110 DAT, in 'Scarlatti' F_1 grafted onto

S. melongena × *S. aethiopicum* gr. *gilo* hybrid rootstocks did not significantly differ from those grafted onto *S. aethiopicum* gr. *gilo* (accession 2).

Rootstock significantly influenced root collar at 50, 80, and 110 DAT (Figure S3). Plants grafted onto *S. aethiopicum* gr. *gilo* (accession 2) and *S. melongena* × *S. aethiopicum* gr. *gilo* hybrid rootstocks had the lowest root collar values, whereas plants grafted on *S. aethiopicum* gr. *gilo* (accession 1) had the highest value (Figure S3).

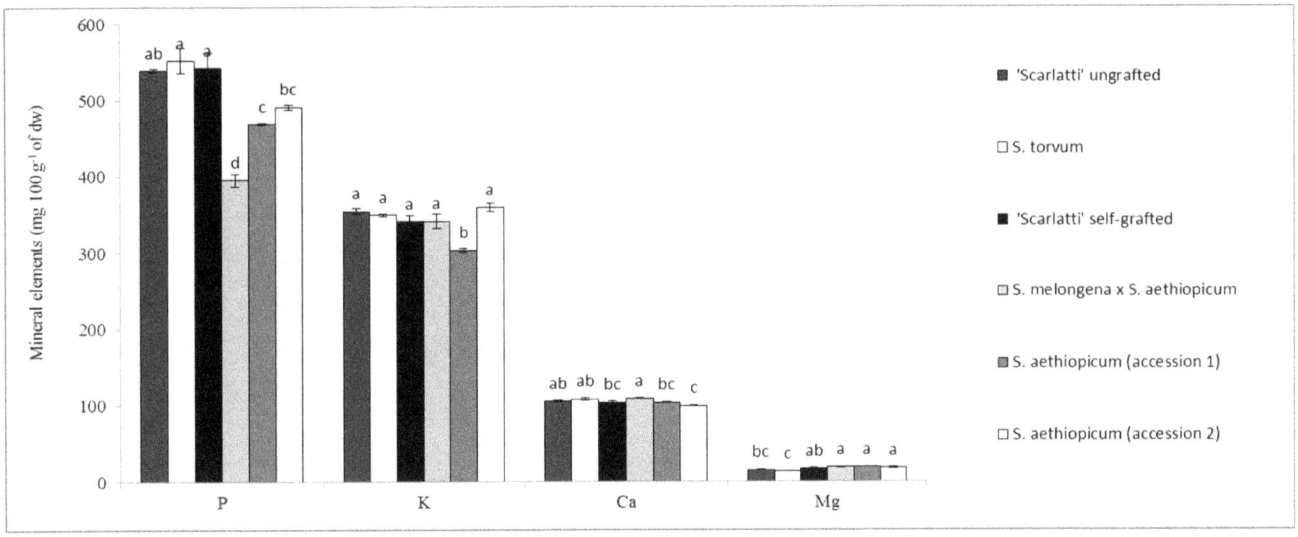

Figure 1. Mineral elements of 'Scarlatti' eggplant fruits produced from ungrafted, self-grafted, and grafted onto *S. torvum*, *S. aethiopicum* gr. *gilo* (accession 1), *S. aethiopicum* gr. *gilo* (accession 2), and *S. melongena* × *S. aethiopicum* gr. *gilo* rootstocks. Bars with different letters are significant by Tukey HSD Test ($p < 0.05$).

3.2. Yield

Rootstock significantly influenced total yield, marketable yield and number of marketable fruits (Table 2).

Table 2. Rootstock effects on yield traits of 'Scarlatti' F_1 scion.

Treatments	Total Yield Plant^{-1} (kg)	Marketable Yield Plant^{-1} (kg)	No. Marketable Fruits Plant^{-1} (No.)	Average Fruit Weight (g)
Rootstock				
'Scarlatti' ungrafted	3.4 ab	3.3 ab	14.5 a	231.3 ns
S. torvum	3.8 a	3.6 a	16.6 a	216.8 ns
'Scarlatti' self-grafted	3.9 a	3.6 a	16.2 a	223.5 ns
Hybrid *S. melongena* × *S. aethiopicum*	3.4 ab	3.2 ab	15.1 a	213.9 ns
S. aethiopicum (accession 1)	2.7 c	2.5 c	9.1 b	330.1 ns
S. aethiopicum (accession 2)	3.1 bc	2.9 bc	14.1 a	207.5 ns
Significance				
Rootstock	***	**	***	NS

Values within a column and a year followed by the same letter are not significantly different at $p \leq 0.05$ (Tukey HSD Test). The significance is designated by asterisks as follows: *, statistically significant differences at p-value below 0.05; **, statistically significant differences at p-value below 0.01; ***, statistically significant differences at p-value below 0.001; NS = not significant.

Total yield ranged between 2.7 kg plant^{-1} for 'Scarlatti' grafted on *S. aethiopicum* gr. *gilo* (accession 1) and 3.9 kg plant^{-1} for those grafted onto *S. melongena* (self-grafted plants) (Table 2). No significant differences were found in the total yield among 'Scarlatti' and *S. torvum* rootstocks, which had total yield that ranged from 3.8 kg plant^{-1} (*S. torvum*) to 3.9 kg plant^{-1} ('Scarlatti' self-grafted). *S. aethiopicum* gr. *gilo* (accessions 1 and 2) had a significantly lower total yield (2.7 and 3.1 kg plant^{-1}, respectively) than the other rootstocks (Table 2). However, plants grafted onto *S. melongena* × *S. aethiopicum* gr. *gilo* rootstock and 'Scarlatti' ungrafted plants did not significantly differ in total yield compared to the plants grafted onto 'Scarlatti' and *S. torvum* rootstocks. Data collected on marketable yield supported the trend established for total yield (Table 2). Plants with *S. aethiopicum* gr. *gilo* (accession 1) as rootstock had a significantly lower fruit number in comparison to those grafted onto 'Scarlatti', *S. torvum*, and *S. melongena* × *S. aethiopicum* gr. *gilo* hybrid, and not different from those grafted onto *S. aethiopicum* gr. *gilo* (accession 2) (Table 2). In addition, no differences were found among plants grafted onto 'Scarlatti', *S. torvum, S. melongena* × *S. aethiopicum* gr. *gilo* hybrid, and *S. aethiopicum* gr. *gilo* (accession 2) for number of marketable fruits plant^{-1} which produced a significantly higher number of fruits than accession 1 of *S. aethiopicum* gr. *gilo*. No significant differences among treatments were found for the average fruit weight (Table 2).

3.3. Apparent Fruit Quality

Rootstock significantly influenced fruit length and fruit calyx length parameters (Table S1). Fruit from plants grafted on *S. melongena* × *S. aethiopicum* gr. *gilo* hybrid rootstock were significantly more elongated than those from plants grafted onto 'Scarlatti' rootstock or from ungrafted 'Scarlatti' plants (Table S1), which in turn had more elongated fruits than those from plants grafted onto *S. aethiopicum* gr. *gilo* (accession 2).

Fruit calyx length in 'Scarlatti' scion grafted onto *S. aethiopicum* gr. *gilo* × *S. melongena* hybrid rootstock was significantly higher than in plants grafted onto *S. torvum*; however, fruits from self-grafted and ungrafted 'Scarlatti' plants did not significantly differ neither from fruits from plants grafted onto *S. melongena* × *S. aethiopicum* gr. *gilo* hybrid nor from those from plants grafted onto *S. torvum*. No significant differences among rootstocks tested were found for fruit width (average of 4.7 cm), fruit length–width ratio (average of 3.9), fruit curvature (average of 1.0), fruit cross-section (average of 7.1), fruit calyx prickles (average of 0.2), or seeds index (average of 1.1) (Table S1).

3.4. Intrinsic and Extrinsic Fruit Quality

Hue color parameter in fruits from plants grafted onto *S.torvum* and *S. aethiopicum* gr. *gilo* (accession 1) had the highest values (360.3 and 360.2, respectively) (Table 3).

Table 3. Rootstock effects on Hue°, fruit dry matter, firmness, and soluble solids content (SSC) in 'Scarlatti' F_1 scion.

Treatments	H°	Fruit Dry Matter (%)	Firmness (N)	SSC (°Brix)
Rootstock				
Scarlatti' ungrafted	360.1 b	7.3 ns	60.0 ns	4.5 b
S. torvum	360.3 a	7.2 ns	43.2 ns	3.9 c
Scarlatti' self-grafted	360.1 b	6.8 ns	44.2 ns	4.1 c
Hybrid *S. melongena* × *S. aethiopicum*	360.0 b	6.8 ns	49.0 ns	4.6 ab
S. aethiopicum (accession 1)	360.2 ab	6.9 ns	50.6 ns	4.6 ab
S. aethiopicum (accession 2)	360.1 b	6.8 ns	38.8 ns	4.9 a
Significance				
Rootstock	***	NS	NS	***

Values within a column and a year followed by the same letter are not significantly different at $p \leq 0.05$ (Tukey HSD Test). The significance is designated by asterisks as follows: *, statistically significant differences at *p*-value below 0.05; **, statistically significant differences at p-value below 0.01; ***, statistically significant differences at *p*-value below 0.001; NS = not significant.

Treatments tested had no effects on fruit dry matter and firmness (Table 3).

Rootstock significantly influenced SSC (Table 3). The highest value of SSC (4.6, 4.9 and 4.6 °Brix, respectively) were found in plants grafted onto *S. aethiopicum* gr. *gilo* (accessions 1 and 2) and *S.*

melongena × *S. aethiopicum* gr. *gilo* hybrid, whereas the lowest value (3.9 and 4.1 °Brix, respectively) were observed in plants grafted onto *S. torvum* and 'Scarlatti'(Table 3).

'Scarlatti' scions grafted onto *S. aethiopicum* gr. *gilo* (accession 1) and *S. melongena* × *S. aethiopicum* gr. *gilo*, had the highest values of L_0 central area (Table 4), while fruits from plants grafted onto *S. aethiopicum* gr. *gilo* (accession 2) rootstocks had the lowest value.

Table 4. Rootstock effect on pulp lightness measured immediately after cutting (L_0) in the central and lateral area and the fruit browning (ΔL_{30}) of the lateral area in 'Scarlatti' F_1 scion.

Treatments	L_0 Central Area	L_0 Lateral Area	ΔL_{30} Lateral Area
Rootstock			
'Scarlatti' ungrafted	82.4 bc	80.6 b	2.5 bc
S. torvum	82.9 b	81.0 b	2.3 c
'Scarlatti' self-grafted	82.5 b	80.6 b	2.5 bc
Hybrid *S. melongena* × *S. aethiopicum*	83.5 ab	81.5 ab	2.7 ab
S. aethiopicum (accession 1)	84.8 a	82.9 a	2.7 ab
S. aethiopicum (accession2)	80.7 c	78.9 c	2.9 a
Significance			
Rootstock	***	***	***

Values within a column and a year followed by the same letter are not significantly different at $p \leq 0.05$ (Tukey HSD Test). The significance is designated by asterisks as follows: *, statistically significant differences at p-value below 0.05; **, statistically significant differences at *p*-value below 0.01; ***, statistically significant differences at *p*-value below 0.001; NS = not significant.

Data collected for L_0 lateral area supported the trend established for L_0 central area (Table 4).

S. aethiopicum gr. *gilo* (accessions 1 and 2) and *S. melongena* × *S. aethiopicum* gr. *gilo* hybrid rootstocks induced the highest ΔL_{30} lateral area values (Table 4). While, *S. torvum* rootstock induced the lowest one.

ANOVA for total anthocyanins was not statistically significant (Table 5).

Table 5. Rootstock effect on total antocyanins, glycoalkaloids, chlorogenic acid, and proteins in 'Scarlatti' F_1 scion.

Treatments	Total Anthocyanins (mg·100 g^{-1} of fw)	Glycoalkaloids (mg·100 g^{-1} of fw)	Chlorogenic Acid (mg·100 g^{-1} of fw)	Proteins (g·100 g^{-1} of fw)
Rootstock				
'Scarlatti' ungrafted	60.6 ns	8.4 a	21.7 ab	1.0 ab
S. torvum	43.0 ns	1.0 b	40.3 a	1.0 ab
'Scarlatti' self-grafted	47.8 ns	2.7 b	34.1 a	1.0 ab
Hybrid *S. melongena* × *S. aethiopicum*	70.8 ns	8.6 a	4.4 b	0.9 bc
S. aethiopicum (accession 1)	80.0 ns	3.7 b	40.0 a	1.1 ab
S. aethiopicum (accession 2)	106.4 ns	1.9 b	17.9 ab	0.8 d
Significance				
Rootstock	NS	*	**	***

Values within a column and a year followed by the same letter are not significantly different at $p \leq 0.05$ (Tukey HSD Test). The significance is designated by asterisks as follows: *, statistically significant differences at *p*-value below 0.05; **, statistically significant differences at *p*-value below 0.01; ***, statistically significant differences at *p*-value below 0.001; NS = not significant; fw = fresh weight.

The highest glycoalkaloids content was observed in fruits from ungrafted plants and in those from plants grafted onto *S. melongena* × *S. aethiopicum* gr. *gilo* rootstocks (Table 5).

Fruits from 'Scarlatti' grafted on *S. torvum*, *S. aethiopicum* gr. *gilo* (accession 1), and 'Scarlatti' self-grafted plants had the highest chlorogenic acid content, whereas, fruits from plants grafted onto *S. melongena* × *S. aethiopicum* gr. *gilo* hybrid rootstock showed the lowest value (Table 5).

We found that protein content in fruits from plants grafted onto *S. aethiopicum* gr. *gilo* (accession 1) was slightly higher than those from 'Scarlatti' ungrafted, self-grafted, and *S. torvum* grafted plants. The lowest protein content was detected in fruits from plants grafted onto *S. aethiopicum* gr. *gilo* (accession 2) (Table 5).

Rootstock significantly affected mineral content (Figure 1).

Ungrafted, self-grafted, and *S. torvum* rootstock grafted plants had the highest fruit P content (538, 551, and 541 mg·100 g^{-1} of dw, respectively). The lowest fruit P content value (395.4 mg·100 g^{-1} of dw) was found in *S.melongena* × *S. aethiopicum* gr. *gilo* hybrid rootstock (Figure 1). Fruits harvested from plants grafted on *S. aethiopicum* gr. *gilo* (accession 1) rootstock had a significantly lower K content (302.9 mg·100 g^{-1} of dw) in comparison to those grafted onto 'Scarlatti', *S. torvum*, *S. melongena* × *S. aethiopicum* gr. *gilo* hybrid, *S. aethiopicum* gr. *gilo* (accession 2), and 'Scarlatti ungrafted (341.0, 349.0, 341.3, 359.1, and 354.4 mg·100 g^{-1} of dw, respectively) (Figure 1). Moreover, no significant differences were found among plants grafted onto 'Scarlatti', *S. torvum*, *S. melongena* × *S. aethiopicum* gr. *gilo* hybrid, *S. aethiopicum* gr. *gilo* (accession 2), and 'Scarlatti ungrafted for fruit K content. 'Scarlatti' grafted onto *S. melongena* × *S. aethiopicum* gr. *gilo* hybrid gave the highest fruit Ca content (108.9 mg·100 g^{-1} of dw), whereas fruits from plants grafted *S. aethiopicum* gr. *gilo* (accession 2) gave the lowest Ca fruit content (99.4 mg·100 g^{-1} of dw) (Figure 1). 'Scarlatti' grafted onto *S. aethiopicum* gr. *gilo* (accessions 1 and 2) and *S. melongena* × *S. aethiopicum* gr. *gilo* hybrid had the highest Mg content (19.3, 18.1, and 18.6 mg 100 g^{-1} of dw, respectively) (Figure 1), whereas the lowest Mg contents were found in fruits from 'Scarlatti' grafted on *S. torvum* (13.8 mg·100 g^{-1} of dw).

4. Discussion

Nowadays, both a high yield performance and fruit quality attributes (intrinsic and extrinsic) are very important due to the functional role of vegetables in the human diet [29].

Along with approaches such as genetic improvement and optimization of the cultivation conditions, herbaceous grafting technique represents a toolbox for securing or increasing yield stability and fruit quality with or without stress conditions [26,34–37]. In this paper, we investigated the effects of *S. aethiopicum* gr. *gilo* (accessions 1 and 2) versus the interspecific hybrid of *S. melongena* × *S. aethiopicum* gr. *gilo* on plant performance and fruit quality attributes of 'Scarlatti' eggplant F_1 hybrid (black cylindrical shape). All the results are compared with those obtained from non-grafted, self-grafted, and *S. torvum* rootstock grafted plants.

Grafting success is an important concern when using new scion–rootstock combinations [37]. Generally, Solanaceous crops are grafted by cleft or tube grafting methods [4,20,38]. In our experiment using the tube grafting technique, grafting success rates of 89.1% for *S. aethiopicum* gr. *gilo* (accession 1), 93.6% for *S. aethiopicum* gr. *gilo* (accession 2), and over 99% for grafting onto 'Scarlatti', *S. torvum*, and *S. melongena* × *S. aethiopicum* gr. *gilo* hybrid, were observed. Our findings are in accord with those obtained by Bletsos et al. [4] and Rahman et al. [39], who reported a high grafting success rate of eggplant cultivars grafted onto *S. torvum*. Although, *Solanum aethiopicum* gr. *gilo* is a close relative species of *S. melongena* [16], from our results it appears that graft incompatibility might exist especially when 'Scarlatti' is grafted onto *S. aethiopicum* gr. *gilo* accession 2. Thus, we may assume that the genotype is a crucial aspect for the identification of a suitable rootstock as significant differences may exist between different accessions of a species. Regarding the *S. melongena* × *S. aethiopicum* gr. *gilo* hybrid rootstock, our outcomes are in accord with those obtained by Gisbert et al. [9], who reported a high graft success rate when interspecific hybrids of eggplants were used as rootstocks. Our results are also consistent with those of Sabatino et al. [10], who reported a high graft success rate of eggplant 'Birgah' F_1 hybrid (violet globose shape) grafted onto *S. melongena* × *S. aethiopicum* rootstock, confirming the capacity of this interspecific hybrid to give good results with two different types of common eggplant. In fact, Birgah is one of the violet F_1 hybrid belonging to a local Sicilian fruit typology that has been subjected to lower breeding improvement efforts when compared to elongated black-purple typelike 'Scarlatti', which is world-wide diffused especially in the most industrialized countries. Our experiment confirms that rootstock might play a major role on plant biometric parameters. Plant height, which may be considered an indicator of vigor was highest in plants with *S. aethiopicum* gr. *gilo* (accession 1), *S. torvum*, and interspecific hybrid *S. melongena* × *S. aethiopicum* gr. *gilo* rootstocks and lowest in those plants with *S. aethiopicum* gr. *gilo* (accession 2) rootstock grafts. Therefore, vigor of the rootstock is important in conferring scion vigor. In the absence of grafting incompatibility signals, grafted plants may also develop more rapidly, hence contributing to earliness. In our work, plants grafted onto *S. torvum* and *S. melongena* × *S. aethiopicum* gr. *gilo* hybrid rootstocks flowered earlier than those grafted on the other rootstocks tested. Our results are consistent with those obtained by Gisbert et al. [9] and Sabatino et al. [10], who reported that plant vigor is positively related to fruit earliness. Increased earliness has also been reported for other fruiting vegetables such as melon grafted onto *Cucurbita* rootstock [40,41]. In our experiment, grafting has proved a useful technique to enhance eggplant 'Scarlatti' F_1 hybrid yield traits. We also found that plants grafted onto *S. melongena* × *S. aethiopicum* gr. *gilo* hybrid rootstock had higher total and marketable yield plant^{-1} than plants grafted onto *S. aethiopicum* gr. *gilo* (accessions 1 and 2), confirming that this latter rootstock has little value for enhancing eggplant yield. Conversely, interspecific hybrid of *S. melongena* × *S. aethiopicum* gr. *gilo* rootstock demonstrated great improvements in agronomic performance due to grafting. However, in our study grafting did not increase the number of marketable fruits. These findings are different from those of Gisbert et al. [9] and Sabatino et al. [10], who reported that eggplants grafted onto interspecific hybrids produced consistently more fruits per plant than ungrafted ones. However, despite in the present work grafting did not significantly improve the number of fruits per plant, the data recorded in our experiment supported the trend established in the experiments by Gisbert et al. [9] and Sabatino et al. [10]. On the other hand, these findings seem to be consistent with those obtained by Maršič et al. [42], who reported that, in the second trial, grafting did not increase the number of fruit per plant in eggplant 'Galine' F_1 (eggplant cylindrical shape). Therefore, it seems that the number of fruit per plant is a yield component influenced by the genetic background of the rootstock scion combination.

Apparent visual quality is an important criterion when consumers decide to purchase eggplant fruits. In our study, rootstocks affect fruit length and fruit calyx length. Our results are in accord with those obtained by Gisbert et al. [9], who hypothesized that fruit shape changes are probably due to changes in the concentration of growth regulators induced by the rootstock. Our findings on apparent visual quality are partially different from those of Sabatino et al. [10], who reported that rootstock significantly influences fruit width but has no significant effect on calyx length. Thus, it seems that, although fruit shape in eggplant is highly heritable and under genetic control [43,44] and even though rootstocks may affect cultivar fruit shape parameters due to changes in the concentration of growth regulators [9], rootstock may differently affect specific fruit shape characters in relation to the scion genotype.

Our results demonstrated that rootstock may enhance H° color parameter of skin eggplant fruit. Our outcomes are in contrast with those of Moncada et al. [45], who revealed that *S. torvum* rootstock decreased H° parameter in 'Black Bell', 'Black Moon', and 'Longo' eggplants. Thus, it seems that fruit skin color parameters are differently affected by scion–rootstock combination. Moreover, in our study, fruits from plants grafted onto *S. torvum* and *S. melongena* × *S. aethiopicum* gr. *gilo* hybrid had the highest H° color parameter value, which is an important commercial and qualitative prerequisite in eggplant fruit. Eggplant ranks among the top 10 vegetables in terms of antioxidant fruit activity. Accordingly to these data, a high fruit pulp browning potential could be expected [46,47]. Mishra et al. [46], King et al. [48], and Prohens et al. [49] revealed that eggplant cultivars differed in their extents of post-cut browning, which could be due to variations in the PPO activity or level of soluble phenolics. Our results on pulp browning are consistent with the findings of Sabatino et al. [10], who reported little or no effect of grafting by using *S. torvum* rootstock. However, we also found that fruits from plant grafted onto *S. aethiopicum* gr. *gilo* (accessions 1 and 2) revealed the highest value in terms of ΔL_{30}.

Mennella et al. [19] revealed that chlorogenic acid is the major monomeric phenolic compound in eggplant fruits. Although few differences were found in fruit composition traits, higher fruit chlorogenic acid content was found in fruits harvested from plants with 'Scarlatti', *S. torvum*, *S. aethiopicum* gr. *gilo* (accessions 1 and 2) rootstocks, and ungrafted plants in comparison to fruits harvested from 'Scarlatti' scions grafted onto *S. melongena* × *S. aethiopicum* gr. *gilo* hybrid. This higher chlorogenic acid concentration may be an additional indication of stress in this rootstock–scion combination, as stress conditions induce accumulation of phenolics [50,51]. Our results are dissimilar from those of Sabatino et al. [10], who reported no differences in terms of fruit chlorogenic acid content among rootstocks tested, but are consistent with those of Sabatino et al. [26] who reported that grafting significantly increased phenolic concentration in fruits from grafted eggplant landraces only in three out of four genotypes. Our outcomes on phenolic content are also consistent with those of Maršič et al. [42] who demonstrated that grafting significantly increased phenolic concentration in fruits from a grafted eggplant landrace as opposed to commercial varieties. Moreover, Toppino et al. [52] discovered environment-specific QTLs associated to the amount of chlorogenic acid. Accordingly, it seems that fruit phenolic content in eggplant is a quality trait also affected by scion–rootstock interactions. Mennella et al. [19], Friedman and McDonald [53], and Friedman [54] observed that genetic, environment, and growing conditions may affect glycoalkaloids concentration in potato, tomato, and eggplant. In addition, Jones and Fenwick [55] and Krits et al. [56] assessed that the level of total glycoalkaloids in potato tubers should not exceed 200 $mg \cdot kg^{-1}$ of fw (or 200 $mg \cdot 100\ g^{-1}$ of dw). Glycoalkaloids concentration ranged from 0.98 to 8.64 $mg \cdot 100\ g^{-1}$ of fresh weight. Rootstock affected glycoalkaloids concentration in eggplant; however, it is important to remark that glycoalkaloids amount in fruits from grafted plant remained below the recommended threshold value. Our findings are in contrast with those of Sabatino et al. [10], who observed no significant differences in terms of glycoalkaloids concentration in 'Birgah' F_1 eggplant fruits from plants grafted onto several potential rootstocks including hybrids and allied species.

Generally, changes in overall fruit composition between grafted and ungrafted plants were reported in tomato [57] and pepper [58]. Our results are in accord with those of Sabatino et al. [10], who found some differences in fruit quality attributes among fruits from ungrafted or self-grafted plants.

5. Conclusions

After the ban of methyl bromide use for soil fumigation, the soilborne diseases arising from continuous cropping are major issues that negatively affect productivity in eggplant. In this scenario, grafting is an effective tool to control vegetable crop stresses. Nowadays, *S. torvum* and tomato hybrids are the main eggplant rootstocks commercially used. However, investigations on the influence of new rootstocks on biometric parameters, yield performance, and fruit quality characteristics of different eggplant groups can provide further advances in this specific grafting field. In the present study, the interspecific hybrid of *S. melongena* × *S. aethiopicum* gr. *gilo* exhibited high grafting compatibility and high yield performance. These findings, together with the absence of negative effects on apparent 'Scarlatti' F_1 hybrid fruit quality and composition, demonstrate that this interspecific hybrid may be a viable rootstock alternative to *S. torvum*.

Supplementary Materials

Figure S1: Monthly meteorological data from May to August of 2014 and 2015 from the meteorological station of the experimental farm of the Department of Agricultural, Alimentary and Forest Sciences, University of Palermo, Italy, Figure S2: Plant height evolution of 'Scarlatti' eggplant plant ungrafted, self-grafted and grafted onto S. torvum, S. aethiopicum gr. gilo (accession 1), S. aethiopicum gr. gilo (accession 2) and S. melongena × S. aethiopicum gr. gilo rootstocks. Bars indicate the standard error of the mean. ANOVA analysis for plant height at 50, 80, and 110 DAT showed the following significance: *** (statistically significant differences at p-value below 0.001), *** and * (statistically significant differences at p-value below 0.05), respectively, Figure S3: Root collar evolution of 'Scarlatti' eggplant plant ungrafted, self-grafted and grafted onto S. torvum, S. aethiopicum gr. gilo (accession 1), S. aethiopicum gr. gilo (accession 2) and S. melongena × S. aethiopicum gr. gilo rootstocks. Bars indicate the standard error of the mean. ANOVA analysis for root collar at 50, 80, and 110 DAT showed the following significance: *** (statistically significant differences at p-value below 0.001), *** and ***, respectively. Table S1: Rootstock effects on apparent fruit quality in 'Scarlatti' F1 scion.

Author Contributions: L.S., G.I., G.L.R., and F.D., conceived and designed the research. L.S. also carried out field work, analyzed the data, and wrote the manuscript. E.P. and G.M. performed laboratory analytical determination. G.I. and G.L.R. helped draft the manuscript. All authors read and approved the manuscript.

Acknowledgments: We thank Carlo Prinzivalli for his assistance in establishing the field trial.

References

1. FAOSTAT. Production. Available online: www.fao.org (accessed on 16 January 2019).
2. Kaushik, P.; Prohens, J.; Vilanova, S.; Gramazio, P.; Plazas, M. Phenotyping of eggplant wild relatives and interspecific hybrids with conventional and phenomics descriptors provides insight for their potential utilization in breeding. *Front. Plant Sci.* **2016**, *7*, 677. [CrossRef]
3. D'Anna, F.; Sabatino, L. Morphological and agronomical characterization of eggplant genetic resources from the Sicily area. *J. Food Agri. Environ.* **2013**, *11*, 401–404.
4. Bletsos, F.; Thanassoulopoulos, C.; Roupakias, D. Effect of grafting on growth, yield and *Verticillium* wilt of eggplant. *HortScience* **2003**, *38*, 183–186. [CrossRef]
5. Colla, G.; Rouphael, Y.; Jawad, R.; Kumar, P.; Rea, E.; Cardarelli, M. The effectiveness of grafting to improve NaCl and $CaCl_2$ tolerance in cucumber. *Sci. Hortic.* **2013**, *164*, 380–391. [CrossRef]
6. Kumar, P.; Rouphael, Y.; Cardarelli, M.; Colla, G. Vegetable grafting as a tool to improve drought resistance and water use efficiency. *Front. Plant Sci.* **2017**, *8*, 1130. [CrossRef] [PubMed]
7. Kumar, P.; Lucini, L.; Rouphael, Y.; Cardarelli, M.; Kalunke, R.M.; Colla, G. Insight into the role of grafting and arbuscular mycorrhiza on cadmium stress tolerance in tomato. *Front. Plant Sci.* **2015**, *6*, 477. [CrossRef]
8. Ginoux, G.; Laterrot, H. Greffage de l'aubergine. Proceedings reflexions du portegreffe. *PHM Revue Hortic.* **1991**, *321*, 49–54.
9. Gisbert, C.; Prohens, J.; Raigón, M.D.; Stommel, J.R.; Nuez, F. Eggplant relatives as sources of variation for developing new rootstocks: Effects of grafting on eggplant yield and fruit apparent quality and composition. *Sci. Hortic.* **2011**, *128*, 14–22. [CrossRef]
10. Sabatino, L.; Iapichino, G.; D'Anna, F.; Palazzolo, E.; Mennella, G.; Rotino, G.L. Hybrids and allied species as potential rootstocks for eggplant: Effect of grafting on vigour, yield and overall fruit quality traits. *Sci. Hortic.* **2018**, *228*, 81–90. [CrossRef]
11. Kyriacou, M.C.; Rouphael, Y.; Colla, G.; Zrenner, R.; Schwarz, D. Vegetable grafting: The implications of a growing agronomic imperative for vegetable fruit quality and nutritive value. *Front. Plant Sci.* **2017**, *8*, 741. [CrossRef]
12. Cappelli, C.; Stravato, V.M.; Rotino, G.L.; Buonaurio, R. Sources of resistance among *Solanum* spp. to an Italian isolate of *Fusarium oxysporum* f sp. *melongenae*. In Proceedings of the 9th EUCARPIA Meeting Genetic and Breeding of Capsicum and Eggplant, Budapest, Hungary, 21–25 August 1995; pp. 221–224.
13. Rizza, F.; Mennella, G.; Collonnier, C.; Shiachakr, D.; Kashyap, V.; Rajam, M.V.; Prestera, M.; Rotino, G.L. Androgenic dihaploids from somatic hybrids between *Solanum melongena* and *S. aethiopicum* group *gilo* as a source of resistance to *Fusarium oxysporum* f. sp. *melongenae*. *Plant Cell Rep.* **2002**, *20*, 1022–1032.
14. Hébert, Y. Comparative resistance of nine species of the genes *Solanum* to bacterial wilt *Psedomonas solanacearum* and the nematode *Meloidogyne incognita*. Implications for the breeding of aubergine (*S. melongena*) in the humid tropical zone. *Agronomie* **1985**, *5*, 27–32. [CrossRef]
15. Prohens, J.; Plazas, M.; Raigón, M.D.; Seguí-Simarro, J.M.; Stommel, J.R.; Vilanova, S. Characterization of interspecific hybrids and backcross generations from crosses between two cultivated eggplants (*Solanum melongena* and *S. aethiopicum* Kumba group) and implications for eggplant breeding. *Euphytica* **2012**, *186*, 517–538. [CrossRef]
16. Sawadogo, B.; Bationo-Kando, P.; Sawadogo, N.; Kiebre, Z.; Kiebre, M.; Nanema, K.R.; Traore, R.E.; Sawadogo, M.; Zongo, J.D. Variation, correlation and heritability of interest characters for selection of African eggplant. *Afr. Crop Sci. J.* **2016**, *24*, 213–221. [CrossRef]
17. Toppino, L.; Mennella, G.; Rizza, F.; D'alessandro, A.; Sihachakr, D.; Rotino, G.L. ISSR and isozyme characterization of androgenetic dihaploids reveals tetrasomic inheritance in tetraploid somatic hybrids

between *Solanum melongena* and *Solanum aethiopicum* group *gilo*. *J. Hered.* **2008**, *99*, 304–315. [CrossRef] [PubMed]

18. Toppino, L.; Vale, G.; Rotino, G.L. Inheritance of fusarium wilt resistance introgressed from *Solanum aethiopicum Gilo* and *Aculeatum* groups into cultivated eggplant (*S. melongena*) and development of associated PCR-based markers. *Mol. Breed.* **2008**, *22*, 237–250. [CrossRef]
19. Mennella, G.; Rotino, G.L.; Fibiani, M.; D'Alessandro, A.; Francese, G.; Toppino, L.; Cavallanti, F.; Acciarri, N.; Lo Scalzo, R. Characterization of health-related compounds in eggplant (*Solanum melongena* L.) lines derived from introgression of allied species. *J. Agric. Food Chem.* **2010**, *58*, 7597–7603. [CrossRef]
20. Lee, J.M. Cultivation of grafted vegetables I: Current status, grafting methods and benefits. *Hort. Sci.* **1994**, *29*, 235–239. [CrossRef]
21. Miceli, A.; Sabatino, L.; Moncada, A.; Vetrano, F.; D'Anna, F. Nursery and field evaluation of eggplant grafted onto unrooted cuttings of *Solanum torvum* Sw. *Sci. Hortic.* **2014**, *178*, 203–210. [CrossRef]
22. Baixauli, C. Berenjena. In *La horticultura espãnola*; Nuez, F., Llácer, G., Eds.; Ediciones de Horticultura: Reus, Spain, 2001.
23. Prohens, J.; Blanca, J.M.; Nuez, F. Morphological and molecular variation in a collection of eggplant from a secondary center of diversity: Implications for conservation and breeding. *J. Am. Soc. Hortic. Sci.* **2005**, *130*, 54–63. [CrossRef]
24. Larrigaudiere, C.; Lentheric, I.; Vendrell, M. Relationship between enzymatic browning and internal disorders in controlled atmosphere stored pears. *J. Sci. Food Agric.* **1998**, *78*, 232–236. [CrossRef]
25. Concellòn, A.; Añón, M.; Chaves, A.R. Effect of low temperature storage on physical and physiological characteristics of eggplant fruit (*Solanum melongena* L.). *LWT* **2007**, *40*, 389–396. [CrossRef]
26. Sabatino, L.; Iapichino, G.; Maggio, A.; D'Anna, E.; Bruno, M.; D'Anna, F. Grafting affects yield and phenolic profile of *Solanum melongena* L. landraces. *J. Integr. Agric.* **2016**, *15*, 1017–1024. [CrossRef]
27. Fogg, D.N.; Wilkinson, N.T. The colorimetric determination of phosphorus. *Analist* **1958**, *83*, 406–414. [CrossRef]
28. Morand, P.; Gullo, J.L. Mineralisation des tissus vegetaux en vue du dosage de P, Ca, Mg, Na, K. *Ann. Agron.* **1970**, *21*, 229–236.
29. Stommel, J.R.; Whitaker, B.D. Phenolic acid content and composition of eggplant fruit in a germplasm core subset. *J. Am. Soc. Hortic. Sci.* **2003**, *128*, 704–710. [CrossRef]
30. Mennella, G.; Lo Scalzo, R.; Fibiani, M.; D'Alessandro, A.; Francese, G.; Toppino, L.; Acciarri, N.; De Almeida, A.E.; Rotino, G.L. Chemical and bioactive quality traits during fruit ripening in eggplant (*S. melongena* L.) and allied species. *J. Agric. Food Chem.* **2012**, *60*, 11821–11831. [CrossRef]
31. Lo Scalzo, R.; Fibiani, M.; Mennella, G.; Rotino, G.L.; Dal Sasso, M.; Culici, M.; Spallino, A.; Braga, P.C. Thermal treatments of eggplant (*Solanum melongena* L.) increases the antioxidant content and the inhibitory effect on human neutrophil burst. *J. Agric. Food Chem.* **2010**, *58*, 3371–3379. [CrossRef]
32. Birner, J. A method for the determination of total steroid bases. *J. Pharm. Sci.* **1969**, *58*, 258–259. [CrossRef]
33. Kuronen, P.; Väänänen, T.; Pehu, E. Reversed-phase liquid chromatographic separation and simultaneous profiling of steroidal glycoalkaloids and their aglycones. *J. Chromatogr. A.* **1999**, *863*, 25–35. [CrossRef]
34. Sabatino, L.; Palazzolo, E.; D'Anna, F. Grafting suitability of Sicilian eggplant ecotypes onto *Solanum torvum*: Fruit composition, production and phenology. *J. Food Agric. Environ.* **2013**, *11*, 1195–1200.
35. Davis, A.R.; Perkins-Veazie, P.; Hassell, R.; Levi, A.; King, S.R.; Zhang, X.P. Grafting effects on vegetable quality. *HortScience* **2008**, *43*, 1670–1672. [CrossRef]
36. Rouphael, Y.; Kyriacou, M.; Colla, G. Vegetable grafting: A toolbox for securing yield stability under multiple stress conditions. *Front. Plant Sci.* **2018**, *8*, 2255. [CrossRef]
37. Kawaguchi, M.; Taji, A.; Backhouse, D.; Oda, M. Anatomy and physiology of graft incompatibility in solanaceous plants. *J. Hortic. Sci. Biotechnol.* **2008**, *83*, 581–588. [CrossRef]
38. Miguel, A.; de la Torre, F.; Baixauli, C.; Maroto, J.V.; Jordá, M.C.; López, M.M.; García- Jiménez, J. *Elinjerto de hortalizas*; Ministerio de Agricultura, Pesca y Alimentación: Madrid, Spain, 2007.
39. Rahman, M.A.; Rashid, M.A.; Hossain, M.M.; Salam, M.A.; Masum, A.S.M.H. Grafting compatibility of cultivated eggplant varieties with wild *Solanum* species. *Pak. J. Biol. Sci.* **2002**, *5*, 755–757.
40. Cohen, R.; Horev, C.; Burger, Y.; Shriber, S.; Hershenhorn, J.; Katanand, J.; Edelstein, M. Horticultural and pathological aspects of *Fusarium* wilt management using grafted melons. *Hort. Sci.* **2002**, *37*, 1069–1073. [CrossRef]

41. Fita, A.; Picó, B.; Roig, C.; Nuez, F. Performance of *Cucumis melo* spp. *agrestis* as a rootstock for melon. *J. Hortic. Sci. Biotechnol.* **2004**, *82*, 184–190.
42. Maršič, N.K.; Mikulič-Petkovšek, M.; Štampar, F. Grafting influences phenolic profile and carpometric traits of fruits of greenhouse-grown eggplant (*Solanum melongena* L.). *J. Agric. Food Chem.* **2014**, *62*, 10504–10514. [CrossRef] [PubMed]
43. Muñoz-Falcón, J.E.; Prohens, J.; Rodríguez-Burruezo, A.; Nuez, F. Potential of local varieties and their hybrids for the improvement of eggplant production in the open field and greenhouse cultivation. *J. Food Agric. Environ.* **2008**, *6*, 83–88.
44. Portis, E.; Barchi, L.; Toppino, L.; Lanteri, S.; Acciarri, N.; Felicioni, N.; Fusari, F.; Barbierato, V.; Cericola, F.; Valè, G.; et al. QTL mapping in eggplant reveals clusters of yield-related loci and orthology with the tomato genome. *PLoS ONE* **2014**, *9*, e89499. [CrossRef] [PubMed]
45. Moncada, A.; Miceli, A.; Vetrano, F.; Mineo, V.; Planeta, D.; D'Anna, F. Effect of grafting on yield and quality of eggplant (*Solanum melongena* L.). *Sci. Hortic.* **2013**, *149*, 108–114. [CrossRef]
46. Mishra, B.B.; Gautam, S.; Sharma, A. Free phenolics and polyphenol oxidase (PPO): The factors affecting post-cut browning in eggplant (*Solanum melongena*). *Food Chem.* **2013**, *139*, 105–114. [CrossRef] [PubMed]
47. Singh, A.P.; Luthria, D.; Wilson, T.; Vorsa, N.; Singh, V.; Banuelos, G.S.; Pasakdee, S. Polyphenols content and antioxidant capacity of eggplant pulp. *Food Chem.* **2009**, *114*, 955–961. [CrossRef]
48. King, S.R.; Davis, A.R.; Zhang, X.; Crosby, K. Genetics, breeding and selection of rootstocks for *Solanaceae* and *Cucurbitaceae*. *Sci. Hortic.* **2010**, *127*, 106–111. [CrossRef]
49. Prohens, J.; Rodriguez-Burruezo, A.; Raigon, M.D.; Nuez, F. Total phenolic concentration and browning susceptibility in a collection of different varietal types and hybrids of eggplant: Implications for breeding for higher nutritional quality and reduced browning. *J. Am. Soc. Hortic. Sci.* **2007**, *132*, 1–9. [CrossRef]
50. Dixon, R.A.; Paiva, N.L. Stress-induced phenylpropanoid metabolism. *Plant Cell* **1995**, *7*, 1085–1097. [CrossRef] [PubMed]
51. Moglia, A.; Lanteri, S.; Comino, C.; Acquadro, A.; de Vos, R.; Beekwilder, J. Stress induced biosynthesis of dicaffeoylquinic acids in globe artichoke. *J. Agric. Food Chem.* **2008**, *56*, 8641–8649. [CrossRef]
52. Toppino, L.; Barchi, L.; Lo Scalzo, R.; Palazzolo, E.; Francese, G.; Fibiani, M.; D'Alessandro, A.; Papa, V.; Laudicina, V.A.; Sabatino, L.; et al. Mapping quantitative trait loci affectingbiochemical and morphologicalfruitproperties in eggplant (*Solanum melongena* L.). *Front. Plant Sci.* **2016**, *7*, 256. [CrossRef]
53. Friedman, M.; McDonald, G.M. Potato glycoalkaloids: Chemistry, analyses, safety and plant physiology. *Crit. Rev. Plant Sci.* **1997**, *16*, 55–132. [CrossRef]
54. Friedman, M. Tomato glycoalkaloids: Role in the plant and in the diet. *J. Agric. Food Chem.* **2002**, *50*, 5751–5780. [CrossRef]
55. Jones, P.G.; Fenwick, G.R. The glycoalkaloid content of some edible solanaceous fruits and potato products. *J. Sci. Food Agric.* **2006**, *32*, 419–421. [CrossRef]
56. Krits, P.; Fogelman, E.; Ginzberg, I. Potato steroidal glycoalkaloid levels and the expression of key isoprenoid metabolic genes. *Planta* **2007**, *227*, 143–150. [CrossRef]
57. Khah, E.M.; Kakava, E.M.A.; Chachalis, D.; Goulas, C. Effect of grafting on growth and yield of tomato (*Lycopersicon esculentum* Mill.) in greenhouse and open field. *J. Appl. Hortic.* **2006**, *8*, 3–7.
58. Gisbert, C.; Sánchez-Torres, P.; Raigón, M.D.; Nuez, F. *Phytophthora capsici* resistance evaluation in pepper hybrids: Agronomic performance and fruit quality of pepper grafted plants. *J. Food Agric. Environ.* **2010**, *8*, 116–121.

14

Rootstock Effects on Yield and Some Consumer Important Fruit Quality Parameters of Eggplant cv. 'Madonna' under Protected Cultivation

Maryam Mozafarian, Nazatul Syaima Binti Ismail and Noémi Kappel *

Department of Vegetable and Mushroom Growing, Faculty of Horticultural Science, Szent István University, Villányi út 29–43, H-1118 Budapest, Hungary; maryammozafariyan67@gmail.com (M.M.); syaima.unibio@gmail.com (N.S.B.I.)

* Correspondence: Kappel.Noemi@szie.hu

Abstract: This study aimed to investigate the effect of different rootstocks on the yield and quality of eggplant cv. 'Madonna' in soilless pot culture in an unheated polyethylene greenhouse. The eggplant was grafted onto several rootstocks, including tomato rootstocks Optifort (O) and Emperador (E), and four *Solanum* rootstocks; *Solanum grandiflorum* × *Solanum melongena* (SH), *Solanum torvum* (ST), *Solanum melongena* × *Solanum integrifolium* (SI), and *Solanum integrifolium* (A) compared with self-grafted (SG) and self-rooted (SR) as control. The results showed that the total marketable yield significantly increased by grafting onto ST (3.94 kg/plant), SH (3.36 kg/plant), and A (3.34 kg/plant) relative to SR (1.65 kg/plant). The chromatics characters of skin and pulp are slightly influenced by rootstocks. Our findings confirmed that grafting eggplant decreased firmness (except SH) of the flesh. Fruit harvested from the Optifort/Madonna combination had the rounded shape, lowest firmness, and Brix value, while the lowest oxidation potential was observed in this combination. The highest seed number was observed in SH/Madonna and SI/Madonna combinations. During the sensory evaluation, the lightest fruit flesh was found in SR, ST, and O, and the sweetest taste was observed in fruits harvested from ST rootstock.

Keywords: eggplant grafting; rootstock; fruit quality; yield; sensory evaluation

---

1. Introduction

The eggplant (*Solanum melongena* L.) is one of the top ten vegetables that originated from Southeast Asia; it has a high antioxidant capability and nutrient value. Soil-borne diseases, and biotic and abiotic stresses have limited the yield in many commercial eggplant plantations. Vegetable grafting is considered to be a rapid alternative way to slow breeding due to the absence of resistance genotype [1]. Grafting has been found effective not only for resistance advantages but also for the improvement of production and some quality traits of the fruit [2,3]. Eggplant grafting was started in the early 1950s by using Scarlet eggplant as rootstocks [4].

Several studies have already investigated different eggplant scion-rootstock combinations [5–8]. For instance, *Solanum paniculatum* increased vigour and fruit yield, but did not have any positive effect on fruit quality and composition [6]. *Solanum incanum* induced tolerance to water and temperature stress [9], while *Solanum aethiopicum* and *Solanum macrocarpon* were tolerant to *Fusarium oxysporum* and resistant to *Ralstonia solanacearum* [10]. *Solanum sisymbriifolium* and *Solanum integrifolium* have been reported effective in controlling bacterial wilt and resulted in higher yield [11]. *Solanum torvum* is the most commonly used rootstock for eggplant, which has been reported to be resistant to soil-borne diseases; but due to lack of rapid and homogeneous seed germination, there is a need to find other alternative rootstocks [12]. Some tomato rootstocks can be valuable, as well, to improve eggplant growth and production. Grafting eggplant onto tomato rootstocks (*Solanum lycopersicum* L. × *Solanum habrochaites*) indicated a positive result on yield and the appearance of fruits, but not equally, and it largely depended on the combination and environment [13]. Overall, some rootstock-scion combinations are moderately compatible, and unfavourable effects may occur; hence, the selection of appropriate rootstock-scion combinations is crucial.

Fruit quality of grafted vegetables can be measured by sensory evaluation and instrumental methods as well [14]. Grieneisen et al. [15] conducted a review of 202 different rootstocks and 1023 experimental treatments related to tomato grafting, and they concluded that fruit quality data based on sensory tests were rare among the published studies. Sensory evaluation of grafted plants of the members of the *Cucurbitaceae* family, i.e., melon [16], watermelon [17,18], and cucumber [19], and those of the *Solanaceae* family [20,21] have been conducted, but there is relatively few data for eggplant.

The majority of previous studies investigated eggplant and tomato rootstock compatibility with eggplant scion individually. The objective of the present study was to compare the effect of tomato and eggplant rootstocks on fruit yield and certain quality parameters of eggplant cv. 'Madonna' in soilless pot culture in an unheated plastic greenhouse. Moreover, current research is aimed at identifying the best rootstock type to maximize yield and fruit quality, conducting not only laboratory testing, but sensory evaluation as well.

2. Materials and Methods

2.1. Experimental Site, Plant Material, and Management Practices

The experiment was conducted from March 2019 until October 2019 at the Experimental and Research Farm, Faculty of Horticultural Science, Szent István University, Budapest, in unheated plastic house (47°23′49′′ N, 19°09′10′′ E, 120 m a.s.l.). The study was arranged based on completely randomized design (CRD) to understand the effect of different rootstocks suitable for eggplant grafting, including *Solanum grandiflorum* × *Solanum melongena* (SH), *Solanum torvum* (ST), *Solanum melongena* × *Solanum integrifolium* (SI), and *Solanum integrifolium* (A) and tomato rootstocks (*Solanum lycopersicum*), including cv. 'Optifort' (O) and cv. 'Emperador' (E). Moreover, self-grafted (SG) cv. 'Madonna' and self-rooted (SR) plants, as control, were used with four replications and five plants in each replication.

The used, popular eggplant cultivar 'Madonna' has evenly coloured, dark-violet fruits. Due to slow seed germination, all four eggplant rootstock seeds (SH, A, ST, and SI) were sown 10 days earlier than cv. 'Madonna' and tomato rootstocks seeds (E and O) (which germinated and developed at the same rate); thus, a uniform stem diameter for grafting was obtained. Eggplant seedlings (with the adequate diameter and 4–5 leaves) were grafted onto the rootstocks by cleft grafting method and kept in high humidity and low light conditions for one week, and then acclimatized to the natural conditions. Three weeks after grafting, all seedlings were transplanted into 10 L pots containing peat substrate, in an unheated polyethylene greenhouse. Temperature and relative humidity inside the greenhouse were recorded by the Flower Power Sensor data logger, as shown in Figure 1. All phytotechnical work recommended for eggplant greenhouse cultivation was performed uniformly. Irrigation was applied twice a week with a commercial fertilizer solution.

Mature fruits were harvested once a week between June and October 2019 (15 times) according to fruit size, colour, and glossiness. Right after picking, the weight of each fruit was recorded by using a digital scale, and the marketable or unmarketable fruits had been sorted. Fruit width (at maximum fruit diameter) and fruit length (from stalk to end of fruit) had been measured by a tape line, and fruit index was calculated based on fruit width/fruit length.

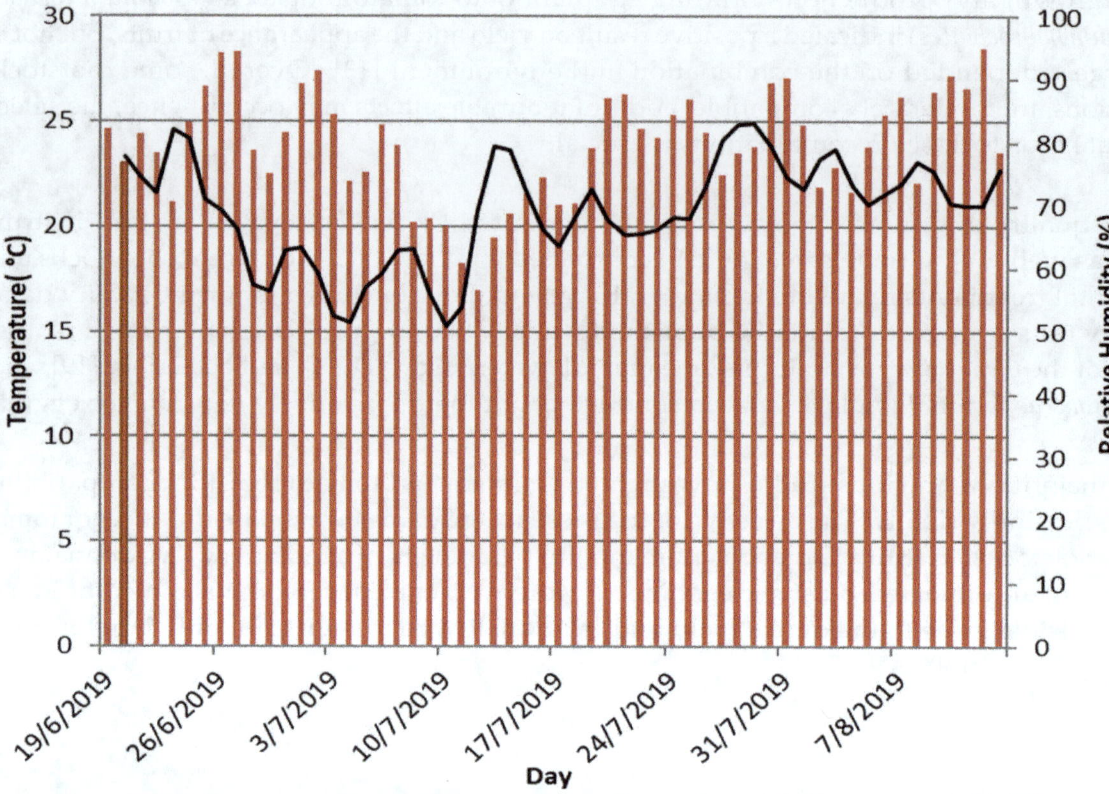

Figure 1. Temperature and relative humidity at the greenhouse during cultivation.

2.2. *Instrumental Measurements*

The fruit flesh acidity (pH) was measured from two sides of each fruit in all combinations by a pH meter (Hanna HI 98128). The pulp of three fruits from each replication were blended using a kitchen blender. The total soluble solids of the fruit (TSS %) was measured by using a refractometer (PAL-1 Brix 0–53% Digital Hand) from the fruit juice. Fruit firmness from each replication (two fruits) were evaluated by a small hand-operated penetrometer. The pressure value was measured in kg/cm^3.

The skin colour of three fruits from each replication was measured on two sides of the skin by using a Minolta Chroma CR-400 colourimeter (Minolta Corporation, Ltd., Osaka, Japan). Fruit chromaticity was expressed in L*, a*, b* colour space coordinates. Chroma (C*) and Hue angles (H°) were calculated according to the following formula 1:

$$C^* = (a^{*2} + b^{*2})^{1/2},\ H^\circ = \tan^{-1}(b^*/a^*). \tag{1}$$

The same colourimeter was used to determine the whitening index (DW), oxidation potential (OP), and colour differences (CD) of the fruit pulp. Two fruits of each replication were cut longitudinally with a straight edge plastic knife. The pulp was measured quickly after being cut (L_0) and after 30 min (L_{30}) in the central and lateral zone. Colour space had been divided into a three-dimensional (L, a and b) so that L (lightness; 0 black and 100 white); a (red to green); and b (blue to yellow). The distance of pure white (DW) was measured as Euclidean distance of the colour coordinates to the pure white coordinates (L* = 100 a* = 0 b = 0) using the formula 2 [22]:

$$DW = ((100 - L)^2 + a^{*2} + b^{*2})^{0.5} \tag{2}$$

The colour differences (CD) were measured as a Euclidean distance between the colour coordinates at 0 and 30 min and calculated by using formula 3:

$$CD = [(L^*30 - L^*0) + (a^*30 - a^*0) + (b^*30 - b^*0)] \quad (3)$$

Moreover, the oxidation potential (OP) was measured by International Commission on Illumination (CIE) L*a*b* values using the following formula 4 [23]:

$$OP = L^*_{30} - L^*_{0} \quad (4)$$

Ten independent slices from the equatorial region of fruits were cut and incubated at 20 °C to induce seed browning and facilitate seed boundaries identification. Images were used to determine the number of seeds by using the software ImageJ (Version 1.8.0_172; Research Services Branch, National Institute of Mental Health, Bethesda, MD, USA).

2.3. Sensory Evaluation

Sensory measurements were made based on questionnaire results. Twelve trained tasting panellists evaluated 10 quality attributes (Table 1) according to ISO 13,299 standard. First, the sensory attributes and their corresponding reference values were determined, to reduce the variation in the resulting dataset. For the sensory test, small pieces of the fruit were boiled for 20 min, placed in numbered plates, and were immediately assessed. The questionnaire results were converted into a grade scale from 0 to 100. The combined results of the properties were plotted on profile diagrams, which were prepared by ProfiSens, a sensory analysis software. Tests were performed according to ISO 8589, and differences between data were evaluated with univariate ANOVA and Fisher least significant difference (LSD) significance level evaluation procedures.

Table 1. Definitions of sensory attributes used in the quantitative descriptive analysis.

Attributes	Characteristic
Flesh Colour	Dark–light
Odour	None–very intensive
Flesh Firmness	Soft–firm
Flesh Juiciness	not very juicy–very juicy
Flavour Intensity	not perceptible–intense
Tart, Bitter Taste	None–very intensive
Pungent Flavour	None–very intensive
Sweet Taste	not very intensive–very intensive
After Taste	None–very intensive
Off-Flavour	None–very intensive

2.4. Statistical Procedures

The experiment was arranged in a completely randomized design (CRD) with four replications and five plants in each replication. Data were statistically analysed using Statistix 8 software (Tallahassee, FL, USA). Data were subjected to the one-way analysis of variance (ANOVA) and means were separated using the least significant difference (LSD) test at $p < 0.05$.

3. Results

As shown in Table 1, grafting cv. 'Madonna' onto 'O and E rootstocks caused the lowest fruit length in comparison with the SR and SG rootstocks. Moreover, it has been observed that self-grafting cv. 'Madonna' significantly increased the width of fruit relative to the fruit harvested from control and other combinations (except ST). Statistical analysis showed that fruit shape index was not significantly ($p < 0.05$) influenced by rootstocks.

Our results showed that the total marketable fruit number significantly increased by grafting in comparison with SR and SG. Average fruit weight decreased by grafting onto O rootstocks in comparison with other treatment (except SG). Grafting cv. 'Madonna' on eggplant rootstocks (SH, ST, SI, and A) significantly increased total marketable fruit yield while tomato rootstocks (E and O) had no significant effect. Total marketable yield of *Solanum torvum* (ST) rootstock was two times higher than control, self-grafted plants (Table 2).

Table 2. Effect of different rootstock combinations on eggplant fruit shape, fruit number, and yield.

Treatment	Fruit Length (cm)	Fruit Width (cm)	Fruit Shape Index	Total Marketable Fruit Number (plant^{-1})	Average Fruit Weight (kg)	Total Marketable Yield (kg plant^{-1})
SR	16.97 a	8.06 bcd	2.13 a	6.35 c	0.28 ab	1.65 c
			Eggplant Rootstocks			
SG	18.10 a	9.38 a	1.92 a	6.70 c	0.26 bc	1.89 c
SH	16.34 ab	7.94 cd	2.06 a	12.13 ab	0.29 ab	3.36 ab
ST	16.58 a	8.36 ab	1.99 a	14.26 a	0.28 ab	3.94 a
SI	16.93 a	8.18 bc	2.08 a	10.80 b	0.31 a	3.23 ab
A	16.91 a	8.19 bc	2.07 a	11.14 b	0.29 ab	3.34 ab
			Tomato Rootstocks			
E	15.66 bc	8.11 bcd	1.95 a	10.71 b	0.28 ab	2.80 abc
O	14.94 c	7.68 d	1.96 a	9.38 b	0.24 c	2.39 bc
CV	10.87	11.19	12.42	15.00	26.67	20.19
p Value	0.000	0.0005	0.003	0.000	0.009	0.0002
Significance	***	***	**	***	**	***

NS = not significant; ** = Significant at $p \leq 0.01$, *** = Significant at $p \leq 0.001$. Different letters indicate significant difference according to the least significant difference (LSD) test ($p < 0.05$). SR = self-root; SG = self-grafting; SH = *S. grandiflorum* × *S. melongena*; ST = *S. torvum*; SI = *S. melongena* × *S. integrifolium*; A = *S. integrifolium*; E = Emperador; O = Optifort.

According to ANOVA, L* value and CIRG index (Color index of red) of fruit skin from all combinations was not significantly ($p < 0.05$) different with the self-rooted (SR) plants. Grafting Madonna onto O, E, A, and SH significantly increased the a* value of skin fruit relative to SR and SG (Table 3). Fruit harvested from O rootstocks had the highest b* value compared with the SR. Our results showed that the highest Hue value was observed at E and O rootstock in comparison with SR and SG. Chroma value of skin fruit increased when O, E, and A was used as rootstock compared with either non-grafted or self-grafted (Table 3).

Table 3. Effect of different rootstock combinations on chromatic characteristics of eggplant fruit skin.

Treatment	L*	a*	b*	Hue	Chroma	CIRG
SR	25.22 a	3.09 d	−0.72 bc	3.18 cd	346.15 cd	6.62 a
			Eggplant Rootstocks			
SG	25.07 a	2.95 d	−0.73 bc	3.05 d	345.13 d	6.69 a
SH	25.36 a	3.62 abc	−0.71 bc	3.70 abc	347.70 abc	6.53 a
ST	25.58 a	3.32 bcd	−0.75 bc	3.40 bcd	346.68 bcd	6.51 a
SI	25.08 a	3.27 cd	−0.79 c	3.72 cd	346.03 cd	6.65 a
A	25.61 a	3.65 abc	−0.69 ab	3.71 abc	348.69 ab	6.47 a
			Tomato Rootstocks			
E	25.53 a	3.89 ab	−0.67 ab	3.95 ab	348.92 ab	6.46 a
O	25.45 a	4.08 a	−0.611 a	4.14 a	349.51 a	6.46 a
CV	4.35	20.15	−22.19	28.66	1.26	5.03
p Value	0.47	0.001	0.006	0.001	0.002	0.084
Significance	NS	***	**	***	**	NS

NS = not significant; ** = Significant at $p \leq 0.01$, *** = Significant at $p \leq 0.001$. Different letters indicate significant difference according to the LSD test ($p < 0.05$). SR = self-root; SG = self-grafting; SH = *S. grandiflorum* × *S. melongena*; ST = *S. torvum*; SI = *S. melongena* × *S. integrifolium*; A = *S. integrifolium*; E = Emperador; O = Optifort.

The results of pulp colour measurements are presented in Table 4. No significant difference was observed in L_0*, a*, CD and DW_0 between grafted and SR plants. Moreover, our results showed that grafting onto tomato rootstocks (E and O) significantly decreased OP value of pulp fruit relative to SR and SG.

Table 4. Effect of different rootstock combinations on chromatic characteristics, colour difference, whiteness degree, and oxidation potential of eggplant pulp.

Treatment	L_0*	a*	b*	CD	DW_0	OP
SR	84.89 ab	−4.10 ab	29.08 a–c	12.16 a–c	33.10 a–c	9.97 a
			Eggplant Rootstocks			
SG	85.43 ab	−3.71 ab	26.84 c	12.62 ab	30.85 c	10.66 a
SH	84.62 ab	−4.01 ab	29.15 abc	13.61 a	33.24 abc	8.67 ab
ST	84.03 b	−2.86 a	30.14 ab	11.46 abc	34.25 ab	9.67 ab
SI	85.26 ab	−4.56 b	29.05 abc	12.48 ab	32.93 abc	9.38 ab
A	84.65 ab	−3.17 ab	30.32 a	13.21 a	34.30 ab	11.03 a
			Tomato Rootstocks			
E	86.36 a	−4.71 b	27.70 bc	7.37 c	31.30 bc	5.81 b
O	83.91 b	−3.68 ab	31.39 a	8.29 bc	35.51 a	5.81 b
CV	2.71	−4.82	8.88	27.58	9.25	16.90
p value	0.4	0.2	0.02	0.1	0.06	0.05
Significance	NS	NS	*	NS	NS	*

NS = not significant; * Significant at $p \leq 0.05$. Different letters indicate significant difference according to LSD's test ($p < 0.05$). SR = self-root; SG = self-grafting; SH = *S. grandiflorum* × *S. melongena*; ST = *S. torvum*; SI = *S. melongena* × *S. integrifolium*; A = *S. integrifolium*; E = Emperador; O = Optifort. CD = colour difference; DW = whiteness degree; OP = oxidation potential.

As shown in Table 5, Brix value of fruit flesh was negatively influenced by grafting onto SI, E, and O rootstocks, while SG resulted the highest Brix value in the pulp. Fruit firmness significantly decreased by grafting (except grafting onto SH rootstock). The pH value of the flesh was not influenced by different rootstocks.

Table 5. Effect of different rootstock combinations on pH, Brix, firmness, and seed number of eggplant fruits.

Treatment	pH	T.S.S (Brix)	Firmness (kg/cm^3)	Seed Number
SR	5.55 a	5.28 b	4.93 a	21.81 c
		Eggplant Rootstocks		
SG	5.57 a	5.80 a	4.18 b	29.50 bc
SH	5.46 ab	5.26 b	4.56 ab	47.30 a
ST	5.53 a	5.34 b	3.51 c	38.70 abc
SI	5.44 ab	4.78 c	3.56 c	44.36 ab
A	5.38 b	5.62 ab	4.04 bc	25.46 c
		Tomato Rootstocks		
E	5.56 a	4.62 c	2.77 d	27.09 bc
O	5.53 a	4.72 c	2.43 d	28.15 bc
CV	2.45	10.59	14.42	14.03
p Value	0.06	0.000	0.000	0.000
Significance	NS	***	***	***

NS = not significant; *** = Significant at $p \leq 0.001$. Different letters indicate significant difference according to the LSD test ($p < 0.05$). SR = self-root; SG = self-grafting; SH = *S. grandiflorum* × *S. melongena*; ST = *S. torvum*; SI = *S. melongena* × *S. integrifolium*; A = *S. integrifolium*; E = Emperador; O = Optifort.

In the current experiment, grafting eggplant onto SH and SI rootstocks sharply increased the seed number of fruits in comparison with SR (Table 4; Figure 2).

Sensory analysis data showed a significant difference in four parameters (flesh colour, firmness, sweet taste, and intensive odour) between selected rootstocks. Fruits harvested from SR, ST, and O had lighter flesh colour than other rootstock combinations. Respectively, SR and E had the lowest and highest flesh firmness between grafting combinations according to panellists' evaluation. Grafting onto ST showed significantly stronger sweet taste as compared to SR fruits. Fruits harvested from SI showed significantly intensive odour among all combinations (Figure 3). Bitter taste, pungent flavour, off flavour, aftertaste, flesh juiciness, and flavour intensity were not significantly influenced by grafting.

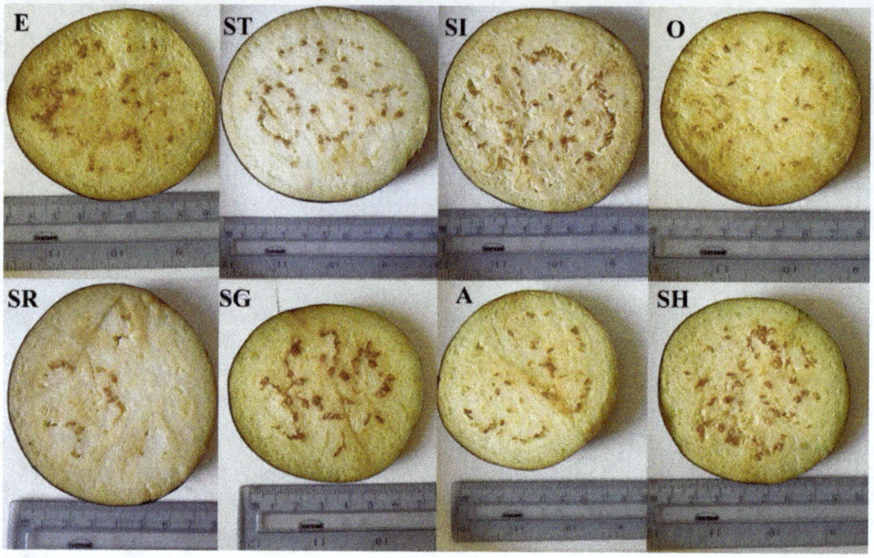

Figure 2. Effect of different rootstock combinations on fruit seed number. SR = self-root; SG = self-grafting; SH = *S. grandiflorum* × *S. melongena*; ST = *S. torvum*; SI = *S. melongena* × *S. integrifolium*; A = *S. integrifolium*; E = Emperador; O = Optifort.

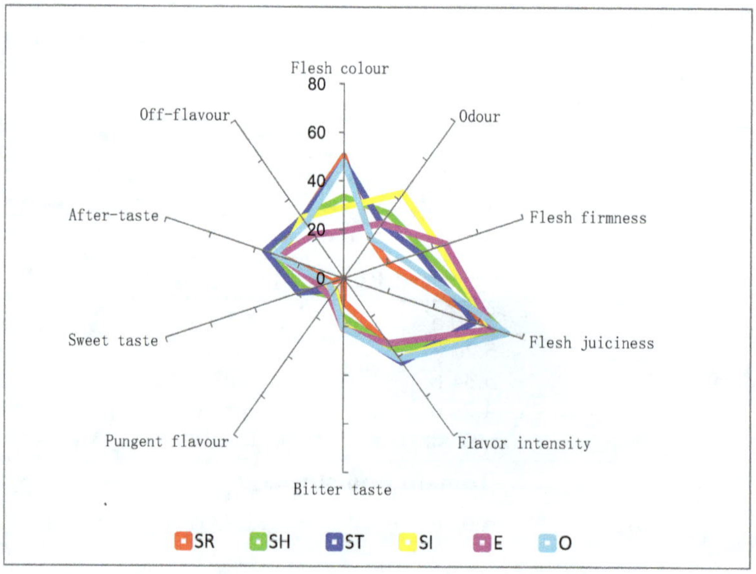

Figure 3. Effect of different rootstock combinations on sensory evaluation of fruit by trained panellists.

4. Discussion

Vegetable quality is defined by external attributes, such as colour, shape, size and freshness, as well as internal features, such as texture, flavour, content of mineral, health-promoting compounds, and sensory parameters. The quality of grafted vegetables can differ and conflicting from the grower and consumer perspectives. However, consumer demand determines economic value and it is an essential aspect in grafted vegetable productions. In the current experiment, we have tested the 'Madonna' eggplant cultivar grafted onto two groups of rootstocks (tomato and other *Solanum* spp.) to find the best combination in terms of yield and quality in both laboratory and consumer evaluation.

Our results showed that the lowest fruit length was observed at tomato rootstock (E and O) and highest fruit width was at SG rootstocks. Fruit shape is controlled genetically; however, Sabatino et al. [6,24] and Gisbert et al. [25] demonstrated that eggplant fruit shape (length and width) is affected by the rootstock. In line with our findings, Passam et al. [26] stated that grafting can increase eggplant fruit size. Moreover, Cassaniti et al. [27] harvested longer fruits from cv. 'Black Bell'—*S. torvum* combination than from self-grafted plants. In cucumber, shape index in grafted plants was higher than non-grafted ones [28], while in watermelon grafting it had no influence on fruit size and shape, according to some studies [29,30]. Therefore, these results are different due to genotype, vigorous rootstocks, and grafting combinations [3]. For instance, grafting tomato onto vigorous rootstocks (i.e., 'Maxifort'; 'Beaufort') increased the fruit size, while grafting onto less vigorous rootstocks (i.e., 'Brigeor'; 'Energy,'; 'Firefly', 'Linea9243,' and 'Nico') reduced tomato fruit size [31–33].

All eggplant rootstocks (SH, ST, A, and SI) improved total fruit yield. Yield, fruit number, and earliness of eggplant cv. 'Cristal F1' grafted onto eggplant rootstocks was higher than tomato rootstocks [25]. In line with our results, Sabatino et al. [19] reported that eggplant cv. 'Scarlatti' F1 grafted onto *S. melongena* × *S. aethiopicum* rootstock had higher total and marketable yield than plants grafted onto *S. aethiopicum*. Earlier, they found that higher marketable yield of grafted eggplant onto *S. torvum* and *S. macrocarpon* can be the result of higher fruit numbers, which associate with increasing water and nutrient absorption [6]. Gisbert et al. [25] confirmed that grafted plants with higher yield had earlier fruit harvesting. In the current experiment, the higher marketable yield is due to a higher harvested fruit number in grafted plants. Our results also correlate with Sabatino et al. [5] who confirmed a higher fruit number of grafted plants than on non-grafted. Higher yield of grafted plants may be related to higher water and nutrient absorption [29].

The darkest skin colour is connected to a high concentration of anthocyanin. Results of Moncada et al. [34] showed that grafting eggplant cv. 'Biragh' onto *S. torvum* caused a darker and less vivid fruit skin colour (lower value of L* and chroma), while other cultivars ('Black Moon' and 'Black Bell') were not influenced by rootstock. Moreover, the highest Hue value was observed when eggplant cv. 'Scarlatti' was grafted onto *S. torvum* and *S. aethiopicum* rootstocks in an experiment by Sabatino et al. [24]. In our study, the most noticeable finding was that grafting cv. 'Madonna' onto O rootstock resulted in low-quality fruit skin, with the Hue value of 4.14, and chroma of 349.51. In tomato, variable results were found in fruit skin colour by grafted plants compared to non-grafted ones [31,35].

A CIRG index of 5 indicates red dark violet skin, and higher than 6 indicates blue–black skin. In the current experiment, 'Madonna' scion normally had dark black skin and grafting did not have any significant influence on CIRG. Kacjan Maršic et al. [13] reported that grafting eggplant cv. 'Blackbell' onto tomato rootstock cv. 'Beaufort' increased CIRG index, while in another cultivar (cv. 'Epic') decreased. It can be explained that scion variety has different responses to grafting according to Kacjan Maršic et al. [13]. Moreover, they concluded that higher vigour of grafted plants negatively influenced the anthocyanins, and the pruning of plants should be carried out during the cultivation to improve the fruit colour [13].

It seems that tomato rootstocks had lower browning degree and colour difference value than self-rooted and self-grafted plants, and fruit harvested from these combinations oxidized less than others (Table 3). High oxidation potential is expected due to the high phenolic compound content of the eggplant. Moreover, it may be influenced by many factors, e.g., scion genotype, rootstocks, and environmental conditions. Moncada et al. [34] and Sabatino et al. [6] reported that grafting eggplant onto *S. torvum* had little or no effect on pulp browning. However, other rootstocks had different reactions; for instance, *S. paniculatum* and *S. macrocarpon* showed lowest [6], while *S. aethiopicum* gave the highest oxidation potential in eggplant [3]. Seed number and fruit size may lead to different colour and oxidization in fruit pulp after cutting. A negative correlation between the browning index and fruit size, and positive relation with seed number, has been reported by Radicetti et al. [36].

Several studies confirmed that fruit quality seldom varied significantly between grafted and non-grafted plants. For instance, previous results by Arvanitoyannis et al. [37] and Cassaniti et al. [27] confirmed a reduction in eggplant flesh firmness by grafting. Similarly, tomato fruits harvested from grafted plants were not firmer than control and self-grafted plants [31,38]. Lower water uptake in non-grafted plant caused lower water content in eggplant fruit and harder texture [37]. The current experiment showed that grafting significantly reduced TSS of eggplant cv. 'Madonna'. In line with our results, lower TSS was observed in eggplant fruits (cv. 'Faselis') obtained from grafted on *S. torvum*, while no significant difference was found in fruits (cv. 'Rima') obtained from tomato hybrid rootstocks compared to the rest of combinations [39]. Lee et al. [40] reported that grafting eggplant onto *S. torvum* had no influence on TSS. Previous studies showed that firmness and SSC of non-grafted and self-grafted melon was higher than grafted fruit in the Guan et al. [14] study, and they explained that it is due to the improved water status of grafted melon by enhancing leaf water potential, leaf stomatal conductance, transpiration rate, and amount of xylem sap, which then can reduce TSS of the fruit [14,41]. Fruits obtained from SH and SI × 'Madonna' combination had the highest seed number and it can be explained that higher nitrogen absorption in the grafted plant may cause a higher seed number in fruit [42].

Combining sensory and instrumental measurements provided a thorough evaluation of the effects of grafting and rootstock combination [14]. Grafting eggplant cv. 'Tsakoniki' onto *S. torvum* resulted in fewer empty spaces and harder fruits more than in case of other combinations [37]. Moreover, they reported that fruits grafted on *S. sisymbriifolium* had fewer seeds and less intensive tart flavour than the rest of the treatments. Another interesting finding was that all grafted plants resulted in less sweet fruits and lower acceptability ratings in the panellist test [37]. In our experiment, fruit harvested from ST had a sweeter taste and lighter colour, and fruit harvested from SI showed intensive odour, according to consumer evaluation. In tomato, according to the experiment of Di Gioia et al. [20], grafting onto cv. 'Beaufort' and 'Maxifort' rootstocks did not influence sweetness, sourness, and the tomato-like taste of the fruits. Barrett et al. [21] and Casals et al. [43] reported that grafting had negative effects on acceptability and the tomato flavour descriptors assessed by a consumer test.

In the *Cucurbitaceae* family, Guan et al. [14] reported that grafting Galia melon onto hybrid squash rootstocks caused lower sensory properties, while Honey Yellow cv. was not influenced by grafting. They explained that lower SSC can be one of the reasons of the lower sensory evaluation results. Similarly, in another experiment, Velkov and Pevicharova [19] confirmed that sensory evaluation of cucumber depended on scion-rootstock combinations and every scion showed a different reaction to the same rootstock.

5. Conclusions

In the present study, we examined the effect of grafting onto different rootstocks in case of an eggplant variety favoured by consumers. Either tomato or other *Solanum* species are recommended for the grafting of eggplant; thus, we also selected the rootstocks to be tested accordingly. Compared to previous studies, we also observed differences between each grafting combination, both in terms of yield and fruit quality.

In our study, *Solanum torvum* × Madonna, as well as *Solanum grandiflorum* × *Solanum melongena* × Madonna combinations, exhibited the highest marketable yield with a higher fruit number per plant and uniform skin colour. Our findings also revealed that tomato rootstocks (O and E) had the lowest pulp colour difference and oxidation potential.

Our results also confirm that it is very important to choose the right rootstock, primarily according to the goal of grafting to be achieved.

Author Contributions: All authors contributed to this research. The design of the experiment was done by N.K., M.M., and N.S.B.I.; M.M. and N.S.B.I. did the recording and processing of the data, as well as the result of the evaluation. The manuscript was written by M.M.; N.K. assisted in writing the paper. N.K. contributed to designing the research and revised the manuscript. The work presented in the paper was conceived within research projects led by N.K. All authors have read and agreed to the published version of the manuscript.

Acknowledgments: This publication was founded by EFOP-3.6.1-16-2016-00016. The specialize of the SZIE Campus of Szarvas research and training profile with intelligent specialization in the themes of water management, hydroculture, precision mechanical engineering, alternative crop production.

References

1. Mozafarian, M.; Kappel, N. The Role of Grafting Vegetable Crops for Reducing Biotic and Abiotic Stresses. In *Handbook of Plant and Crop Stress*, 4th ed.; Pessarakli, M., Ed.; CRC Press: Boca Raton, FL, USA, 2019; pp. 855–863. [CrossRef]
2. Mozafarian, M.; Kappel, N. Grafting Plants to Improve Abiotic Stress Tolerance. In *Plant Ecophysiology and Adaptation under Climate Change: Mechanisms and Perspectives II*; Hasanuzzaman, M., Ed.; Springer: Singapore, 2020; pp. 477–490. [CrossRef]
3. Kyriacou, M.C.; Rouphael, Y.; Colla, G.; Zrenner, R.; Schwarz, D. Vegetable grafting: The implications of a growing agronomic imperative for vegetable fruit quality and nutritive value. *Front. Plant Sci.* **2017**, *8*, 1–23. [CrossRef] [PubMed]
4. Boyaci, H.F.; Ellialtioglu, S.S. Rootstock usage in eggplant: Actual situation and recent advances. *Acta Hortic.* **2020**, *1271*, 403–410. [CrossRef]
5. Sabatino, L.; Iapichino, G.; Maggio, A.; D'Anna, E.; Bruno, M.; D'Anna, F. Grafting affects yield and phenolic profile of *Solanum melongena* L. landraces. *J. Integr. Agric.* **2016**, *15*, 1017–1024. [CrossRef]
6. Sabatino, L.; Iapichino, G.; D'Anna, F.; Palazzolo, E.; Mennella, G.; Rotino, G.L. Hybrids and allied species as potential rootstocks for eggplant: Effect of grafting on vigour, yield and overall fruit quality traits. *Sci. Hortic.* **2018**, *228*, 81–90. [CrossRef]
7. Pebriana, E.; Dawam Maghfoer, M.O.C.H.; Widaryanto, E. Effect of grafting using wild eggplant as rootstock on growth and yield of four eggplant (*Solanum melongena* L.) cultivars. *Biosci. Res.* **2018**, *15*, 337–347.
8. Giuffrida, F.; Cassaniti, C.; Agnello, M.; Leonardi, C. Growth and ionic concentration of eggplant as influenced by rootstocks under saline conditions. *Acta Hortic.* **2015**, *1086*, 161–166. [CrossRef]
9. Daunay, M.C. Eggplant. In *Handbook of Plant Breeding*; Prohens, J., Nuez, F., Eds.; Vegetables II; Springer: New York, NY, USA, 2008; pp. 163–220. [CrossRef]
10. Furini, A.; Wunder, J. Analysis of eggplant (*Solanum melongena*) related germplasm: Morphological and AFLP data contribute to phylogenetic interpretations and germplasm utilization. *Theor. Appl. Genet* **2004**, *108*, 197–208. [CrossRef]
11. Rahman, M.A.; Rashid, M.A.; Hossain, M.M.; Salam, M.A.; Masum, A.S.M.H. Grafting compatibility of cultivated eggplant varieties with wild Solanum species. *Pak. J. Biol. Sci.* **2002**, *5*, 755–757. [CrossRef]
12. King, S.R.; Davis, A.R.; Zhang, X.; Crosby, K. Genetics, breeding and selection of rootstocks for Solanaceae and Cucurbitaceae. *Sci. Hort.* **2010**, *127*, 106–111. [CrossRef]
13. Kacjan Maršič, N.; Mikulič-Petkovšek, M.; Stampar, F. Grafting influences phenolic profile and carpometric traits of fruits of greenhouse-grown eggplant (*Solanum melongena* L.). *J. Agric. Food Chem.* **2014**, *62*, 10504–10514. [CrossRef]
14. Guan, W.; Zhao, X.; Huber, D.J.; Sims, C.A. Instrumental and sensory analyses of quality attributes of grafted specialty melons. *J. Sci. Food Agric.* **2015**, *95*, 2989–2995. [CrossRef] [PubMed]
15. Grieneisen, M.L.; Aegerter, B.J.; Scott Stoddard, C.; Zhang, M. Yield and fruit quality of grafted tomatoes, and their potential for soil fumigant use reduction. A meta-analysis. *Agron. Sustain. Dev.* **2018**, *38*, 29. [CrossRef]
16. Németh, D.; Balázs, G.; Daood, H.; Kovács, Z.; Bodor, Z.; Zinia Zaukuu, J.-L.; Szentpéteri, V.; Kókai, Z.; Kappel, N. Standard Analytical Methods, Sensory Evaluation, NIRS and Electronic Tongue for Sensing Taste Attributes of Different Melon Varieties. *Sensors* **2019**, *19*, 5010. [CrossRef]
17. Fekete, D.; Balázs, G.; Bőhm, V.; Várvölgyi, E.; Kappel, N. Sensory evaluation and electronic tongue for sensing grafted and non-grafted watermelon taste attributes. *Acta Alimentaria* **2018**, *47*, 487–494. [CrossRef]

18. Liu, Q.; Zhao, X.; Brecht, J.K.; Sims, C.A.; Sanchez, T.; Dufault, N.S. Fruit quality of seedless watermelon grafted onto squash rootstocks under different production systems. *J. Sci. Food Agric.* **2017**, *97*, 4704–4711. [CrossRef]
19. Velkov, N.; Pevivharova, G. Effects of cucumber grafting on yield and fruit sensory characteristics. *Zemdirbyste Agric.* **2016**, *103*, 405–410. [CrossRef]
20. Di Gioia, F.; Serio, F.; Buttaro, D.; Ayala, O.; Santamaria, P. Influence of rootstock on vegetative growth, fruit yield and quality in 'Cuore di Bue', an heirloom tomato. *J. Hort. Sci. Biotechnol.* **2010**, *854*, 77–482. [CrossRef]
21. Barrett, C.E.; Zhao, X.; Hodges, A.W. Cost benefit analysis of using grafted transplants for root-knot nematode management in organic heirloom tomato production. *HortTechnology* **2012**, *22*, 252–257. [CrossRef]
22. Larrigaudiere, C.; Lentheric, I.; Vendrell, M. Relationship between enzymatic browning and internal disorders in controlled atmosphere stored pears. *J. Sci. Food Agric.* **1998**, *78*, 232–236. [CrossRef]
23. Prohens, J.; Rodriguez-Burruezo, A.; Raigon, M.D.; Nuez, F. Total phenolics concentration and browning susceptibility in a collection of different varietal types and hybrids of eggplant: Implications for breeding for higher nutritional quality and reduced browning. *J. Am. Soc. Hort. Sci.* **2007**, *132*, 638–646. [CrossRef]
24. Sabatino, L.; D'Anna, F.; D'Anna, F.; Iapichino, G.; Moncada, A.; D'Anna, E.; De Pasquale, C. Interactive effects of genotype and molybdenum supply on yield and overall fruit quality of tomato. *Front. Plant Sci.* **2019**, *9*, 19–22. [CrossRef] [PubMed]
25. Gisbert, C.; Prohens, J.; Raigón, M.D.; Stommel, J.R.; Nuez, F. Eggplant relatives as sources of variation for developing new rootstocks: Effects of grafting on eggplant yield and fruit apparent quality and composition. *Sci. Hortic.* **2011**, *128*, 14–22. [CrossRef]
26. Passam, H.C.; Stylianou, M.; Kotsiras, A. Performance of eggplant grafted on tomato and eggplant rootstocks. *Eur. J. Hort. Sci.* **2005**, *70*, 130–134.
27. Cassaniti, C.; Giuffrida, F.; Scuderi, D.; Leonardi, C. Effect of rootstock and nutrient solution concentration on eggplant grown in a soilless system. *J. Food Agric. Environ.* **2011**, *9*, 252–256.
28. Colla, G.; Rouphael, Y.; Jawad, R.; Kumar, P.; Rea, E.; Cardarelli, M. The effectiveness of grafting to improve NaCl and $CaCl_2$ tolerance in cucumber. *Sci. Hortic.* **2013**, *164*, 380–391. [CrossRef]
29. Colla, G.; Rouphael, Y.; Cardarelli, M.; Rea, E. Effect of salinity on yield, fruit quality, leaf gas exchange, and mineral composition of grafted watermelon plants. *HortScience* **2006**, *41*, 622–627. [CrossRef]
30. Soteriou, G.A.; Kyriacou, M.C. Rootstock mediated effects on watermelon field performance and fruit quality characteristics. *Intern. J. Vegetable. Sci.* **2014**, *21*, 344–362. [CrossRef]
31. Schwarz, D.; Öztekin, G.B.; Tüzel, Y.; Brückner, B.; Krumbein, A. Rootstocks can enhance tomato growth and quality characteristics at low potassium supply. *Sci. Hortic.* **2013**, *149*, 70–79. [CrossRef]
32. Turhan, A.; Ozmen, N.; Serbeci, M.S.; Seniz, V. Effects of grafting on different rootstocks on tomato fruit yield and quality. *Hortic. Sci.* **2011**, *38*, 142–149. [CrossRef]
33. Romano, D.; Paratore, A.; Vindigni, G. Caratteristiche dei fruttidi pomodoro in rapporto a diversi portinnesti. In Proceedings of the Workshop on 'Applicazione di Tecnologie Innovative per il Miglioramentodell' Orticoltura Meridionale', Consiglio Nazionale delle Ricerche, Rome, Italy, 18 July 2000; pp. 113–114.
34. Moncada, A.; Miceli, A.; Vetrano, F.; Mineo, V.; Planeta, D.; D'Anna, F. Effect of grafting on yield and quality of eggplant (*Solanum melongena* L.). *Sci. Hortic.* **2013**, *149*, 108–114. [CrossRef]
35. Ulukapi, K.; Onus, A.N. Comparison of the productivity and quality of the grafted and ungrafted tomato plants grown in the greenhouse with mycorrhiza application. *X Int. Symp. Process. Tomato* **2007**. [CrossRef]
36. Radicetti, E.; Massantini, R.; Campigliaa, E.; Mancinelli, R.; Ferri, S.; Moscetti, R. Yield and quality of eggplant (*Solanum melongena* L.) as affected by cover crop species and residues. *Sci. Hortic.* **2016**, *204*, 161–171. [CrossRef]
37. Arvanitoyannis, I.S.; Khah, E.; Christakou, E.C.; Bletsos, F.A. Effect of grafting and modified atmosphere packaging on eggplant quality parameters during storage. *Int. J. Food Sci. Tech.* **2005**, *40*, 311–322. [CrossRef]
38. Khah, E.M.; Kakava, E.; Mavromatis, A.; Chachalis, D.; Goulas, C. Effect of grafting on growth and yield of tomato (*Lycopersicon esculentum* Mill.) in greenhouse and open-field. *J. Appl. Hortic.* **2006**, *8*, 3–7. [CrossRef]
39. Khah, E.M. Effect of grafting on growth, performance and yield of aubergine (*Solanum melongena* L.) in greenhouse and open-field. *Intern. J. Plant Prod.* **2011**, *5*, 359–366. [CrossRef]
40. Lee, J.M.; Kubota, C.; Tsao, S.J.; Biel, Z.; Hoyos Echevaria, P.; Morra, L. Current status of vegetable grafting: Diffusion, grafting techniques, automation. *Sci. Hortic.* **2010**, *127*, 93–105. [CrossRef]

41. Rouphael, Y.; Schwarz, D.; Krumbein, A.; Colla, G. Impact of grafting on product quality of fruit vegetables. *Sci. Hortic.* **2010**, *127*, 172–179. [CrossRef]
42. Akanbi, W.B.; Togun, A.O.; Olaniran, O.A.; Akinfasoye, J.O.; Tairu, F.M. Physico-chemical properties of eggplant (*Solanum melongena* L.) fruit in response to nitrogen fertilizer and fruit size. *Agric. J.* **2007**, 2, 140–148.
43. Casals, J.; Rull, A.; Bernal, M.; González, R.; del Castillo, R.R.; Simó, J. Impact of grafting on sensory profile of tomato landraces in conventional and organic management systems. *Hortic. Environ. Biotech.* **2018**, *59*, 597–696. [CrossRef]

15

Rootstock and Arbuscular Mycorrhiza Combinatorial Effects on Eggplant Crop Performance and Fruit Quality under Greenhouse Conditions

Leo Sabatino [1,*], Giovanni Iapichino [1,*], Beppe Benedetto Consentino [1], Fabio D'Anna [1] and Youssef Rouphael [2]

1 Dipartimento Scienze Agrarie, Alimentari e Forestali (SAAF), Università di Palermo, 90128 Palermo, Italy; beppebenedetto.consentino@unipa.it (B.B.C.); fabio.danna@unipa.it (F.D.)

2 Dipartimento di Agraria, Università degli Studi di Napoli Federico II, 80055 Portici, Italy; youssef.rouphael@unina.it

* Correspondence: leo.sabatino@unipa.it (L.S.); giovanni.iapichino@unipa.it (G.I.)

Abstract: The herbaceous grafting of fruiting vegetables is considered a toolbox for safeguarding yield stability under various distresses and for improving fruit quality. Inoculation with arbuscular mycorrhiza (AM) fungi seems also to be an efficient tool for increasing the assimilation, uptake and translocation of macroelements and microelements, for modulating plant secondary metabolism and for overcoming several forms of plant distress. The present work evaluated the combined effect of grafting the "Birgah" (B) eggplant onto its wild/allied relatives' rootstocks (*Solanum torvum* (T), *S. macrocarpon* (M) and *S. paniculatum* (P)) and AM fungi (*R. irregularis*) on the yield, fruit quality, nitrogen use efficiency, mineral profile, and nutritional and functional quality. The B/T, B/M and B/P grafting combinations significantly increased the marketable fruit and fruit number compared with those in the ungrafted control. Furthermore, irrespective of the grafting combinations, AM fungi significantly enhanced the marketable fruit, fruit number and nitrogen use efficiency (NUE) by 13.3%, 12.7% and 13.3%, respectively compared to those in the untreated control. Exposing the B/T and B/P grafted plants to the +AM treatment significantly increased the ascorbic acid contents by 17.2% and 10.4%, respectively, compared with those in the ungrafted control. Fruits from the combination B/P × +AM had a higher chlorogenic acid content than fruits from the ungrafted control plots. Finally, the B/T × +AM and B/P × +AM combinations decreased glycoalkaloids by 58.7% and 63.7%, respectively, compared with those in the ungrafted control, which represents a highly important target for eggplant fruit healthiness.

Keywords: vegetable grafting; *Solanum melongena* L.; grafting combinations; arbuscular micorrhizal fungi; yield traits; NUE; mineral profile; functional properties

---

1. Introduction

Eggplant (*Solanum melongena* L.) is a major solanaceous vegetable crop cultivated worldwide [1]. Eggplant originated in the Old World, evolved from *S. insanum*, and it was self-sufficiently domesticated in India and China [2,3]. With a global production around 49.5 Mt, predominantly in Asia [4], eggplant is a relevant source of minerals and vitamins, and in terms of total nutritional value, it has been compared to tomato [1]. *S. macrocarpon* L., also known as African eggplant, and *S. aethiopicum* L., also known as Aethiopicum eggplant, are closely related to *S. melongena* and are mainly distributed in the African continent. In the eggplant cultivation scenario, the Mediterranean Basin represents an important area and Sicily is counted as a secondary eggplant diversification zone [5]. In Italy, eggplant is grown under either greenhouse or outdoor conditions, and due to the intensive cropping systems commonly used in the vegetable production farms, soilborne diseases and pests widely spread, limiting the yield and growth traits in eggplant [6]. At present, the absence of resistant genotypes, together with the ban on methyl bromide [7], has increased interest in the use of grafted eggplant. There are reports underlining the advantages derived from adopting grafting both in terms of plant biotic/abiotic stress tolerance and yield stability [8–10]. *Solanum torvum* Sw. is the most used rootstock for eggplant as it allows several soilborne diseases to be overcome, such as *Verticilium dahliae* Klebahn, *Ralstonia solanacearum* (Smith) Yabuuchiet al., *Fusarium oxysporum* (Schlechtend:Fr.) f. sp. *Melongenae* Matuoand Ishigami, and *Meloidogyne* spp. root-knot nematodes [6,11]. However, both eggplant's wild and allied relatives and interspecific hybrids are proposed as alternative eggplant rootstocks [12–14]. The use of tomato rootstock is also suggested to improve yield, fruit visual quality and plant vigor in eggplant [15].

Crop rotations and biological diversity long have been the key factors in successful traditional agricultural production systems. Rotation is essential for minimizing the build up of pest and soil borne disease problems. However, intensive greenhouse production often precludes vegetable growers from applying rotation. The lack of rotation and tendency towards monocropping in intensive protected vegetable production systems not only increase the prevalence of insects, soil plant pathogens and nematodes but, along with high nutrient levels, can suppress arbuscular mycorrhizal fungi (AMF). AMF are obligate symbionts between plants and fungi belonging to the monophyletic phylum *Glomeromycota* [16]. AMF symbiosis is extremely important for enhancing the assimilation of key macroelements and microelements (P, Cu and Zn) and improving nutrient uptake and efficiency due to its ability to develop an extended external hypha up to 40–50 times their length [17]. Another prominent and sustainable means of boosting yield is the inoculation of beneficial AMF in specific environments such as soil greenhouses where these fungi are generally suppressed.

Moreover, recent findings demonstrated that AM may modulate plant secondary metabolism, enhancing antioxidant activity as well as the accumulation of antioxidant molecules known as phytochemicals (i.e., carotenoids, flavonoids, glucosinolates and phenolic acids) with health-promoting properties [18]. AM fungi have been reported to not only enhance nutritional status and the quality of the produce but also to be able to reduce several forms of plant distress such as thermal stress [19], salinity stress [20], drought stress [21] and soil heavy metal stress [22–25]. Furthermore, there are reports showing an increased resistance to certain diseases in mycorrhized plants [26]. Although there are several studies on the synergistic effects of grafting and other agronomical or chemical means commonly applied in vegetable crop production [27,28], few experiments have been conducted on the interactive effects of grafting and AM on plant performance and the nutritive value of vegetables. In particular, Kumar et al. [24], studying the combined role of grafting and AM on Cd stress tolerance in tomato, observed that vigorous rootstock such as Maxifort successfully alleviate Cd stress symptoms via several physiological and biochemical mechanisms, such as (i) better nutrient absorption and translocation, (ii) higher synthesis of pigments linked to photosynthetic activity and (iii) better capability of producing enzymes (CAT and APX), proline and key metabolites (phytochelatin, fructans and inulins). Furthermore, Oztekin et al. [29], after studying the effect of grafting and AM on the performance of tomato plants cultivated under different salinity conditions in two growing seasons, conclude that grafting and AM have synergistic effects on tomato plant salinity tolerance. Nevertheless, no specified research has been conducted to study the co-operative effects between using wild and allied eggplant relatives as rootstocks and AM application in improving eggplant crop performance and nutritive value. On the aforesaid basis, the aims of our investigation were to appraise the concerted action of eggplant wild/allied relatives' rootstocks and AM inoculation on the yield, fruit quality and nitrogen use efficiency of "Birgah" eggplant grown under greenhouse conditions. The outcome of this study should provide useful information on the performance of new eggplant rootstocks and on their response to AMF inoculation.

2. Materials and Methods

2.1. Trial Site, Nursery Production and Growing Settings

The investigation was conducted in 2016 at the experimental farm of the University of Palermo. Seeds of the *Solanum torvum*, *Solanum macrocarpon* and *Solanum paniculatum* rootstocks were sown in 44-cell seedling trays, containing pasteurized peat moss (FAP, Padova, Italia), in a protected environment with a day/night temperature cycle of 25 °C/18 °C. Twenty days after rootstock sowing, seeds of the "Birgah" eggplant cultivar were seeded in 104-cell trays and maintained under the same climatic conditions and with the same planting method as the rootstocks. "Birgah" is an eggplant F_1 hybrid belonging to the violet round shape group, and it is one of the most cultivated eggplants, both in open fields and in protected environments, in Sicily. Seventy-five days after planting, all seedlings attained a hypocotyl diameter of at least 2 mm (the minimum recommended for the tube grafting method). The grafting of "Birgah" scions was performed as reported by Sabatino et al. [12,13]. Briefly, the rootstock was cut off, 0.5 cm below the cotyledons, at a 45° angle, and a similar cut was performed on the scion. Care was taken to ensure that the cut surfaces fitted perfectly. To complete the grafting procedure, a plastic clip was attached at the grafting point to guarantee that the correct amount of pressure was applied. The grafted plantlets were placed in a greenhouse equipped with a fog system in order to maintain a humidity of 95% and an air temperature of 20 °C for 7 days. After that, the grafted transplants were subjected, for 3 days, to a slow dropping of the humidity, useful for plant acclimatization. Then, they were ready for transplantation. Ungrafted and self-grafted eggplants were also included.

Prior to transplanting, half of the ungrafted, self-grafted and grafted plantlets were treated with 10 g plant^{-1} of micorrhizal inoculum carrying 40 spores g^{-1} of *Rhizophagus irregularis* (formerly *Glomus intraradices*). All plantlets were transplanted on 9 February, 2016 and maintained till the end of May, 2016 in an unheated greenhouse in Typic Rhodoxeralf soil, characterized by the following soil texture: 46.5% sand, 22.3% silt and 31.2% clay. The soil pH was 7.2. Before the experiment, a brassica crop was cultivated. Mulching with a black polyethylene film of 20 μm was installed. Eggplant plantlets were spaced in order to obtain a plant density of 2 plants m^{-2}. Plants were periodically drip irrigated, receiving 250 kg of nitrogen ha^{-1}, 150 kg of phosphorous pentoxide ha^{-1} and 250 kg of potassium oxide ha^{-1}. The cultivation practices reported by Baixauli [30], suggested for eggplant growth in Mediterranean conditions, were applied regularly subject to crop requirements.

2.2. Yield, Nitrogen Use Efficiency, Nutritional Traits and Functional Compounds

Immediately after harvest, the fruits were weighed and separated into marketable and waste production categories. The number of marketable fruits per plant was recorded, and average fruit weight was also calculated. In total, ten harvests were realized, starting on 21 March, 2016.

A digital penetrometer (Trsnc, Forlì, Italy) was used to determine fruit firmness. Fruit firmness was measured based on three replicates of five fruits per scion/rootstock combination. Each fruit was perforated, using a 6 mm diameter stainless steel cylinder probe, on two opposite sides of the fruit equatorial zone. Firmness was expressed in newtons (N).

Soluble solids content (SSC) was recorded based on three replicates of five fruits and was determined using a refractometer (MTD-045nD, Three-In-OneEnterprises Co. Ltd. Taiwan). Prior to the SSC determinations, the fruit juice was adequately clarified.

Three to five commercially mature fruits for each replicate, from the second and third harvests, were used for the analytical determinations. Each sample consisted of the same quantity of apical, equatorial and distal parts of the fruits. Nutritional and functional determinations were performed on fruits harvested from labelled flowers at the fruit set stage, and all the fruits were harvested 35 days after labelling (fruit commercial maturity stage).

Fruit dry weight was measured via the dehydration of the sample in a heater at 80 °C until a constant weight was achieved. The fruit dry matter percentage was calculated using the fruit fresh and dry weights. For protein determination, the Kjeldal method was applied. Specifically, sample was exposed to acid-catalyzed mineralization to convert the organic nitrogen into ammoniacal nitrogen. Subsequently, the ammoniacal nitrogen was distilled under alkaline pH. The ammonia produced via the distillation was collected in a boric acid solution and measured through titrimetric dosage. The ammoniacal nitrogen value was multiplied by 6.25.

For Ca, Mg, K and Fe determinations, atomic absorption spectroscopy following wet mineralization was adopted as reported by Morand and Gullo [31]. Phosphorus values were assessed using colorimetry as reported by Fogg and Wilkinson [32].

The nitrogen use efficiency (NUE = yield/N application rate) was calculated and expressed as $t{\cdot}kg^{-1}$.

The ascorbic acid content was determined as described by Sabatino et al. [33] for tomato. Therefore, the determinations were made with a reflectometer (Merck RQflex* 10 m) using Reflectoquant Ascorbic Acid Test Strips. Thus, one gram of eggplant fruit juice was mixed with distilled water to a 10 mL total volume. Subsequently, an appropriate test strip was immersed into the prepared sample then inserted into the meter. The results were expressed as mg of ascorbic acid per 100 g of fresh weight.

The anthocyanin extraction procedure applied was that described by Mennella et al. [34]. The determination was conducted on a lyophilized and powdered sample of 200 mg. The flow rate and absorbance units full scale adopted were 0.8 mL min^{-1} and at 0.1, respectively. For RP-HPLC analyses, purified delphinidin-3-rutinoside (D3R, Polyphenols Laboratories AS, Sandnes, Norway) was used as an external standard with a dissimilar retention time (23.9 min) compared to delphinidin-3-(p-coumaroyl rutinoside)-5-glucoside (nasunin), which was eluted with a longer retention time (25.8 min for cis-nasunin and 26.1 min for trans-nasunin, respectively). Regarding the nasunin determination, in agreement with Lo Scalzo et al. [35], a partially purified standard was used. Total anthocyanins were expressed as mg·100 g^{-1} dry weight (dw), highlighting that the detection limit was 2.00 mg 100 g^{-1} of dw.

For chlorogenic acid determination, the extraction and analysis procedure described by Stommel and Whitaker [36], with slight variations, was applied. Therefore, a binary mobile phase gradient of methanol in 0.01% aqueous phosphoric acid was provided according to this procedure: 0–15 min, linear increase from 5% to 25% methanol; 15–28 min, linear increase from 25% to 50% methanol; 28–30 min, linear increase from 50% to 100% methanol; 30–32 min, 100% methanol; 32–36 min, linear decrease from 100% to 5% methanol; 36–43 min, 5% methanol. The flow rate was 0.8 $mL{\cdot}min^{-1}$. The chlorogenic acid quantification, subsequently conducted by RP-HPLC separation, was based on the absorbance at 325 nm relative to the external standard of chlorogenic acid (Sigma-Aldrich, St.Louis, MO) and expressed as mg·100 g^{-1} of dw.

For glycoalkaloids, the extraction method adopted was that reported by Birner [37] with some adjustments. Thus, 0.5 g of lyophilized and powdered flesh tissue sample was extracted with 95% ethanol. For glycoalkaloids (expressed as mg·100 g^{-1}dw) analyses, the method described by Kuronen et al. [38] was applied. Therefore, the analyses were conducted by means of RP-HPLC using purified solasonine and solamargine as the external standards. The detection limit was 0.03 mg 100 g^{-1} of dw.

2.3. Root Colonization Assessment

To evaluate mycorrhizal colonization, the method described by Phillips and Haymann [39] and revised by Torta et al. [40] was adopted. Briefly, three lateral root samples from mychorrhized plants were collected and marked with acid fuchsin. As reported by Kormanik and McGraw [41], mycorrhizal colonization (Mycorrhization Index (MI = % of marked tissue, with respect to the hyaline portion, on the unit of the length of the root)) was evaluated on three fragments, attaining the average value.

2.4. Experimental Design and Statistical Analysis

The trial was organized in a randomized complete block design with 3 replicates per treatment. Each replicate consisted of 10 plants. Consequently, the total number of plants was 300 (10 treatments × 30 plants per treatment = 300 plants). The data were subjected to GLM (General Linear Model) analysis using the SPSS software package version 20.0. For data expressed in percentages, an arcsin transformation before ANOVA analysis (Ø = arcsin(p/100)$^{1/2}$) was performed. Tukey's HSD test ($p < 0.05$) was used to compare means. Principal component analysis (PCA) was conducted on the whole yield, fruit quality and nitrogen use efficiency data set.

3. Results

3.1. Root Colonization by AM Fungi

AM fungi root colonization was significantly affected by mycorrhizal inoculation (M), whereas the grafting combination (G) did not significantly influence AM root colonization. ANOVA showed no significant interaction (G × M) (Figure S1). Irrespective of the grafting combination, the percentage of root colonization was higher with the inoculated treatment (71.8%) compared to with the non-inoculated one (2.5%) (Figure S1).

3.2. Effect of Grafting Combination and Mycorrhizal Inoculation on Yield, Yield Components and NUE

The marketable yield, number of marketable fruit per plant and percentage of unmarketable production were significantly influenced by the G × M interaction (Table 1).

Table 1. Main effects of the grafting combination and arbuscular mycorrhiza (AM) inoculation on yield and the yield components of greenhouse eggplant.

Treatment	Marketable Yield (t ha^{-1})		Fruits Number (n. plant^{-1})		Fruit Mean Weight (g)		Discarded Production (%)	
Grafting combination (G)								
Birgah ungrafted (B)	36.5	c	6.1	c	505.5	a	6.3	b
Birgah self-grafted (B/B)	37.1	c	6.2	c	501.6	a	6.3	b
Birgah/*S. torvum* (B/T)	48.4	a	8.1	a	500.8	a	5.2	b
Birgah/*S. macrocarpon* (B/M)	40.1	b	6.7	b	500.6	a	10.3	a
Birgah/*S. paniculatum* (B/P)	38.5	bc	6.5	bc	494.7	a	6.3	b
AM fungi (M)								
−AM	37.6	b	6.3	b	499.8	a	9.4	a
+AM	42.6	a	7.1	a	501.4	a	4.4	b
Significance								
G	***		***		NS		***	
M	***		***		NS		***	
G × M	NS		NS		NS		NS	

NS, *** non-significant or significant at $p \leq 0.001$. Data represent mean values of three replicates. Values within a column followed by the same letter are not significantly different at $p \leq 0.05$ according to Tukey's HSD Test. NS = not significant. +AM, −AM = mycorrhizal and non-mycorrhizal eggplants, respectively.

On the other hand, neither the grafting combination nor mycorrhizal treatment had a significant effect on fruit mean weight (avg. 500.6 g fruit^{-1}). When averaged across mycorrhizal treatments, the marketable yield production reached a maximum value with Brigah grafted onto *S. torvum* (B/T), followed by both the B/M and B/P grafting combinations, while the lowest crop productivity was recorded for ungrafted and self-grafted crops (Table 1). Interestingly, the higher marketable fresh yield observed for eggplants grafted onto *S. torvum* and to a lesser extent onto *S. macrocarpon* and *S. paniculatum*, in comparison to both the ungrafted and self-grafted plants, was associated with an increase in the number of eggplant fruits per plant and not with a change in the mean fruit weight (Table 1). Moreover, the highest discarded production was observed with the B/M grafting combination (Table 1). A significant effect of mycorrhizal treatment on the yield and yield components was also observed (Table 1). Irrespective of grafting combinations, the inoculation of eggplant at the transplantation stage with *R. irregularis* elicited significant increases in the marketable yield and number of fruits per plant of 13.3% and 12.6%, respectively, compared to for the non-inoculated control (Table 1).

Similarly to the effects on the yield and yield components, the nitrogen use efficiency (NUE) for the B/T grafting combination (0.242 t kg^{-1}) was significantly higher, by 31.2%, than that obtained from the ungrafted and self-grafted eggplant (avg. 0.184 t kg^{-1}; Table 1). Finally, the inoculation of eggplant with *R. irregularis*, when averaging across grafting combinations, induced a significant increase in the NUE, which was 13.3% higher than for the non-inoculated control treatment (Figure 1).

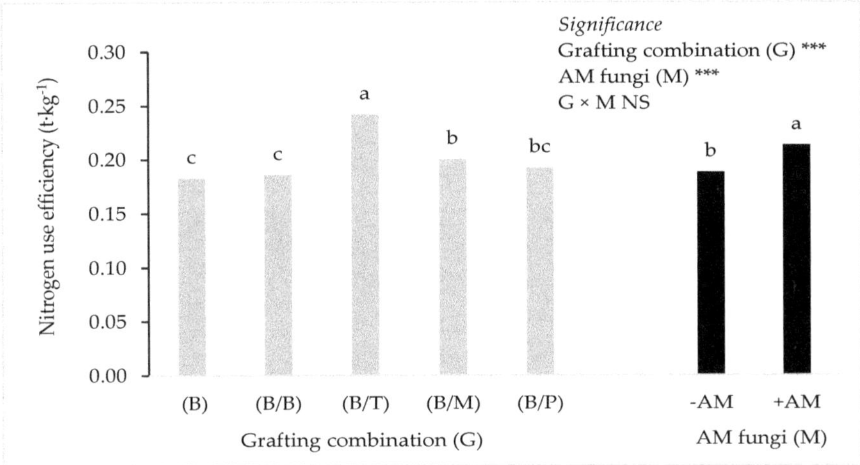

Figure 1. Main effects of grafting combination and AM inoculation on the nitrogen use efficiency (NUE) of greenhouse eggplant. (B): Birgah ungrafted; (B/B): Birgah self-grafted; (B/T): Birgah/*S. torvum*; (B/M): Birgah/*S. macrocarpon*; (B/P): Birgah/*S. paniculatum*.

3.3. Effects of Grafting Combination and Mycorrhizal Inoculation on the Mineral Profile and Nutritional and Functional Quality

Neither the grafting combination nor mycorrhizal inoculation had a significant effect on the fruit dry matter content (DM). The soluble solids content (SSC) and K content of eggplant fruits in the B/T and B/M plots, averaged across mycorrhizal inoculation status, were higher than those for the other grafting combinations (Table 2).

Table 2. Main effects of the grafting combination and AM inoculation on the fruit dry matter (DM) percentage, soluble solids content (SSC) and macromineral content of greenhouse eggplant.

Treatment	Dry Matter (%)		SSC (°Brix)		K (mg 100 g^{-1} dw)		Ca (mg 100 g^{-1} dw)		Mg (mg 100 g^{-1} dw)	
Grafting combination (G)										
Birgah ungrafted (B)	6.0	a	4.3	b	4.3	b	309.2	c	18.0	a
Birgah self-grafted (B/B)	5.9	a	4.3	b	4.3	b	311.5	c	17.0	a
Birgah/*S. torvum* (B/T)	6.0	a	5.1	a	5.1	a	311.4	c	12.3	c
Birgah/*S. macrocarpon* (B/M)	5.9	a	5.1	a	5.1	a	335.5	b	16.8	a
Birgah/*S. paniculatum* (B/P)	5.9	a	4.4	b	4.4	b	347.5	a	14.5	b
AM fungi (M)										
−AM	5.9	a	4.4	b	4.4	b	324.0	a	16.1	a
+AM	6.0	a	4.9	a	4.9	a	322.0	a	15.6	a
Significance										
G	NS		***		***		***		***	
M	NS		***		***		NS		NS	
G × M	NS		NS		NS		NS		NS	

NS, *** non-significant or significant at $p \leq 0.001$. Data represent mean values of three replicates. Values within a column followed by the same letter are not significantly different at $p \leq 0.05$ according to Tukey's HSD Test. NS = not significant. +AM, −AM = mycorrhizal and non-mycorrhizal eggplants, respectively.

The highest and lowest Ca and mg fruit contents were recorded in the B/P and B/T plots, respectively. Regardless of the grafting combination, the SSC and K fruit contents in the inoculated plots were 11.4% and 11.4% higher than in the non-inoculated ones, respectively. No significant interaction was observed between the grafting combination and mycorrhizal inoculation in terms of the DM percentage; SSC; and K, Ca and mg fruit contents (Table 2).

The grafting combination and mycorrhizal inoculation significantly affected the P and Fe fruit contents, protein content, firmness, ascorbic acid, total anthocyanins, chlorogenic acid and glycoalkaloids (Table 3).

Table 3. Effects of the grafting combination and AM inoculation on the phosphorus (P), iron (Fe), protein, fruit firmness, ascorbic acid, total anthocyanins, chlorogenic acid and glycoalkaloid content of greenhouse eggplant.

Treatment	P (mg 100 g^{-1} dw)		Fe (µg g^{-1})		Proteins (g 100 g^{-1} dw)		Firmness (N)		Ascorbic Acid (mg 100 g^{-1} fw)		Total Anthocyanins (mg 100 g^{-1} dw)		Chlorogenic Acid (mg 100 g^{-1} dw)		Glycoalkaloids (mg 100 g^{-1} dw)	
Birgah ungrafted (B) × −AM	418.07	e	21.13	e	10.77	e	43.5	c	6.51	de	7462.03	d	982.37	b	89.38	b
Birgah ungrafted (B) × +AM	431.43	e	22.73	e	12.93	d	53.2	a	7.27	b	7770.40	c	992.60	b	74.16	c
Birgah self-grafted (B/B) × −AM	419.27	e	25.17	d	10.60	e	43.1	c	6.43	e	7428.13	d	982.71	b	89.13	b
Birgah self-grafted (B/B) × +AM	435.53	e	27.40	c	12.70	d	52.2	a	7.19	bc	7772.43	c	984.99	b	73.27	c
Birgah/*S. torvum* (B/T) × −AM	494.90	d	30.17	b	14.03	c	54.7	a	7.27	b	7154.77	e	757.15	e	44.13	d
Birgah/*S. torvum* (B/T) × +AM	525.70	c	30.70	ab	15.23	a	53.8	a	7.63	a	7393.67	de	823.98	d	36.93	e
Birgah/*S. macrocarpon* (B/M) × −AM	383.67	f	31.43	a	9.87	f	46.4	b	6.71	d	7241.60	e	955.27	c	97.20	a
Birgah/*S. macrocarpon* (B/M) × +AM	382.67	f	31.30	a	9.83	f	52.2	a	7.09	c	7222.50	e	948.57	c	90.33	b
Birgah/*S. paniculatum* (B/P) × −AM	553.23	b	29.33	b	13.07	d	41.5	c	6.67	d	9858.37	b	989.16	b	41.94	d
Birgah/*S. paniculatum* (B/P) × +AM	584.30	a	31.33	a	14.90	b	47.3	b	7.19	bc	11,162.96	a	1027.86	a	32.43	e
Significance																
G	***		***		***		***		***		***		***		***	
M	***		***		***		***		***		***		***		***	
G × M	***		***		***		**		***		***		**		***	

, * significant at $p \leq 0.01$ or significant at $p \leq 0.001$. Data represent mean values of three replicates. Values within a column followed by the same letter are not significantly different at $p \leq 0.05$ according to Tukey's HSD Test. NS = not significant. +AM, −AM = mycorrhizal and non-mycorrhizal eggplants, respectively.

The inoculation treatment elicited a significant increase in the P fruit content in the B/P and B/T plots compared to in the non-inoculated ones. However, this effect was not apparent with the other grafting combinations.

A similar positive effect was also observed in eggplant grafted onto *S. paniculatum*, since mycorrhizal inoculation increased Fe concentration in inoculated plants compared to in the non-inoculated ones (Table 3). Moreover, eggplant "Birgah" grafted onto *S. torvum* and inoculated with AM fungi displayed the highest total protein value (Table 3). Except for the B/T grafting combination where no significant changes in fruit firmness were observed, the inoculation of eggplant with AM fungi in all the other grafting combinations (ungrafted, B/B, B/M and B/P) incurred a significant increase in fruit firmness (Table 3). Contrarily, for the latter quality parameters (P, Fe, protein and firmness), the beneficial effect of AM fungi inoculation on ascorbic acid in eggplant fruits was apparent with all the grafting combinations tested, with the highest values recorded for "Birgah" grafted onto *S. torvum* (Table 3).

The total anthocyanin and chlorogenic acid contents ranged from 7154 to 11,162 and 757 to 1027 mg 100 g^{-1} dw, respectively. Interestingly, the B/P grafting combination inoculated with AM fungi produced a major amplification of total anthocyanins and chlorogenic acid in comparison to the rest of the treatments (Table 3). Finally, the results indicated that the highest glycoalkaloid content was observed in the non-inoculated B/M grafting combination, whereas the lowest values were recorded in the B/P and B/T combinations inoculated with *R. irregularis* (Table 3).

3.4. *Principal Component Analysis of All Agronomic and Qualitative Parameters*

The principal component analysis (PCA) on the agronomic and qualitative traits of greenhouse eggplant in response to the grafting combination and mycorrhizal inoculation is reported in Figure 2.

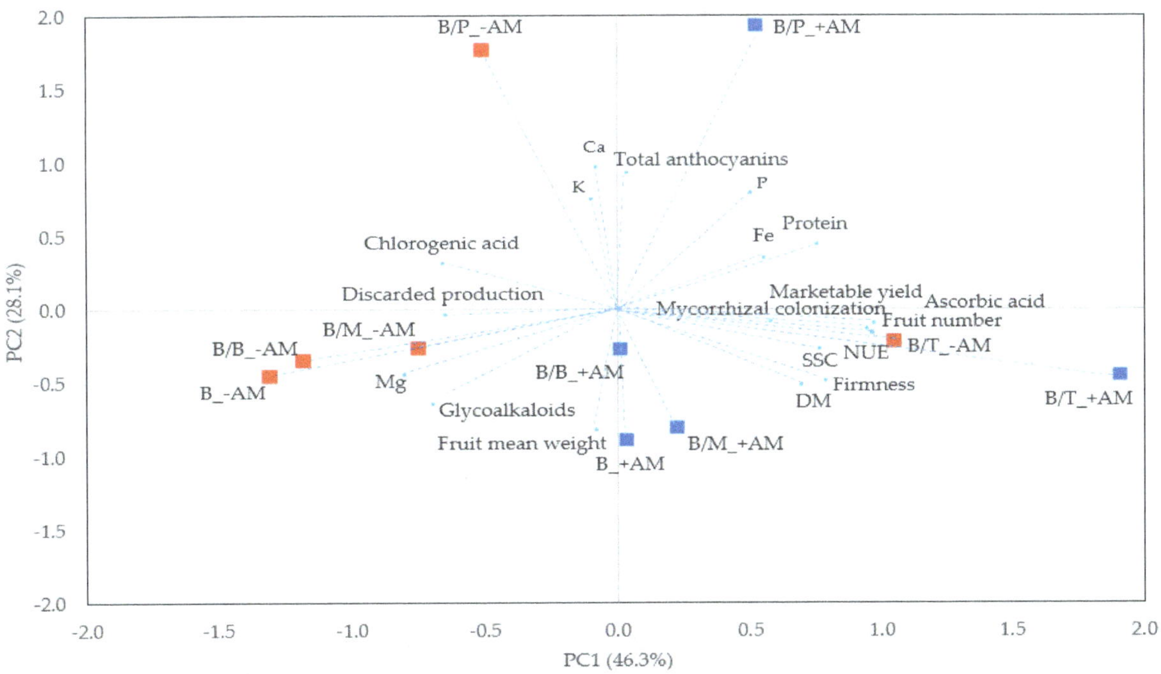

Figure 2. Principal component loading plot and scores of the principal component analysis (PCA) of the yield, yield components, nitrogen use efficiency (NUE), dry matter percentage (DM), protein, firmness, soluble solids content (SSC), minerals (P, K, Ca, mg and Fe), bioactive molecules (ascorbic acid, anthocyanins and chlorogenic acid) and glycoalkaloids of eggplant cv. "Birgah" as a function of grafting combination (ungrafted B and self-grafted B/B, B/T, B/M and B/P) and AM inoculation (+AM and −AM).

The first four PCs (i.e., principal components) were related with eigenvalues higher than 1 and explained 94.7% of the total variance (Table 4).

Table 4. Correlation coefficients for each agronomic and qualitative traits, eigenvalues, variance and cumulative proportions of total variance of the four principal components (PCs).

Variable	PC1	PC2	PC3	PC4
Marketable yield	0.961	−0.157	−0.129	−0.003
Fruit number	0.969	−0.090	−0.156	−0.012
Fruit mean weight	−0.080	−0.813	0.377	−0.041
Discarded production	−0.657	−0.029	−0.727	−0.082
Protein	0.747	0.444	0.360	−0.295
K	−0.097	0.752	−0.394	0.463
P	0.498	0.798	0.193	−0.261
Ca	−0.083	0.972	0.078	0.046
Mg	−0.809	−0.435	0.243	0.290
Fe	0.550	0.359	−0.644	0.345
NUE	0.959	−0.155	−0.130	−0.017
Fruit dry matter	0.691	−0.509	−0.001	−0.001
Firmness	0.780	−0.488	0.088	0.146
SSC	0.757	−0.262	−0.365	0.425
Ascorbic acid	0.940	−0.125	0.210	0.162
Total anthocyanins	0.034	0.936	0.268	0.189
Chlorogenic acid	−0.659	0.319	0.431	0.517
Glycoalkaloids	−0.698	−0.635	−0.155	0.285
Mycorrhizal colonization	0.571	−0.077	0.500	0.636
Eigenvalue	8.793	5.348	2.247	1.601
Variance (%)	46.281	28.148	11.825	8.427
Cumulative (%)	46.281	74.429	86.254	94.681

In the current greenhouse experiment, the loading matrix indicates that variation in Ca and K was most closely aligned with the total anthocyanins, whereas the variation in the glycoalkaloids was not correlated to the SSC (Figure 2). The score plot of the PCA superimposed on the agronomic and qualitative parameters demonstrated a strong clustering of the two mycorrhizal treatments along PC1 (with the exception of *S. torvum* × −AM), with +AM plants concentrating marketable yield, NUE, DM, SSC, mineral composition and ascorbic acid (Figure 2). Particularly, the B/T grafting combination inoculated with AM fungi was positioned on the positive side of PC1 in the lower right quadrant of the PCA score plot, and it exhibited the highest crop performance (high yield, number of fruits and NUE) with premium quality due to high concentrations of DM, SSC and ascorbic acid (Figure 2). Moreover, eggplant grafted onto *S. paniculatum* and inoculated with AM was characterized by higher P, Fe and total anthocyanin content (Figure 2). Finally, non-inoculated ungrafted and self-grafted plants were positioned in the lower left quadrant, characterized by higher glycoalkaloid content (Figure 2).

4. Discussion

In the present work we tested the use of various eggplant grafting combinations and AM fungi, alone or in combination, to boost crop performance and fruit quality in "Birgah" eggplant cultivated in a protected environment. Our results showed that enhancements in terms of yield, yield related traits, NUE, and fruit nutritional and functional quality can be accomplished by harnessing the combined effects of specific grafting combinations and AM fungi. These results are in line with those reported by Sabatino et al. [12,13], who—exploring the effects of using eggplant hybrids and allied species as rootstocks on eggplant yield, plant vigor and overall fruit quality—found that *S. paniculatum* and the interspecific hybrid of *S. melongena* × *S. aethiopicum* gr. *Gilo* provided a higher yield performance and a better fruit quality compared to the ungrafted or self-grafted plants. Similarly, our results are in line with those observed by Oztekin et al. [29], who, studying the influence of AM fungi on salinity tolerance in tomato grafted plants, stated that AM fungi significantly increased the total and marketable yields as well as the average fruit weight. Several authors [42,43] report that the phytostimulating effect of the AM fungi could be attributed to numerous mechanisms such as enhancing the uptake and translocation of major and trace elements, inducing a more developed root system, improving the water status and photosynthetic efficiency, bolstering the antioxidative defense system, balancing plant hormones, upregulating nutrient transporter action and promoting the production of enzymes such as phosphatases. Therefore, we might hypothesize that the higher yield and yield related traits of mycorrhized plants could have similar explanations. Our outcomes regarding the NUE are in agreement with those reported by Djidonou et al. [44], who—investigating the yield, water status and NUE in field-grown, grafted tomato—found that grafting the "Florida 47" tomato onto "Beaufort" or "Multifort" significantly increased N use. Our findings are also consistent with those of Colla et al. [45], who, studying the influence of selected rootstocks on the plant performance and nitrogen use efficiency of the "Proteo" melon, report that the use of melon grafted onto designated rootstocks would represent a prospective plan for increasing crop productivity and NUE. In the present study, AM fungi increased the NUE of eggplant plants. This accords with the results of Zhu et al. [46], who—evaluating the effects of AM fungi on growth, nitrogen uptake and NUE in wheat—found that AM symbiosis increased NUE. Our findings are also in line with those presented by Liu et al. [47], who, studying the influence of AM on NUE in soybean, concluded that AM fungi play an imperative function in increasing NUE. Likewise, Rouphael et al. [43], in a review, report that plant biostimulants, including microbial plant biostimulants, can enhance NUE. Our study showed that both the *S. torvum* and *S. macrocarpon* rootstocks increased SSC values. This is in congruence with the results of Sabatino et al. [12], who used the "Birgah" eggplant F_1 hybrid as a scion. However, these findings are in contrast to the outcomes reported by Sabatino et al. [13], who tested the "Scarlatti" eggplant F_1 hybrid as an eggplant scion. Furthermore, our results for SSC revealed that AM fungi increase the SSC level in eggplant. However, this is in contrast with previous reports [29] indicating that AM symbiosis does not affect SSC in tomato fruit. Our results on firmness are corroborated by the results of Sabatino et al. [12], who found that *S. torvum* rootstock increased fruit firmness in "Birgah" hybrid and by those reported by Miceli et al. [48], who, assessing the effect of AM fungi and grafting on yield and fruit quality in mini-watermelon, noted that AM symbiosis increases pulp firmness. In our study, at the end of the experiment, the mean mycorrhizal colonization rates were 71.8% and 2.5% in +AM and –AM plants, respectively. There are reports that mycorrhizal hyphae extending into the soil or substrate increase nutrient uptake [43,49] thanks to the mycorrhizal hyphae's capacity to penetrate into very small particles [50]. Our results on fruit mineral composition showed that the B/T, B/M and B/P scion/rootstock combinations improved the K, Ca and Fe fruit contents. Meanwhile, fruits from the B/M grafts revealed a decrease in terms of P content compared to the fruits from the control plots (ungrafted or self-grafted plants). Moreover, the present study showed that AM fungi treatment increased the K fruit content, without affecting the Ca and mg fruit content. In addition, our results highlighted that the B/T and B/P grafts benefit significantly from AM fungi in terms of P and Fe fruit concentrations. These outcomes are in accord with those of Kaya et al. [51] and Oztekin et al. [29], who supposed that a higher fruit mineral (macro and micro) concentration in grafted and mycorrhized plants could be attributed to a positive impact on water and nutrient uptake. Our trial showed that the *S. torvum* and *S. paniculatum* rootstocks increased fruit protein concentrations. This is in agreement with the results of Sabatino et al. [12] but in contrast to those of Sabatino et al. [13], who found no significant differences in terms of fruit protein content between ungrafted, self-grafted or *S. torvum* grafted plants. These results mark the important role played by the scion in terms of nutritional fruit quality in grafted plants. In addition, our results showed that AM fungi augmented the protein concentration in eggplant fruits. This seems to be in accord with the study of Baum et al. [52], who asserted that mycorrhizal inoculation induces a higher accumulation of proteins in onion. Contemporary outcomes exposed that AM symbioses are capable to adjust host plant primary and secondary metabolism, encouraging the synthesis of phytochemicals in the root system and shoots of mycorrhized plants [53]. Our data showed that the *S. torvum* rootstock significantly increased ascorbic acid in eggplant fruit, whereas fruits from the B/M and B/P grafts did not show significant differences in terms of ascorbic acid compared to the fruits from the control plots, demonstrating that the rootstock plays an imperative role in eggplant fruit quality. This results are partially in accord with those reported by Oztekin et al. [29], who observed that the rootstock factor did not affect vitamin C in tomato fruits, and with those of Miceli et al. [48], who reported that grafting increases fruit ascorbic acid content in mini-watermelon. Conversely, our results on the positive effect of the AM fungi on fruit ascorbic acid content are fully in accord with those of Oztekin et al. [29] and Miceli et al. [48]. There is evidence that the increase in secondary metabolites in mycorrhized plants could be correlated to the higher contents of mineral nutrients [25]. In this regard, overall, our study indicated that AM fungi positively affected total anthocyanins and chlorogenic acid, with the sole exception of with the B/M scion/rootstock combination. Our findings seem to be in line with those reported by Walter et al. [54], who revealed that some graminaceous crops inoculated with AM fungi produce undergo important biochemical modulation leading to phenolic acid accumulation. Our results are in accord with those of other authors [55–57], who, investigating the influence of AM symbiosis on the production of phytochemicals in *Ocimum basilicum*, found higher concentrations of antioxidants when plants were treated with diverse AM fungi species. Baslam et al. [58] remarked that mycorrhized lettuce leaves display higher anthocyanin, carotenoid and phenolic contents than plants from control plots. Castellanos-Morales et al. [59] found that the strawberry mycorrhized plants produced fruits with a higher content of anthocyanidin cyanidin-3-glucoside. Ceccarelli et al. [60] state that artichoke, when treated with two AM fungi species, have increased leaf polyphenolic concentrations and antioxidant activity. Our results on glycoalkaloids support the findings reported by Sabatino et al. [12], who, although not showing a statistically significant effect of the rootstock on glycoalkaloids, highlighted a certain effect of rootstocks on glycoalkaloid content, remarking on the effectiveness of *S. torvum* and *S. paniculatum* in reducing glycoalkaloid concentrations. Furthermore, our results evidenced that +AM treatment had a positive effect on glycoalkaloid reduction which in turn could positively influence human health.

The effectiveness of PCA in interpreting species/cultivar differences across several agronomic and qualitative attributes in response to a wide range of pre-harvest factors such as the genetic material and agricultural practices (i.e., AM inoculation) has been reported previously by several researchers [61–63]. This was also evident in the present study, conducted under protected environment conditions, as the score plot of PCA integrated information on the yield, yield components, NUE, mineral profile and nutritive value of eggplant from five grafting combinations with or without inoculation with AM.

5. Conclusions

In the present work, eggplant wild/allied relatives' rootstocks and AMF significantly interacted, improving the crop performance and quality traits of "Birgah" eggplant. However, the B/T and B/P grafts combined with AMF stood out as producing the best results in terms of the yield traits, NUE, mineral profile, and nutritional and functional quality. Our findings could be useful information for vegetable crop nurseries interested in introducing new eggplant rootstocks that successfully respond when combined with AMF.

Author Contributions: Conceptualization, L.S., F.D. and G.I.; methodology, L.S. and Y.R.; validation, L.S. and G.I.; formal analysis, B.B.C. and L.S.; investigation, L.S.; resources, F.D.; data curation, B.B.C. and L.S.; writing—original draft preparation, L.S.; writing—review and editing, L.S., G.I. and Y.R.; visualization, L.S., G.I. and Y.R.; supervision, F.D., G.I. and Y.R.; project administration, F.D.; funding acquisition, F.D. All authors have read and agreed to the published version of the manuscript.

Abbreviations

AM	Arbuscular Mycorrhizal
ANOVA	Analysis of Variance

APX	Ascorbate Peroxidase
B	Birgah
B/B	Birgah self-grafted
B/M	Birgah/S. *macrocarpon*
B/P	Birgah/S. *paniculatum*
B/T	Birgah/S. *torvum*
Ca	Calcium
CAT	Catalase
DM	Dry Matter
Fe	Iron
K	Potassium
Mg	Magnesium
NUE	Nitrogen Use Efficiency
P	Phosphorus
PCA	Principal Component Analysis
RP-HPLC	Reversed Phase High Performance Liquid Chromatography
SSC	Soluble Solids Content

References

1. Kalloo, G. Eggplant *Solanum melongena* L. In *Genetic Improvement of Vegetable Crops*; Kalloo, G., Bergh, B.O., Eds.; Pergamon Press: Oxford, UK, 1993; pp. 587–604.
2. Vavilov, N.I. The Origin, Variation, Immunity and Breeding of Cultivated Plants. *Soil Sci.* **1951**, *72*, 482. [CrossRef]
3. Cericola, F.; Portis, E.; Toppino, L.; Barchi, L.; Acciarri, N.; Ciriaci, T.; Sala, T.; Rotino, G.L.; Lanteri, S. The Population Structure and Diversity of Eggplant from Asia and the Mediterranean Basin. *PLoS ONE* **2013**, *8*, e73702. [CrossRef] [PubMed]
4. Toppino, L.; Barchi, L.; Scalzo, R.L.; Palazzolo, E.; Francese, G.; Fibiani, M.; D'Alessandro, A.; Papa, V.; Laudicina, V.A.; Sabatino, L.; et al. Mapping Quantitative Trait Loci Affecting Biochemical and Morphological Fruit Properties in Eggplant (Solanum melongena L.). *Front. Plant Sci.* **2016**, *7*, 3017. [CrossRef] [PubMed]
5. D'Anna, F.; Sabatino, L. Morphological and agronomical characterization of eggplant genetic resources from the Sicily area. *J. Food Agri. Environ.* **2013**, *11*, 401–404.
6. Bletsos, F.; Thanassoulopoulos, C.; Roupakias, D. Effect of Grafting on Growth, Yield, and Verticillium Wilt of Eggplant. *HortScience* **2003**, *38*, 183–186. [CrossRef]
7. Maier, J. UNEP–United Nations Environment Programme. In *A Concise Encyclopedia of the United Nations*; Brill: Caton, The Netherlands, 2010; pp. 712–714.
8. Daunay, M.-C. Eggplant. In *Vegetables II*; Springer Science and Business Media LLC: Berlin, Germany, 2008; pp. 163–220.
9. King, S.R.; Davis, A.R.; Zhang, X.; Crosby, K. Genetics, breeding and selection of rootstocks for Solanaceae and Cucurbitaceae. *Sci. Hortic.* **2010**, *127*, 106–111. [CrossRef]
10. Rouphael, Y.; Kyriacou, M.C.; Colla, G. Vegetable Grafting: A Toolbox for Securing Yield Stability under Multiple Stress Conditions. *Front. Plant Sci.* **2018**, *8*, 2255. [CrossRef]
11. Bogoescu, M.; Doltu, M.; Sora, D. Hybrids and allied species as potential rootstocks for eggplant: Effect of grafting on vigour, yield and overall fruit quality traits. *Acta Hortic.* **2014**, 331–336. [CrossRef]
12. Sabatino, L.; Iapichino, G.; D'Anna, F.; Palazzolo, E.; Mennella, G.; Rotino, G.L. Hybrids and allied species as potential rootstocks for eggplant: Effect of grafting on vigour, yield and overall fruit quality traits. *Sci. Hortic.* **2018**, *228*, 81–90. [CrossRef]
13. Sabatino, L.; Iapichino, G.; Rotino, G.L.; Palazzolo, E.; Mennella, G.; Anna, D.; D'Anna, F. Solanum aethiopicum gr. gilo and Its Interspecific Hybrid with S. melongena as Alternative Rootstocks for Eggplant:

Effects on Vigor, Yield, and Fruit Physicochemical Properties of Cultivar 'Scarlatti'. *Agronomy* **2019**, *9*, 223. [CrossRef]
14. Gisbert, C.; Prohens, J.; Raigón, M.D.; Stommel, J.R.; Nuez, F. Eggplant relatives as sources of variation for developing new rootstocks: Effects of grafting on eggplant yield and fruit apparent quality and composition. *Sci. Hortic.* **2011**, *128*, 14–22. [CrossRef]
15. Maršić, N.K.; Mikulič-Petkovšek, M.; Štampar, F. Grafting Influences Phenolic Profile and Carpometric Traits of Fruits of Greenhouse-Grown Eggplant (Solanum melongenaL.). *J. Agric. Food Chem.* **2014**, *62*, 10504–10514. [CrossRef] [PubMed]
16. Schüβler, A.; Schwarzott, D.; Walker, C. A new fungal phylum, the Glomeromycota: Phylogeny and evolution. *Mycol. Res.* **2001**, *105*, 1413–1421. [CrossRef]
17. Rouphael, Y.; Cardarelli, M.; Colla, G. Role of arbuscular mycorrhizal fungi in alleviating the adverse effects of acidity and aluminium toxicity in zucchini squash. *Sci. Hortic.* **2015**, *188*, 97–105. [CrossRef]
18. Avio, L.; Turrini, A.; Giovannetti, M.; Sbrana, C. Designing the Ideotype Mycorrhizal Symbionts for the Production of Healthy Food. *Front. Plant Sci.* **2018**, *9*, 9. [CrossRef] [PubMed]
19. Zhu, X.-C.; Song, F.-B.; Liu, S.-Q.; Liu, T.-D. Effects of arbuscular mycorrhizal fungus on photosynthesis and water status of maize under high temperature stress. *Plant Soil* **2011**, *346*, 189–199. [CrossRef]
20. Cantrell, I.C.; Linderman, R.G. Preinoculation of lettuce and onion with VA mycorrhizal fungi reduces deleterious effects of soil salinity. *Plant Soil* **2001**, *233*, 269–281. [CrossRef]
21. Püschel, D.; Rydlová, J.; Vosátka, M. Can mycorrhizal inoculation stimulate the growth and flowering of peat-grown ornamental plants under standard or reduced watering? *Appl. Soil Ecol.* **2014**, *80*, 93–99. [CrossRef]
22. Ouziad, F.; Hildebrandt, U.; Schmelzer, E.; Bothe, H. Differential gene expressions in arbuscular my-corrhizal-colonized tomato grown under heavy metal stress. *J. Plant Physiol.* **2005**, *162*, 634–649. [CrossRef]
23. Shahabivand, S.; Maivan, H.Z.; Goltapeh, E.M.; Sharifi, M.; Aliloo, A.A. The effects of root endophyte and arbuscular mycorrhizal fungi on growth and cadmium accumulation in wheat under cadmium toxicity. *Plant Physiol. Biochem.* **2012**, *60*, 53–58. [CrossRef]
24. Kumar, P.; Lucini, L.; Rouphael, Y.; Cardarelli, M.; Kalunke, R.M.; Colla, G. Insight into the role of grafting and arbuscular mycorrhiza on cadmium stress tolerance in tomato. *Front. Plant Sci.* **2015**, *6*, 477. [CrossRef] [PubMed]
25. Rouphael, Y.; Franken, P.; Schneider, C.; Schwarz, D.; Giovannetti, M.; Agnolucci, M.; De Pascale, S.; Bonini, P.; Colla, G. Arbuscular mycorrhizal fungi act as biostimulants in horticultural crops. *Sci. Hortic.* **2015**, *196*, 91–108. [CrossRef]
26. Dugassa, G.D.; Von Alten, H.; Schönbeck, F. Effects of arbuscular mycorrhiza (AM) on health ofLinum usitatissimum L. infected by fungal pathogens. *Plant Soil* **1996**, *185*, 173–182. [CrossRef]
27. Ioannou, N. Integrating soil solarization with grafting on resistant rootstocks for management of soil-borne pathogens of eggplant. *J. Hortic. Sci. Biotechnol.* **2001**, *76*, 396–401. [CrossRef]
28. Katan, J. Physical and cultural methods for the management of soil-borne pathogens. *Crop. Prot.* **2000**, *19*, 725–731. [CrossRef]
29. Oztekin, G.B.; Tuzel, Y.; Tuzel, I.H. Does mycorrhiza improve salinity tolerance in grafted plants? *Sci. Hortic.* **2013**, *149*, 55–60. [CrossRef]
30. Baixauli, C. Berenjena. In *La Horticultura Espãnola*; Nuez, F., Llácer, G., Eds.; Ediciones de Horticultura: Reus, Spain, 2001; pp. 104–108.
31. Morand, P.; Gullo, J.L. Mineralisation des tissus vegetaux en vue du dosage de P, Ca, Mg, Na, K. *Ann. Agron.* **1970**, *21*, 229–236.
32. Fogg, D.N.; Wilkinson, N.T. The colorimetric determination of phosphorus. *Analist* **1958**, *83*, 406. [CrossRef]
33. Sabatino, L.; D'Anna, F.; Iapichino, G.; Moncada, A.; D'Anna, E.; De Pasquale, C. Interactive Effects of Genotype and Molybdenum Supply on Yield and Overall Fruit Quality of Tomato. *Front. Plant Sci.* **2019**, *9*, 9. [CrossRef]
34. Mennella, G.; Scalzo, R.L.; Fibiani, M.; D'Alessandro, A.; Francese, G.; Toppino, L.; Acciarri, N.; De Almeida, A.E.; Rotino, G.L. Chemical and Bioactive Quality Traits During Fruit Ripening in Eggplant (*S. melongena* L.) and Allied Species. *J. Agric. Food Chem.* **2012**, *60*, 11821–11831. [CrossRef]

35. Scalzo, R.L.; Fibiani, M.; Mennella, G.; Rotino, G.L.; Sasso, M.D.; Culici, M.; Spallino, A.; Braga, P.C. Thermal Treatment of Eggplant (Solanum melongenaL.) Increases the Antioxidant Content and the Inhibitory Effect on Human Neutrophil Burst. *J. Agric. Food Chem.* **2010**, *58*, 3371–3379. [CrossRef] [PubMed]
36. Stommel, J.R.; Whitaker, B.D. Phenolic Acid Content and Composition of Eggplant Fruit in a Germplasm Core Subset. *J. Am. Soc. Hortic. Sci.* **2003**, *128*, 704–710. [CrossRef]
37. Birner, J. A method for the determination of total steroid bases. *J. Pharm. Sci.* **1969**, *58*, 258–259. [CrossRef] [PubMed]
38. Kuronen, P.; Väänänen, T.; Pehu, E. Reversed-phase liquid chromatographic separation and simultaneous profiling of steroidal glycoalkaloids and their aglycones. *J. Chromatogr. A* **1999**, *863*, 25–35. [CrossRef]
39. Phillips, J.; Hayman, D. Improved procedures for clearing roots and staining parasitic and vesicular-arbuscular mycorrhizal fungi for rapid assessment of infection. *Trans. Br. Mycol. Soc.* **1970**, *55*, 158–168. [CrossRef]
40. Torta, L.; Mondello, V.; Burruano, S. Valutazione delle caratteristiche morfo-anatomiche di alcune simbiosi micorriziche mediante tecniche colorimetriche usuali e innovative. *Micol. Ital.* **2003**, *2*, 53–59.
41. Kormanik, P.P.; McGraw, A.C. Quantification of vesicular-arbuscular mycorrhizae in plant roots. In *Methods and Principles of Mycorrhizal Research*; Schenck, N.C., Ed.; APS Press: St. Paul, MN, USA, 1991; pp. 37–45.
42. De Pascale, S.; Rouphael, Y.; Colla, G. Plant biostimulants: Innovative tool for enhancing plant nutrition in organic farming. *Eur. J. Hortic. Sci.* **2018**, *82*, 277–285. [CrossRef]
43. Rouphael, Y.; Colla, G. Synergistic Biostimulatory Action: Designing the Next Generation of Plant Biostimulants for Sustainable Agriculture. *Front. Plant Sci.* **2018**, *9*, 9. [CrossRef]
44. Djidonou, D.; Zhao, X.; Simonne, E.H.; Koch, K.E.; Erickson, J.E. Yield, Water-, and Nitrogen-use Efficiency in Field-grown, Grafted Tomatoes. *HortScience* **2013**, *48*, 485–492. [CrossRef]
45. Colla, G.; Suárez, C.M.C.; Cardarelli, M.; Rouphael, Y. Improving Nitrogen Use Efficiency in Melon by Grafting. *HortScience* **2010**, *45*, 559–565. [CrossRef]
46. Zhu, X.; Song, F.; Liu, S.; Liu, F. Arbuscular mycorrhiza improve growth, nitrogen uptake, and nitrogen use efficiency in wheat grown under elevated CO_2. *Mycorrhiza* **2015**, *26*, 133–140. [CrossRef]
47. Liu, H.; Song, F.; Liu, S.; Li, X.; Liu, F.; Zhu, X. Arbuscular mycorrhiza improves nitrogen use efficiency in soybean grown under partial root-zone drying irrigation. *Arch. Agron. Soil Sci.* **2018**, *65*, 269–279. [CrossRef]
48. Miceli, A.; Romano, C.; Moncada, A.; Piazza, G.; Torta, L.; D'Anna, F.; Vetrano, F. Yield and quality of mini-watermelon as affected by grafting and mycorrhizal inoculum. *J. Agric. Sci. Technol.* **2016**, *18*, 505–516.
49. Colla, G.; Rouphael, Y.; Di Mattia, E.; El-Nakhel, C.; Cardarelli, M. Co-inoculation of Glomus intraradices and Trichoderma atroviride acts as a biostimulant to promote growth, yield and nutrient uptake of vegetable crops. *J. Sci. Food Agric.* **2015**, *95*, 1706–1715. [CrossRef] [PubMed]
50. Al-Karaki, G.N. Nursery inoculation of tomato with arbuscular mycorrhizal fungi and subsequent performance under irrigation with saline water. *Sci. Hortic.* **2006**, *109*, 1–7. [CrossRef]
51. Kaya, C.; Ashraf, M.; Sonmez, O.; Aydemir, S.; Tuna, A.L.; Cullu, M.A. The influence of arbuscular mycorrhizal colonisation on key growth parameters and fruit yield of pepper plants grown at high salinity. *Sci. Hortic.* **2009**, *121*, 1–6. [CrossRef]
52. Baum, C.; El-Tohamy, W.; Gruda, N. Increasing the productivity and product quality of vegetable crops using arbuscular mycorrhizal fungi: A review. *Sci. Hortic.* **2015**, *187*, 131–141. [CrossRef]
53. Sbrana, C.; Avio, L.; Giovannetti, M. Beneficial mycorrhizal symbionts affecting the production of health-promoting phytochemicals. *Electrophoresis* **2014**, *35*, 1535–1546. [CrossRef]
54. Walter, M.; Fester, T.; Strack, D. Arbuscular mycorrhizal fungi induce the non-mevalonate methylerythritol phosphate pathway of isoprenoid biosynthesis correlated with accumulation of the 'yellow pigment' and other apocarotenoids. *Plant J.* **2000**, *21*, 571–578. [CrossRef]
55. Copetta, A.; Lingua, G.; Berta, G. Effects of three AM fungi on growth, distribution of glandular hairs, and essential oil production in Ocimum basilicum L. var. Genovese. *Mycorrhiza* **2006**, *16*, 485–494. [CrossRef]
56. Toussaint, J.P.; Smith, F.A.; Smith, S.E. Arbuscular mycorrhizal fungi caninduce the production of phytochemicals in sweet basil irrespective of phosphorus nutrition. *Mycorrhiza* **2007**, *17*, 291–297. [CrossRef] [PubMed]
57. Rasouli-Sadaghiani, M.; Hassani, A.; Barin, M.; Rezaee Danesh, Y.; Sefidkon, F. Effects of arbuscular mycorrhizal (AM) fungi on growth, essential oil production and nutrients uptake in basil. *J. Med. Plant Res.* **2010**, *4*, 2222–2228.

58. Baslam, M.; Garmendia, I.; Goicoechea, N. Arbuscular Mycorrhizal Fungi (AMF) Improved Growth and Nutritional Quality of Greenhouse-Grown Lettuce. *J. Agric. Food Chem.* **2011**, *59*, 5504–5515. [CrossRef] [PubMed]
59. Castellanos-Morales, V.; Villegas, J.; Wendelin, S.; Vierheilig, H.; Eder, R.; Cardenas-Navarro, R. Root colonization by the arbuscular mycorrhizal fungus Glomus intraradices alters the quality of strawberry fruit (Fragaria × ananassa Duch) at different nitrogen levels. *J. Sci. Food Agric.* **2010**, *90*, 1774–1782. [PubMed]
60. Ceccarelli, N.; Curadi, M.; Martelloni, L.; Sbrana, C.; Picciarelli, P.; Giovannetti, M. Mycorrhizal colonization impacts on phenolic content and antioxidant properties of artichoke leaves and flower heads two years after field transplant. *Plant Soil* **2010**, *335*, 311–323. [CrossRef]
61. Colonna, E.; Rouphael, Y.; Barbieri, G.; De Pascale, S. Nutritional quality of ten leafy vegetables harvested at two light intensities. *Food Chem.* **2016**, *199*, 702–710. [CrossRef]
62. Kyriacou, M.C.; Soteriou, G.A.; Rouphael, Y.; Siomos, A.S.; Gerasopoulos, D. Configuration of watermelon fruit quality in response to rootstock-mediated harvest maturity and postharvest storage. *J. Sci. Food Agric.* **2015**, *96*, 2400–2409. [CrossRef]
63. Carillo, P.; Colla, G.; El-Nakhel, C.; Bonini, P.; D'Amelia, L.; Dell'Aversana, E.; Pannico, A.; Giordano, M.; Sifola, M.I.; Kyriacou, M.C.; et al. Biostimulant application with a tropical plant extract enhances *Corchorus olitorius* adaptation to sub-optimal nutrient regimens by improving physiological parameters. *Agronomy* **2019**, *9*, 249. [CrossRef]

Permissions

All chapters in this book were first published in MDPI; hereby published with permission under the Creative Commons Attribution License or equivalent. Every chapter published in this book has been scrutinized by our experts. Their significance has been extensively debated. The topics covered herein carry significant findings which will fuel the growth of the discipline. They may even be implemented as practical applications or may be referred to as a beginning point for another development.

The contributors of this book come from diverse backgrounds, making this book a truly international effort. This book will bring forth new frontiers with its revolutionizing research information and detailed analysis of the nascent developments around the world.

We would like to thank all the contributing authors for lending their expertise to make the book truly unique. They have played a crucial role in the development of this book. Without their invaluable contributions this book wouldn't have been possible. They have made vital efforts to compile up to date information on the varied aspects of this subject to make this book a valuable addition to the collection of many professionals and students.

This book was conceptualized with the vision of imparting up-to-date information and advanced data in this field. To ensure the same, a matchless editorial board was set up. Every individual on the board went through rigorous rounds of assessment to prove their worth. After which they invested a large part of their time researching and compiling the most relevant data for our readers.

The editorial board has been involved in producing this book since its inception. They have spent rigorous hours researching and exploring the diverse topics which have resulted in the successful publishing of this book. They have passed on their knowledge of decades through this book. To expedite this challenging task, the publisher supported the team at every step. A small team of assistant editors was also appointed to further simplify the editing procedure and attain best results for the readers.

Apart from the editorial board, the designing team has also invested a significant amount of their time in understanding the subject and creating the most relevant covers. They scrutinized every image to scout for the most suitable representation of the subject and create an appropriate cover for the book.

The publishing team has been an ardent support to the editorial, designing and production team. Their endless efforts to recruit the best for this project, has resulted in the accomplishment of this book. They are a veteran in the field of academics and their pool of knowledge is as vast as their experience in printing. Their expertise and guidance has proved useful at every step. Their uncompromising quality standards have made this book an exceptional effort. Their encouragement from time to time has been an inspiration for everyone.

The publisher and the editorial board hope that this book will prove to be a valuable piece of knowledge for researchers, students, practitioners and scholars across the globe.

List of Contributors

Roberta Spanò, Donato Gallitelli and Tiziana Mascia
Department of Soil, Plant and Food Sciences, University of Bari "Aldo Moro", 70126 Bari, Italy

Massimo Ferrara
Institute of Sciences of Food Production (ISPA) – CNR, 70126 Bari, Italy

Ibrahim Musa
Institute of Tropical Agriculture and Food Security, Universiti Putra Malaysia, Serdang 43400, Selangor, Malaysia
Department of Agronomy, Faculty of Agriculture, Federal University of Kashere, Gombe 0182, Gombe State, Nigeria

Mohd Y. Rafii
Institute of Tropical Agriculture and Food Security, Universiti Putra Malaysia, Serdang 43400, Selangor, Malaysia
Department of Crop Science, Faculty of Agriculture, Universiti Putra Malaysia, Serdang 43400, Selangor, Malaysia

Khairulmazmi Ahmad
Department of Plant Protection, Faculty of Agriculture, Universiti Putra Malaysia, Serdang 43400, Selangor, Malaysia

Shairul Izan Ramlee and Jamilu Halidu
Department of Crop Science, Faculty of Agriculture, Universiti Putra Malaysia, Serdang 43400, Selangor, Malaysia

Muhammad Asyraf Md Hatta
Department of Agriculture Technology, Faculty of Agriculture, Universiti Putra Malaysia, Serdang 43400, Selangor, Malaysia

Yusu Oladosu, Isma'ila Muhammad, Samuel Chibuike Chukwu, Nur Nadzirah Mat Sulaiman and Arolu Fatai Ayanda
Institute of Tropical Agriculture and Food Security, Universiti Putra Malaysia, Serdang 43400, Selangor, Malaysia

Mariateresa Cardarelli and Catello Pane
Consiglio per la Ricerca in Agricoltura e L'analisi Dell'economia Agraria, Centro di Ricerca Orticoltura e Florovivaismo, 84098 Pontecagnano Faiano, Italy

Marios C. Kyriacou
Department of Vegetable Crops, Agricultural Research Institute, Nicosia 1516, Cyprus

Giuseppe Colla
Department of Agriculture and Forest Sciences, University of Tuscia, 01100 Viterbo, Italy

Hira Singh
Department of Vegetable Science, Punjab Agricultural University, Ludhiana 141004, India

Pradeep Kumar
ICAR-Central Arid Zone Research Institute, Jodhpur 342003, India

Ashwani Kumar
ICAR-Central Soil Salinity Research Institute, Karnal 132001, India

Hao Wei, Jin Zhao and Jiangtao Hu
Department of Horticulture, Division of Applied Life Science (BK21 Plus Program), Graduate School of Gyeongsang National University, Jinju 52828, Korea

Byoung Ryong Jeong
Department of Horticulture, Division of Applied Life Science (BK21 Plus Program), Graduate School of Gyeongsang National University, Jinju 52828, Korea
Institute of Agriculture & Life Science, Gyeongsang National University, Jinju 52828, Korea
Research Institute of Life Science, Gyeongsang National University, Jinju 52828, Korea

José-Luis Pardo-Alonso, Ángel Carreño-Ortega and Ángel-Jesús Callejón-Ferre
Department of Engineering, University of Almería, Agrifood Campus of International Excellence (CeiA3), La Cañada de San Urbano, Almería 04120, Spain

Tian Gong, Zachary T. Ray, Kylee E. Butcher, Zachary E. Black, Xin Zhao and Jeffrey K. Brecht
Horticultural Sciences Department, University of Florida, Gainesville, FL 32611, USA

Ángel Carreño-Ortega and Marta Gómez Galán
Department of Engineering, University of Almería, Research Center CIMEDES, Agrifood Campus of International Excellence (CeiA3), La Cañada de San Urbano, 04120 Almería, Spain

Carolina-Clara Martínez-Gaitán
Tecnova, Technological Center: Foundation for Auxiliary Technologies for Agriculture; Parque Tecnológico de Almería, Avda. de la Innovación, 23, 04131 Almería, Spain

Iacopo Golasi
DIAEE-Area Fisica Tecnica, Università degli Studi di Roma "Sapienza", 00184 Rome, Italy

Rana Shahzad Noor
Department of Agriculture, Biological, Environment and Energy Engineering, College of Engineering, Northeast Agricultural University, Harbin 150030, China
Faculty of Agricultural Engineering & Technology, PMAS-Arid Agriculture University, Rawalpindi 46000, Pakistan

Zhi Wang and Yong Sun
Department of Agriculture, Biological, Environment and Energy Engineering, College of Engineering, Northeast Agricultural University, Harbin 150030, China

Muhammad Umair
Faculty of Agricultural Engineering & Technology, PMAS-Arid Agriculture University, Rawalpindi 46000, Pakistan

Muhammad Yaseen
Department of Agricultural Extension, College of Agriculture, University of Sargodha, Sargodha 40100, Pakistan

Muhammad Ameen
College of Engineering, Nanjing Agricultural University, Nanjing 210031, China

Shoaib-Ur Rehman and Muzammil Usman Khan
Institute of Horticultural Sciences, University of Agriculture, Faisalabad 38040, Pakistan

Muhammad Imran
Faculty of Water Resources and Hydroelectric Engineering, Xi'an University of Technology, Xi'an 710048, China

Waqar Ahmed
Key Laboratory of Soybean Biology, Chinese Ministry of Education, Northeast Agricultural University, Harbin 150030, China

Branimir Urlić, Marko Runjić, Marija Mandušić, Katja Žanić and Ana Mateškovič
Department of Applied Sciences, Institute for Adriatic Crops and Karst Reclamation, Put Duilova 11, 21000 Split, Croatia

Gabriela Vuletin Selak and Gvozden Dumičić
Department of Plant Sciences, Institute for Adriatic Crops and Karst Reclamation, Put Duilova 11, 21000 Split, Croatia

Enrica Allevato and Rosita Marabottini
DIBAF, University of Tuscia, Via S.C. De Lellis snc, 01100 Viterbo, Italy

Rosario Paolo Mauro, Cherubino Leonardi and Francesco Giuffrida
Di3A, University of Catania, Via Valdisavoia 5, 95123 Catania, Italy

Silvia Rita Stazi
DSCF, University of Ferrara, Via Borsari 46, 44121 Ferrara, Italy

Anita Ierna
CNR-IBE, Via P. Gaifami 18, 95126 Catania, Italy

Giuseppe Carlo Modarelli and Stefania De Pascale
Department of Agricultural Sciences, University of Naples Federico II, 80055 Portici, Italy

Gölgen Bahar Öztekin and Yüksel Tüzel
Department of Horticulture, Faculty of Agriculture, Ege University, 35100 Bornova-Izmir, Turkey

Francesco Orsini and Giorgio Gianquinto
DISTAL—Department of Agricultural and Food Sciences, Alma Mater Studiorum—University of Bologna, 40127 Bologna, Italy

Giovanni Iapichino and Eristanna Palazzolo
Dipartimento Scienze Agrarie, Alimentari e Forestali, Università di Palermo, Viale delle Scienze, 90128 Palermo, Italy

Giuseppe Leonardo Rotino
Consiglio per la Ricerca in Agricoltura e l'Analisi dell'Economia Agraria CREA-GB, Centro di Ricerca Genomica e Bioinformatica, Via Paullese, 26836 Montanaso Lombardo (LO), Italy

Giuseppe Mennella
Consiglio per la Ricerca in Agricoltura e l'Analisi dell'Economia Agraria CREA-OF, Centro di Ricerca Orticoltura e Florovivaismo, Via Cavalleggeri, 84098 Pontecagnano-Faiano, Italy

Maryam Mozafarian, Nazatul Syaima Binti Ismail and Noémi Kappel
Department of Vegetable and Mushroom Growing, Faculty of Horticultural Science, Szent István University, Villányi út 29-43, H-1118 Budapest, Hungary

Leo Sabatino, Giovanni Iapichino, Beppe Benedetto Consentino and Fabio D'Anna
Dipartimento Scienze Agrarie, Alimentari e Forestali (SAAF), Università di Palermo, 90128 Palermo, Italy

Youssef Rouphael
Dipartimento di Agraria, Università degli Studi di Napoli Federico II, 80055 Portici, Italy

Index